D1664441

Godfrey Gumbs and
Danhong Huang

Properties of Interacting Low-Dimensional Systems

Related Titles

Yan, D., Wang, H., Du, B.

Introduction to Organic Semiconductor Heterojunctions

2010

ISBN: 978-0-470-82594-5

Guldi, D. M., Martín, N. (eds.)

Carbon Nanotubes and Related Structures
Synthesis, Characterization, Functionalization, and Applications

2010

ISBN: 978-3-527-32406-4

Butt, H.-J., Kappl, M.

Surface and Interfacial Forces

2010

ISBN: 978-3-527-40849-8

Heinzel, T.

Mesoscopic Electronics in Solid State Nanostructures

2010

ISBN: 978-3-527-40932-7

Wolf, E. L.

Nanophysics and Nanotechnology
An Introduction to Modern Concepts in Nanoscience

2006

ISBN: 978-3-527-40651-7

Godfrey Gumbs and Danhong Huang

Properties of Interacting Low-Dimensional Systems

WILEY-VCH Verlag GmbH & Co. KGaA

The Authors

Dr. Godfrey Gumbs
University of New York
Department of Physics
695, Park Avenue
New York, NY 10065
USA

Dr. Danhong Huang
USAF Research Lab (AFRL/RVSS)
Adv. E/O Space Sensors Group
3550, Aberdeen Ave, SE Bldg 426
Kirtland AFB, NM 87117
USA

Library of Congress Card No.:
applied for

British Library Cataloguing-in-Publication Data:
A catalogue record for this book is available from the British Library.

Bibliographic information published by the Deutsche Nationalbibliothek
The Deutsche Nationalbibliothek lists this publication in the Deutsche Nationalbibliografie; detailed bibliographic data are available on the Internet at http://dnb.d-nb.de.

Composition le-tex publishing services GmbH, Leipzig
Printing and Binding Fabulous Printers Pte Ltd, Singapore
Cover Design Schulz Grafik-Design, Fußgönheim

Printed in Singapore
Printed on acid-free paper

ISBN 978-3-527-40894-8

Contents

Preface

There has been a considerable amount of research on "mesoscopic" structures whose sizes are intermediate, that is, between the macroscopic and the atomic scale. These include semiconductor heterojunctions, quantum dots and wires as well as carbon nanotubes and atomic layers of graphene. One is unable to explain the properties of these systems simply in terms of a single-particle Schrödinger equation since many-body effects cannot be neglected. Therefore, there is a need to combine an introduction to some typical topics of interest and the methods and techniques needed to handle them in a single volume. This book tries to achieve that goal by carefully presenting a number of topics concerned with the optical response and transport properties of low-dimensional structures. This material is supplemented by a selection of problems at the end of each chapter to give the reader a chance to apply the ideas and techniques in a challenging manner. There are several excellent textbooks which already deal with electron–electron interaction effects. However, the material we cover supplements those publications by covering more recently studied topics, that is, especially in semiconductors.

Since the aim of the book is to be self-contained, we first present some background diagrammatic methods. This is based on standard field theoretic techniques. (See, for example, the early work in [1–17].) However, here we give several examples which have useful applications to the topics covered in later chapters. We evaluate the Green's function expansion and the polarization function which is a necessary ingredient in our investigations employing the linear response formalism which can then be used in the study of the collective plasma excitations in quantum dots and wires, electron transport, light absorption, electron energy loss spectroscopy for nanotubes, graphene containing massless Dirac fermions or layered semiconductor structures, just to name a few examples which are covered in this book. We have developed some of the methods introduced here in collaboration with co-workers. By learning the formalism and getting introduced to the novel physical properties of low-dimensional systems, the reader should be able to understand scientific papers in condensed matter physics dealing with the effects arising from many-particle interactions.

This book is inspired by a collection of lectures which were given over the years at the Graduate Center of the City University of New York. This course is usually taken by graduate students who have had some exposure to basic quantum mechanics, statistical mechanics and introductory solid state physics at the undergraduate level. What makes this book different from previously published ones on the many-body theory of solids is that it presents a variety of current topics of interest in the field of mesoscopic systems and it also provides basic formal theory which is relevant to these systems but not available in previously published books. This makes it suitable for either a "Special Topics" course in solid state physics in which a few of the chapters may be selected, or as a textbook for an advanced solid state physics course in which the methodology is taught. In any event, problem solving would be an integral part of the course. A number of problems and the references have been given at the end of each chapter for students wanting to become more familiar with the topics and their background.

We would like to express our gratitude to Dr. Oleksiy Roslyak for his generous time in helping us write the chapters on graphene, nonlinear Green's function theory as well as on excitons and biexcitons in quantum dots. This contribution as well as his helpful comments on the manuscript are gratefully acknowledged. Our thanks also to Dr. Paula Fekete, Dr. Tibab McNeish, Dr. Oleg Berman Dr. Antonios Balassis, Andrii Lurov, Hira Ghumman, and Alisa Dearth for their helpful comments and criticisms of the manuscript.

Hunter College of the City University of New York — *Godfrey Gumbs*
AFRL, New Mexico, March 2011 — *Danhong Huang*

References

1 Martin, P.C. and Schwinger, J. (1959) Theory of Many-Particle Systems. *Phys. Rev.*, **115**, 1342.

2 Abraham, M. and Becker, R. (1949) *Classical Theory of Magnetism*, Hafner, NY.

3 Ambegaokar, V. and Baratoff, A. (1963) Tunneling Between Superconductors. *Phys. Rev. Letts.*, **10**, 486.

4 Anderson, P.W. (1964) *Lectures on the Many-Body Problem*, Vol. 2 (ed. E.R. Caianello), Academic Press.

5 Anderson, P.W. (1958) Random-Phase Approximation in the Theory of Superconductivity. *Phys. Rev.*, **112**, 1900.

6 Ashcroft, N., Mermin, D., and Bardeen, J. (1956) Theory of Superconductivity, in *Handbuch der Phys.*, Vol. 15 (ed. S. Flugge), 274.

7 Bardeen, J. (1961) Tunnelling from a Many-Particle Point of View. *Phys. Rev. Lett.*, **6**, 57.

8 Bardeen, J. and Schrieffer, J.R. (1961) in *Prog. in Low Temp. Phys.* (ed. C.J. Gorter), North Holland, Vol. B, p. 170.

9 Bardeen, J., Cooper, L.N., and Schrieffer, J.R. (1957) Theory of Superconductivity. *Phys. Rev.*, **108**, 1175.

10 Blatt, J.M. and Butler, S.T. (1954) Superfluidity of a Boson Gas. *Phys. Rev.*, **96**, 1149.

11 Bogoliubov, N.N. (1958) On a New Method in the Theory of Superconductivity. *Nuovo Cim.*, **7**, 794.

12 Bogoliubov, N.N. (1959) *Us. Fiz. Nauk*, **67**, 549.

13 Cohen, M.H., Falicov, L.M., and Phillips, J.C. (1962) Superconductive Tunneling. *Phys. Rev. Lett.*, **8**, 316.

14 Ferrell, R.A. and Prange, R.E. (1963) Self-Field Limiting of Josephson Tunneling of Superconducting Electron Pairs. *Phys. Rev. Lett*, **10**, 479.

15 Feynman, R.P. (1972) *Statistical Mechanics*, W.A Benjamin, Reading, MA.

16 Fock, V. (1932) Konfigurationsraum und zweite Quantelung. *Z. Phys.*, **75**, 622.

17 Callaway, J. (1974) *Quantum Theory of the Solid State*, Ch. 7, 2nd edn, Academic Press Inc., San Diego.

Part One Linear Response of Low Dimensional Quantum Systems

1
Introduction

1.1
Second-Quantized Representation for Electrons

The use of a Schrödinger equation to describe one or more electrons already treats the electron quantum mechanically and is sometimes referred to as first quantization. As long as electrons are neither created nor destroyed, such a description is complete. However, an electron that is transferred from state n to state m is often described as the destruction of an electron in state n and creation in state m by an operator obeying an algebra of the form $c_m^\dagger c_n$. It is convenient therefore to further refine the algebra of such operators analogous to the operators $b_q^\dagger$ and b_q that create and destroy phonons of wave-vector $\boldsymbol{q}$. However, electrons are fermions rather than bosons and the state occupancy number $c_n^\dagger c_n$ should only be permitted to take the values zero or one. This aim is achieved by using anti-commutation rules {described by braces} or by square brackets with a + subscript, that is, $[\cdots]_+$, instead of commutation rules described by brackets or square brackets with a − subscript, that is, $[\cdots]_-$.

In this book, we will be primarily concerned with low-dimensional systems such as quantum wells, dots and wires. A typical band structure of the valence and conduction bands for a heterostructure like GaAs/AlGaAs is shown in Figure 1.1. However, the formulation in this chapter and in some of the others is independent of dimensionality.

The phrase "second quantization" is descriptive of the notion that the Schrödinger wave function $\Psi(\boldsymbol{r})$ is to be quantized, that is, treated as an operator. In terms of any complete set of states $\phi_k(\boldsymbol{r})$, we can write:

$$\Psi(\boldsymbol{r}) = \sum_k c_k \phi_k(\boldsymbol{r}) , \quad \int d^3 r \phi_k^*(\boldsymbol{r}) \phi_{k'}(\boldsymbol{r}) = \delta_{kk'} , \tag{1.1}$$

where the anti-commutation rules are given by

$$\left[c_k, c_{k'}^\dagger\right]_+ = \delta_{kk'} , \tag{1.2}$$

and

$$[c_k, c_{k'}]_+ = \left[c_k^\dagger, c_{k'}^\dagger\right]_+ = 0 . \tag{1.3}$$

Properties of Interacting Low-Dimensional Systems, First Edition. G. Gumbs and D. Huang.

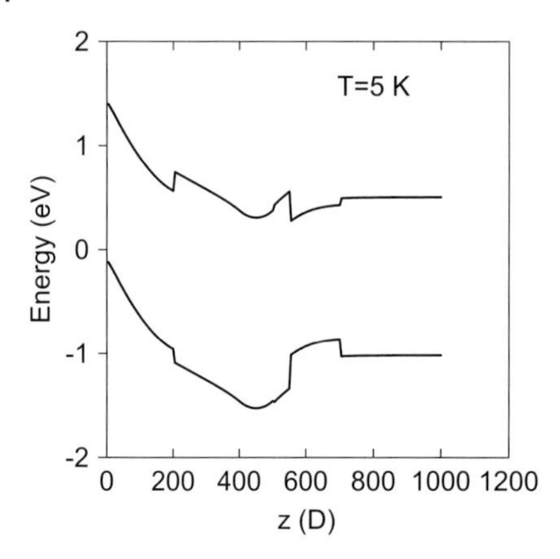

Figure 1.1 Valence (lower curve) and conduction (upper curve) bands of electrons in a semiconductor heterostructure.

For a single state, we can omit the subscripts and examine the consequences. Equation (1.3) implies that

$$c^2 = 0\,, \quad \left(c^\dagger\right)^2 = 0\,. \tag{1.4}$$

Let $N = c^\dagger c$, then

$$\begin{aligned} N^2 = \left(c^\dagger c\right)^2 = c^\dagger c c^\dagger c &= c^\dagger\left(1 - c^\dagger c\right) c \\ &= c^\dagger c - \left(c^\dagger\right)^2 c^2 = c^\dagger c - 0 = N\,. \end{aligned} \tag{1.5}$$

Thus, $N = 1$ or 0.

Consider the two eigenstates of N:

$$N\Psi_0 = 0\Psi_0\,, \quad \text{and} \quad N\Psi_1 = 1\Psi_1\,. \tag{1.6}$$

Then, is $c^\dagger\Psi_0$ also an eigenstate of N?

$$\begin{aligned} \left(c^\dagger c\right) c^\dagger\Psi_0 &= c^\dagger\left(1 - c^\dagger c\right)\Psi_0 \\ &= c^\dagger\Psi_0 - \left(c^\dagger\right)^2 c\Psi_0 = c^\dagger\Psi_0\,. \end{aligned} \tag{1.7}$$

Therefore,

$$N\left(c^\dagger\Psi_0\right) = 1\left(c^\dagger\Psi_0\right)\,, \tag{1.8}$$

that is, $c^\dagger\Psi_0$ is proportional to Ψ_1. Evaluate the normalization:

$$\begin{aligned} \int d^3\boldsymbol{r}\left(c^\dagger\Psi_0\right)^*\left(c^\dagger\Psi_0\right) &= \int d^3\boldsymbol{r}\,\Psi_0^* c c^\dagger\Psi_0 \\ &= \int d^3\boldsymbol{r}\,\Psi_0^*\left(1 - c^\dagger c\Psi_0\right) = 1\,. \end{aligned} \tag{1.9}$$

Therefore, $c^\dagger \Psi_0$ is normalized and we can simply choose

$$c^\dagger \Psi_0 = \Psi_1 \,. \tag{1.10}$$

Similarly,

$$cc^\dagger (c\Psi_1) = c\left(1 - cc^\dagger\right) \Psi_1 = c\Psi_1 \,. \tag{1.11}$$

Therefore, $c\Psi_1$ is an eigenvector of $cc^\dagger = 1 - c^\dagger c$ with eigenvalue 1 or eigenvector of $c^\dagger c$ with eigenvector 0. By a similar procedure to Eq. (1.9), $c\Psi_1$ is normalized and we can write

$$c\Psi_1 = \Psi_0 \,. \tag{1.12}$$

Note that

$$c\Psi_0 = c^2 \Psi_1 = 0 \,, \tag{1.13}$$

$$c^\dagger \Psi_1 = \left(c^\dagger\right)^2 \Psi_0 = 0 \,. \tag{1.14}$$

We begin by rewriting the Schrödinger equation in second quantized form. In most cases, the Hamiltonian has the form

$$H = \sum_{i=1}^{N} T(x_i) + \frac{1}{2} \sum_{i \neq j = 1}^{N} V\left(x_i, x_j\right) \,, \tag{1.15}$$

where T is the kinetic energy and V is the potential energy of interaction of the particles. Here, $x_i = (\boldsymbol{x}_i, t_i, s_i)$ is the space-time-spin point. The potential energy term represents the interaction between every pair of particles counted once, precisely why we have the factor of 1/2. We will not give the details for reformulating Eq. (1.15) in second quantized form since it can be found in many textbooks on quantum mechanics and will simply quote the results along with some others. For the Hamiltonian in Eq. (1.15), we have

$$\hat{H} = \int d^3\boldsymbol{x},\, \hat{\Psi}^\dagger(x)\, T(x)\, \hat{\Psi}(x) + \frac{1}{2} \int d^3\boldsymbol{x} \int d^3\boldsymbol{x}'\, \hat{\Psi}^\dagger(x)\, \hat{\Psi}^\dagger(x')\, V(\boldsymbol{x}, \boldsymbol{x}')\, \hat{\Psi}(x')\, \hat{\Psi}(x) \,. \tag{1.16}$$

The single-particle operator

$$J = \sum_{i=1}^{N} J(\boldsymbol{x}_i) \tag{1.17}$$

in second quantized form becomes

$$\hat{J} = \int d^3\boldsymbol{x}\, \hat{\Psi}^\dagger(x)\, J(x)\, \hat{\Psi}(x) \,, \tag{1.18}$$

and, in particular, the number density $n(\boldsymbol{x}) = \sum_{i=1}^{N} \delta(\boldsymbol{x} - \boldsymbol{x}_i)$ is given by

$$\hat{n}(x) = \hat{\Psi}^\dagger(x)\, \hat{\Psi}(x) \,, \tag{1.19}$$

where $\hat{\Psi}^\dagger(x)$ and $\hat{\Psi}(x)$ are creation and annihilation operators, respectively.

1.2
Second Quantization and Fock States

For a system in which the number of particles is variable, it is essential to introduce creation and destruction operators. However, it is also possible to do so when the number of particles is conserved. In that case, of course, the perturbation operators will contain an equal number of creation and destruction operators. In that case, it is customary to describe the procedure as "second quantization". First quantization replaces classical mechanical equations of motion, and second quantization replaces a Hamiltonian containing one-body forces, two-body forces, and so on by a Hamiltonian that is bilinear in creation and destruction operators, quadratic in creation and destruction operators, and so on. Nothing new is added, but the commutation rules of the creation and destruction operators make the bookkeeping of the states simpler than using permanents or determinants for the Schrödinger wave functions. For Bose particles, second quantization was developed by Dirac [1], and extended to Fermi particles by Wigner and Jordan [2]. A more detailed discussion is given by Fock [3] and by Landau and Lifshitz [4].

1.3
The Boson Case

It is simplest to describe the relation between the Schrödinger description and the second quantized description by assuming that we have a set of N non-interacting particles that can occupy any one of a set of orthonormal states $\phi_n(x_i)$. Besides, if the particles do not interact, the wave function can be a product of the ϕ_n that are occupied. In addition, (for Boson statistics) the wave function must be symmetric with respect to exchange of any two particles. If there are N_1 particles in state ϕ_1, N_2 in ϕ_2 for a total $N = \sum_i N_i$ of particles, the wave function can be written in the form

$$\Psi(N_1, N_2, \cdots) = \sqrt{\frac{N_1! N_2! \cdots}{N!}} \sum \phi_{p_1}(x_1)\, \phi_{p_2}(x_2) \cdots \phi_{p_N}(x_N) \,, \qquad (1.20)$$

where $p_1, p_2, \ldots, p_N$ is any set of occupied states (such as $p_1 = 1$, $p_2 = 3$, $p_3 = 4$, etc.). These indices must not all be different since some states can be multiply occupied. However, where they are distinct, the sum must be taken over all permutations of the distinct indices. Since the number of ways of placing N_1 particles (out of N) in one box, N_2 in a second and so on is given by $N!/(N_1!N_2!\cdots)$. The prefactor in Eq. (1.20) is added to preserve normalization. We can refer to $\Psi(N_1, N_2, \cdots)$ as a Fock state with N_1 particles in $\phi_1(x)$, N_2 in $\phi_2(x)$, and so on.

We then introduce creation and destruction operators $B_i^\dagger$ and B_j defined by

$$B_i^\dagger \Psi(N_1, \cdots, N_i, \cdots) = \sqrt{N_i + 1}\, \Psi(N_1, \cdots, N_i + 1, \cdots) \,, \qquad (1.21)$$

and

$$B_j \Psi(N_1, \cdots, N_j, \cdots) = \sqrt{N_j}\, \Psi(N_1, \cdots, N_j - 1, \cdots) \,. \qquad (1.22)$$

The principal simplification of second quantization is that a one-body operator

$$V = \sum_a V(\boldsymbol{r}_a) , \tag{1.23}$$

which can take $\Psi(N_i - 1, N_k)$ into $\Psi(N_i, N_k - 1)$ by "destroying" a particle in state k and creating one in state i in the Schrödinger permanent wave functions can be much more easily calculated when the operator

$$\hat{V} = V_{ik} B_i^\dagger B_k \tag{1.24}$$

acts on the Fock states. In particular, the matrix element

$$\begin{aligned}
&\int d^3\boldsymbol{r}\Psi^*(N_i, N_k - 1)\,\hat{V}\Psi(N_i - 1, N_k) \\
&= \int d^3\boldsymbol{r}\Psi^*(N_i, N_k - 1)\,V_{ik} B_i^\dagger B_k \Psi(N_i - 1, N_k) \\
&= \sqrt{N_i N_k}\,V_{ik} ,
\end{aligned} \tag{1.25}$$

where the first integral is over the Schrödinger space $dx_l dx_2 \cdots dx_N$. The second one is thought of in terms of creation and destruction operators in a space described by the number set $\{N_j\}$, and the matrix element

$$V_{ik} = \int d^3\boldsymbol{r}\phi_i^*(\boldsymbol{r})\,V(\boldsymbol{r})\,\phi_k(\boldsymbol{r}) \tag{1.26}$$

is the usual one-body matrix element in the Schrödinger representation.

Landau and Lifshitz [4] do not derive this result. They merely state that "The calculation of these matrix elements is, in principle, very simple, it is easier to do it oneself than to follow an account of it." It would be unfair to leave the matter there: Landau knows how to do it; let it be an exercise for the reader.

We can make the answer plausible by showing that the right-hand side of Eq. (1.25) is a product of four factors:

$$P \equiv \left[\frac{(N_i - 1)!N_k!}{(N_i + N_k - 1)!}\right]^{1/2} \left[\frac{N_i!(N_k - 1)!}{(N_i + N_k - 1)!}\right]^{1/2} V_{ik} M . \tag{1.27}$$

The first factor comes from the normalization factor of the initial state, and the second from the normalization of the final state. Factors involving N_j for $j \neq i$ or k are ignored since they merely contribute to the normalization of the remaining states. A single product of ϕs for the initial state, a product for the final state and one $V(\boldsymbol{r}_a)$ give rise to a term V_{ik} or zero. The factor M is simply the number of such non-vanishing terms. Since each $V(\boldsymbol{r}_a)$ for $a = 1, 2, \ldots, N$ makes the same contribution, M contains a factor $N = (N_i - 1) + N_k = N_i + (N_k - 1)$, the total number of particles in these two states (in either the final or the initial states). For a given $V(\boldsymbol{r}_a)$, the factor $\int d^3\boldsymbol{r}_a\phi_i^*(\boldsymbol{r}_a)V(\boldsymbol{r}_a)\phi_k(\boldsymbol{r}_a)$ appears and one particle is used up. The remaining $N - 1 = N_i + N_k - 2$ particles must now be distributed with $N_i - 1$ in the ith state and $N_k - 1$ in the kth state. Thus,

$$M = N\frac{(N - 1)!}{(N_i - 1)!(N_k - 1)!} , \tag{1.28}$$

where the first factor, $N = N_i + N_k - 1$, is the number of terms in the sum $\sum_a V(\boldsymbol{r}_a)$ and the second factor is the number of ways of distributing $N-1$ particles in two states with $N_i - 1$ in the first state and $N_k - 1$ in the second state. Finally,

$$P = \sqrt{N_i N_k} \frac{(N_i - 1)!\,(N_k - 1)!}{(N_i + N_k - 1)!} V_{ik} \frac{(N_i + N_k - 1)!}{(N_i - 1)!\,(N_k - 1)!}$$
$$= \sqrt{N_i N_k} V_{ik} \,. \tag{1.29}$$

The matrix elements in Eqs. (1.21) and (1.22) are such as to insure the commutations rules

$$[B_i, B_j]_- = [B_i^\dagger, B_j^\dagger]_- = 0\,, \quad [B_i, B_j^\dagger]_- = \delta_{ij}\,. \tag{1.30}$$

A natural generalization of Eq. (1.25) to two-body operators implies the replacement:

$$\sum_{a>b} W(\boldsymbol{r}_a, \boldsymbol{r}_b) \to \sum W_{lm}^{ik} B_i^\dagger B_k^\dagger B_l B_m \,. \tag{1.31}$$

A compact statement of these commutation rules in Eq. (1.30) can be obtained by introducing an operator $\psi(\boldsymbol{r})$ in the form

$$\psi(\boldsymbol{r}) = \sum_i B_i \phi_i(\boldsymbol{r})\,, \quad \psi^\dagger(\boldsymbol{r}) = \sum_i B_i^\dagger \phi_i^*(\boldsymbol{r})\,. \tag{1.32}$$

Then, the "Schrödinger operators", $\psi(\boldsymbol{r})$ and $\psi^\dagger(\boldsymbol{r}')$, obey the commutation rules

$$\left[\psi(\boldsymbol{r}), \psi(\boldsymbol{r}')\right]_- = \left[\psi^\dagger(\boldsymbol{r}), \psi^\dagger(\boldsymbol{r}')\right]_- = 0\,, \tag{1.33}$$

$$\left[\psi(\boldsymbol{r}), \psi^\dagger(\boldsymbol{r}')\right]_- = \sum_{i,j} \phi_i(\boldsymbol{r})\,\phi_j^*(\boldsymbol{r}')\,\delta_{ij} = \delta(\boldsymbol{r} - \boldsymbol{r}')\,, \tag{1.34}$$

where the δ_{ij} arises from Eq. (1.30) and the Dirac delta function follows from the completeness.

The second quantized Hamiltonian of a boson system with one and two-body forces can be written in the form

$$\hat{H} = \int d^3\boldsymbol{r} \left[\frac{\hbar^2}{2m} \nabla\psi^\dagger(\boldsymbol{r}) \cdot \nabla\psi(\boldsymbol{r}) + V(\boldsymbol{r})\,\psi^\dagger(\boldsymbol{r})\,\psi(\boldsymbol{r})\right]$$
$$+ \frac{1}{2} \int d^3\boldsymbol{r} \int d^3\boldsymbol{r}'\,\psi^\dagger(\boldsymbol{r})\,\psi^\dagger(\boldsymbol{r}')\,W(\boldsymbol{r}, \boldsymbol{r}')\,\psi(\boldsymbol{r}')\,\psi(\boldsymbol{r})\,. \tag{1.35}$$

In addition to the correspondence

$$\sum_a V(\boldsymbol{r}_a) \to \sum_{i,k} V_{ik} B_i^\dagger B_k \tag{1.36}$$

for one-body forces, we have a similar correspondence for two-body forces:

$$\sum_{a>b} W(\boldsymbol{r}_a, \boldsymbol{r}_b) \to \sum_{i,k;l,m} W_{lm}^{ik} B_i^\dagger B_k^\dagger B_l B_m\,, \tag{1.37}$$

where

$$W_{lm}^{ik} = \int d^3r \int d^3r' \phi_i^*(\mathbf{r})\phi_k^*(\mathbf{r}') W(\mathbf{r},\mathbf{r}')\phi_l(\mathbf{r})\phi_m(\mathbf{r}') \,. \tag{1.38}$$

We note that the commutation rules, Eqs. (1.30), (1.33), and (1.34) for bosons are the same as the ones we are familiar with for harmonic oscillators and phonons, which are of course bosons.

1.4
The Fermion Case

In the fermion case, the Pauli principle requires that the wave function be antisymmetric. The simplest example of a set of independent fermions is then described by a determinant

$$\Psi(N_1, N_2, \cdots) = \begin{vmatrix} \phi_{p_1}(\mathbf{r}_1) & \phi_{p_1}(\mathbf{r}_2) & \cdots \phi_{p_1}(\mathbf{r}_N) \\ \phi_{p_2}(\mathbf{r}_1) & \phi_{p_2}(\mathbf{r}_2) & \cdots \phi_{p_2}(\mathbf{r}_N) \\ \cdots & \cdots & \cdots\cdots \\ \phi_{p_N}(\mathbf{r}_1) & \phi_{p_N}(\mathbf{r}_2) & \cdots \phi_{p_N}(\mathbf{r}_N) \end{vmatrix} \tag{1.39}$$

in terms of the set of functions $\phi_i(\mathbf{r})$. The latter are usually taken as members of a complete set of eigen-functions of the one-body Hamiltonian. Here, N is the total number of eigen-functions appearing in the determinant, that is, the total number of occupied states. The set of numbers $p_1, p_2, \cdots, p_N$ are some chosen ordering of the set $\{i\}$. To make the sign of the determinant unique, a fixed order must be chosen. It is conventional to choose the ordering

$$p_1 < p_2 < p_3 < \cdots < p_N \,. \tag{1.40}$$

This is not necessary, but a fixed choice must be maintained in the ensuing discussion.

The result of second quantization for fermions will look similar to that for bosons in the sense that Eq. (1.32) is replaced by

$$\psi(\mathbf{r}) = \sum_i F_i \phi_i(\mathbf{r}) \,, \quad \psi^\dagger(\mathbf{r}) = \sum_i F_i^\dagger \phi_i^*(\mathbf{r}) \,, \tag{1.41}$$

where the boson operators B_i have been replaced by the fermion operators F_i. With this change, Eqs. (1.35)–(1.38) remain valid. However, fermion states can have occupancies only of $N_i = 0$ or $N_i = 1$. This is accomplished by the use of anti-commutation rules

$$\left[F_i, F_j^\dagger\right]_+ = \delta_{ij} \,, \quad \left[F_i, F_j\right]_+ = 0 \,, \quad \left[F_i^\dagger, F_j^\dagger\right]_+ = 0 \tag{1.42}$$

rather than the commutation rules used in the boson case. In particular, Eqs. (1.33) and (1.34) are replaced by

$$[\psi(\mathbf{r}), \psi(\mathbf{r}')]_+ = [\psi^\dagger(\mathbf{r}), \psi^\dagger(\mathbf{r}')]_+ = 0 \,, \tag{1.43}$$

$$[\psi(\mathbf{r}), \psi^\dagger(\mathbf{r}')]_+ = \delta(\mathbf{r} - \mathbf{r}') \,, \tag{1.44}$$

which follows directly from Eq. (1.42).

To see that the anti-commutation rules, Eq. (1.42), accomplish the desired objectives, we first consider a single state $\phi_i(\mathbf{r})$ with operators F_i and $F_i^\dagger$, and omit the index i.

$$F^2 = 0 \,, \quad \left(F^\dagger\right)^2 = 0 \,. \tag{1.45}$$

Let $N = F^\dagger F$. Then,

$$\begin{aligned} N^2 = \left(F^\dagger F\right)^2 &= F^\dagger F F^\dagger F = F^\dagger \left(1 - F^\dagger F\right) F \\ &= F^\dagger F - \left(F^\dagger\right)^2 F^2 = F^\dagger F - 0 = N \end{aligned} \tag{1.46}$$

so that

$$N = 1 \quad \text{or} \quad N = 0 \,. \tag{1.47}$$

Consider the eigenstates of N:

$$N \Psi_0 = 0 \Psi_0 \quad \text{or} \quad N \Psi_1 = \Psi_1 \,. \tag{1.48}$$

Then, is $F^\dagger \Psi_0$ also an eigenstate of N?

$$\begin{aligned} \left(F^\dagger F\right) F^\dagger \Psi_0 &= F^\dagger \left(1 - F^\dagger F\right) \Psi_0 \\ &= F^\dagger \Psi_0 - \left(F^\dagger\right)^2 F \Psi_0 \,. \end{aligned} \tag{1.49}$$

Therefore,

$$N F^\dagger \Psi_0 = 1 \left(F^\dagger \Psi_0\right) \,, \tag{1.50}$$

that is, $F^\dagger \Psi_0$ is proportional to Ψ_1. Evaluate the normalization:

$$\begin{aligned} \int d^3\mathbf{r} \left(F^\dagger \Psi_0\right)^* \left(F^\dagger \Psi_0\right) &= \int d^3\mathbf{r}, \Psi_0^* F F^\dagger \Psi_0 \\ &= \int d^3\mathbf{r} \Psi_0^* \left(1 - F^\dagger F\right) \Psi_0 = 1 \,. \end{aligned} \tag{1.51}$$

Therefore, $F^\dagger \Psi_0$ is normalized and we can choose

$$F^\dagger \Psi_0 = \Psi_1 \,. \tag{1.52}$$

Similarly,

$$F F^\dagger \left(F \Psi_1\right) = F \left(1 - F F^\dagger\right) \Psi_1 = F \Psi_1 \,. \tag{1.53}$$

Therefore, $F \Psi_1$ is an eigenvector of $F F^\dagger = 1 - F^\dagger F$ with eigenvalue one or eigenvector of $F^\dagger F$ with eigenvector zero. By a similar procedure to Eq. (1.28), $F \Psi_1$ is normalized and we choose

$$F \Psi_1 = \Psi_0 \,. \tag{1.54}$$

Note that

$$F\Psi_0 = F^2\Psi_1 = 0 \,, \tag{1.55}$$

$$F^\dagger \Psi_1 = \left(F^\dagger\right)^2 \Psi_0 = 0 \,. \tag{1.56}$$

The above discussion has established that the anti-commutation rules generate a set of states with occupancies zero and one. The full correspondence between first and second quantization requires that we establish the analogue of Eq. (1.36):

$$V = \sum_a V(\boldsymbol{r}_a) \to \sum_{ik} V_{i,k} F_i^\dagger F_k \,. \tag{1.57}$$

This involves the matrix element of the one-body operator V between two determinantal states. In effect, a transition in which a fermion in state k is destroyed and one in state i is created was found to have the matrix element

$$\int d^3\boldsymbol{r}\Psi^*(N_i, N_k - 1)\, V\Psi(N_i - 1, N_k) = V_{ik} \tag{1.58}$$

between determinantal states.

[4, Eq. (61.3)] allege (without proof) that the result in Eq. (1.58) should instead be

$$\int d^3\boldsymbol{r}\Psi^*(N_i, N_k - 1)\, V\Psi(N_i - 1, N_k) = V_{ik}(-1)^{\Sigma} \,, \tag{1.59}$$

where the symbol

$$\Sigma = \sum_{j=i+1}^{k-1} N_j \,. \tag{1.60}$$

This discrepancy can be resolved as follows. In our calculation, we obtained the final wave function (before anti-symmetrization) from the initial wave function simply by replacing $\phi_k(\boldsymbol{r}_a)$ with $\phi_i(\boldsymbol{r}_b)$. However, this procedure does not preserve the chosen ordering, Eq. (1.40). To restore the chosen ordering, one must interchange row i and k in the final determinant. These gain a factor $(-1)^{\Sigma}$ where Σ is the number of occupied states between i and k.

To maintain the validity of Eq. (1.57), the operator F_k takes the state $N_k = 1$ into $N_k = 0$ with an extra factor

$$F_k\Psi(N_k = 1) = \eta_k\Psi(N_k = 0) \,, \tag{1.61}$$

and

$$F_k^\dagger\Psi(N_k = 0) = \eta_k^*\Psi(N_k = 1) \,. \tag{1.62}$$

The anti-commutator $[F_k, F_k^\dagger]_+$ is unchanged as long as $|\eta_k|^2 = 1$. If we define

$$\eta_k = \prod_{j=1}^{k-1} (-1)^{N_j} \,, \tag{1.63}$$

then $F_i^\dagger F_k$ acquires just the extra factor $(-1)^{\sum(i+1,k-1)}$ demanded by Eq. (1.59). Moreover, it is easy to see why

$$F_m F_n = -F_n F_m \tag{1.64}$$

because one of m, n (say m) is higher in the sequence of states. Then, the matrix for F_m in $F_m F_n$ is reversed in sign because F_n has acted and eliminated the state n below m. In the reverse order, $F_n F_m$, F_n is unaffected by the elimination of state m above it. Hence, the two orders differ by a factor -1 to yield the desired anti-commutation rule. These remarks are stated clearly in [5].

1.5
The Hamiltonian of Electrons

We first consider the case of a single electron, or of a set of non-interacting electrons. The Hamiltonian can be written in the form:

$$H = \int d^3 r \psi^\dagger(\mathbf{r})\, U(\mathbf{r})\, \psi(\mathbf{r}) \; . \tag{1.65}$$

Here, $\psi(\mathbf{r})$ and $\psi^\dagger(\mathbf{r})$ are regarded as operators that can be expanded in an arbitrary orthonormal set $\phi_n(\mathbf{r})$:

$$\psi(\mathbf{r}) = \sum_n c_n \phi_n(\mathbf{r}) \; , \tag{1.66}$$

$$\psi^\dagger(\mathbf{r}) = \sum_m c_m^\dagger \phi_m^*(\mathbf{r}) \; . \tag{1.67}$$

The Hamiltonian then takes the form

$$\begin{aligned} H &= \sum_{m,n} c_m^\dagger \int d^3 r \phi_m^*(\mathbf{r})\, U(\mathbf{r})\, \phi_n(\mathbf{r})\, c_n \\ &= \sum_{m,n} c_m^\dagger \langle m|U|n\rangle c_n \; , \end{aligned} \tag{1.68}$$

where $\langle m|U|n\rangle$ represents the matrix for destruction of electrons in n and its creation in m. If the original Schrödinger equation is

$$\left[-\frac{\hbar^2}{2m^*}\nabla^2 + V(\mathbf{r})\right]\phi = E\phi \; , \tag{1.69}$$

where m^* represents the mass of an electron, then U is the operator defined by

$$U = -\frac{\hbar^2}{2m}\nabla^2 + V(\mathbf{r}) \; , \tag{1.70}$$

whose matrix element is

$$\langle m|U|n\rangle = \int d^3 r \phi_m^*(\mathbf{r}) \left[-\frac{\hbar^2}{2m^*}\nabla^2 + V(\mathbf{r})\right] \phi_n(\mathbf{r}) = U_{mn} \; . \tag{1.71}$$

To best understand the eigenstates of the operator

$$H = \sum_{m,n} U_{mn} c_m^\dagger c_n \ , \qquad (1.72)$$

we can choose the ϕ_n to be the eigenstates of U with eigenvalues E_n. Then,

$$U_{mn} = E_m \delta_{mn} \ , \qquad (1.73)$$

$$H = \sum_n E_n c_n^\dagger c_n \ . \qquad (1.74)$$

The Hilbert vector Ψ that is an eigenvector of H will then simply be described by a set of occupancies zero or one of each $N = c_n^\dagger c_n$, for example, $|0,1,0,0,1,1,0,1,\ldots\rangle$. To solve the Schrödinger equation $H\Psi = E\Psi$, we can write

$$\sum_{m,n} \left(c_m^\dagger c_n U_{mn}\right) \Psi = E\Psi \ . \qquad (1.75)$$

1.6 Electron–Phonon Interaction

Following Callaway [6] the Hamiltonian is written as an electron energy, plus a phonon energy, plus an electron–phonon interaction:

$$H = \sum_{k,\sigma} E_k c_{k\sigma}^\dagger c_{k\sigma} + \sum_q \hbar\omega_q a_q^\dagger a_q + \sum_{k,q,\sigma} \left[D(q) c_{k+q,\sigma}^\dagger c_{k\sigma} a_q + D(-q) c_{k-q,\sigma}^\dagger c_{k\sigma} a_q^\dagger \right] . \qquad (1.76)$$

Quasi-momentum conservation is built into the above expression and σ is the index for electron spin. The original form for phonon absorption was

$$V_{k'k} \delta_{k',k+q} c_{k'\sigma}^\dagger c_{k\sigma} a_q \ . \qquad (1.77)$$

We assume that the potential is local so that $V_{k'k} = V_{k'-k}$. There are also screened Coulomb electron–electron terms that we ignore here. These have the form

$$\int d^3 \boldsymbol{r}_1 \cdots \int d^3 \boldsymbol{r}_n \Psi^* (\boldsymbol{r}_1, \cdots, \boldsymbol{r}_n) \frac{1}{2} \sum_{i \neq j = 1}^{n} W\left(\boldsymbol{r}_i - \boldsymbol{r}_j\right) \Psi\left(\boldsymbol{r}_1, \cdots, \boldsymbol{r}_n\right) , \qquad (1.78)$$

corresponding to the many-body wave-function energy. If we insert $\psi(\boldsymbol{r}) = \sum_j c_j \phi_j(\boldsymbol{r})$ which expresses the operators $\psi(\boldsymbol{r})$ in terms of the c_j operators, we get

$$H_{\text{two-body}} = \frac{1}{2} \sum_{i,k;l,m} c_i^\dagger c_k^\dagger \langle kl | W | ml \rangle c_l c_m \ , \qquad (1.79)$$

which is a representation of two-body interactions in second quantized form.

1.7
Effective Electron–Electron Interaction

In an electromagnetic field, the charge-1 acts on the field and the field acts on charge-2. If we can eliminate the field, we obtain a direct interaction between charge-1 and charge-2. Here, the field is the phonon field. After we eliminate the electron–phonon interaction, we should obtain an effective electron–electron interaction.

Let

$$H = H_0 + H_1 \,, \tag{1.80}$$

where H_1 is the interaction Hamiltonian. The transformed Hamiltonian

$$\begin{aligned} H_\mathrm{T} &= e^{-iS} H e^{iS} \\ &= H + i[H, S] - \frac{1}{2}[[H, S], S] + \cdots \\ &= H_0 + H_1 + i[H_0, S] + i[H_1, S] - \frac{1}{2}[[H_0, S], S] + \cdots \end{aligned} \tag{1.81}$$

to second order of S. To dispose of H_1 to lowest order, we set

$$H_1 + i[H_0, S] = 0 \,, \quad \text{i.e.,} \quad [H_0, S] = iH_1 \,. \tag{1.82}$$

Then,

$$H_\mathrm{T} = H_0 + i[H_1, S] - \frac{i}{2}[H_1, S] = H_0 + \frac{i}{2}[H_1, S] \,. \tag{1.83}$$

Let $|m\rangle$ and $|n\rangle$ be energy eigenstates of the complete system of electrons and phonons. Then, we get

$$\begin{aligned} \langle m|[H_0, S]|n\rangle &= i\langle m|H_1|n\rangle \,, \quad \text{and} \quad \langle m|H_1|m\rangle = 0 \,, \\ (E_m - E_n)\, S_{mn} &= i\langle m|H_1|n\rangle \,, \\ S_{mn} &= \frac{\langle m|H_1|n\rangle}{E_m - E_n} \,, \quad \text{for} \quad m \neq n \,. \end{aligned} \tag{1.84}$$

However,

$$\langle n_q - 1|a_q|n_q\rangle = n_q^{1/2} \,, \quad \langle n_q + 1|a_q^\dagger|n_q\rangle = \left(n_q + 1\right)^{1/2} \,. \tag{1.85}$$

Writing S as operators cs (electronic part) and as a matrix in the vibrational part, we obtain

$$\langle n_q - 1|S|n_q\rangle = i\sum_{k,\sigma} \frac{D(q) c_{k+q,\sigma}^\dagger c_{k\sigma} n_q^{1/2}}{E(k+q) - E(k) - \hbar\omega_q} \,. \tag{1.86}$$

Here, a phonon of wave vector q is absorbed and an electron is scattered from k to $k + q$ at the same time, that is,

$$\langle n_q + 1|S|n_q\rangle = i\sum_{k,\sigma} \frac{D(-q)c^\dagger_{k-q,\sigma}c_{k\sigma}\left(n_q + 1\right)^{1/2}}{E\,(k - q) - E(k) - \hbar\omega_q} . \tag{1.87}$$

In Eq. (1.87), a phonon of wave vector q is created and an electron is scattered from k to $k - q$ at the same time.

We are concerned with the effective second-order interaction

$$\frac{i}{2}[H_1, S]$$

and by this we mean the part diagonal in the phonon numbers. (The off-diagonal elements can be transformed away to give still higher-order interactions). We can write

$$\begin{aligned}\langle n_q|\frac{i}{2}[H_1, S]|n_q\rangle = \frac{i}{2}\sum &\{\langle n_q|H_1|n_q \pm 1\rangle\langle n_q \pm 1|S|n_q\rangle \\ &- \langle n_q|S|n_q \pm 1\rangle\langle n_q \pm 1|H_1|n_q\rangle\} ,\end{aligned} \tag{1.88}$$

where the intermediate states $|n_q \pm 1\rangle$ are summed over. We write out one term explicitly:

$$\begin{aligned}&\frac{i}{2}\langle n_q\,|H_1|\,n_q - 1\rangle\langle n_q - 1\,|S|\,n_q\rangle \\ &= \frac{i}{2}\sum_{k',\sigma'} D(-q)n_q^{1/2}c^\dagger_{k'-q\sigma'}c_{k'\sigma'}\sum_{k,\sigma}\frac{i\,D(q)c^\dagger_{k+q,\sigma}c_{k\sigma}n_q^{1/2}}{E(k + q) - E(k) - \hbar\omega_q} \\ &= -\frac{1}{2}\sum_{k,k';\sigma,\sigma'}\frac{|D(q)|^2 c^\dagger_{k'-q,\sigma'}c_{k'\sigma'}c^\dagger_{k+q\sigma}c_{k\sigma}}{E(k + q) - E(k) - \hbar\omega_q} ,\end{aligned} \tag{1.89}$$

where $D(-q) = D^*(q)$. Callaway's Eq. (7.8.5) [6] states that the four terms combine to give

$$\begin{aligned}H_1 = \sum_{k,k',q;\sigma,\sigma'} &\frac{|D(q)|^2\hbar\omega_q}{[E(k) - E(k - q)]^2 - (\hbar\omega_q)^2} \\ &\times c^\dagger_{k-q\sigma}c^\dagger_{k'+q\sigma'}c_{k'\sigma'}c_{k\sigma} .\end{aligned} \tag{1.90}$$

With the replacement $k \to k + q$, we get

$$\begin{aligned}H_1 = \sum_{k,k',q;\sigma,\sigma'} &\frac{|D(q)|^2\hbar\omega_q}{[E(k + q) - E(k)]^2 - (\hbar\omega_q)^2} \\ &\times c^\dagger_{k\sigma}c^\dagger_{k'+q\sigma'}c_{k'\sigma'}c_{k+q\sigma} .\end{aligned} \tag{1.91}$$

Note that the effective Hamiltonian is independent of temperature. For $E(k) - E(k - q) < \hbar\omega_q$, the interaction becomes attractive.

An alternate derivation of the above H_1 is obtained in a semiclassical way by Rickayzen [7, p. 117–121], by considering an electron fluid and an ion fluid and retarded interactions between the two components.

We will be dealing with systems of many interacting particles and, as a result, we need to include the inter-particle potential in the Schrödinger equation. This problem is the basis of the present book. The N-particle wave function in configuration space contains all the possible information. However, a direct solution of the Schrödinger equation is not practical. We therefore need other techniques which involve (a) second quantization, (b) quantum field theory, and (c) Green's functions.

Second quantization describes the creation and annihilation of particles and quantum statistics as well as simplifying the problem of many interacting particles. This approach reformulates the Schrödinger equation. The advantage it has is that we avoid the awkward use of symmetrized and anti-symmetrized product of single-particle wave functions. With the method of quantum field theory, we avoid dealing with the wave functions and thus the coordinates of all the particles – bosons and fermions.

Green's functions can be used to calculate many physical quantities such as (1) the ground state energy, (2) thermodynamic functions, (3) the energy and lifetime of excited states, and (4) linear response to external perturbations. The exact Green's functions are also difficult to calculate which means we must use perturbation theory. This is presented with the use of Feynman diagrams. This approach allows us to calculate physical quantities to any order of perturbation theory. We use functional derivative techniques in the perturbation expansion of the Green's function determined by the Dyson equation and show that only linked diagrams contribute. Wick's theorem which forms all possible pairs of the field operators is not used in this approach.

1.8 Degenerate Electron Gases

We now illustrate the usefulness of the second quantization representation by applying it to obtain some qualitative results for a metal. The simple model we use is that of an interacting electron gas with a uniform positive background so that the total system is neutral. We ignore the motion of the ions/positive charge. We do not consider any surface effects by restricting our attention to the bulk medium. We insert the system into a large box of side L and apply periodic boundary conditions; this ensures invariance under spatial translations of all physical quantities. The single-particle states are plane waves

$$\varphi_{k,\lambda}(x) = \frac{1}{L^{3/2}} e^{ik\cdot x} \eta_\lambda \,, \quad \eta\uparrow = \begin{pmatrix} 1 \\ 0 \end{pmatrix}, \quad \eta\downarrow = \begin{pmatrix} 0 \\ 1 \end{pmatrix}, \tag{1.92}$$

where $k_i = 2\pi n_i/L$, $n_i = 0, \pm 1, \pm 2, \ldots$ The total Hamiltonian is

$$H = H_e + H_{e-b} + H_b \,, \tag{1.93}$$

where

$$H_e = \sum_{i=1}^{N} \frac{p_i^2}{2m^*} + \frac{e^2}{2} \sum_{i \neq j=1}^{N} \frac{e^{-\kappa |x_i - x_j|}}{|x_i - x_j|} , \tag{1.94}$$

$$H_b = \frac{e^2}{2} \int d^3x \int d^3x' \frac{n(x)\, n(x')}{|x - x'|} e^{-\kappa |x - x'|} , \tag{1.95}$$

$$H_{e-b} = -e^2 \sum_{i=1}^{N} \int d^3x\, n(x) \frac{e^{-\kappa |x - x_i|}}{|x - x_i|} . \tag{1.96}$$

Here, κ is the inverse screening length required for convergence of the integrals. Individual integrals diverge in the thermodynamic limit $N \to \infty$, $\mathcal{V} \to \infty$ but $n = N/\mathcal{V}$ is a constant. The sum of the three terms must however remain meaningful in this limit. For a uniform positive background $n(x) = N/\mathcal{V}$, we have

$$\begin{aligned} H_b &= \frac{e^2}{2} \left(\frac{N}{\mathcal{V}} \right)^2 \int d^3x \int d^3x' \frac{e^{-\kappa |x - x'|}}{|x - x'|} \\ &= \frac{e^2}{2} \left(\frac{N}{\mathcal{V}} \right)^2 \int d^3x \int d^3x'' \frac{e^{-\kappa x''}}{x''} \\ &= \frac{e^2}{2} \frac{N^2}{\mathcal{V}} \frac{4\pi}{\kappa^2} , \end{aligned} \tag{1.97}$$

$$\begin{aligned} H_{e-b} &= -e^2 \sum_{i=1}^{N} \left(\frac{N}{\mathcal{V}} \right) \int d^3x \frac{e^{-\kappa |x - x_i|}}{|x - x_i|} \\ &= -e^2 \sum_{i=1}^{N} \left(\frac{N}{\mathcal{V}} \right) \int d^3x \int d^3x'' \frac{e^{-\kappa x''}}{x''} \\ &= -e^2 \frac{N^2}{\mathcal{V}} \frac{4\pi}{\kappa^2} . \end{aligned} \tag{1.98}$$

Therefore, the total Hamiltonian is

$$H = \underbrace{-\frac{e^2}{2} \frac{N^2}{\mathcal{V}} \frac{4\pi}{\kappa^2}}_{\text{a C-number}} + H_e . \tag{1.99}$$

Forming a linear combination of the creation and destruction operators as

$$\hat{\psi}(x) = \sum_{k,\lambda} \varphi_{k,\lambda}(x)\, \hat{a}_{k,\lambda} , \quad \hat{\psi}^\dagger(x) = \sum_{k,\lambda} \varphi^*_{k,\lambda}(x)\, \hat{a}^\dagger_{k,\lambda} , \tag{1.100}$$

we rewrite H_e in second quantized form and the total Hamiltonian is

$$\begin{aligned} H = &-\frac{e^2}{2} \frac{N^2}{\mathcal{V}} \frac{4\pi}{\kappa^2} + \sum_{k,\lambda} \frac{\hbar^2 k^2}{2m^*} \hat{a}^\dagger_{k,\lambda} \hat{a}_{k,\lambda} \\ &+ \frac{e^2}{2\mathcal{V}} \sum_{k_1,\lambda_1} \sum_{k_2,\lambda_2} \sum_{k_3,\lambda_3} \sum_{k_4,\lambda_4} \delta_{\lambda_1,\lambda_3} \delta_{\lambda_2,\lambda_4} \delta_{k_1+k_2,k_3+k_4} \\ &\times \frac{4\pi}{|k_1 - k_3|^2 + \kappa^2} \hat{a}^\dagger_{k_1,\lambda_1} \hat{a}^\dagger_{k_2,\lambda_2} \hat{a}_{k_4,\lambda_4} \hat{a}_{k_3,\lambda_3} . \end{aligned} \tag{1.101}$$

By changing variables in the potential energy term to $\boldsymbol{k}$, $\boldsymbol{p}$ and $\boldsymbol{q}$, where $\boldsymbol{k}_1 = \boldsymbol{k}+\boldsymbol{q}$, $\boldsymbol{k}_2 = \boldsymbol{p}-\boldsymbol{q}$, $\boldsymbol{k}_3 = \boldsymbol{k}$ and $\boldsymbol{k}_4 = \boldsymbol{p}$, it becomes

$$\begin{aligned}
\text{P.E.} &= \frac{e^2}{2\mathcal{V}} \sum_{\boldsymbol{k},\boldsymbol{p},\boldsymbol{q}} \sum_{\lambda_1,\lambda_2} \frac{4\pi}{q^2+\kappa^2} \hat{a}^\dagger_{\boldsymbol{k}+\boldsymbol{q},\lambda_1} \hat{a}^\dagger_{\boldsymbol{p}-\boldsymbol{q},\lambda_2} \hat{a}_{\boldsymbol{p},\lambda_2} \hat{a}_{\boldsymbol{k},\lambda_1} \\
&= \frac{e^2}{2\mathcal{V}} \sum_{\boldsymbol{k},\boldsymbol{p},\boldsymbol{q}\neq 0} \sum_{\lambda_1,\lambda_2} \frac{4\pi}{q^2+\kappa^2} \hat{a}^\dagger_{\boldsymbol{k}+\boldsymbol{q},\lambda_1} \hat{a}^\dagger_{\boldsymbol{p}-\boldsymbol{q},\lambda_2} \hat{a}_{\boldsymbol{p},\lambda_2} \hat{a}_{\boldsymbol{k},\lambda_1} \\
&\quad + \underbrace{\frac{e^2}{2\mathcal{V}} \sum_{\boldsymbol{k},\boldsymbol{p}} \sum_{\lambda_1,\lambda_2} \frac{4\pi}{\kappa^2} \hat{a}^\dagger_{\boldsymbol{k},\lambda_1} \hat{a}^\dagger_{\boldsymbol{p},\lambda_2} \hat{a}_{\boldsymbol{p},\lambda_2} \hat{a}_{\boldsymbol{k},\lambda_1}}_{\frac{e^2}{2\mathcal{V}} \frac{4\pi}{\kappa^2} \sum_{\boldsymbol{k},\lambda_1} \sum_{\boldsymbol{p},\lambda_2} \hat{a}^\dagger_{\boldsymbol{k},\lambda_1} \hat{a}_{\boldsymbol{k},\lambda_1} \left(\hat{a}^\dagger_{\boldsymbol{p},\lambda_2} \hat{a}_{\boldsymbol{p},\lambda_2} - \delta_{\boldsymbol{k}\boldsymbol{p}} \delta_{\lambda_1\lambda_2}\right)} ,
\end{aligned} \tag{1.102}$$

where we separated the potential energy term into two terms corresponding to $\boldsymbol{q} = 0$ and $\boldsymbol{q} \neq 0$. The $\boldsymbol{q} = 0$ term can be further simplified as

$$\frac{e^2}{2\mathcal{V}} \frac{4\pi}{\kappa^2} \left(\hat{\mathrm{N}}^2 - \hat{\mathrm{N}}\right) , \tag{1.103}$$

where $\hat{\mathrm{N}}$ is the number operator. The ground state expectation value of Eq. (1.103) is

$$\frac{e^2}{2} \frac{\mathrm{N}^2}{\mathcal{V}} \frac{4\pi}{\kappa^2} - \frac{e^2}{2} \frac{\mathrm{N}}{\mathcal{V}} \frac{4\pi}{\kappa^2} , \tag{1.104}$$

where the first term in Eq. (1.104) cancels the first term of the Hamiltonian in Eq. (1.99) and the second term in Eq. (1.104) gives $-(e^2/2)(4\pi/\mathcal{V}\kappa^2)$ as an energy per particle. This second term vanishes when the thermodynamic limit is taken first. Therefore, the Hamiltonian for a bulk electron gas in a uniform positive background is

$$H = \sum_{\boldsymbol{k},\lambda} \frac{\hbar^2 k^2}{2m^*} \hat{a}^\dagger_{\boldsymbol{k},\lambda} \hat{a}_{\boldsymbol{k},\lambda} + \frac{2\pi e^2}{\mathcal{V}} \sum_{\boldsymbol{k},\boldsymbol{p},\boldsymbol{q}=\neq 0} \sum_{\lambda_1,\lambda_2} \frac{1}{q^2} \hat{a}^\dagger_{\boldsymbol{k}+\boldsymbol{q},\lambda_1} \hat{a}^\dagger_{\boldsymbol{p}-\boldsymbol{q},\lambda_2} \hat{a}_{\boldsymbol{p},\lambda_2} \hat{a}_{\boldsymbol{k},\lambda_1} , \tag{1.105}$$

where we have now safely set $\kappa = 0$.

1.9 Ground-State Energy in the High Density Limit

Let us denote the Bohr radius by $a_0 = \hbar^2/m^* e^2$ and the inter-particle spacing by r_0 so that $4/3\pi r_0^3 = \mathcal{V}/N$. Also, set $r_s = r_0/a_0$ so that $r_s \to 0$ in the high density limit.

Setting $\bar{V} = \mathcal{V}/r_0^3$ ($\bar{V}$ is fixed for given N) and $\bar{k} = k r_0$, we rewrite Eq. (1.105) as

$$H = \frac{e^2}{a_0 r_s^2} \left\{ \frac{1}{2} \sum_{\bar{k},\lambda} \bar{k}^2 \hat{a}^\dagger_{\bar{k},\lambda} \hat{a}_{\bar{k},\lambda} + \frac{2\pi r_s}{\bar{V}} \sum_{\bar{k},\bar{p},\bar{q}\neq 0} \sum_{\lambda_1,\lambda_2} \frac{1}{\bar{q}^2} \hat{a}^\dagger_{\bar{k}+\bar{q},\lambda_1} \hat{a}^\dagger_{\bar{p}-\bar{q},\lambda_2} \hat{a}_{\bar{p},\lambda_2} \hat{a}_{\bar{k},\lambda_1} \right\}. \quad (1.106)$$

Therefore,

1. The potential energy is a small perturbation of the kinetic energy in the high density limit, that is, $r_s \to 0$, of an electron gas.
2. The leading term of the interaction energy of a high density electron gas can be obtained using first order perturbation theory even though the potential is not weak or short-ranged.
3. The ground state energy is given by

$$E_{\text{GS}} = \frac{e^2}{a_0 r_s^2} \left\{ a + b r_s + c r_s^2 \ln r_s + d r_s^2 + \cdots \right\}, \quad (1.107)$$

where a, b and c are numerical constants. As a matter of fact, the "a" term corresponds to the ground state energy $E^{(0)}$ of a free Fermi gas, the "b" term gives the first-order energy shift $E^{(1)}$.

It is fairly straightforward to obtain $E^{(0)}$ and $E^{(1)}$, though we need advanced techniques to obtain the coefficients c and d. Denote the Fermi wave vector by $k_{\text{F}} = (3\pi^2 N/\mathcal{V})^{1/3} = (9\pi/4)^{1/3} r_s^{-1}$ so that

$$\begin{aligned} E^{(0)} &= \frac{\hbar^2}{2m^*} \sum_{k,\lambda} k^2 \theta\left(k_{\text{F}} - k\right) = \frac{\hbar^2}{2m^*} \cdot 2 \cdot \frac{\mathcal{V}}{(2\pi)^3} \int d^3 k k^2 \theta\left(k_{\text{F}} - k\right) \\ &= \frac{e^2}{2a_0} \frac{N}{r_s^2} \frac{3}{5} \left(\frac{9\pi}{4}\right)^{2/3}, \end{aligned} \quad (1.108)$$

where $\theta(x)$ is the unit step function. Thus, for a free Fermi gas, the ground state energy per particle is $E^{(0)}/N = 2.21/r_s^2$ Ry where $e^2/2a_0 = 13.6$ eV is 1 Ry. We now calculate the first-order correction to $E^{(0)}$, that is,

$$E^{(1)} = \frac{2\pi e^2}{\mathcal{V}} \sum_{k,p,q\neq 0} \sum_{\lambda_1,\lambda_2} \frac{1}{q^2} \langle g | \hat{a}^\dagger_{k+q,\lambda_1} \hat{a}^\dagger_{p-q,\lambda_2} \hat{a}_{p,\lambda_2} \hat{a}_{k,\lambda_1} | g \rangle, \quad (1.109)$$

where $|g\rangle$ is the ground state for non-interacting electrons. The states $(\boldsymbol{k}, \lambda_1)$ and $(\boldsymbol{p}, \lambda_2)$ must be occupied, and the states $(\boldsymbol{k} + \boldsymbol{q}, \lambda_1)$ and $(\boldsymbol{p} - \boldsymbol{q}, \lambda_2)$ must also be occupied. Therefore, we must have either

$$\text{either (a)} \begin{cases} (\boldsymbol{k}+\boldsymbol{q}, \lambda_1) = (\boldsymbol{k}, \lambda_1) \\ (\boldsymbol{p}-\boldsymbol{q}, \lambda_2) = (\boldsymbol{p}, \lambda_2) \end{cases} \text{or (b)} \begin{cases} (\boldsymbol{k}+\boldsymbol{q}, \lambda_1) = (\boldsymbol{p}, \lambda_2) \\ (\boldsymbol{p}-\boldsymbol{q}, \lambda_2) = (\boldsymbol{k}, \lambda_1) \end{cases}. \quad (1.110)$$

The choice given as (a) is forbidden since $\boldsymbol{q} \neq 0$ and the matrix element in Eq. (1.109) is

$$\begin{aligned} \langle g | \cdots | g \rangle &= \delta_{\boldsymbol{p},\boldsymbol{k}+\boldsymbol{q}} \delta_{\lambda_1 \lambda_2} \langle g | \hat{a}^{\dagger}_{\boldsymbol{k}+\boldsymbol{q},\lambda_1} \hat{a}^{\dagger}_{\boldsymbol{k},\lambda_1} \hat{a}_{\boldsymbol{k}+\boldsymbol{q},\lambda_1} \hat{a}_{\boldsymbol{k},\lambda_1} | g \rangle \\ &= -\delta_{\boldsymbol{p},\boldsymbol{k}+\boldsymbol{q}} \delta_{\lambda_1 \lambda_2} \langle g | \hat{n}_{\boldsymbol{k}+\boldsymbol{q},\lambda_1} \hat{n}_{\boldsymbol{k},\lambda_1} | g \rangle \\ &= -\delta_{\boldsymbol{p},\boldsymbol{k}+\boldsymbol{q}} \delta_{\lambda_1 \lambda_2} \theta \left(k_{\mathrm{F}} - k \right) \theta \left(k_{\mathrm{F}} - |\boldsymbol{k} + \boldsymbol{q}| \right) , \end{aligned} \tag{1.111}$$

so that

$$\begin{aligned} E^{(1)} &= -\frac{2\pi e^2}{\mathcal{V}} \sum_{\lambda} \sum_{\boldsymbol{k},\boldsymbol{q} \neq 0} \frac{1}{q^2} \theta \left(k_{\mathrm{F}} - k \right) \theta \left(k_{\mathrm{F}} - |\boldsymbol{k} + \boldsymbol{q}| \right) \\ &= -\frac{e^2}{2a_0} \frac{N}{r_s} \frac{3}{2\pi} \left(\frac{9\pi}{4} \right)^{1/3} . \end{aligned} \tag{1.112}$$

Thus, by combining the results for $E^{(0)}$ and $E^{(1)}$, we obtain the energy per particle in the limit as $r_s \to 0$ to be given by

$$\lim_{r_s \to 0} \frac{E_{\mathrm{GS}}}{N} \approx \frac{e^2}{2a_0} \left\{ \frac{2.21}{r_s^2} - \frac{0.916}{r_s} + \cdots \right\} . \tag{1.113}$$

The first term is the kinetic energy of the Fermi gas of electrons and dominates in the high density limit. The second term is the exchange energy term. It is negative and arises from the antisymmetry of the wave function. The direct part arises from the $\boldsymbol{q} = 0$ part of the Hamiltonian and cancels the $H_b + H_{e-b}$ terms as a result of charge neutrality.

The exchange term is not the total that arises from the electron–electron interaction. All that is left out is called the correlation energy. The leading contribution to the correlation energy of the degenerate electron gas will be obtained using Feynman graph techniques. However, we note that E_{GS}/N has a minimum at a negative value of the energy, that is, the system is bound, as shown in Figure 1.2. The Rayleigh–Ritz variational principle tells us that the exact ground state energy of a quantum mechanical system always has a lower energy than that evaluated using a normalized state for the expectation value of the Hamiltonian. The exact solution must also be that for a bound system with energy below our approximate solution and the binding energy is that of vaporization for metals.

1.10 Wigner Solid

The energy of the Fermi gas can be lowered if the electrons crystallize into a Wigner solid. The range of values of r_s for metals is $1.8 \lesssim r_s \lesssim 6.0$. At low densities, Wigner suggested that the electrons will become localized and form a regular lattice. This lattice could be a closed packed structure such as bcc, fcc or hcp. The electrons would vibrate around their equilibrium positions and the positive charge

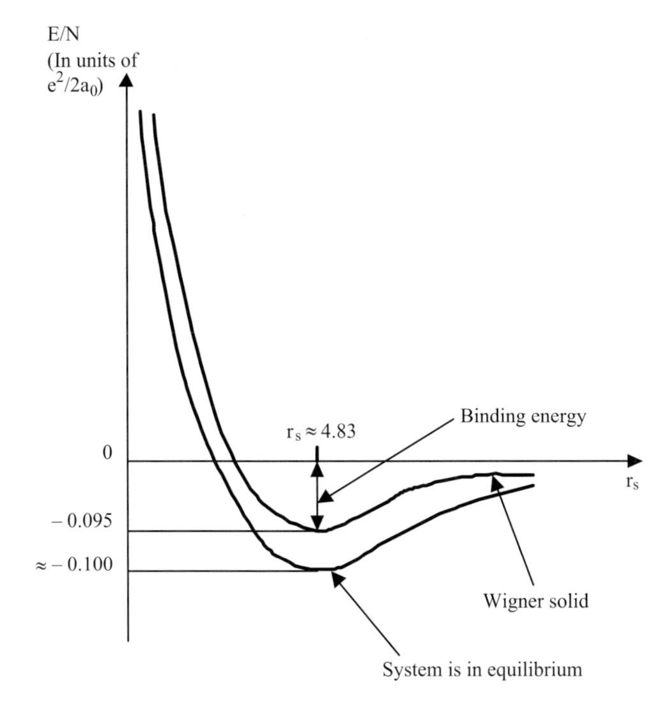

Figure 1.2 The energy per particle as a function of a dimensionless density parameter r_s, where $r_s \to 0$ corresponds to the high density limit, while $r_s \to \infty$ corresponds to the low density limit.

is still spread out in the system. The vibrational modes of the electrons would be at the plasmon frequency. For large r_s, the potential energy is much larger than the kinetic energy and there could be localization. In our discussion, the unit cell is taken as a sphere of radius $r_s a_0$ with the electron at the center. The total charge within the sphere is zero. Outside each sphere, the electric field is zero and consequently the spheres do not exert any electric fields on each other.

The potential energy between the electron and the uniform positive background is

$$E_{e-b} = n \int d^3\mathbf{r} \left(\frac{-e^2}{r} \right) = -\frac{3e^2}{r_s^3 a_0^3} \int_0^{r_s a_0} r\,dr = -\frac{3}{r_s} \left(\frac{e^2}{2a_0} \right) . \tag{1.114}$$

The potential energy due to the interaction of the positive charge with itself is obtained as follows. Let $V(r)$ be the potential energy from the positive charge at distance r from the center. The electric field is $E(r)$ where

$$e\,E(r) = -\frac{\partial V(r)}{\partial r} = \frac{e^2}{r^2} n \left(\frac{4}{3} \pi r^3 \right) = \left(\frac{e^2}{r_s^3 a_0^3} \right) r \,. \tag{1.115}$$

Integrating to obtain $V(r)$ gives a constant of integration. This is obtained by observing that the total potential from the electron and positive charge must vanish on the surface of the sphere and we obtain

$$V(r) = \frac{1}{r_s}\left[3 - \left(\frac{r}{r_s a_0}\right)^2\right]\frac{e^2}{2a_0} . \tag{1.116}$$

The interaction of the positive charge with itself is found by using

$$E_{b-b} = \frac{n}{2}\int d^3\mathbf{r}\,V(r) = \frac{6}{5r_s}\left(\frac{e^2}{2a_0}\right) . \tag{1.117}$$

Therefore, the total potential energy for the Wigner lattice in the Wigner–Seitz approximation is

$$E_{e-b} + E_{b-b} = -\frac{1.8}{r_s}\left(\frac{e^2}{2a_0}\right) . \tag{1.118}$$

This is larger than the exchange contribution for the free particle system. This system has gained energy by the localization of the electrons. Stroll has calculated the actual energy for several lattices. His results, expressed as $-A/r_s$, in unit of $(e^2/2a_0)$ are given as follows:

Lattice	A
sc	1.76
fcc	1.79175
bcc	1.79186
hcp	1.79168

1.11 The Chemical Potential of an Ideal Bose Gas and Bose–Einstein Condensation

For non-interacting bosons of energy $\varepsilon_k = \hbar^2 k^2/2m^*$, the total number at temperature T ($\beta = 1/(k_B T)$) is

$$\begin{aligned}\frac{N}{\mathcal{V}} &= \frac{g}{(2\pi)^3}\int_0^\infty dk 4\pi k^2 \frac{1}{e^{\beta(\varepsilon_k-\mu)} - 1} \\ &= \frac{g}{4\pi^2}\left(\frac{2m^*}{\hbar^2}\right)^{3/2}\int_0^\infty d\varepsilon \frac{\varepsilon^{1/2}}{e^{\beta(\varepsilon-\mu)} - 1} ,\end{aligned} \tag{1.119}$$

where g is the degeneracy and μ is the chemical potential. We must have $\varepsilon - \mu \geq 0$ since the mean occupation number must be positive for all energies. However, since we can have $\varepsilon = 0$, then $\mu \leq 0$. If a classical limit is taken with

$$\frac{\mu}{k_B T} \to -\infty , \tag{1.120}$$

then we get

$$\frac{N}{\mathcal{V}} \to \frac{g}{4\pi^2}\left(\frac{2m^*}{\hbar^2}\right)^{3/2} \int_0^\infty d\varepsilon\, \varepsilon^{1/2} e^{\beta(\mu-\varepsilon)}$$
$$= g e^{\beta\mu}\left(\frac{m^* k_B T}{2\pi\hbar^2}\right)^{3/2}, \tag{1.121}$$

which is the result of the Boltzmann distribution, where $\int_0^\infty dx x^2 e^{-x^2} = \sqrt{\pi}/4$ is used. Solving this equation for μ, we obtain

$$\frac{\mu}{k_B T} = \ln\left[\frac{N}{g\mathcal{V}}\left(\frac{2\pi\hbar^2}{m^* k_B T}\right)^{3/2}\right]. \tag{1.122}$$

A plot of this classical result is shown in Figure 1.3.

If T_0 is the temperature where $\mu = 0$, then Eq. (1.119) gives

$$\frac{N}{\mathcal{V}} = \frac{g}{4\pi^2}\left(\frac{2m^*}{\hbar^2}\right)^{3/2} \int_0^\infty d\varepsilon \frac{\varepsilon^{1/2}}{e^{\varepsilon/(k_B T_0)} - 1}. \tag{1.123}$$

The question which we now answer is what is the value of μ for $T < T_0$. If $\mu = 0$ for $T < T_0$, the integral in Eq. (1.119) is less than $N/\mathcal{V}$ in Eq. (1.123) because the value of the denominator is increased relative to its value at T_0 and the full value of $N/\mathcal{V}$ will not be reproduced. This can be rectified if we treat the system as follows. Below T_0, the system consists of two components: (1) particles occupying the zero momentum state with a mean occupation number N_0, and (2) particles occupying the excited state. This leads to

$$\frac{N}{\mathcal{V}} = \frac{N_0}{\mathcal{V}} + \frac{g}{4\pi^2}\left(\frac{2m^*}{\hbar^2}\right)^{3/2} \int_{0+}^\infty d\varepsilon \frac{\varepsilon^{1/2}}{e^{\varepsilon/k_B T} - 1}, \tag{1.124}$$

which gives $N_0/\mathcal{V} = (N/\mathcal{V})[1 - (T/T_0)^{3/2}]$ for $T < T_0$. Experimentally, it has been found that liquid He^4 has a phase transition at 2.2 K. Below that temperature, it acts

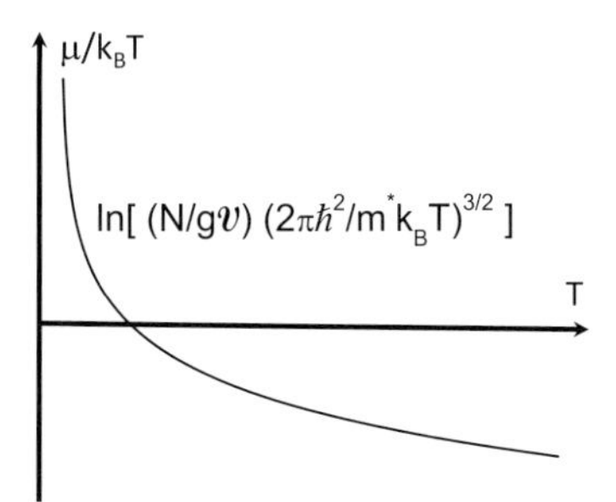

Figure 1.3 $\mu/(k_B T) = \ln[(N/g\mathcal{V})(2\pi\hbar^2/(m^* k_B T))^{3/2}]$ of a Bose gas as a function of T for fixed $N/\mathcal{V}$ in the classical limit.

like a mixture of superfluid and normal fluid. This discussion illustrates the Bose–Einstein condensation of the ideal Bose gas and thus gives a qualitative description of He^4. Inter-particle interactions play a key role in the properties of quantum fluids such as liquid He^4.

1.12 Problems

1. Show that

$$\left[\psi(x), \psi^\dagger(x')\,\psi(x'')\right] = \delta(x - x')\,\psi(x'')$$
$$\left[\psi^\dagger(x), \psi^\dagger(x')\,\psi(x'')\right] = -\delta(x - x'')\,\psi^\dagger(x')\,,$$

where $\psi(x)$ is a Boson field or a fermion field.

2. Show that for a fermion field

$$\begin{aligned}&\left[\psi(x), \psi^\dagger(x')\,\psi^\dagger(x'')\,\psi(x''')\,\psi(x'''')\right]\\ &= \delta(x - x')\,\psi^\dagger(x'')\,\psi(x''')\,\psi(x'''')\\ &\quad - \delta(x - x'')\,\psi^\dagger(x')\,\psi(x''')\,\psi(x'''')\end{aligned}$$

and

$$\begin{aligned}&\left[\psi^\dagger(x), \psi^\dagger(x')\,\psi^\dagger(x'')\,\psi(x''')\,\psi(x'''')\right]\\ &= \delta(x - x''')\,\psi^\dagger(x')\,\psi^\dagger(x'')\,\psi(x'''')\\ &\quad - \delta(x - x'''')\,\psi^\dagger(x')\,\psi^\dagger(x'')\,\psi(x''')\,.\end{aligned}$$

3. Show that for fermions

$$[a_{k'}a_{k''}, a^\dagger_{k'''}a_{k''''}]_- = \delta(k'' - k''')\,a_{k'}a_{k''''} - \delta(k' - k''')\,a_{k''}a_{k''''}\,.$$

4. a) Starting with the commutation relation $[a, a^\dagger] = 1$ for bosonic creation $a^\dagger$ and annihilation a operators, show that

$$\left[a^\dagger a, a\right] = -a\,, \quad \left[a^\dagger a, a^\dagger\right] = a^\dagger\,.$$

Using this result, show that if $|\alpha\rangle$ represents an eigenstate of the operator $a^\dagger a$ with eigenvalue α, $a|\alpha\rangle$ is also an eigenstate with eigenvalue $\alpha - 1$ (unless $a|\alpha\rangle = 0$).

b) If $|\alpha\rangle$ represents a normalized eigenstate of the operator $a^\dagger a$ with eigenvalue α for all $\alpha \geq 0$, show that

$$a|\alpha\rangle = \sqrt{\alpha}|\alpha - 1\rangle\,, \quad a^\dagger|\alpha\rangle = \sqrt{\alpha + 1}|\alpha + 1\rangle\,.$$

Defining the normalized vacuum state $|\Omega\rangle$ as the normalized state that is annihilated by the operator a, show that $|n\rangle = (1/\sqrt{n!})\,(a^\dagger)^n|\Omega\rangle$ is a normalized eigenstate of $a^\dagger a$ with eigenvalue n.

c) Assuming the operators a and $a^\dagger$ obey Fermionic anti-commutation relations, repeat parts (a) and (b).

5. Starting from first principles, show that the second quantized representation of the one-body kinetic energy operator is given by

$$\hat{T} = \int_0^L dx a^\dagger(x) \frac{p^2}{2m^*} a(x) .$$

Hint: Remember that the representation is most easily obtained from the basis in which the operator is diagonal.

6. Transforming to the Fourier basis, diagonalize the non-interacting three-dimensional cubic lattice tight-binding Hamiltonian

$$\hat{\mathcal{H}}^{(0)} = -\sum_{(m,n)} t_{mn,\sigma} c^\dagger_{m\sigma} c_{n\sigma} ,$$

where the matrix elements t_{mn} take the positive real value t between neighboring sites and zero otherwise. Comment on how this result compares with the spectrum of the Heisenberg ferromagnet.

7. Making use of the Pauli matrix identity $\sigma_{\alpha\beta} \cdot \sigma_{\gamma\delta} = 2\delta_{\alpha\delta}\delta_{\beta\gamma} - \delta_{\alpha\beta}\delta_{\gamma\delta}$, where "·" denotes the scalar or dot product, prove that

$$\hat{\mathbf{S}}_m \cdot \hat{\mathbf{S}}_n = -\frac{1}{2}\sum_{\alpha,\beta} c^\dagger_{m\alpha} c^\dagger_{n\beta} c_{m\beta} c_{n\alpha} - \frac{1}{4}\hat{n}_m \hat{n}_n ,$$

where $\hat{\mathbf{S}}_m = 1/2\sum_{\alpha,\beta} c^\dagger_{m\alpha} \sigma_{\alpha\beta} c_{m\beta}$ denotes the spin operator and $\hat{n}_m = \sum_\alpha c^\dagger_{m\alpha} c_{m\beta}$ represents the total number operator on site m. (Here, assume that lattice sites m and n are distinct.)

8. Starting with the definition

$$\hat{S}^- = (2S)^{1/2} a^\dagger \left(1 - \frac{a^\dagger a}{2S}\right)^{1/2} ,$$

confirm the validity of the Holstein–Primakoff transformation by explicitly checking the commutation relations of spin raising and lowering operators.

9. **Frustration**: On a bipartite lattice (i.e., one in which the neighbors of one sublattice belong to the other sublattice), the ground state (known as a Néel state) of a classical antiferromagnet can adopt a staggered spin configuration in which the exchange energy is maximized. Lattices which cannot be classified in this way are said to be frustrated – the maximal exchange energy associated with each bond cannot be recovered. Using only symmetry arguments, specify one of the possible ground states of a classical three-site triangular lattice antiferromagnet. (Note that the invariance of the Hamiltonian under a global

rotation of the spins means that there is a manifold of continuous degeneracy in the ground state.) Using the result, construct one of the classical ground states of the infinite triangular lattice.

10. Confirm that the bosonic commutation relations of the operators a and $a^\dagger$ are preserved by the Bogoliubov transformation,

$$\begin{pmatrix} \alpha \\ \alpha^\dagger \end{pmatrix} = \begin{pmatrix} \cosh\theta & \sinh\theta \\ \sinh\theta & \cosh\theta \end{pmatrix} \begin{pmatrix} a \\ a^\dagger \end{pmatrix} .$$

11. a) Making use of the spin commutation relation, $[S_i^\alpha, S_j^\beta] = i\delta_{ij}\varepsilon^{\alpha\beta\gamma} S_i^\gamma$, apply the identity to express the equation of motion of a spin in a nearest-neighbor spin S one-dimensional Heisenberg ferromagnet as a difference equation.
    b) Interpreting the spins as classical vectors and taking the continuum limit, show that the equation of motion of the *hydrodynamic modes* takes the form

$$\hbar \dot{\mathbf{S}} = J a^2 \mathbf{S} \times \partial^2 \mathbf{S} ,$$

    where a denotes the lattice spacing. (Hint: Going to the continuum limit, apply a Taylor expansion to the spins, i.e., $S_{i+1} = S_i + a\partial S_i + a^2 \partial^2 S_i/2 + \cdots$)
    c) Parameterizing the spin as

$$\mathbf{S} = \left(C\cos(kx - \omega t), C\sin(kx - \omega t), \sqrt{S^2 - C^2} \right) ,$$

    solve the equation. Sketch a "snapshot" configuration of the spins in a spin chain.

12. **Valence bond solid**: Starting with the spin-1/2 Majumdar–Ghosh Hamiltonian

$$\hat{H}_{\mathrm{MG}} = \frac{4|J|}{3} \sum_{n=1}^{N} \left(\hat{\mathbf{S}}_n \cdot \hat{\mathbf{S}}_{n+1} + \frac{1}{2} \hat{\mathbf{S}}_n \cdot \hat{\mathbf{S}}_{n+2} \right) + \frac{N}{2} ,$$

where the total number of states N is even and $\hat{\mathbf{S}}_{N+1} = \hat{\mathbf{S}}_1$, show that the two-dimer or valence bond states

$$|0_+\rangle = \prod_{n=1}^{N/2} \frac{1}{\sqrt{2}} \left(|\uparrow_{2n}\rangle \otimes |\downarrow_{2n\pm1}\rangle - |\downarrow_{2n}\rangle \otimes |\uparrow_{2n\pm1}\rangle \right)$$

are exact eigenstates. (Hint: Try to recast the Hamiltonian in terms of the total spin of a triad $\hat{\mathbf{J}}_n = \hat{\mathbf{S}}_{n+1} + \hat{\mathbf{S}}_n + \hat{\mathbf{S}}_{n-1}$ and consider what this representation implies.) In fact, these states represent the ground states of the Hamiltonian. Suggest what would happen if the total number of states was odd.

13. **Su–Schrieffer–Heeger model** of conducting polymers: Polyacetylene consists of bonded CH groups forming an isomeric long chain polymer. According to molecular orbital theory, the carbon atoms are expected to be sp^2 hybridized

suggesting a planar configuration of the molecule. An unpaired electron is expected to occupy a single p-orbital which points out of the plane. The weak overlap of the p-orbitals delocalizes the electrons into a π-conduction band. Therefore, according to the nearly-free electron theory, one might expect the half-filled conduction band of a polyacetylene chain to be metallic. However, the energy of a half-filled band of a one-dimensional system can always be lowered by applying a periodic lattice distortion known as the *Peierls instability*. One can think of an enhanced probability of finding the π electron on the short bond where the orbital overlap is stronger than the "the double bond". The aim of this problem is to explore the instability.

a) At its simplest level, the conduction band of polyacetylene can be modeled by a simple Hamiltonian, due to Su, Schrieffer and Heeger, in which the hopping matrix elements of the electrons are modulated by the lattice distortion of the atoms. Taking the displacement of the atomic sites from the equilibrium from the equilibrium separation $a \equiv 1$ to be unity, and treating their dynamics as classical, the effective Hamiltonian takes the form

$$\hat{H} = -t \sum_{n=1}^{N} \sum_{\sigma} (1 + u_n) \left[c^{\dagger}_{n\sigma} c_{n+1\sigma} + \text{h.c.} \right] + \sum_{n=1}^{N} \frac{k_s}{2} \left(u_{n+1} - u_n \right)^2 ,$$

where, for simplicity, the boundary conditions are taken to be periodic. The first term describes the hopping of electrons between neighboring sites with a matrix element modulated by the periodic distortion of the bond length, while the last term represents the associated increase in the elastic energy. Taking the lattice distortion to be periodic, $u_n = (-1)^n \alpha$, and the number of sites to be even, diagonalize the Hamiltonian. (Hint: The lattice distortion lowers the symmetry of the lattice. The Hamiltonian is most easily diagonalized by distinguishing the two sites of the sublattice, i.e., doubling the size of the elementary unit cell, and transforming to the Fourier representation.) Show that the Peierls distortion of the lattice opens a gap in the spectrum at the Fermi level of the half-filled system.

b) By estimating the total electronic and elastic energy of the half-filled band, that is, an average of one electron per lattice site, show that the one-dimensional system is always unstable towards the Peierls distortion. To do this calculation, you will need the approximate formula for the elliptic integral,

$$\int_{-\pi/2}^{\pi/2} dk \sqrt{1 - (1 - \alpha^2) \sin^2 k} \approx 2 + \left(a_1 - b_1 \ln \alpha^2 \right) \alpha^2 ,$$

where a_1 and b_1 are unspecified numerical constants.

c) For an even number of sites, the Peierls instability has two degenerate configurations – ABABAB$\cdots$ and BABABA$\cdots$ Comment on the qualitative form of the ground state lattice configuration if the number of sites is odd. Explain why such configurations give rise to mid-gap states.

14. In the Schwinger boson representation, the quantum mechanical spin is expressed in terms of two bosonic operators a and b in the form

$$\hat{S}^{+} = a^{\dagger} b \,, \quad \hat{S}^{-} = \left(\hat{S}^{+}\right)^{\dagger} \,, \quad \hat{S}^{z} = \frac{1}{2}\left(a^{\dagger} a - b^{\dagger} b\right) \,.$$

a) Show that this definition is consistent with the commutation relations for spin: $[\hat{S}^{+}, \hat{S}^{-}] = 2\hat{S}^{z}$.

b) Using the bosonic commutation relations, show that

$$|S, m\rangle = \frac{\left(a^{\dagger}\right)^{S+m}}{\sqrt{(S+m)!}} \frac{\left(b^{\dagger}\right)^{S-m}}{\sqrt{(S-m)!}} |\Omega\rangle$$

is compatible with the definition of an eigenstate of the total spin operator $\mathbf{S}^2$ and S^z. Here, $|\Omega\rangle$ denotes the vacuum of the Schwinger bosons, and the total spin S defines the physical subspace

$$\{|n_a, n_b\rangle : n_a + n_b = 2S\} \,.$$

15. **The Jordan–Wigner transformation**: So far, we have shown how the algebra of quantum mechanical spin can be expressed using boson operators, c.f., the Holstein–Primakoff transformation and the Schwinger representation. In this problem, we show that a representation for spin 1/2 can be obtained in terms of Fermion operators. Specifically, let us formally represent an up-spin as a particle and a down-spin as the vacuum $|0\rangle$, namely,

$$|\uparrow\rangle \equiv |1\rangle = f^{\dagger}(0) \,,$$
$$|\downarrow\rangle \equiv |0\rangle = f(1) \,.$$

In this representation, the spin raising and lowering operators are expressed in the form $\hat{S}^{+} = f^{\dagger}$ and $\hat{S}^{-} = f$, while $\hat{S}^{z} = f^{\dagger} f - 1/2$.

a) With this definition, confirm that the spins obey the algebra $[\hat{S}^{+}, \hat{S}^{-}] = 2\hat{S}^{z}$. However, there is a problem, that is, spins on different sites commute while fermion operators anticommute, for example,

$$\hat{S}_i^{+} \hat{S}_j^{+} = \hat{S}_j^{+} \hat{S}_i^{+} \,, \quad \text{but} \quad f_i^{\dagger} f_j^{\dagger} = -f_j^{\dagger} f_i^{\dagger} \,.$$

To obtain a faithful spin representation, it is necessary to cancel this unwanted sign. Although a general procedure is hard to formulate in one dimension, this can be achieved by a nonlinear transformation, that is,

$$\hat{S}_l^{+} = f_l^{\dagger} e^{i\pi \sum_{j<l} \hat{n}_j} \,, \quad \hat{S}_l^{-} = e^{-i\pi \sum_{j<l} \hat{n}_j} f_l \,, \quad \hat{S}_l^{z} = f_l^{\dagger} f_l - \frac{1}{2} \,.$$

Operationally, this seemingly complicated transformation is straightforward; in one dimension, the particles can be ordered on the line. By counting the number of particles "to the left", we can assign an overall phase of $+1$ or -1 to a given configuration and thereby transmute the particles into fermions. (Put differently, the exchange to two fermions induces a sign change which is compensated by the factor arising from the phase – the "Jordan–Wigner string".)

b) Using the Jordan–Wigner representation, show that

$$\hat{S}_m^+ \hat{S}_{m+1}^- = f_m^\dagger f_{m+1} \, .$$

c) For the spin 1/2 anisotropic quantum Heisenberg spin chain, the spin Hamiltonian assumes the form

$$\hat{H} = -\sum_n \left[J_z \hat{S}_n^z \hat{S}_{n+1}^z + \frac{J_\perp}{2} \left(\hat{S}_n^+ \hat{S}_{n+1}^- + \hat{S}_n^- \hat{S}_{n+1}^+ \right) \right] .$$

Turning to the Jordan–Wigner representation, show that the Hamiltonian can be cast in the form

$$\hat{H} = \sum_n \left[\frac{J_\perp}{2} \left(f_n^\dagger f_{n+1} + \text{h.c.} \right) + J_z \left(\frac{1}{4} - f_n^\dagger f_n + f_n^\dagger f_n f_{n+1}^\dagger f_{n+1} \right) \right] .$$

d) The mapping above shows that the one-dimensional quantum spin-1/2 XY-model, that is, $J_z = 0$, can be diagonalized as a non-interacting theory of spinless fermions. In this case, show that the spectrum assumes the form

$$\epsilon(k) = -J_\perp \cos(ka) \, .$$

References

1 Dirac, P.A.M. (1927) The Quantum Theory of the Emission and Absorption of Radiation, *Proc. R. Soc. Lond. A*, **114**, 243.

2 Jordan, P. and Wigner, E. (1928) *Über Paulisches Äquivalenzverbot*, Z. Phys., **47**, 631.

3 Fock, V. (1932) *Konfigurationsraum und zweite Quantelung*, Z. Phys., **75**, 622.

4 Landau, L.D. and Lifshitz, E.M. (1958) *Statistical Physics*, Pergamon Press, London.

5 Wentzel, G. (1949) *Quantum Theory of Fields*, Interscience, New York, Chap. 1.

6 Callaway, J. (1974) *Quantum Theory of the Solid State*, 2nd edn, Academic Press Inc., San Diego, Chap. 7.

7 Rickayzen, G. (1965) *Theory of Superconductivity*, 1st edn, John Wiley & Sons, Inc., New York.

2
The Kubo–Greenwood Linear Response Theory

In this chapter, we present the linear response theory of Kubo and Greenwood. It lays the foundations for calculating the electrical transport coefficients of conductivity and mobility as well as the density–density response function and its relationship to fluctuations in the system. This formalism is adopted from the pioneering papers listed in the references which is not specific to any dimensionality [1–14].

2.1
Fluctuations and Dissipation

Let us consider a gas or liquid which contains a density $n_i(z)$ of impurities, for example, ions. There will be a net flow current of ions in the condensed matter if either (a) a density gradient exists or (b) there is an external force, such as, an electric field, acting on the impurities. The current due to a density gradient in terms of the diffusion coefficient D is

$$J_z^{(D)} = -D\frac{dn_i}{dz} , \tag{2.1}$$

while the external force F will produce a net drift velocity μF, where μ is the mobility and consequently a net current

$$J_z^{(F)} = n_i \mu F . \tag{2.2}$$

Suppose now that *both* a density gradient and an external field exist, and act against each other so that there is no net flow of the impurities. In this case,

$$J_z^{(D)} + J_z^{(F)} = 0 , \tag{2.3}$$

or

$$n_i \mu F = D\frac{dn_i}{dz} . \tag{2.4}$$

Now, when this condition is satisfied, the system is in equilibrium, and, therefore, the probability of finding an impurity at z is $e^{-\beta E(z)} = e^{-\beta[E_0 + U(z)]}$, where $U(z)$ is

Properties of Interacting Low-Dimensional Systems, First Edition. G. Gumbs and D. Huang.

the potential energy due to the external force, that is, $F = -dU(z)/dz$. Therefore, we have

$$\frac{dn_i}{dz} = -\beta n_i \frac{dU}{dz} = \beta F n_i \ . \tag{2.5}$$

From this, we obtain

$$\mu = \beta D \ . \tag{2.6}$$

This relation was derived by Nernst [1] and independently by Einstein [2]. It is the earliest form of the fluctuation–dissipation theorem, giving a link between the diffusive and the drift motion of the impurities. The impurities diffuse because of the fluctuations in the forces exerted on them by an external force. This drift is dissipative, generating Joule heat by doing work against the internal forces. The diffusive motion is not.

2.2 Nyquist's Relation

In 1928, Johnson [3] discovered the existence of a noise voltage in conductors, and attributed it to thermal motion of the electrons. The spectrum of the fluctuating voltage was related by Nyquist [4–6] to the frequency-dependent impedance of the conductor as well as the temperature. The argument presented by Nyquist is as follows.

Let us consider two conductors with resistance R in series and at the same temperature T. The voltage V_a due to charge fluctuations in R_a makes a current $V/2R$ flow. This current transfers power from "a" to "b". In the same way, the charge fluctuations in R_b transfer power from "b" to "a". Since R_b and R_a are at the same temperature, the average power flowing in one direction must be exactly equal to the average power flowing in the other direction, according to the Second Law of thermodynamics. This equality must hold not only for total power exchange, but also for the powers exchanged in any given frequency range. For, if there were a frequency range in which R_a delivered more power than it received, then by connecting a non-dissipative network (shown below), the Second Law could be violated. It follows that the voltage developed due to current fluctuations is a universal function of the temperature, resistance and frequency, and only of these variables. Since it is a universal function, it may be determined by solving any particular model. Nyquist took the following: Two resistors R connected by a long loss-less line of length ℓ, with inductance and capacitance per unit length, chosen so that the characteristic impedance of the line is equal to R. The line is then matched at each end, so that all energy traveling down the line will be absorbed without reflection. The number of waves in wave number range dk is $\ell dk/2\pi$, each with energy ε_k. The average rate of flow of energy in each direction in the wave number range dk is therefore

$$(\text{average energy}) \times \left(\frac{\text{velocity}}{\text{length of line}}\right) \times (\text{number of modes})$$

$$= \bar{\varepsilon}_k \left(\frac{c}{\ell}\right)\left(\frac{\ell}{2\pi}\right)\left(\frac{d\omega}{c}\right) \equiv \varepsilon_k d f\,. \tag{2.7}$$

However, this must be equal to

$$I_f^2 R_f = \left(\frac{V_f}{2R_f}\right)^2 R_f = \left(\frac{V_f^2}{4R_f}\right). \tag{2.8}$$

Therefore, we have

$$V_f^2 = 4R_f \bar{\varepsilon}_k d f\,. \tag{2.9}$$

The mean energy of the mode of frequency ω at temperature T is

$$\bar{\varepsilon}_\omega = \frac{1}{2}\hbar\omega + \frac{\hbar\omega}{e^{\beta\hbar\omega}-1} = \frac{1}{2}\hbar\omega \coth\left(\frac{1}{2}\beta\hbar\omega\right). \tag{2.10}$$

This has the high-temperature limit $k_\mathrm{B} T$, and Nyquist's relation becomes

$$V_f^2 = 4R_f k_\mathrm{B} T d f\,. \tag{2.11}$$

Equations (2.9) and (2.11) are another form of the fluctuation–dissipation theorem, relating the voltage fluctuations at frequency ω to the dissipative properties given by R_f at the same frequency.

2.3 Linear Response Theory

2.3.1 Generalized Susceptibility

There is an important class of physical problems where the effect of an external force $F(t)$ is to change the Hamiltonian of the system by an operator of the form

$$\hat{H}' = -X F(t) \int_0^\infty d\tau \phi(\tau) F(t-\tau) \tag{2.12}$$

and where, furthermore, the steady state of the system after the force is switched on is close enough to equilibrium that it is sufficient to consider terms *linear* in F in calculating the expectation value of the generalized displacement operator X. The most general way of describing the linear response of the system, consistent with the principle of causality, is to write

$$\langle X(t)\rangle_\beta = \int_{-\infty}^{t} dt' \phi(t-t')F(t') = \int_0^{\infty} d\tau \phi(\tau) F(t-\tau) \tag{2.13}$$

where the response function ϕ depends on the temperature, and the properties of the system such as spatial variables. We note that Eq. (2.13) is written in such a way that $\langle X(t)\rangle_\beta$ depends on values of F at times less than t, that is, the response is *causal*.

It is useful to resolve both the force and the displacement into frequency components

$$\begin{aligned} F(t) &= \sum_\omega \tilde{F}(\omega) e^{-i\omega t} \\ \langle X(t)\rangle_\beta &= \sum_\omega \tilde{X}(\omega) e^{-i\omega t} \end{aligned} \tag{2.14}$$

and, we obtain

$$\langle X(t)\rangle_\beta = \int_0^{\infty} d\tau \phi(\tau) F(t-\tau) = \int_0^{\infty} d\tau \phi(\tau) \sum_\omega \tilde{F}(\omega) e^{-i\omega(t-\tau)} . \tag{2.15}$$

Since the response is linear, different frequency components will not mix and we obtain

$$\tilde{X}(\omega) = \chi(\omega)\tilde{F}(\omega) , \tag{2.16}$$

where

$$\chi(\omega) = \int_0^{\infty} d\tau \phi(\tau) e^{i\omega\tau} \tag{2.17}$$

is called the *generalized susceptibility* or *admittance*. We note that since ϕ is real because X is a Hermitian operator corresponding to a real observable, we have $\chi(-\omega) = \chi^*(\omega)$, or

$$\mathrm{Re}\left[\chi(-\omega)\right] = \mathrm{Re}\left[\chi(\omega)\right], \quad \mathrm{Im}\left[\chi(-\omega)\right] = -\mathrm{Im}\left[\chi(-\omega)\right]. \tag{2.18}$$

Thus, the real and imaginary parts of $\chi(\omega)$ are respectively even and odd functions of frequency. These relations express the fact that $\langle X\rangle$ must be real for any real $F(t)$. For example, take the real force

$$F(t) = \frac{1}{2}\left(F_0 e^{-i\omega t} + F_0^* e^{i\omega t}\right). \tag{2.19}$$

Then,

$$\langle X\rangle = \frac{1}{2}\left[\chi(\omega) F_0 e^{-i\omega t} + \chi(-\omega) F_0^* e^{i\omega t}\right] \tag{2.20}$$

and this expression is real if $\chi(-\omega) = \chi^*(\omega)$.

2.3.2
Kronig–Kramers Relations

Let us consider the behavior of the generalized susceptibility defined in Eq. (2.17) as a function of complex frequency $\omega = \omega_1 + i\omega_2$. For $\omega_2 > 0$, the integration contains the factor $e^{-\omega_2 t}$ and so goes to zero in the upper half-plane when $t > 0$. Also, $\phi(t)$ is finite everywhere, being the response of a physical system to an impulse force at $t = 0$. It follows that $\chi(\omega)$ is finite and single-valued throughout the upper half-plane, that is, it has no singularities there. We also note that we used $t \geq 0$, that is, analyticity of $\chi(\omega)$ in the upper half-plane is a consequence of the principle of causality.

Now, consider the integral

$$\int\limits_C d\omega \frac{\chi(\omega)}{\omega - \Omega} , \tag{2.21}$$

where C is the contour shown in Figure 2.1.

Since $\chi(\omega)$ has no singularities in the upper half-plane, and the point Ω is excluded from inside the contour, the integral is zero. We have

$$\begin{aligned} 0 &= -i\pi\chi(\Omega) + \lim_{\gamma\to 0}\left\{\int\limits_{-\infty}^{\Omega-\gamma} d\omega \frac{\chi(\omega)}{\omega - \Omega} + \int\limits_{\Omega+\gamma}^{\infty} d\omega \frac{\chi(\omega)}{\omega - \Omega}\right\} \\ &= -i\pi\chi(\Omega) + \mathcal{P}\int\limits_{-\infty}^{\infty} d\omega \frac{\chi(\omega)}{\omega - \Omega} , \end{aligned} \tag{2.22}$$

where $\mathcal{P}$ denotes the principal part. If $\chi(\omega)$ has a pole at the origin of the form iA/ω, a term $-\pi A/\Omega$ has to be added to Eq. (2.22). Therefore,

$$i\chi(\Omega) = \frac{\mathcal{P}}{\pi}\int\limits_{-\infty}^{\infty} d\omega \frac{\chi(\omega)}{\omega - \Omega} - \frac{A}{\Omega} . \tag{2.23}$$

Separating out the real and imaginary parts of $\chi(\Omega)$, we obtain

$$\begin{aligned} \mathrm{Re}\left[\chi(\Omega)\right] &= \frac{\mathcal{P}}{\pi}\int\limits_{-\infty}^{\infty} d\omega \frac{\mathrm{Im}\left[\chi(\omega)\right]}{\omega - \Omega} \\ \mathrm{Im}\left[\chi(\Omega)\right] &= -\frac{\mathcal{P}}{\pi}\int\limits_{-\infty}^{\infty} d\omega \frac{\mathrm{Re}\left[\chi(\omega)\right]}{\omega - \Omega} + \frac{A}{\Omega} . \end{aligned} \tag{2.24}$$

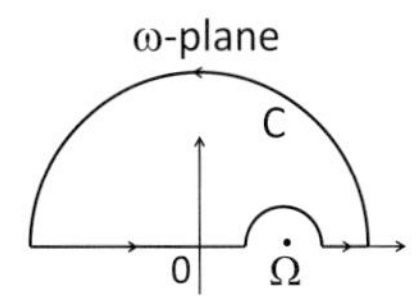

Figure 2.1 Plot for the contour C in the upper complex ω-plane, where a point at $\omega = \Omega$ on the real axis is excluded.

These are the Kronig–Kramers relations. They are a direct consequence of causality. $\mathrm{Re}\left[\chi(\omega)\right]$ and $\mathrm{Im}\left[\chi(\omega)\right]$ are Hilbert transforms of each other.

2.3.3
Dielectric Function in Three Dimensions

Consider a system consisting of n electrons per unit volume. For an electric field F ($F = F_0 e^{-i\omega t}$), the equation of motion for each electron of mass m^* and charge e is

$$\ddot{x} + \frac{\dot{x}}{\tau} = \frac{eF}{m^*} , \tag{2.25}$$

where τ is a phenomenological relaxation time. Assuming that $x \sim e^{-i\omega t}$, we obtain

$$\left(\omega^2 + i\frac{\omega}{\tau}\right) x = -\frac{eF}{m^*} . \tag{2.26}$$

However, in three dimensions, the dielectric function is defined by

$$\epsilon_{\mathrm{r}}(\omega) = 1 + \frac{4\pi P}{\epsilon_{\mathrm{s}} F} , \quad P = nex , \tag{2.27}$$

where P is the polarization and $\epsilon_{\mathrm{s}} = 4\pi\epsilon_0\epsilon_{\mathrm{b}}$ with ϵ_{b} denoting the background dielectric constant. This definition only holds for a uniform electric field and must be modified accordingly when this field is spatially nonuniform. This issue will be addressed in a later chapter. It follows from Eqs. (2.26) and (2.27) that

$$\epsilon_{\mathrm{r}}(\omega) = 1 - \frac{\Omega_{\mathrm{p}}^2}{\omega\left(\omega + \frac{i}{\tau}\right)} , \quad \Omega_{\mathrm{p}} = \sqrt{\frac{4\pi n e^2}{\epsilon_{\mathrm{s}} m^*}} , \tag{2.28}$$

where Ω_{p} is the bulk plasma frequency. Also, since the polarization is given by

$$P(\omega) = \epsilon_{\mathrm{s}} \left[\frac{\epsilon_{\mathrm{r}}(\omega) - 1}{4\pi}\right] F(\omega) , \tag{2.29}$$

it means that $\epsilon_{\mathrm{r}}(\omega) - 1$ is a frequency-dependent susceptibility which satisfies the Kronig–Kramers relations. The example given in Eq. (2.28) does.

2.4
The Density Matrix and Quantum Statistics

2.4.1
The von Neumann Density Matrix

The density matrix for a dynamical system of fermions which, at a given instant of time, is in one or another of a possible set of normalized states $|\alpha\rangle$ with probability

p_α for being in that state is defined by

$$\hat{\rho}(t) = \sum_\alpha |\alpha\rangle\langle\alpha| p_\alpha \,. \tag{2.30}$$

In equilibrium, p_α is the Fermi–Dirac distribution function and the equilibrium density matrix is

$$\hat{\rho}_0 = \sum_\alpha |\alpha\rangle\langle\alpha| f_0(\varepsilon_\alpha) \,. \tag{2.31}$$

When the system is in the state $|\alpha\rangle$, the average value of any observable $\hat{B}$ is $\langle\alpha|\hat{B}|\alpha\rangle$. Taking an ensemble average of these systems, distributed over the various states $|\alpha\rangle$ according to the probability law p_α, the average value over many measurements is

$$\langle\hat{B}\rangle = \sum_\alpha p_\alpha \langle\alpha|\hat{B}|\alpha\rangle \,. \tag{2.32}$$

Introducing a representation of base vectors $|\xi\rangle$ with $|\xi\rangle\langle\xi| = \hat{I}$, we obtain

$$\langle\hat{B}\rangle = \sum_\xi \langle\xi|\hat{\rho}\hat{B}|\xi\rangle \equiv \mathrm{Tr}\left(\hat{\rho}\hat{B}\right), \tag{2.33}$$

involving the trace over the complete set of states $|\xi\rangle$. In a straightforward way, it can be shown that the equation of motion of $\hat{\rho}$ is given by

$$i\hbar\frac{d\hat{\rho}}{dt} = \left[\hat{H}, \hat{\rho}\right]. \tag{2.34}$$

2.4.2
Entropy

In quantum statistical mechanics, entropy is defined in terms of the density matrix as

$$S = -\mathrm{Tr}(\hat{\rho}\ln\hat{\rho}) \tag{2.35}$$

in conjunction with the constraint $\mathrm{Tr}(\hat{\rho}) = 1$. The entropy is a measure of the lack of knowledge about the states of the system in the ensemble. If we maximize S subject to the constraint on $\hat{\rho}$, we have, for all $\delta\hat{\rho}$,

$$-\mathrm{Tr}\,(\delta\hat{\rho}\ln\hat{\rho} + \delta\hat{\rho} + \lambda\delta\hat{\rho}) = 0 \,, \tag{2.36}$$

or $\hat{\rho}$ is constant and λ is a variational multiplier. That is, when the entropy is a maximum, the probability of finding the system in any of its possible states is the same.

Now suppose, as is usually the case, that the system has a fixed total energy

$$\langle E\rangle = \mathrm{Tr}\left(\hat{\rho}\hat{H}\right). \tag{2.37}$$

By maximizing the entropy subject to Eq. (2.37) and introducing another multiplier β, we obtain

$$-\mathrm{Tr}\left(\delta\hat{\rho}\ln\hat{\rho} + \delta\hat{\rho} + \lambda\delta\hat{\rho} + \beta\hat{H}\delta\hat{\rho}\right) = 0\,, \tag{2.38}$$

which has the solution

$$\hat{\rho} = e^{-(1+\lambda)}e^{-\beta\hat{H}}\,. \tag{2.39}$$

Since $\mathrm{Tr}(\hat{\rho}) = 1$, we must have $e^{(1+\lambda)} = \mathrm{Tr}(e^{-\beta\hat{H}})$. Therefore,

$$\hat{\rho} = \frac{e^{-\beta\hat{H}}}{\mathrm{Tr}(e^{-\beta\hat{H}})} \equiv \frac{e^{-\beta\hat{H}}}{Z}\,. \tag{2.40}$$

The average energy can therefore be written as

$$\langle E\rangle = \frac{\mathrm{Tr}\left(\hat{H}e^{-\beta\hat{H}}\right)}{\mathrm{Tr}\left(e^{-\beta\hat{H}}\right)} = -\frac{\partial}{\partial\beta}\ln Z\,. \tag{2.41}$$

Defining the free energy F by $\beta F = -\ln Z$ or $Z = e^{-\beta F}$, we have the following result for the density matrix of a system in thermodynamic equilibrium:

$$\hat{\rho} = e^{\beta(F-\hat{H})}\,. \tag{2.42}$$

We note that in equilibrium $\hat{\rho}$ is a function of $\hat{H}$ and therefore commutes with it, that is, $d\hat{\rho}/dt = [\hat{H},\hat{\rho}]/i\hbar = 0$ in equilibrium.

2.5 Kubo's Theory

Consider systems close to thermal equilibrium, perturbed by some external force. Assume that the external force is weak enough so that the difference between the expectation value of any physical quantity and its equilibrium value (usually taken to be zero) is small and linear in the force. The perturbing Hamiltonian due to the external force is

$$\hat{H}' = -\hat{A}F(t)\,, \tag{2.43}$$

where $\hat{A}$ is said to be conjugate to F. Examples are $\hat{H}' = -e\hat{x}\,E(t)$, $\hat{H}' = -\hat{M}\,H(t)$ where the displacement $\hat{x}$ and the magnetization $\hat{M}$ are conjugate to the external forces $eE(t)$ and the magnetic field $H(t)$. Now, we suppose that we are interested in the observed deviation $\langle\Delta\hat{B}(t)\rangle$ of a quantity $\hat{B}$ (not necessarily the conjugate operator $\hat{A}$) from its equilibrium value (taken to be zero). Assume that $\langle\Delta\hat{B}(t)\rangle$ is linear in F, and define a (causal) response function $\phi_{\mathrm{BA}}(t,t')$ by

$$\left\langle\Delta\hat{B}(t)\right\rangle = \int_{-\infty}^{t} dt'\,\phi_{\mathrm{BA}}(t,t')F(t')\,. \tag{2.44}$$

This response function gives the effect of a delta function force at later times for $F(t) = \delta(t - t_0)$, as shown in Figure 2.2, that is,

$$\left\langle \Delta \hat{B}(t) \right\rangle = \begin{cases} 0\,, & t < t_0 \\ \phi_{\mathrm{BA}}(t, t_0)\,, & t > t_0 \end{cases}. \tag{2.45}$$

If the deviations from equilibrium are small, the system is basically stationary so the response functions only depend on the difference between the times of the pulse force and the measurement of its effect:

$$\phi_{\mathrm{BA}}(t, t') = \phi_{\mathrm{BA}}\left(t - t'\right). \tag{2.46}$$

For a periodic force $F(t) = \mathrm{Re}(F_0 e^{-i\omega t})$, the response can be written as

$$\left\langle \Delta \hat{B}(t) \right\rangle = \mathrm{Re}\left[\chi_{\mathrm{BA}}(\omega) F_0 e^{-i\omega t}\right], \tag{2.47}$$

where

$$\chi_{\mathrm{BA}}(\omega) = \int_0^\infty d\tau e^{i\omega\tau} \phi_{\mathrm{BA}}(\tau) \tag{2.48}$$

is the generalized susceptibility or admittance. The real and imaginary parts of $\chi(\omega)$ obey the Kramers–Kronig relations.

The response function $\phi_{\mathrm{BA}}(\tau)$, giving the response of the observable $\hat{B}$ to a pulse force conjugate to $\hat{A}$, is a functional of $\hat{A}$, $\hat{B}$ and the equilibrium properties of the system. Kubo showed that

$$\phi_{\mathrm{BA}}(t) = \mathrm{Tr}\left\{\hat{\rho}_0 \frac{1}{i\hbar}\left[\hat{A}, \hat{B}(t)\right]\right\} \equiv \frac{1}{i\hbar}\left\langle\left[\hat{A}(0), \hat{B}(t)\right]\right\rangle_\beta, \tag{2.49}$$

where $\hat{\rho}_0$ is the density matrix of the equilibrium state.

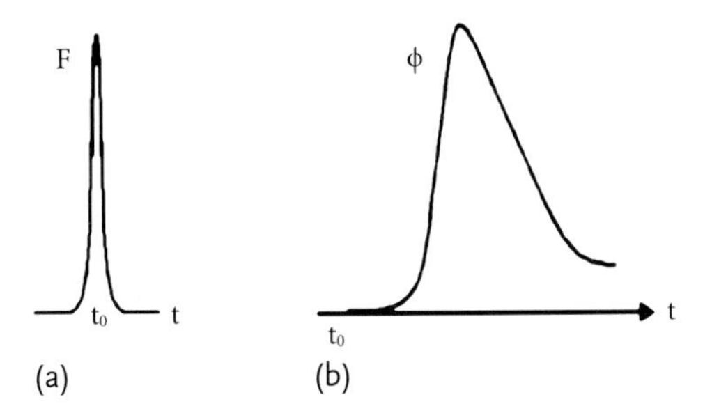

Figure 2.2 (a) The applied force F and (b) linear response function ϕ.

2.6 The Kubo Equation

We move on to the proof of the Kubo equation (2.49). The expectation value $\langle \Delta \hat{B}(t) \rangle$ may be obtained either in terms of the response ϕ_{BA} and given by

$$\left\langle \Delta \hat{B}(t) \right\rangle = \int_{-\infty}^{t} dt' \phi_{\mathrm{BA}}(t-t') F(t'), \tag{2.50}$$

or as the trace of $\hat{B}$ times the density matrix,

$$\left\langle \Delta \hat{B}(t) \right\rangle = \mathrm{Tr}\left(\hat{\rho}_1(t) \hat{B}(t)\right). \tag{2.51}$$

However, $\hat{\rho}(t)$ evolves according to Eq. (2.34), where

$$\hat{H} = \hat{H}_0 + \hat{H}_{\mathrm{ext}}(t) = \hat{H}_0 - \hat{A} F(t). \tag{2.52}$$

We now expand $\hat{\rho}$ as $\hat{\rho}_0 + \hat{\rho}_1(t)$ and only keep first-order terms in Eq. (2.34). This gives

$$i\hbar \frac{d\hat{\rho}_1}{dt} \approx -\left[\hat{\rho}_0, \hat{H}_1\right] - \left[\hat{\rho}_1(t), \hat{H}_0\right], \tag{2.53}$$

which, assuming that $\hat{\rho}_1(t = -\infty) = 0 = F(-\infty)$, has the solution

$$\hat{\rho}_1(t) = \frac{1}{i\hbar} \int_{-\infty}^{t} dt' e^{-i(t-t')\hat{H}_0/\hbar} \left[\hat{\rho}_0, \hat{A}\right] e^{i(t-t')\hat{H}_0/\hbar} F(t'). \tag{2.54}$$

The response of the observable $\hat{B}$ is therefore

$$\begin{aligned} \left\langle \Delta \hat{B}(t) \right\rangle &= \mathrm{Tr}\left[\hat{\rho}(t) \hat{B}\right] = \mathrm{Tr}\left[\hat{\rho}_1(t) \hat{B}\right] \\ &= \frac{1}{i\hbar} \mathrm{Tr} \left\{ \int_{-\infty}^{t} dt' \left[\hat{\rho}_0, \hat{A}\right] \hat{B}(t-t') F(t') \right\}, \end{aligned} \tag{2.55}$$

where the time-dependent operator is defined as $\hat{B}(t) = e^{it\hat{H}_0/\hbar} \hat{B} e^{-it\hat{H}_0/\hbar}$. Comparing Eqs. (2.50) and (2.55), we obtain the linear response function

$$\begin{aligned} \phi_{\mathrm{BA}}(t) &= Tr \left\{ \frac{1}{i\hbar} \left[\hat{\rho}_0, \hat{A}\right] \hat{B}(t) \right\} \\ &= \mathrm{Tr} \left\{ \hat{\rho}_0 \frac{1}{i\hbar} \left[\hat{A}, \hat{B}(t)\right] \right\} \\ &\equiv \frac{1}{i\hbar} \left\langle \left[\hat{A}(0), \hat{B}(t)\right] \right\rangle_{\beta}. \end{aligned} \tag{2.56}$$

2.7 Fluctuation–Dissipation Theorem

The Fourier spectrum of the time correlation functions $\langle \hat{A}(0)\hat{B}(t)\rangle$ and $\langle \hat{B}(t)\hat{A}(0)\rangle$ is related to the spectrum of the response function $\phi_{BA}(t)$ by virtue of the Kubo equation (2.49). This in turn gives the spectrum of the dissipation through the generalized susceptibility

$$\chi_{BA}(\omega) = \int_0^\infty dt e^{i\omega t}\phi_{BA}(t) \,. \tag{2.57}$$

These relations lead to the generalized fluctuation–dissipation theorem. Making use of

$$\mathrm{Tr}\left[e^{-\beta \hat{H}_0}\hat{A}(0)\hat{B}(t)\right] = \mathrm{Tr}\left[e^{-\beta \hat{H}_0}\hat{B}(t - i\hbar\beta)\hat{A}(0)\right], \tag{2.58}$$

it follows that

$$\int_{-\infty}^{\infty} dt e^{i\omega t}\left\langle \hat{A}(0)\hat{B}(t)\right\rangle_\beta = e^{-\beta\hbar\omega}\int_{-\infty}^{\infty} dt e^{i\omega t}\left\langle \hat{B}(t)\hat{A}(0)\right\rangle_\beta \tag{2.59}$$

provided that $\langle \hat{B}(t)\hat{A}(0)\rangle_\beta$ has no singularities in the lower half-plane down to $t = -i\hbar\beta$ and $\langle \hat{B}(t)\hat{A}(0)\rangle_\beta \to 0$ as $t \to \pm\infty$. The contour of integration is given in Figure 2.3.

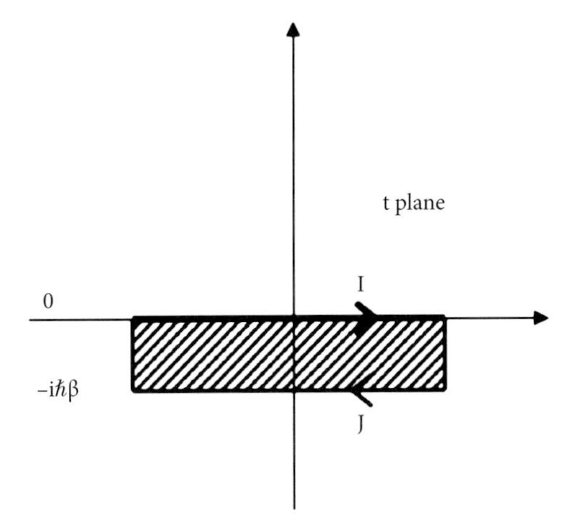

Figure 2.3 The contour of integration for Eq. (2.59).

Therefore, the Fourier transform of $\phi_{\mathrm{BA}}(t)$ is

$$\int_{-\infty}^{\infty} dt e^{i\omega t}\phi_{\mathrm{BA}}(t) = \frac{1}{i\hbar}\left(1 - e^{\beta\hbar\omega}\right)\int_{-\infty}^{\infty} dt e^{i\omega t}\left\langle \hat{A}(0)\,\hat{B}(t)\right\rangle_{\beta}$$

$$= \frac{1}{i\hbar}\frac{\left(1 - e^{\beta\hbar\omega}\right)}{\left(1 + e^{\beta\hbar\omega}\right)}\int_{-\infty}^{\infty} dt e^{i\omega t}\left\langle [\hat{A}(0), \hat{B}(t)]_{+}\right\rangle_{\beta} . \tag{2.60}$$

If $\hat{A}$ and $\hat{B}$ are Hermitian operators corresponding to real observables, then ϕ_{BA} is real. Furthermore, ϕ_{BA} is usually either even in time (Re(χ) dissipative) or odd in time (Im(χ) dissipative). From the definition of χ in Eq. (2.48), we find that

$$\int_{-\infty}^{\infty} dt e^{i\omega t}\phi_{\mathrm{BA}}(t) = \chi_{\mathrm{BA}}(\omega) \pm \chi^{*}_{\mathrm{BA}}(\omega)$$

$$= 2\begin{cases}\mathrm{Re}\left[\chi_{\mathrm{BA}}(\omega)\right], & \text{if } \phi_{\mathrm{BA}} \text{ is even}\\ \mathrm{Im}\left[\chi_{\mathrm{BA}}(\omega)\right], & \text{if } \phi_{\mathrm{BA}} \text{ is odd}\end{cases} . \tag{2.61}$$

This means that

$$\mathrm{Re}\left[\chi_{\mathrm{BA}}(\omega)\right] = -\frac{1}{2i\hbar}\tanh\left(\frac{1}{2}\beta\hbar\omega\right)$$
$$\times \int_{-\infty}^{\infty} dt e^{i\omega t}\left\langle [\hat{A}(0), \hat{B}(t)]_{+}\right\rangle_{\beta}, \quad \text{if } \phi_{\mathrm{BA}} \text{ is even} \tag{2.62}$$

or

$$\mathrm{Im}\left[\chi_{\mathrm{BA}}(\omega)\right] = \frac{1}{2\hbar}\tanh\left(\frac{1}{2}\beta\hbar\omega\right)$$
$$\times \int_{-\infty}^{\infty} dt e^{i\omega t}\left\langle [\hat{A}(0), \hat{B}(t)]_{+}\right\rangle_{\beta}, \text{if } \phi_{\mathrm{BA}} \text{ is odd} . \tag{2.63}$$

The inverse of these generalized fluctuation–dissipation relations gives the time evolution of the correlation functions in terms of the frequency spectrum of the susceptibility, that is,

$$\left\langle [\hat{A}(0), \hat{B}(t)]_{+}\right\rangle_{\beta} = -\frac{i\hbar}{\pi}\int_{-\infty}^{\infty} d\omega e^{-i\omega t}\coth\left(\frac{1}{2}\beta\hbar\omega\right)$$
$$\times\, \mathrm{Re}\left[\chi_{\mathrm{BA}}(\omega)\right], \quad \text{if } \phi_{\mathrm{BA}} \text{ is even} \tag{2.64}$$

$$\left\langle [\hat{A}(0), \hat{B}(t)]_{+}\right\rangle_{\beta} = \frac{\hbar}{\pi}\int_{-\infty}^{\infty} d\omega e^{-i\omega t}\coth\left(\frac{1}{2}\beta\hbar\omega\right)$$
$$\times\, \mathrm{Im}\left[\chi_{\mathrm{BA}}(\omega)\right], \quad \text{if } \phi_{\mathrm{BA}} \text{ is odd} . \tag{2.65}$$

2.8 Applications

2.8.1 Mobility and the Nernst–Einstein Relation

For a system containing an impurity, for example, an ion, with position operator $\hat{X}$ acted on by an external force $F(t)$, the perturbing Hamiltonian is

$$\hat{H}_{\text{ext}} = -\hat{X} F(t) . \tag{2.66}$$

We wish to calculate the mobility, defined by

$$\langle \dot{X} \rangle = \mu F(t) . \tag{2.67}$$

The appropriate operators in Kubo's formalism are therefore $\hat{A} = \hat{X}$ and $\hat{B} = \dot{X}$, and the response function is

$$\phi(t) = \frac{1}{i\hbar} \left\langle \left[\hat{X}, \dot{X}(t) \right] \right\rangle_{\beta} . \tag{2.68}$$

Therefore, the susceptibility or admittance is now the frequency-dependent mobility

$$\mu(\omega) = \chi_{\dot{X}X}(\omega) = \frac{1}{i\hbar} \int_{0}^{\infty} dt e^{i\omega t} \left\langle \left[\hat{X}, \dot{X}(t) \right] \right\rangle_{\beta} , \tag{2.69}$$

as shown in Figure 2.4.

Writing Eq. (2.68) as

$$\phi(t) = \frac{1}{(i\hbar)^2} \left\langle \left[\hat{X}, \left[\hat{X}(t), \hat{H}_0 \right] \right] \right\rangle_{\beta} , \tag{2.70}$$

and using the stationary property

$$\left\langle \hat{X}(0) \hat{X}(-t) \hat{H}_0 \right\rangle_{\beta} = \left\langle \hat{X}(t) \hat{X}(0) \hat{H}_0 \right\rangle_{\beta} , \tag{2.71}$$

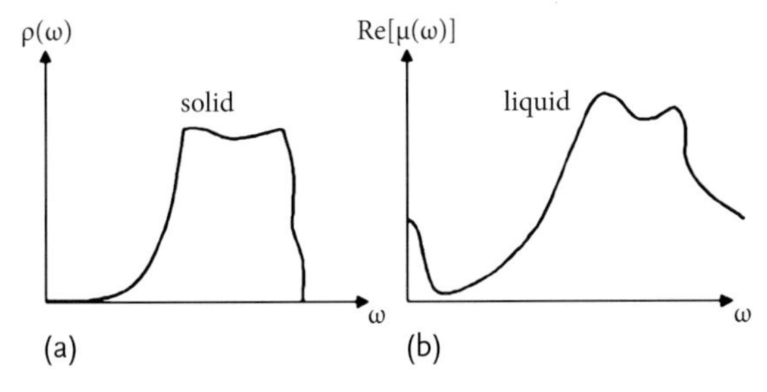

Figure 2.4 Schematic representation of the frequency spectrum $\rho(\omega)$ of the phonon modes as a function of frequency for (a) a solid and the real part of the frequency-dependent mobility spectrum $\text{Re}[\mu(\omega)]$ for (b) a liquid.

it may be shown that ϕ is even in t. The fluctuation–dissipation theorem therefore gives

$$\left\langle \left[\hat{X}(0), \dot{X}(t)\right]_{+} \right\rangle_{\beta} = -\frac{i\hbar}{\pi} \int_{-\infty}^{\infty} d\omega e^{-i\omega t} \coth\left(\frac{1}{2}\beta\hbar\omega\right) \mathrm{Re}\left[\mu(\omega)\right]$$

$$= -\frac{2\hbar}{\pi} \int_{0}^{\infty} d\omega \sin(\omega t) \coth\left(\frac{1}{2}\beta\hbar\omega\right) \mathrm{Re}\left[\mu(\omega)\right] \quad (2.72)$$

since $\mathrm{Re}\left[\mu(\omega)\right]$ is even in ω. Integrating with respect to time, we obtain

$$\left\langle \left(\hat{X}(t) - \hat{X}(0)\right)^2 \right\rangle_{\beta} = \frac{2\hbar}{\pi} \int_{0}^{\infty} \frac{d\omega}{\omega} \left[1 - \cos(\omega t)\right] \coth\left(\frac{1}{2}\beta\hbar\omega\right) \mathrm{Re}\left[\mu(\omega)\right]. \quad (2.73)$$

Also, it can be shown that

$$\left\langle \left[\hat{X}(0), \hat{X}(t)\right] \right\rangle_{\beta} = \frac{2i\hbar}{\pi} \int_{0}^{\infty} \frac{d\omega}{\omega} \sin(\omega t) \mathrm{Re}\left[\mu(\omega)\right]. \quad (2.74)$$

Using $\lim_{t\to\infty} \sin(\omega t)/\omega = \pi\delta(\omega)$, we have

$$\mu_0 \equiv \lim_{\omega\to 0} \mathrm{Re}\left[\mu(\omega)\right] = \frac{1}{i\hbar} \lim_{t\to\infty} \left\langle \left[\hat{X}(0), \hat{X}(t)\right] \right\rangle_{\beta}. \quad (2.75)$$

These relations are due to Josephson and Lekner [12]. As a matter of fact, the results for the mean-square displacement are the same as for an atom in a harmonic solid if we make the replacement

$$\frac{2}{\pi} \mathrm{Re}\left[\mu(\omega)\right] \to \frac{\rho(\omega)}{M}, \quad (2.76)$$

where $\rho(\omega)$ is the frequency spectrum of the phonon modes and M is the mass of the atom.

Let us now define the diffusion constant D by

$$\lim_{t\to\infty} \left\langle \left(\hat{X}(t) - \hat{X}(0)\right)^2 \right\rangle_{\beta} \to 2Dt + \text{constant}. \quad (2.77)$$

Then, the exact relation (2.73) gives

$$D = \lim_{t\to\infty} \frac{\hbar}{2\pi} \int_{-\infty}^{\infty} d\omega \left(\frac{\sin \omega t}{\omega}\right) \left[\frac{\omega}{\tanh \frac{\beta\hbar\omega}{2}}\right] \mathrm{Re}\left[\mu(\omega)\right]$$

$$= \mu_0 k_\mathrm{B} T. \quad (2.78)$$

Thus, the Nernst–Einstein relation between the mobility and the diffusion constant is exact and its range of validity is $t \gg \hbar\beta$.

2.8.2
Electrical Conductivity and the Nyquist Relation

The perturbation due to an external electric field $\boldsymbol{E}(t)$ on the system is

$$\hat{H}_{\text{ext}} = -\sum_i q_i \boldsymbol{r}_i \cdot \boldsymbol{E}(t) \tag{2.79}$$

and we wish to calculate the current

$$\boldsymbol{I} = \sum_i q_i \dot{\boldsymbol{r}}_i \,, \tag{2.80}$$

where q_i is the charge on the ith carrier with coordinate $\boldsymbol{r}_i$. Therefore, we have for the current response function from Eq. (2.56)

$$\begin{aligned} \phi_{\mu\nu} &= \frac{1}{i\hbar}\left\langle\left[\sum_i e_i r_{i\nu}, I_\mu(t)\right]\right\rangle_\beta \\ &= \int_0^\beta d\lambda \left\langle I_\nu(-i\hbar\lambda) I_\mu(t)\right\rangle_\beta . \end{aligned} \tag{2.81}$$

This is the response of the μth component of the current at time t when a pulse field is applied at time $t = 0$ in the νth direction. It is given by current-current correlations at real time t and at complex times up to $-i\hbar\beta$. Thus, the conductivity tensor for a periodic field ($\sim e^{-i\omega t}$) is

$$\sigma_{\mu\nu}(\omega) = \frac{1}{\mathcal{V}}\int_0^\infty dt e^{i\omega t}\int_0^\beta d\lambda \left\langle I_\nu(-i\hbar\lambda) I_\mu(t)\right\rangle_\beta , \tag{2.82}$$

where $\mathcal{V}$ is the volume of the sample.

With $\hat{A} = \sum_i q_i X_i$ and $\hat{B} = \sum_i q_i \dot{X}_i = I$, Eq. (2.64) gives

$$\left\langle\left[\hat{A}(0), \hat{B}(t)\right]_+\right\rangle_\beta = -\frac{i\hbar}{\pi}\int_{-\infty}^{\infty} d\omega e^{-i\omega t}\coth\left(\frac{1}{2}\beta\hbar\omega\right)\mathrm{Re}\left[\sigma(\omega)\right]. \tag{2.83}$$

However, the left-hand side is $\langle[\hat{A}(-t), \hat{B}(0)]_+\rangle_\beta$. Therefore, differentiating with respect to $-t$ gives

$$\left\langle\left[\hat{I}(0), \hat{I}(t)\right]_+\right\rangle_\beta = \frac{\hbar}{\pi}\int_{-\infty}^{\infty} d\omega\omega e^{-i\omega t}\coth\left(\frac{1}{2}\beta\hbar\omega\right)\mathrm{Re}\left[\sigma(\omega)\right]. \tag{2.84}$$

As $t \to \infty$, the right-hand side tends to zero by the Riemann–Lebesgue theorem. At $t = 0$, we have spontaneous fluctuations and

$$\left\langle\hat{I}^2(0)\right\rangle_\beta = \frac{\hbar}{2\pi}\int_{-\infty}^{\infty} d\omega\omega\coth\left(\frac{1}{2}\beta\hbar\omega\right)\mathrm{Re}\left[\sigma(\omega)\right]. \tag{2.85}$$

2.8.3 Magnetic Susceptibility

When a magnetic field $\boldsymbol{B}(t)$ is applied to a sample, the perturbing Hamiltonian is

$$\hat{H}_{\text{ext}} = -\boldsymbol{M} \cdot \boldsymbol{B}(t) . \tag{2.86}$$

We look for the response in M_μ when the magnetic field is in the ν direction. The response function is

$$\phi_{\mu\nu}(t) = \frac{1}{i\hbar} \langle [M_\nu, M_\mu(t)] \rangle , \tag{2.87}$$

and the magnetic susceptibility is

$$\begin{aligned} \chi_{\mu\nu}(\omega) &= \frac{1}{i\hbar} \int_0^\infty dt e^{i\omega t} \langle [M_\nu, M_\mu(t)] \rangle \\ &= \int_0^\infty dt e^{i\omega t} \int_0^\beta d\lambda \left\langle \dot{M}_\nu(-i\hbar\lambda) M_\mu(t) \right\rangle . \end{aligned} \tag{2.88}$$

For systems with permanent magnetization $\boldsymbol{M}^0$, we should subtract M_ν^0 from M_ν and M_μ^0 from M_μ.

2.8.4 The Langevin Equation

For a system without permanent magnetization, the static susceptibility is

$$\chi_{xx}(\omega = 0) = \frac{1}{i\hbar} \int_0^\infty dt \, \langle M_x(0) M_x(t) - M_x(t)) M_x(0) \rangle . \tag{2.89}$$

By using $\text{Tr}[e^{-\beta H} \hat{A}(0) \hat{B}(t)] = \text{Tr}[e^{-\beta H} \hat{B}(t - i\hbar\beta) \hat{A}(0)]$, Eq. (2.89) becomes

$$\chi_{xx}(0) = \frac{1}{i\hbar} \int_0^\infty dt \, \langle (M_x(t - i\hbar\beta) - M_x(t)) \, M_x(0) \rangle . \tag{2.90}$$

In the classical limit, when $\beta \to 0$, we obtain

$$\chi_{xx}(0) \to -\beta \int_0^\infty dt \left\langle \dot{M}_x(t) M_x(0) \right\rangle = \beta \left\langle M_x^2(0) \right\rangle , \tag{2.91}$$

where we used $\langle M_x(\infty) M_x(0) \rangle = 0$. Now, let us set $M_x = \sum_i \mu_{ix}$, where the μ_{ix} are individual dipoles. If the dipoles are non-interacting, that is, uncorrelated, the average gives

$$\left\langle M_x^2 \right\rangle = \sum_i \left\langle \mu_{ix}^2 \right\rangle = \frac{N}{3} \mu^2 , \tag{2.92}$$

where μ is the magnitude of each dipole moment. Therefore,

$$\chi_{xx}(0) = \frac{N\mu^2}{3k_B T} , \tag{2.93}$$

which is valid for large T and in the absence of interactions.

2.8.5
Stochastic Model of Magnetic Resonance

For the magnetic moment $\boldsymbol{m}$ of a given nucleus, we write the equation of motion as

$$\frac{d\boldsymbol{m}}{dt} = \left[\boldsymbol{\omega}_0 + \boldsymbol{\omega}'(t)\right] \times \boldsymbol{m} , \tag{2.94}$$

where $\boldsymbol{\omega}_0 \times \boldsymbol{m} = (ge\boldsymbol{B}_0/2M_p) \times \boldsymbol{m}$ is the torque due to the external field and $\boldsymbol{\omega}' \times \boldsymbol{m}$ is produced by the surroundings of the magnetic moment in question. In this notation, g is the gyromagnetic ratio. Assume that $\boldsymbol{\omega}'(t)$ is a stochastic variable with $\langle \boldsymbol{\omega}'(t) \rangle = 0$ and that $\boldsymbol{\omega}'$ is always parallel to $\boldsymbol{\omega}_0$. The problem is then reduced to the following. Consider a stochastic variable $x(t)$ with the equation of motion

$$\frac{dx(t)}{dt} = i\left[\omega_0 + \omega'(t)\right] x(t) , \tag{2.95}$$

which has the solution

$$x(t) = x(0) e^{i\omega_0 t + i\int_0^t dt' \omega'(t')} . \tag{2.96}$$

The correlation function of $x(t)$ is defined as

$$\langle x(t)x(0)\rangle = \left\langle x^2(0)\right\rangle e^{i\omega_0 t}\phi(t) , \tag{2.97}$$

where

$$\phi(t) \equiv \left\langle e^{i\int_0^t dt' \omega'(t')} \right\rangle . \tag{2.98}$$

By the fluctuation–dissipation theorem of Kubo, this gives the resonance absorption spectrum

$$I(\omega - \omega_0) = \frac{1}{2\pi} \int_{-\infty}^{\infty} dt e^{-i(\omega-\omega_0)t} \phi(t) . \tag{2.99}$$

Now, define the correlation time τ_c by

$$\tau_c = \frac{1}{\omega'^2} \int_0^{\infty} dt' \left\langle \omega'(t)\omega'\left(t + t'\right)\right\rangle \tag{2.100}$$

and the amplitude of the modulation Δ by

$$\Delta^2 = \int_{-\infty}^{\infty} d\omega' \omega'^2 P(\omega') = \langle \omega'^2 \rangle , \tag{2.101}$$

where $P(\omega')$ is the distribution function of the random variable.

Extreme cases include:

1. $\tau_c \gg 1/\Delta$ for which $\omega'(t)$ changes slowly. The absorption spectrum then directly gives the distribution of the local field ω' with

$$I(\omega - \omega_0) = P(\omega - \omega_0) \, .$$

2. $\tau_c \ll 1/\Delta$ for which $\omega'(t)$ changes rapidly. Then, the effect of the perturbation will just average out and the linewidth approaches a sharp spike, that is,

$$I(\omega - \omega_0) \to \delta(\omega - \omega_0) \, .$$

A decrease in width or sharpening of the resonance line due to a rapid change of the perturbation is called *motional narrowing* (Bloembergen, *et al.* [15]).

2.8.6 Gaussian Process

We may take the distribution law of the variables $\omega'(t_1), \ldots, \omega'(t_N)$ for an arbitrary number N of time points. Physically, this is a good approximation, that is, the field ω' consists of a large number of small components

$$\omega' = \sum_{i=1}^{N} \Delta\omega'_i \, , \quad \text{for } N \gg 1 \, .$$

Under this assumption, ϕ is given by

$$\begin{aligned} \phi(t) &= \left\langle e^{i \int_0^t dt' \omega'(t')} \right\rangle \\ &= \exp\left[-\int_0^t d\tau (t - \tau) \left\langle \omega'(0)\omega'(\tau) \right\rangle \right] \\ &\equiv \exp\left[-\Delta^2 \int_0^t d\tau (t - \tau) \Psi_{\omega'}(\tau) \right] , \end{aligned} \tag{2.102}$$

where the correlation function

$$\Psi_{\omega'}(\tau) = \frac{\langle \omega'(t)\omega'(t+\tau) \rangle}{\langle \omega'^2 \rangle} \, . \tag{2.103}$$

2.9 Kinetic Equation for Elastic Processes

2.9.1 Boltzmann's Transport Equation

One may develop a theory for the kinetic properties in terms of a phenomenological mean free path and relaxation time without giving precise definitions. In a given system, the particles contributing to the transport properties do not have a single velocity, but distribution of velocities. Maxwell introduced a function $f(\boldsymbol{v}, t)$ or $f(\boldsymbol{p}, t)$ defined such that at time t, the number of particles having momenta in $d\boldsymbol{p}$ about $\boldsymbol{p}$ is proportional to $f(\boldsymbol{p}, t)\, d\boldsymbol{p}$. Consider a system where the distribution of momenta of the particles is being changed by external field via collisions of the particles with scattering centers, and by the time evolution of the system. Then,

$$\frac{df}{dt} = \frac{\partial f}{\partial t} + \left(\frac{\partial f}{\partial t}\right)_{\text{field}} + \left(\frac{\partial f}{\partial t}\right)_{\text{collisions}} . \tag{2.104}$$

In a steady state, a metal or semiconductor in a uniform time-independent electric field, f is constant in time so that

$$\left(\frac{\partial f}{\partial t}\right)_{\text{field}} + \left(\frac{\partial f}{\partial t}\right)_{\text{collisions}} = 0 . \tag{2.105}$$

The field term is straightforward to calculate. In the presence of an electric field $\boldsymbol{E}$, all momenta increase at the same rate since $d\boldsymbol{p}/dt = e\boldsymbol{E}$, or in a lattice $dk_z/dt = eE/\hbar$ for a field in the z-direction. The whole distribution function is, therefore, shifted uniformly so that

$$f(k_x, k_y, k_z; t + \Delta t) = f\left(k_x, k_y, k_z - \frac{eE}{\hbar}\Delta t; t\right) . \tag{2.106}$$

This means that those particles having momentum $\hbar k_z$ at $t + \Delta t$ had momentum $\hbar k_z - eE\Delta t$ at t and

$$\left(\frac{\partial f}{\partial t}\right)_{\text{field}} = -\frac{eE}{\hbar}\frac{\partial f}{\partial k_z} . \tag{2.107}$$

Generalization of this equation to include variation of f with space and magnetic field B is given by

$$\left(\frac{\partial f}{\partial t}\right)_{\text{field}} = -e\,(\boldsymbol{E} + \boldsymbol{v} \times \boldsymbol{B}) \cdot \frac{\partial f}{\partial \boldsymbol{p}} - \boldsymbol{v} \cdot \nabla f . \tag{2.108}$$

2.9.2 The Collision Term

The collision term $(\partial f/\partial t)_{\text{collision}}$ gives the net rate at which the number of electrons with momentum $\hbar\boldsymbol{k}$ is being increased by collisions. The rate at which electrons are being scattered out from point $\boldsymbol{k}$ in $\boldsymbol{k}$-space is

$$n_s \int d\Omega' \sigma(\boldsymbol{k}, \boldsymbol{k}') v(\boldsymbol{k}') + f(\boldsymbol{k}) \left[1 - f(\boldsymbol{k}')\right], \tag{2.109}$$

where n_s is the number of scattering centers per unit volume, $\sigma(\boldsymbol{k}, \boldsymbol{k}')$ is the differential cross-section for scattering from $\boldsymbol{k}$ to $\boldsymbol{k}'$, $v(\boldsymbol{k})$ is the velocity and the factor $f(\boldsymbol{k})[1 - f(\boldsymbol{k}')]$ gives the probability of there being a state available for scattering at $\boldsymbol{k}$ and a state available to be scattered into at $\boldsymbol{k}'$. The differential probability is integrated over all angles of the state $\boldsymbol{k}'$. Similarly, the rate of scattering into $\boldsymbol{k}$ is given by

$$n_s \int d\Omega' \sigma\left(\boldsymbol{k}', \boldsymbol{k}\right) v(\boldsymbol{k}') + f\left(\boldsymbol{k}'\right) \left[1 - f\left(\boldsymbol{k}\right)\right]. \tag{2.110}$$

For elastic processes, $\boldsymbol{k}' = \boldsymbol{k}$. Also, $\sigma(\boldsymbol{k}', \boldsymbol{k}) = \sigma(\boldsymbol{k}, \boldsymbol{k}')$ by time-reversal invariance and isotropy, and can be written as $\sigma(k, \theta)$ where θ is the angle between $\boldsymbol{k}$ and $\boldsymbol{k}'$, Therefore,

$$\begin{aligned}\left(\frac{\partial f}{\partial t}\right)_{\text{collision}} &= n_s v(k) \int d\Omega' \sigma(k, \theta) \left[f\left(\boldsymbol{k}'\right)(1 - f\left(\boldsymbol{k}\right)) - f\left(\boldsymbol{k}\right)\left(1 - f\left(\boldsymbol{k}'\right)\right)\right] \\ &= n_s v(k) \int d\Omega' \sigma(k, \theta) \left[f\left(\boldsymbol{k}'\right) - f\left(\boldsymbol{k}\right)\right].\end{aligned} \tag{2.111}$$

Thus, the Pauli exclusion principle has no effect on the collision term for elastic scattering of Fermions by any foreign scatterers. The effect of Fermi statistics does not cancel out in inelastic collisions or fermion–fermion collisions when the collision term contains the factor

$$f\left(\boldsymbol{k}_1'\right) f\left(\boldsymbol{k}_2'\right) \left[1 - f\left(\boldsymbol{k}_1\right)\right] \left[1 - f\left(\boldsymbol{k}_2\right)\right] - f\left(\boldsymbol{k}_1\right) f\left(\boldsymbol{k}_2\right) \left[1 - f\left(\boldsymbol{k}_1'\right)\right] \left[1 - f\left(\boldsymbol{k}_2'\right)\right]. \tag{2.112}$$

2.9.3 Solution in the Ohmic Regime

To solve Boltzmann's equation, we will write

$$f\left(\boldsymbol{k}\right) = f_0\left(\varepsilon_k\right) + \frac{k_z}{k} f_1\left(\varepsilon_k\right). \tag{2.113}$$

For an isotropic system, these are the first two terms of an expansion in Legendre polynomials about the direction of the field. When the field is zero, f_1 is also zero, and

$$f_0\left(\varepsilon_k\right) = \frac{1}{e^{\beta\left(\varepsilon_k - \mu_c\right) + 1}}, \tag{2.114}$$

which is the equilibrium distribution. Restricting ourselves to the ohmic regime, Boltzmann's equation gives

$$\begin{aligned} 0 &= \left(\frac{\partial f}{\partial t}\right)_{\text{field}} + \left(\frac{\partial f}{\partial t}\right)_{\text{collision}} \\ &= -\frac{eE}{\hbar}\frac{\partial f}{\partial k_z} + n_s v(k) \int d\Omega' \sigma(k,\theta) \left[f(\boldsymbol{k}') - f(\boldsymbol{k}) \right], \end{aligned} \tag{2.115}$$

or,

$$\frac{eE}{\hbar}\frac{d f_0(\varepsilon_k)}{dk} = -n_s v(k) f_1(\varepsilon_k) \int d\Omega' \sigma(k,\theta) \left(1 - \frac{k_z'}{k_z}\right). \tag{2.116}$$

Take $\boldsymbol{k}$ as the polar axis. Then,

$$\begin{aligned} \boldsymbol{k}' &= k\left(\sin\theta\cos\phi', \sin\theta\sin\phi', \cos\theta\right) \\ \hat{z}' &= (\sin\xi\cos\phi, \sin\xi\sin\phi, \cos\xi) \\ k_z' &= \boldsymbol{k}'\cdot\hat{z} = k\left[\sin\theta\sin\xi\cos(\phi'-\phi) + \cos\theta\cos\xi\right]. \end{aligned} \tag{2.117}$$

Therefore, $k_z'/k_z = \cos\theta$ plus a term in $\cos(\phi' - \phi)$. The latter term gives zero on integration over the azimuthal angle. So, we have

$$\begin{aligned} \frac{eE}{\hbar}\frac{d f_0(\varepsilon_k)}{dk} &= -n_s v(k) f_1(\varepsilon_k) \int_0^\pi d\theta \sin\theta\, \sigma(k,\theta) \int_0^{2\pi} d\phi' \left(1 - \frac{k_z'}{k_z}\right) \\ &= -2\pi n_s v(k) f_1(\varepsilon_k) \int_0^\pi d\theta \sin\theta\,(1 - cos\theta)\,\sigma(k,\theta)\,. \end{aligned} \tag{2.118}$$

Using $v(k) = 1/\hbar d\varepsilon_k/dk$ and introducing a mean free path Λ_k by

$$\frac{1}{\Lambda(k)} \equiv 2\pi n_s \int_0^\pi d\theta \sin\theta\,(1 - cos\theta)\sigma(k,\theta)\,, \tag{2.119}$$

we have

$$f_1(\varepsilon_k) = -eE\Lambda(k)\frac{d f_0(\varepsilon_k)}{d\varepsilon_k}\,. \tag{2.120}$$

We note that Eq. (2.119) gives a precise definition of the transport cross-section via $1/\Lambda(k) = n_s\sigma_k$ as

$$\sigma_k = 2\pi \int_0^\pi d\theta \sin\theta\,(1 - \cos\theta)\sigma(k,\theta)\,. \tag{2.121}$$

The effect of the Pauli exclusion principle cancels out in the collision term. Thus, the above derivation of Eq. (2.120) is valid for nondegenerate collisions as well. Since $d f_0(\varepsilon_k)/d\epsilon_k \approx -\delta(\varepsilon_k - E_F)$ for a Fermi distribution with $E_F \gg k_B T$, the only contribution to the transport coefficients comes from the neighborhood of the Fermi surface.

2.9.4
Conductivity and Mobility

The current per electron is $ev(\boldsymbol{k})$ and the number of electrons per unit volume in $d\boldsymbol{k}$ is $2f(\boldsymbol{k})d\boldsymbol{k}/(2\pi)^3$. Therefore, the total current is

$$\begin{aligned} J &= \frac{2}{(2\pi)^3}\int d\boldsymbol{k}ev_z(\boldsymbol{k})f(\boldsymbol{k}) \\ &= \frac{2e}{(2\pi)^3}\int d\boldsymbol{k}\frac{1}{\hbar}\frac{d\varepsilon_k}{dk_z}\left[f_0(\varepsilon_k)+\frac{k_z}{k}f_1(\varepsilon_k)\right] \\ &= \frac{2e}{(2\pi)^3\hbar}\int d\boldsymbol{k}\frac{d\varepsilon_k}{dk_z}\frac{k_z}{k}f_1(\varepsilon_k) \\ &= \frac{2e}{(2\pi)^3\hbar}\int d\boldsymbol{k}\frac{k_z}{k}\frac{d\varepsilon_k}{dk_z}\left[-eE\Lambda(k)\frac{df_0(\varepsilon_k)}{d\varepsilon_k}\right]. \end{aligned} \tag{2.122}$$

For $E_F \gg k_B T$, $f_0(\varepsilon_k)$ is nearly the unit step function at $\varepsilon_k = E_F$ and we have $df_0(\varepsilon_k)/d\varepsilon_k = -\delta(\varepsilon_k - E_F)$. Therefore,

$$\begin{aligned} J &= \frac{2e^2E}{(2\pi)^3\hbar}\left(\frac{4\pi}{3}\right)\int_0^\infty dkk^2\frac{d\varepsilon_k}{dk}\Lambda(k)\delta(\varepsilon_k - E_F) \\ &= \left[\frac{e^2E\Lambda(k_F)}{\hbar}\right]\left(\frac{k_F^2}{3\pi^2}\right) = \left[\frac{n_0e^2\Lambda(k_F)}{\hbar k_F}\right]E\,, \end{aligned} \tag{2.123}$$

where $n = k_F^3/3\pi^2$ is the electron density. The conductivity is

$$\sigma = \rho^{-1} = n_0e^2\frac{\Lambda(k_F)}{\hbar k_F}\,, \tag{2.124}$$

which may be obtained using qualitative arguments and is known as the Drude conductivity. We note that the velocity factors have canceled out. Thus, the effective mass of the electron does not appear in the formula. However, it can enter in $\Lambda(k_F)$ as the density of final states in the scattering.

2.10
Problems

1. Using the fact that Re[$\chi(\omega)$] and Im[$\chi(\omega)$] are respectively even and odd, show that the Kronig–Kramers relations may be written as

$$\begin{aligned} \mathrm{Re}\left[\chi(\Omega)\right] &= \frac{2\mathcal{P}}{\pi}\int_0^\infty d\omega\frac{\omega\,\mathrm{Im}\left[\chi(\omega)\right]}{\omega^2-\Omega^2} \\ \mathrm{Im}\left[\chi(\Omega)\right] &= -\frac{2\Omega\mathcal{P}}{\pi}\int_0^\infty d\omega\frac{\mathrm{Re}\left[\chi(\omega)\right]}{\omega^2-\Omega^2}\,. \end{aligned} \tag{2.125}$$

2. Verify that $\epsilon_r(\omega) - 1$, with the dielectric function given by Eq. (2.28), satisfies the Kronig–Kramers relations.

3. Verify that

$$\left[e^{-\beta\hat{H}}, \hat{A}\right] = e^{-\beta\hat{H}} \int_0^\beta d\lambda e^{\lambda\hat{H}}[\hat{A}, \hat{H}]e^{-\lambda\hat{H}} .$$

4. Using the notation defined in the text, show that

$$\phi_{\rm BA}(t) = -\mathrm{Tr}\left[\hat{\rho}_0 \int_0^\beta d\lambda \hat{A}(-i\hbar\lambda)\dot{\hat{B}}(t)\right] .$$

5. For a system containing an impurity/ion with position vector X acted on by an external force $F(t)$, the perturbing Hamiltonian is $H' = -X F(t)$. We wish to calculate the mobility defined in the static case by

$$\langle \dot{X} \rangle = \mu F .$$

The appropriate formulas in Kubo's formalism are therefore $A = X$ and $B = \dot{X}$, and the response function is

$$\phi(t) = \left\langle \frac{1}{i\hbar}\left[X, \dot{X}(t)\right]\right\rangle .$$

Thus, the susceptibility or frequency-dependent mobility is

$$\mu(\omega) = \chi_{\dot{X}X} = \int_0^\infty dt e^{i\omega t}\left\langle \frac{1}{i\hbar}\left[X, \dot{X}(t)\right]\right\rangle .$$

a) Writing

$$\phi(t) = \left\langle \left(\frac{1}{i\hbar}\right)^2 \left[X, \left[X(t), H\right]\right]\right\rangle ,$$

show that $\phi(t)$ is an even function in t.

b) Using the fluctuation–dissipation theorem

$$\left\langle \left[X(0), \dot{X}(t)\right]_+ \right\rangle = -\frac{i\hbar}{\pi}\int_{-\infty}^{\infty} d\omega e^{-i\omega t} \coth\left(\frac{1}{2}\beta\hbar\omega\right) \mathrm{Re}\left[\mu(\omega)\right] ,$$

where $\mathrm{Re}[\mu(\omega)]$ is even in ω, show that

$$\langle (X(t) - X(0))^2 \rangle = \frac{2\hbar}{\pi}\int_{-\infty}^{\infty} \frac{d\omega}{\omega}(1-\cos\omega t)\coth\left(\frac{1}{2}\beta\hbar\omega\right) \mathrm{Re}\left[\mu(\omega)\right] .$$

Deduce the Nernst–Einstein relation $\beta D = \mathrm{Re}[\mu(0)]$.

6. For the frequency-dependent mobility $\mu(\omega)$, show that

$$\frac{1}{\pi}\int_{-\infty}^{\infty} d\omega \mathrm{Re}\left[\mu(\omega)\right] = \frac{1}{m^*}$$

and

$$\frac{1}{\pi}\int_{-\infty}^{\infty} d\omega\omega^2 \mathrm{Re}\left[\mu(\omega)\right] = \frac{1}{m^{*2}}\left\langle\frac{\partial^2 V}{\partial x^2}\right\rangle$$

for particles with effective mass m^* interacting with its environment via a momentum-independent potential V.

7. Show that the zero-frequency polarizability of an atom with eigenstates $|n\rangle$, $H|n\rangle = E_n|n\rangle$ is

$$\alpha(0) = 2e^2\sum_{n\neq 0}\frac{|\langle < 0|\hat{X}|n\rangle|^2}{E_n - E_0} .$$

8. Define the conductivity tensor $\sigma_{\mu\nu}(\omega)$ and hence prove the sum rule:

$$\frac{1}{\pi}\int_{-\infty}^{\infty} d\omega \mathrm{Re}\left[\sigma_{\mu\nu}(\omega)\right] = \frac{ne^2}{m^*}\delta_{\mu\nu} . \tag{2.126}$$

Furthermore, show that

$$\lim_{\omega\to\infty}\omega\mathrm{Im}\left[\sigma_{\mu\nu}\right] = \frac{ne^2}{m^*}\delta_{\mu\nu} ,$$
$$\lim_{\omega\to\infty}\omega\mathrm{Re}\left[\sigma_{\mu\nu}\right] = 0 . \tag{2.127}$$

9. Define the magnetic susceptibility tensor $\chi_{ij}(\omega)$. Show that for a system of N non-interacting dipoles $\boldsymbol{\mu}$, the classical limit of $\chi_{ij}(0)$ is diagonal with elements

$$\chi_{xx}(0) = \frac{1}{3}N\beta\mu^2 . \tag{2.128}$$

10. **Relaxation time**: Boltzmann's equation is usually solved by introducing a phenomenological time τ_k, defined so that the collision term

$$\left(\frac{\partial f}{\partial t}\right)_{\text{collision}} = -\frac{\Delta f}{\tau_k} \tag{2.129}$$

where $\Delta f = f(\boldsymbol{k}) - f_0(\varepsilon_k)$. Show that these assumptions lead to results equivalent to those above with

$$\tau_k = \frac{\Lambda(k)}{v(k)} . \tag{2.130}$$

11. Show that in the relaxation-time approximation, Boltzmann's equation for electrons in an electromagnetic field can be written as

$$eE\frac{d f_0(\varepsilon_k)}{d\varepsilon_k} = -\left[\frac{1}{\tau_k} + \frac{e}{m^*}(\boldsymbol{v} \times \boldsymbol{B}) \cdot \frac{\partial}{\partial \boldsymbol{v}}\right](f - f_0) \tag{2.131}$$

to terms that are first order in $\boldsymbol{E}$. It is assumed that the energy ε_k of an electron is a function of $|\boldsymbol{v}|$ only. Solve this equation by assuming for f the form

$$f = \left(1 - \boldsymbol{v} \cdot \boldsymbol{P}\frac{d}{d\varepsilon_k}\right) f_0 \, . \tag{2.132}$$

References

1 Nernst, W. (1884) Über Elektrostriktion durch freie Ionen. *Z. Phys. Chem.*, **9**, 613.

2 Einstein, A. (1905) *Ann. Phys.*, **17**, 549 Translation in *Theory of the Brownian Movement* (Sover).

3 Johnson, J.B. (1928) Thermal Agitation of Electricity in Conductors. *Phys. Rev.*, **32**, 97.

4 Nyquist, H. (1928) Thermal Agitation of Electric Charge in Conductors. *Phys. Rev.*, **32**, 110.

5 Callen, H.B. and Welton, T.A. (1951) Irreversibility and Generalized Noise. *Phys. Rev.*, **83**, 34.

6 Landau, L.D. and Lifshitz, E.M. (1958) *Statistical Physics*, Pergamon Press, London, Chap. 12.

7 de L. Kronig, R. (1926) On the Theory of Dispersion of X-rays. *J. Opt. Soc. Am.*, **12**, 547.

8 Kramers, H.A. (1927) *Atti. Congr. Int. fis. Como*, **2**, 545.

9 Kubo, R. (1957) Statistical-Mechanical Theory of Irreversible Processes. *J. Phys. Soc. Jpn.*, **12**, 570.

10 Kubo, R. (1958) *Lectures in Theoretical Physics*, Vol. I, Interscience, Boulder, p. 120.

11 Kubo, R. (1966) *Rep. Prog. Phys.*, **29**, 255.

12 Josephson, B.D. and Lekner, J. (1969) Mobility of an Impurity in a Fermi Liquid. *Phys. Rev. Lett.*, **23**, 111.

13 Thornber, K.K. and Feynman, R.P. (1970) Velocity Acquired by an Electron in a Finite Electric Field in a Polar Crystal. *Phys. Rev. B*, **1**, 4099.

14 Purcell, E.M., Torrey, H.C., and Pound, R.V. (1946) Resonance Absorption by Nuclear Magnetic Moments in a Solid. *Phys. Rev.*, **69**, 37.

15 Bloembergen, N. (1965) *Nonlinear Optics*, 1st edn, Benjamin, New York.

3
Feynman Diagrammatic Expansion

In this chapter, we deal with the diagrammatic expansion formalism for calculating the contributions to the Green's functions perturbatively. Several monographs and journal articles where the Feynman diagrammatic method is employed [1–14] already exist. We use the method of taking functional derivatives in evaluating the perturbation expansion of the Green's function which satisfies the Dyson equation and show that only linked diagrams contribute. In this approach, we avoid using Wick's theorem which forms all possible pairs of the field operators.

3.1
General Formalism

We now consider a system of particles (bosons or fermions) each with effective mass m^* interacting with each other through a spin and time-independent potential $V(\boldsymbol{r}_1 - \boldsymbol{r}_2)$. In addition, there is an externally applied spin-independent potential, $W(\boldsymbol{r}, t)$. Denoting the chemical potential by μ, we introduce the operator $\hat{K}(t)$ defined by

$$\begin{aligned}\hat{K}(t) &= \sum_{\sigma_1} \int d^3\boldsymbol{r}_1 \hat{\psi}^\dagger_{\sigma_1 S}(\boldsymbol{r}_1)\left(-\frac{\hbar^2\nabla_1^2}{2m^*} - \mu\right)\hat{\psi}_{\sigma_1 S}(\boldsymbol{r}_1)\\ &+ \frac{1}{2}\sum_{\sigma_1}\sum_{\sigma_2}\int d^3\boldsymbol{r}_1 \int d^3\boldsymbol{r}_2 \hat{\psi}^\dagger_{\sigma_1 S}(\boldsymbol{r}_1)\hat{\psi}^\dagger_{\sigma_2 S}(\boldsymbol{r}_2) V(\boldsymbol{r}_1 - \boldsymbol{r}_2)\hat{\psi}_{\sigma_2 S}(\boldsymbol{r}_2)\hat{\psi}_{\sigma_1 S}(\boldsymbol{r}_1)\\ &+ \sum_{\sigma_1}\int d^3\boldsymbol{r}_1 \hat{\psi}^\dagger_{\sigma_1 S}(\boldsymbol{r}_1) W(\boldsymbol{r}_1, t)\hat{\psi}_{\sigma_1 S}(\boldsymbol{r}_1)\,,\end{aligned} \tag{3.1}$$

where σ is the spin index, and the subscript S means that the operators are in the Schrödinger representation. One of our goals for this system is to calculate the single-particle thermal Green's function which is defined by

$$\mathcal{G}(1,2) = -i\left\langle \mathcal{T}\left[\hat{\psi}_K(1)\hat{\psi}^\dagger_K(2)\right]\right\rangle. \tag{3.2}$$

Properties of Interacting Low-Dimensional Systems, First Edition. G. Gumbs and D. Huang.

The ensemble average 2 of an operator $\hat{A}$ is defined by

$$\left\langle \hat{A} \right\rangle \equiv \frac{\mathrm{Tr}\left[e^{-i\int_0^{-i\beta\hbar} dt'\hat{K}(t')/\hbar} \hat{A} \right]}{\mathrm{Tr}\left[e^{-i\int_0^{-i\beta\hbar} dt'\hat{K}(t')/\hbar} \right]} = e^{\beta\Omega}\mathrm{Tr}\left[e^{-i\int_0^{-i\beta\hbar} dt'\hat{K}(t')/\hbar} \hat{A} \right]. \tag{3.3}$$

Where $n = 1, 2$ is the space-time-spin point $(r_n, 1t_n, \sigma_n)$. The above equation introduces the grand canonical potential Ω by

$$e^{-\beta\Omega} = \mathrm{Tr}\left[e^{-i\int_0^{-i\beta\hbar} dt'\hat{K}(t')/\hbar} \right]. \tag{3.4}$$

This definition of $\mathcal{G}$ is the Martin–Schwinger generating Green's function used also by Kadanoff and Baym. In Eq. (3.2), the times are purely imaginary and lie in the interval $(0, -i\beta\hbar)$. The time-ordered operator $\mathcal{T}$ for imaginary times places the operator with time argument closest to zero on the right, and the operator with time argument closest to $-i\beta\hbar$ on the left with a factor of $\varepsilon(1,2) = \pm 1$, where $+1$ applies to bosons and -1 is for fermions, that is,

$$\varepsilon(1,2) = \begin{cases} 1\,, & \text{for bosons} \\ 1\,, & \text{for fermions if } t_1 > t_2 \\ -1\,, & \text{for fermions if } t_2 > t_1 \end{cases}. \tag{3.5}$$

The reason for using imaginary times is that the properties of $\mathcal{G}$ are particularly simple in this case. Since observable quantities can be calculated knowing $\mathcal{G}$ for imaginary times, such times are decidedly not unphysical. If the real-time Green's function is required, it can be obtained from the thermal Green's function by analytic continuation. The subscript K on the operators in Eq. (3.2) means that eigenfunctions are in the K-representation defined by

$$\hat{\psi}_K(1) = e^{i\int_0^{t_1} dt'\hat{K}(t')/\hbar} \hat{\psi}_{\sigma_1 S}(\boldsymbol{r}_1) e^{-i\int_0^{t_1} dt'\hat{K}(t')/\hbar}, \tag{3.6}$$

$$\hat{\psi}_K^\dagger(1) = e^{i\int_0^{t_1} dt'\hat{K}(t')/\hbar} \hat{\psi}_{\sigma_1 S}^\dagger(\boldsymbol{r}_1) e^{-i\int_0^{t_1} dt'\hat{K}(t')/\hbar}. \tag{3.7}$$

We note that $\hat{\psi}_K^\dagger(1)$ is not the Hermitian conjugate of $\hat{\psi}_K(1)$ unless $\hat{K}$ is Hermitian and the time is real. In general, an operator in the K-representation is given in terms of its S-representation by

$$\hat{O}_K(t) = e^{i\int_0^{t} dt'\hat{K}(t')/\hbar} \hat{O}_S e^{-i\int_0^{t} dt'\hat{K}(t')/\hbar}. \tag{3.8}$$

In our treatment below, we shall also need to express operators in the interaction representation in terms of the corresponding operators in the S-representation. These are defined by

$$\hat{O}_{\text{I}}(t) = e^{i\hat{K}_0 t/\hbar}\hat{O}_{\text{S}} e^{-i\hat{K}_0 t/\hbar} , \tag{3.9}$$

where $\hat{K}_0 = \hat{K}(t) - \hat{K}_1(t)$, that is, $\hat{K}_0$ is the time-independent contribution to $\hat{K}(t)$ and given by the first two terms on the right-hand side of Eq. (3.1).

As we mentioned above, we only need imaginary times in the interval $(0, -i\beta\hbar)$. The reason for this is that $\mathcal{G}$ is periodic for bosons (or antiperiodic for fermions) for imaginary times. The periodicity (or antiperiodicity) properties are

$$\mathcal{G}(\mathbf{r}_1, 0, \sigma_1; \mathbf{r}_2, t_2, \sigma_2) = \pm\mathcal{G}(\mathbf{r}_1, -i\beta\hbar, \sigma_1; \mathbf{r}_2, t_2, \sigma_2) , \tag{3.10}$$

$$\mathcal{G}(\mathbf{r}_1, t_1, \sigma_1; \mathbf{r}_2, 0, \sigma_2) = \pm\mathcal{G}(\mathbf{r}_1, t_1, \sigma_1; \mathbf{r}_2, -i\beta\hbar, \sigma_2) . \tag{3.11}$$

Similar properties are obtained for the multi-particle thermal Green's function defined by

$$\begin{aligned} &\mathcal{G}_n(1, \ldots, n; 1', \ldots, n') \\ &= (-i)^n \left\langle \mathcal{T}\left[\hat{\psi}_K(1)\ldots\hat{\psi}_K(n)\hat{\psi}_K^\dagger(n')\ldots\hat{\psi}_K^\dagger(1')\right]\right\rangle . \end{aligned} \tag{3.12}$$

We have

$$\mathcal{G}_n(\ldots, t_i = 0, \ldots) = \pm\mathcal{G}_n(\ldots, t_i = -i\beta\hbar, \ldots) . \tag{3.13}$$

The periodicity (or antiperiodicity) of $\mathcal{G}_n$ allows us to obtain a Fourier series representation for bosons (or fermions) as follows, that is,

$$\mathcal{G}_n(\ldots, t, \ldots) = \frac{1}{-i\beta\hbar}\sum_\nu e^{-iz_\nu t}\mathcal{G}_n(\ldots, z_\nu, \ldots) , \tag{3.14}$$

where $z_\nu = (\pi\nu)/(-i\beta\hbar)$ and $\nu = \begin{pmatrix}\text{even}\\ \text{odd}\end{pmatrix}$ integer for $\nu = \begin{pmatrix}\text{bosons}\\ \text{fermions}\end{pmatrix}$. The inverse Fourier transform of Eq. (3.14) is

$$\mathcal{G}_n(\ldots, z_\nu, \ldots) = \int_0^{-i\beta\hbar} dt e^{iz_\nu t}\mathcal{G}_n(\ldots, t, \ldots) . \tag{3.15}$$

We now introduce the time-dependent operator $\hat{U}(t_1, t_2)$ defined as

$$\hat{U}(t_1, t_2) = e^{i\hat{K}_0 t_1/\hbar} e^{-i\int_{t_2}^{t_1} dt' \hat{K}(t')/\hbar} e^{-i\hat{K}_0 t_2/\hbar} . \tag{3.16}$$

In terms of $\hat{U}(t_1, t_2)$, the operators $\hat{O}_{\text{K}}$ and $\hat{O}_{\text{I}}$ are related by

$$\hat{O}_{\text{K}}(t) = \hat{U}(0, t)\hat{O}_{\text{I}}(t)\hat{U}(t, 0) , \tag{3.17}$$

where $\hat{U}$ has the properties

$$\hat{U}(t,t) = 1 \,, \tag{3.18}$$

$$\hat{U}(t_1,t_2)\,\hat{U}(t_2,t_3) = \hat{U}(t_1,t_3) \,, \tag{3.19}$$

and the time-derivative of $\hat{U}$ satisfies

$$i\hbar\frac{\partial}{\partial t}\hat{U}(t,t_0) = \hat{K}_{1\mathrm{I}}(t)\,\hat{U}(t,t_0) \,, \tag{3.20}$$

where

$$\hat{K}_{1\mathrm{I}}(t) = e^{i\hat{K}_0 t/\hbar}\,\hat{K}_1(t)e^{-i\hat{K}_0 t/\hbar} \,. \tag{3.21}$$

Solving Eq. (3.20), we obtain an explicit expression for $\hat{U}(t,t_0)$ in terms of $\hat{K}_{1\mathrm{I}}(t)$ as

$$\hat{U}(t,t_0) = \left\{\exp\left[-\frac{i}{\hbar}\int_{t_0}^{t} dt'\,\hat{K}_{1I}(t')\right]\right\}_+ \,, \tag{3.22}$$

where the subscript "+" means that the operators are to be ordered with the earlier time on the right and the later time on the left. An alternate expression is given by

$$\begin{aligned}\hat{U}(t,t_0) &= \sum_{n=0}^{\infty}\frac{\left(-\frac{i}{\hbar}\right)^n}{n!}\int_{t_0}^{t} dt_1\ldots\int_{t_0}^{t} dt_n\mathcal{T}\left[\hat{K}_{1\mathrm{I}}(t_1)\ldots\hat{K}_{1\mathrm{I}}(t_n)\right]\\ &\equiv \sum_{n=0}^{\infty}\frac{\left(-\frac{i}{\hbar}\right)^n}{n!}\int_{t_0}^{t} dt_1\ldots\int_{t_0}^{t} dt_n\left[\hat{K}_{1\mathrm{I}}(t_1)\ldots\hat{K}_{1\mathrm{I}}(t_n)\right]_+ \,.\end{aligned} \tag{3.23}$$

It is convenient to express the Green's function $\mathcal{G}$ defined in Eq. (3.2) in terms of operators in the interaction representation. From Eq. (3.16), it follows that

$$e^{-i\int_0^t dt'\hat{K}(t')/\hbar} = e^{-it\hat{K}_0/\hbar}\,\hat{U}(t,0) \,. \tag{3.24}$$

Making use of this result in Eq. (3.2) and introducing the notation that $\hat{\psi}_{\mathrm{K}}^{(+)}(t_>)$ stands for $\hat{\psi}_{\mathrm{K}}(1)$ or $\hat{\psi}_{\mathrm{K}}^{\dagger}(2)$ depending on whether t_1 is less or greater than t_2, ($t_1 \lessgtr t_2$) we have

$$\begin{aligned}\mathcal{G}(1,2) &= -i\varepsilon(1,2)e^{\beta\Omega}\mathrm{Tr}\left[e^{-i\int_0^{-i\beta\hbar} dt'\hat{K}(t')/\hbar}\,\hat{\psi}_{\mathrm{K}}^{(+)}(t_>)\hat{\psi}_{\mathrm{K}}^{(+)}(t_<)\right]\\ &= -i\varepsilon(1,2)e^{\beta\Omega}\mathrm{Tr}\left[e^{-\beta\hat{K}_0}\,\hat{U}(-i\beta\hbar,0)\,\hat{U}(0,t_>)\hat{\psi}_{\mathrm{I}}^{(+)}(t_>)\right.\\ &\quad\left.\times\;\hat{U}(t_>,0)\,\hat{U}(0,t_<)\hat{\psi}_{\mathrm{I}}^{(+)}(t_<)\,\hat{U}(t_<,0)\right]\\ &= -i\varepsilon(1,2)e^{\beta\Omega}\mathrm{Tr}\left[e^{-\beta\hat{K}_0}\,\hat{U}(-i\beta\hbar,t_>)\hat{\psi}_{\mathrm{I}}^{(+)}(t_>)\,\hat{U}(t_>,t_<)\hat{\psi}_{\mathrm{I}}^{(+)}(t_<)\,\hat{U}(t_<,0)\right]\\ &= -i\varepsilon(1,2)e^{\beta\Omega}\mathrm{Tr}\left[e^{-\beta\hat{K}_0}\left\{\hat{U}(-i\beta\hbar,0)\hat{\psi}_{\mathrm{I}}(1)\hat{\psi}_{\mathrm{I}}^{\dagger}(2)\right\}_+\right] \,,\end{aligned} \tag{3.25}$$

where we employed Eq. (3.17) as well as Eq. (3.19). The notation used here is defined as follows

$$\begin{aligned}\left[\hat{U}(-i\beta\hbar,0)\hat{\psi}_{\mathrm{I}}(1)\hat{\psi}_{\mathrm{I}}^{\dagger}(2)\right]_{+} &= \left[\hat{U}\left(-i\beta\hbar,t_{>}\right)\hat{U}\left(t_{>},t_{<}\right)\hat{U}\left(t_{<},0\right)\hat{\psi}_{\mathrm{I}}(1)\hat{\psi}_{\mathrm{I}}^{\dagger}(2)\right]_{+} \\ &= \hat{U}\left(-i\beta\hbar,t_1\right)\hat{\psi}_{\mathrm{I}}(1)\hat{U}\left(t_1,t_2\right)\hat{\psi}_{\mathrm{I}}^{\dagger}(2)\hat{U}\left(t_2,0\right),\ t_1>t_2 \\ &= \hat{U}\left(-i\beta\hbar,t_2\right)\hat{\psi}_{\mathrm{I}}^{\dagger}(2)\hat{U}\left(t_2,t_1\right)\hat{\psi}_{\mathrm{I}}(1)\hat{U}\left(t_1,0\right),\ t_2>t_1\ .\end{aligned} \tag{3.26}$$

However, a trivial application of Eq. (3.24) gives

$$e^{-\beta\Omega} = \mathrm{Tr}\left[e^{-\beta\hat{K}_0/\hbar}\,\hat{U}\left(-i\beta\hbar,0\right)\right]. \tag{3.27}$$

When this relation is substituted into Eq. (3.25), we have the desired result for the Green's function when the field operators are expressed in the interaction representation, that is,

$$\mathcal{G}(1,2) = -i\varepsilon(1,2)\frac{\mathrm{Tr}\left[e^{-\beta\hat{K}_0}\left\{\hat{U}\left(-i\beta\hbar,0\right)\hat{\psi}_{\mathrm{I}}(1)\hat{\psi}_{\mathrm{I}}^{\dagger}(2)\right\}_{+}\right]}{\mathrm{Tr}\left[e^{-\beta\hat{K}_0}\hat{U}\left(-i\beta\hbar,0\right)\right]}\ . \tag{3.28}$$

We thus have two forms for $\mathcal{G}$, that is, Eq. (3.2) in terms of field operators in the K-representation, and Eq. (3.28) in terms of field operators in the interaction representation. The first form is used when we derive the equation of motion for $\mathcal{G}$. However, we use the second form when we take functional derivatives and, subsequently, obtain the Feynman diagrammatic expansion.

For a system of bosons or fermions interacting with each other and with an external potential $W(\boldsymbol{r},t)$, we have, from Eq. (3.1),

$$\hat{K}_1(t_1) = \sum_{\sigma_1}\int d^3\boldsymbol{r}_1\hat{\psi}_{\sigma_1 S}^{\dagger}(\boldsymbol{r}_1)\hat{\psi}_{\sigma_1 S}(\boldsymbol{r}_1)\,W\left(\boldsymbol{r}_1,t_1\right). \tag{3.29}$$

Then, in the interaction representation, this becomes $\hat{K}_{1\mathrm{I}}(t_1)$, which is obtained as follows

$$\hat{K}_{1\mathrm{I}}(t_1) = \sum_{\sigma_1}\int d^3\boldsymbol{r}_1\hat{\psi}_{\mathrm{I}}^{\dagger}(1)\hat{\psi}_{\mathrm{I}}(1)\,W\left(\boldsymbol{r}_1,t_1\right), \tag{3.30}$$

where "1″ is the space-time-spin point $(\boldsymbol{r}_1,t_1,\sigma_1)$. The equation of motion of $\mathcal{G}$ is easily obtained from Eq. (3.2). Since

$$\begin{aligned} i\hbar\frac{\partial\hat{\psi}_{\mathrm{K}}(1)}{\partial t_1} &= \left[\hat{\psi}_{\mathrm{K}}(1),\,\hat{K}(t_1)\right] \\ &= \left[e^{i\int_0^{t_1}dt'\hat{K}(t')/\hbar}\hat{\psi}_{\sigma_1 S}(\boldsymbol{r}_1)e^{-i\int_0^{t_1}dt'\hat{K}(t')/\hbar},\,\hat{K}(t_1)\right] \\ &= e^{i\int_0^{t_1}dt'\hat{K}(t')/\hbar}\left[\hat{\psi}_{\sigma_1 S}(\boldsymbol{r}_1),\,\hat{K}(t_1)\right]e^{-i\int_0^{t_1}dt'\hat{K}(t')/\hbar}, \end{aligned} \tag{3.31}$$

where $\hat{K}(t_1)$ is in the Schrödinger representation, we have

$$\left[i\hbar\frac{\partial}{\partial t_1} + \frac{\hbar^2\nabla_1^2}{2m^*} + \mu - W(\boldsymbol{r}_1, t_1)\right]\mathcal{G}(1,2)$$
$$= \delta(1-2) \mp i\int d^3\boldsymbol{r}_3 V(\boldsymbol{r}_1 - \boldsymbol{r}_3)$$
$$\times \sum_{\sigma_3}\mathcal{G}_2\left(1; \boldsymbol{r}_3, t_1, \sigma_3; 2; \boldsymbol{r}_3, t_1^+, \sigma_3\right), \tag{3.32}$$

where the upper (lower) sign is for bosons (fermions), $\delta(1-2) = \delta(\boldsymbol{r}_1 - \boldsymbol{r}_2)\delta(t_1 - t_2)\delta_{\sigma_1,\sigma_2}$ and the two-particle Green's function $\mathcal{G}_2(1,2;3,4)$ is given explicitly by the equation

$$\mathcal{G}_2(1,2;3,4) = (-i)^2\left\langle \mathcal{T}\left[\hat{\psi}_K(1)\hat{\psi}_K(2)\hat{\psi}_K^\dagger(4)\hat{\psi}_K^\dagger(3)\right]\right\rangle. \tag{3.33}$$

We may rewrite Eq. (3.32) more conveniently by introducing the notation

$$v(1,2) = V(\boldsymbol{r}_1 - \boldsymbol{r}_2)\,\delta(t_1 - t_2) \tag{3.34}$$

which is independent of σ_1 and σ_2. Denoting

$$\int (d3)\ldots = \int d^3\boldsymbol{r}_3 \int_0^{-i\beta\hbar} dt_3 \sum_{\sigma_3}\ldots, \tag{3.35}$$

we have

$$\left[i\hbar\frac{\partial}{\partial t_1} + \frac{\hbar^2\nabla_1^2}{2m^*} + \mu - W(\boldsymbol{r}_1, t_1)\right]\mathcal{G}(1,2)$$
$$= \delta(1-2) \mp i\int (d3) v(1,3)\mathcal{G}_2(1,3;2,3^+), \tag{3.36}$$

where the upper (lower) sign is for bosons (fermions) and the space-time-spin point 3^+ is $(\boldsymbol{r}_3, t_3^+, \sigma_3)$. Equation (3.36) can be rewritten with the use of a function $\mathcal{G}_0(1,2)$ defined by

$$\left[i\hbar\frac{\partial}{\partial t_1} + \frac{\hbar^2\nabla_1^2}{2m^*} + \mu - W(\boldsymbol{r}_1, t_1)\right]\mathcal{G}_0(1,2) = \delta(1-2), \tag{3.37}$$

and we have

$$\mathcal{G}(1,2) = \mathcal{G}_0(1,2) \pm i\int (d3)\int (d4)\mathcal{G}_0(1,4)v(4,3)\mathcal{G}_2(4,3;2,3^+). \tag{3.38}$$

If we know $\mathcal{G}_2$, Eq. (3.38) would be the solution to the problem of calculating $\mathcal{G}$. Unfortunately, we do not know $\mathcal{G}_2$. Moreover, we cannot derive an explicit expression for $\mathcal{G}_2$ in terms of $\mathcal{G}$ in order to obtain a closed set of equations. The equation of motion for $\mathcal{G}_2$ involves even higher-order Green's functions. We thus have a never-ending hierarchy of equations to deal with.

There are two approaches to obtain approximations for $\mathcal{G}$. One is to truncate the hierarchy of equations by approximating the higher-order Green's functions as sums and products of lower ordered Green's functions. For example, we may make the Hartree–Fock approximation

$$\mathcal{G}_2(4,3;2,3^+) = \mathcal{G}(4,2)\mathcal{G}(3,3^+) \pm \mathcal{G}(4,3)\mathcal{G}(3,2) , \tag{3.39}$$

which is exact for non-interacting systems. However, it is hard to go much farther using this approach. Instead, we use a systematic diagrammatic technique. In this approach, we generate a diagrammatic expansion and then select a subset of diagrams from the ensuing series. We shall employ the method involving Feynman diagrams. However, instead of using Wick's theorem, we shall generate the diagrams by functional derivative techniques.

3.2 Functional Derivative Techniques

If $F[u]$ is the functional

$$F[u] = \int d^3 r_1 \int_{t_0}^{t} dt_1 \mathcal{F}(\boldsymbol{r}_1, t_1) u(\boldsymbol{r}_1, t_1) \tag{3.40}$$

of the function $u(\boldsymbol{r}, t)$, then the functional derivative of F with respect to u is

$$\frac{\delta F[u]}{\delta u(\boldsymbol{r}, t)} = \begin{cases} \mathcal{F}(\boldsymbol{r}, t) , & \text{if } (\boldsymbol{r}, t) \text{ is within the range of integration} \\ 0 , & \text{otherwise} \end{cases} . \tag{3.41}$$

Here, we take the integral over $\boldsymbol{r}$ to extend over all space whereas the time-integral is within the time interval (t_0, t). In consideration of our evaluation of the Green's function, we evaluate the functional derivative of $\hat{U}(t, t_0)$ with respect to $W(\bar{\boldsymbol{r}}_1, \bar{t}_1)$. From Eq. (3.23), we have

$$\begin{aligned} &\frac{\delta \hat{U}(t, t_0)}{\delta W(\bar{\boldsymbol{r}}_1, \bar{t}_1)} \\ &= \frac{\delta}{\delta W(\bar{\boldsymbol{r}}_1, \bar{t}_1)} \left[1 + \sum_{n=1}^{\infty} \frac{\left(-\frac{i}{\hbar}\right)^n}{n!} \int_{t_0}^{t} dt_1 \ldots \int_{t_0}^{t} dt_n \right. \\ &\quad \left. \times \left\{ \hat{K}_{11}(t_1) \ldots \hat{K}_{11}(t_n) \right\}_+ \right] . \end{aligned} \tag{3.42}$$

Substituting Eq. (3.30) into Eq. (3.42), we obtain

$$
\begin{aligned}
\frac{\delta \hat{U}(t,t_0)}{\delta W(\bar{r}_1,\bar{t}_1)} &= \frac{-\frac{i}{\hbar}}{1!}\frac{\delta}{\delta W(\bar{r}_1,\bar{t}_1)}\int_{t_0}^{t} dt_1 \sum_{\sigma_1}\int d\boldsymbol{r}_1 \hat{\psi}_{\mathrm{I}}^{\dagger}(1)\hat{\psi}_{\mathrm{I}}(1)\,W(\boldsymbol{r}_1,t_1) \\
&+ \frac{\delta}{\delta W(\bar{r}_1,\bar{t}_1)}\sum_{n=2}^{\infty}\frac{\left(-\frac{i}{\hbar}\right)^n}{n!}\int_{t_0}^{t} dt_1\ldots\int_{t_0}^{t} dt_n \sum_{\sigma_1}\ldots\sum_{\sigma_n}\int d^3\boldsymbol{r}_1\ldots\int d^3\boldsymbol{r}_n \\
&\times\left\{\hat{\psi}_{\mathrm{I}}^{\dagger}(1)\hat{\psi}_{\mathrm{I}}(1)\ldots\hat{\psi}_{\mathrm{I}}^{\dagger}(n)\hat{\psi}_{\mathrm{I}}(n)\right\}_{+} W(\boldsymbol{r}_1,t_1)\ldots W(\boldsymbol{r}_n,t_n) \\
&= \sum_{\sigma_1}\hat{\psi}_{\mathrm{I}}^{\dagger}(\bar{1})\hat{\psi}_{\mathrm{I}}(\bar{1}) + \sum_{n=2}^{\infty}\frac{\left(-\frac{i}{\hbar}\right)^n}{n!}\int_{t_0}^{t} dt_1\ldots\int_{t_0}^{t} dt_{n-1} \\
&\times\sum_{\sigma_1}\ldots\sum_{\sigma_{n-1}}\int d^3\boldsymbol{r}_1\ldots\int d^3\boldsymbol{r}_{n-1} \\
&\times n\sum_{\bar{\sigma}_1}\left\{\hat{\psi}_{\mathrm{I}}^{\dagger}(\bar{1})\hat{\psi}_{\mathrm{I}}(\bar{1})\hat{\psi}_{\mathrm{I}}^{\dagger}(1)\hat{\psi}_{\mathrm{I}}(1)\ldots\hat{\psi}_{\mathrm{I}}^{\dagger}(n-1)\hat{\psi}_{\mathrm{I}}(n-1)\right\}_{+} \\
&\times W(\boldsymbol{r}_1,t_1)\ldots W(\boldsymbol{r}_{n-1},t_{n-1})\,,
\end{aligned}
\tag{3.43}
$$

provided that $t \le \bar{t}_1 \le t_0$. Thus, from Eq. (3.43) we have

$$
\frac{\delta \hat{U}(t,t_0)}{\delta W(\bar{r}_1,\bar{t}_1)} = \begin{cases} -i\left\{\hat{U}(t,t_0)n_{\mathrm{I}}(\bar{r}_1,\bar{t}_1)\right\}_{+}, & t_0 \le \bar{t}_1 \le t \\ 0\,, & \text{otherwise} \end{cases}\,,
\tag{3.44}
$$

where the density operator at the space-time point $(\boldsymbol{r}_1, t_1)$ is

$$
n_{\mathrm{I}}(\boldsymbol{r}_1,t_1) = \sum_{\sigma_1}\hat{\psi}_{\mathrm{I}}^{\dagger}(1)\hat{\psi}_{\mathrm{I}}(1)\,.
\tag{3.45}
$$

We now use this result to calculate the functional derivative of the Green's function, that is, $(\delta\mathcal{G}(1,2))/(\delta W(\boldsymbol{r}_3,t_3))$ from Eq. (3.28). Since the dependence of $\mathcal{G}$ on the potential W is entirely in $\hat{U}(-i\beta\hbar,0)$, we have

$$
\begin{aligned}
\frac{\delta\mathcal{G}(1,2)}{\delta W(\boldsymbol{r}_3,t_3)} &= \left\{\frac{-i\varepsilon(1,2)}{\mathrm{Tr}\left[e^{-\beta\hat{K}_0}\hat{U}(-i\beta\hbar,0)\right]}\right\}\frac{\delta}{\delta W(\boldsymbol{r}_3,t_3)} \\
&\times \mathrm{Tr}\left\{e^{-\beta\hat{K}_0}\left[\hat{U}(-i\beta\hbar,0)\hat{\psi}_{\mathrm{I}}(1)\hat{\psi}_{\mathrm{I}}^{\dagger}(2)\right]_{+}\right\} - \frac{\mathcal{G}(1,2)}{\mathrm{Tr}\left[e^{-\beta\hat{K}_0}\hat{U}(-i\beta\hbar,0)\right]} \\
&\times\frac{\delta}{\delta W(\boldsymbol{r}_3,t_3)}\mathrm{Tr}\left[e^{-\beta\hat{K}_0}\hat{U}(-i\beta\hbar,0)\right],
\end{aligned}
\tag{3.46}
$$

which can be rewritten using Eq. (3.44) as

$$\begin{aligned}\frac{\delta \mathcal{G}(1,2)}{\delta W(\boldsymbol{r}_3,t_3)} &= -\varepsilon(1,2)e^{\beta\Omega} \\ &\quad\times \mathrm{Tr}\left\{e^{-\beta\hat{K}_0}\left[\hat{U}(-i\beta\hbar,t_3)n_{\mathrm{I}}(\boldsymbol{r}_3,t_3)\,\hat{U}(t_3,0)\hat{\psi}_{\mathrm{I}}(1)\hat{\psi}_{\mathrm{I}}^{\dagger}(2)\right]_+\right\} \\ &\quad+ i\mathcal{G}(1,2)e^{\beta\Omega}\mathrm{Tr}\left[e^{-\beta\hat{K}_0}\,\hat{U}(-i\beta\hbar,t_3)n_{\mathrm{I}}\,(\boldsymbol{r}_3,t_3)\;\hat{U}(t_3,0)\right].\end{aligned} \tag{3.47}$$

We now discuss each term in Eq. (3.47) separately. The second term can be expressed as

$$\begin{aligned}&i\mathcal{G}(1,2)\sum_{\sigma_3}\frac{\mathrm{Tr}\left\{e^{-\beta\hat{K}_0}\left[\hat{U}(-i\beta\hbar,0)\hat{\psi}_{\mathrm{I}}(3)\hat{\psi}_{\mathrm{I}}^{\dagger}(3^+)\right]_+\right\}}{\mathrm{Tr}\left[e^{-\beta\hat{K}_0}\,\hat{U}(-i\beta\hbar,0)\right]} = -\mathcal{G}(1,2)\varepsilon(3,3^+) \\ &\quad\times\sum_{\sigma_3}\mathcal{G}(3,3^+) = \mp\mathcal{G}(1,2)\sum_{\sigma_3}\mathcal{G}(3,3^+)\,,\end{aligned} \tag{3.48}$$

by employing Eq. (3.28) and the definition of $\varepsilon(3,3^+)$. We now calculate the first term in Eq. (3.47). For this, we look at the definition of $\mathcal{G}_2$ in Eq. (3.33) and write it in terms of operators in the interaction representation. We have

$$\begin{aligned}\mathcal{G}_2(1,3;2,3^+) &= -e^{\beta\Omega}\mathrm{Tr}\left\{e^{-i\int_0^{-i\beta\hbar}dt\,\hat{K}(t)/\hbar}\right. \\ &\quad\left.\times\,\mathcal{T}\left[\hat{\psi}_{\mathrm{K}}(1)\hat{\psi}_{\mathrm{K}}(3)\hat{\psi}_{\mathrm{K}}^{\dagger}(3^+)\hat{\psi}_{\mathrm{K}}^{\dagger}(2)\right]\right\} \\ &= -e^{\beta\Omega}\,\varepsilon(1,3;3^+,2) \\ &\quad\times\mathrm{Tr}\left\{e^{-i\int_0^{-i\beta\hbar}dt\,\hat{K}(t)/\hbar}\left[\hat{\psi}_{\mathrm{K}}(1)\hat{\psi}_{\mathrm{K}}(3)\hat{\psi}_{\mathrm{K}}^{\dagger}(3^+)\hat{\psi}_{\mathrm{K}}^{\dagger}(2)\right]_+\right\},\end{aligned} \tag{3.49}$$

where

$$\varepsilon\left(1,3;3^+,2\right) = \begin{cases} 1\,, & \text{for bosons} \\ \pm 1\,, & \text{for fermions and the number of permutations}\end{cases}\,. \tag{3.50}$$

Explicit calculation shows that

$$\varepsilon\left(1,3;3^+,2\right) = \begin{cases} 1\,, & \text{for bosons} \\ +1\,, & \text{for fermions, } t_2 > t_1 \\ -1\,, & \text{for fermions, } t_1 > t_2\end{cases}\,. \tag{3.51}$$

Therefore, we have

$$\varepsilon(1,3;3^+,2) = \varepsilon(2,1) = \pm\varepsilon(1,2) . \tag{3.52}$$

This means that

$$\mathcal{G}_2(1,3;2,3^+) = -e^{\beta\Omega}(\pm 1)\varepsilon(1,2) \times \mathrm{Tr}\left\{e^{-\beta\hat{K}_0}\left[\hat{U}(-i\beta\hbar,0)\hat{\psi}_{\mathrm{I}}(1)\hat{\psi}_{\mathrm{I}}(3)\hat{\psi}_{\mathrm{I}}^{\dagger}(3^+)\hat{\psi}_{\mathrm{I}}^{\dagger}(2)\right]_+\right\} . \tag{3.53}$$

Therefore, the first term in Eq. (3.47) is

$$-e^{\beta\Omega}\varepsilon(1,2)\sum_{\sigma_3}\mathrm{Tr}\left\{e^{-\beta\hat{K}_0}\left[\hat{U}(-i\beta\hbar,0)\hat{\psi}_{\mathrm{I}}(1)\hat{\psi}_{\mathrm{I}}(3)\hat{\psi}_{\mathrm{I}}^{\dagger}(3^+)\hat{\psi}_{\mathrm{I}}^{\dagger}(2)\right]_+\right\} = \pm\sum_{\sigma_3}\mathcal{G}_2(1,3;2,3^+) . \tag{3.54}$$

Combining these results in Eq. (3.47) and rearranging the terms, we obtain

$$\pm\sum_{\sigma_3}\mathcal{G}_2(1,3;2,3^+) = \left[\frac{\delta}{\delta W(\mathbf{r}_3,t_3)} \pm \sum_{\sigma_3}\mathcal{G}\left(3,3^+\right)\right]\mathcal{G}(1,2) . \tag{3.55}$$

The result in Eq. (3.55) allows Eq. (3.38) to be expressed in the form

$$\mathcal{G}(1,2) = \mathcal{G}_0(1,2) + i\int d^3\mathbf{r}_3 \int_0^{-i\beta\hbar} dt_3 \int (d4)\mathcal{G}_0(1,4)v(4,3) \times \left[\frac{\delta}{\delta W(\mathbf{r}_3,t_3)} \pm \sum_{\sigma_3}\mathcal{G}(3,3^+)\right]\mathcal{G}(4,2) . \tag{3.56}$$

However, to generate a perturbation expansion in the interaction potential v, we need to evaluate $\delta/(\delta W(\mathbf{r}_3,t_3))\mathcal{G}_0(1,2)$. We will make use of the result

$$\frac{\delta\mathcal{G}_0(1,2)}{\delta W(\mathbf{r}_3,t_3)} = \pm\sum_{\sigma_3}[\mathcal{G}_2(1,3;2,3^+)]_0 \mp \mathcal{G}_0(1,2)\sum_{\sigma_3}\mathcal{G}_0(3,3^+), \tag{3.57}$$

which follows from Eq. (3.55). Furthermore, rewriting Eq. (3.37) as

$$\int_0^{-i\beta\hbar}(d2)\mathcal{G}_0^{-1}(1,2)\mathcal{G}_0(2,3) = \delta(1-3) , \tag{3.58}$$

where

$$\mathcal{G}_0^{-1}(1,2) = \left[i\hbar\frac{\partial}{\partial t_1} + \frac{\hbar^2\nabla_1^2}{2m^*} + \mu - W(\mathbf{r}_1,t_1)\right]\delta(1-2) , \tag{3.59}$$

we have $\delta[\mathcal{G}_0^{-1}\mathcal{G}_0] = \delta\mathcal{G}_0^{-1}\mathcal{G}_0 + \mathcal{G}_0^{-1}\delta\mathcal{G}_0 = 0$ or $\delta\mathcal{G}_0 = -\mathcal{G}_0\delta\mathcal{G}_0^{-1}\mathcal{G}_0$. From this, we deduce that

$$\frac{\delta\mathcal{G}_0(1,2)}{\delta W(\mathbf{r}_3, t_3)} = \sum_{\sigma_3} \mathcal{G}_0(1,3)\mathcal{G}_0(3,2) . \tag{3.60}$$

Combining Eqs. (3.57) and (3.60), we obtain

$$[\mathcal{G}_2(1,3;2,3^+)]_0 \equiv \mathcal{G}_0(1,2)\mathcal{G}_0(3,3^+) \pm \mathcal{G}_0(1,3)\mathcal{G}_0(3,2) , \tag{3.61}$$

which is an identity for non-interacting particles. Based on this result, one makes the Hartree–Fock approximation Eq. (3.39).

Equation (3.60) allows us to generate a perturbation expansion from Eq. (3.56). In addition, since $\mathcal{G}_0$ and $\mathcal{G}$ are diagonal in the spin indices, the functional derivative term in Eq. (3.56) can, without error, be summed over σ_3, that is,

$$\mathcal{G}(1,2) = \mathcal{G}_0(1,2) + i\int(d3)\int(d4)\mathcal{G}_0(1,4)v(4,3) \times\left[\frac{\delta}{\delta W(\mathbf{r}_3,t_3)} \pm \mathcal{G}(3,3^+)\right]\mathcal{G}(4,2) . \tag{3.62}$$

3.3 Unrenormalized Expansion for $\mathcal{G}$ and Σ

We begin this section by iterating Eq. (3.62) for the one-particle Green's function, that is, by replacing $\mathcal{G}(1,2) \to \mathcal{G}_0(1,2)$ on the right-hand side in order to obtain

$$\mathcal{G}(1,2) = \mathcal{G}_0(1,2) + i\int(d3)\int(d4)\mathcal{G}_0(1,4)v(4,3) \times\left[\mathcal{G}_0(4,3)\mathcal{G}_0(3,2) \pm \mathcal{G}_0(3,3^+)\mathcal{G}_0(4,2)\right] + \mathcal{O}(v^2) . \tag{3.63}$$

Expressing this equation in terms of diagrams, we have the results in Figure 3.1.

Here, the rules for constructing the Feynman diagrams are defined in Figure 3.2. We also specify the notation used in the diagrams.

The complete set of second-order diagrams are given in Figure 3.3. The first four are iterations; the next four are dressed first order diagrams. Only the last two are quite distinct. We have thus demonstrated that the functional derivative

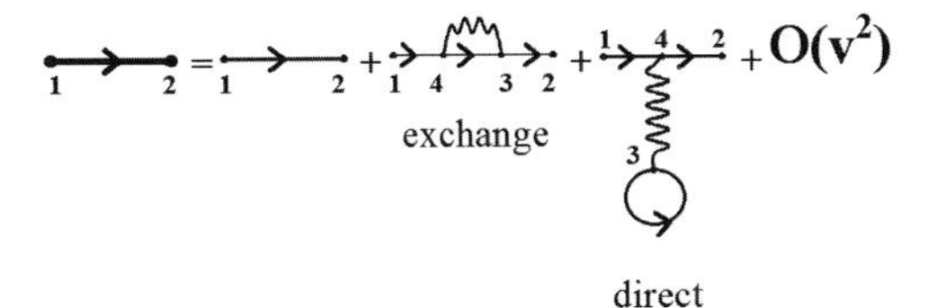

Figure 3.1 Feynman diagrams for the single-particle Green's function $\mathcal{G}(1,2)$, to the lowest order in the Coulomb potential v.

1. 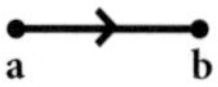 a ——→—— b $= G\,(\mathrm{a}, \mathrm{b})$

2. 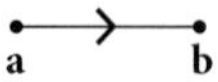 a ——→—— b $- G_0\,(\mathrm{a}, \mathrm{b})$

3. 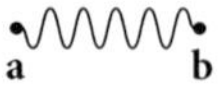 a ∿∿∿∿ b $= t\, v(\mathrm{a}, \mathrm{b}) = t\, v(\mathrm{b}, \mathrm{a})$

4. Closed loop: ±1

5. G_0 line ending at its starting point 3: $G_0\,(3, 3^+)$

6. "Integrate" over each intermediate space-time-spin point i according to:

$$\int d^3r_i \int_0^{-i\beta} dt_i \sum_{S_i}$$

7. Remember that the interaction does not change the spin at the vertices, and that G_0 and G are diagonal in spin

Figure 3.2 Feynman rules for constructing the perturbation expansion and constructing a diagrammatic expansion for the Green's function $\mathcal{G}(1,2)$.

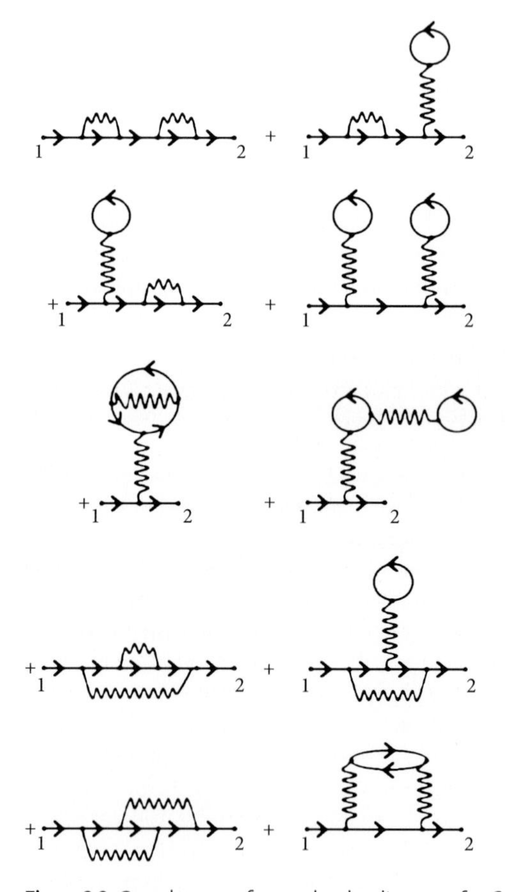

Figure 3.3 Complete set of second-order diagrams for $\mathcal{G}(1,2)$.

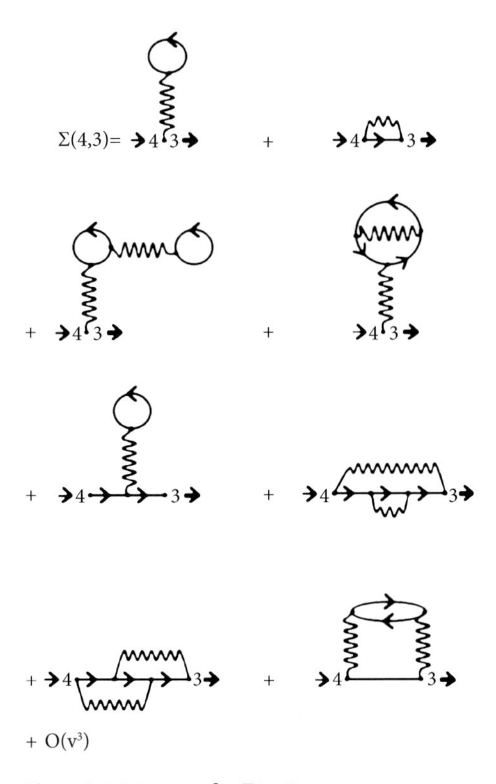

Figure 3.4 Diagrams for $\Sigma(4,3)$.

technique yields Feynman graphs without ever having to use Wick's theorem. All the graphs in the diagrammatic expansion are connected which means that there was no need to show that the disconnected graphs cancel as is the case when Wick's theorem is used. So far, $\mathcal{G}_0$ is the Green's function for non-interacting particles in a non-zero external field W. We can now set $W = 0$ and $\mathcal{G}_0$ will correspond to the Green's function for non-interacting particles in zero external field, and $\mathcal{G}$ will be the Green's function for interacting particles when this external field is switched off. This means that we may safely set the external potential W equal to zero in our discussion which now follows.

We introduce the proper self-energy $\Sigma(3,4)$ by means of

$$\mathcal{G}(1,2) = \mathcal{G}_0(1,2) + \int (d3) \int (d4) \mathcal{G}_0(1,4) \Sigma(4,3) \mathcal{G}(3,2) , \tag{3.64}$$

where each term in the above expansion for $\mathcal{G}(1,2)$ must be produced only once via our selection of the self-energy. Thus far we have generated an *unrenormalized* expansion for $\mathcal{G}$ and Σ and this is accomplished if we employ $\Sigma(4,3)$ in Figure 3.4.

The Hartree–Fock (HF) approximation for the self-energy is given in Figure 3.5.

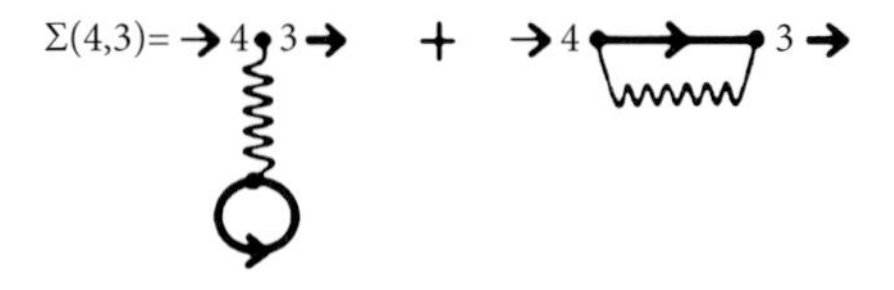

Figure 3.5 Feynman Diagrams for $\Sigma(4, 3)$ in the Hartree–Fock approximation.

Applying the Feynman rules described above, we obtain an expression for the self-energy as

$$\Sigma_{\mathrm{HF}}\left(x_1, x_1'\right) = \pm i\delta\left(t_1 - t_1'\right)\Big[\delta\left(\boldsymbol{x}_1 - \boldsymbol{x}_1'\right)\int d^3\boldsymbol{x}_2 G\left(\boldsymbol{x}_2, t_2; \boldsymbol{x}_2, t_2^+\right) v\left(\boldsymbol{x}_1, -\boldsymbol{x}_2\right) - v\left(\boldsymbol{x}_1 - \boldsymbol{x}_1'\right) G\left(\boldsymbol{x}_1, t_1; \boldsymbol{x}_1', t_1^+\right)\Big]. \tag{3.65}$$

By Fourier transforming this equation in time, we obtain

$$\Sigma_{\mathrm{HF}}(\boldsymbol{x}_1, \boldsymbol{x}_1') = \pm i\delta(\boldsymbol{x}_1 - \boldsymbol{x}_1') \int d^3\boldsymbol{x}_2 v\left(\boldsymbol{x}_1 - \boldsymbol{x}_2\right) \int_{-\infty}^{\infty} \frac{d\omega}{2\pi} e^{i\omega\eta^+} G\left(\boldsymbol{x}_2, \boldsymbol{x}_2; \omega\right) + iv\left(\boldsymbol{x}_1 - \boldsymbol{x}_1'\right) \int_{-\infty}^{\infty} \frac{d\omega}{2\pi} e^{i\omega\eta^+} G\left(\boldsymbol{x}_1, \boldsymbol{x}_1'; \omega\right), \tag{3.66}$$

which gives us the first-order terms in the expansion for the single-particle Green's function $\mathcal{G}$. It also gives all second-order terms except the last two terms in Figure 3.3. Additionally, it produces an infinite number of other diagrams besides those in Figure 3.3. Thus, this Hartree–Fock approximation results in an integral equation for $\mathcal{G}$ which must be solved self-consistently. The appearance of a positive infinitesimal quantity η^+ allows us to close the contour in the upper half of the complex ω-plane.

3.4 Renormalized Expansion for Self-Energy Σ

In our presentation so far, we have used the functional derivative technique to reproduce well-known results that have been obtained using Wick's theorem. We now use the functional derivative method to obtain results which, to our knowledge, cannot be obtained using Wick's theorem. Namely, we derive a set of equations which, when iterated, yield a perturbation expansion for the proper self-energy in which only renormalized propagator lines appear, that is, a given diagram occurs only once.

To carry out this calculation, we reintroduce the external potential W and take the functional derivative of Eq. (3.64)

$$\begin{aligned}\frac{\delta \mathcal{G}(1,2)}{\delta W(\mathbf{r}_3,t_3)} &= \frac{\delta \mathcal{G}_0(1,2)}{\delta W(\mathbf{r}_3,t_3)} + \int (d5) \int (d4) \frac{\delta \mathcal{G}_0(1,4)}{\delta W(\mathbf{r}_3,t_3)} \Sigma(4,5) \mathcal{G}(5,2) \\ &+ \int (d5) \int (d4) \mathcal{G}_0(1,4) \frac{\delta \Sigma(4,5)}{\delta W(\mathbf{r}_3,t_3)} \mathcal{G}(5,2) \\ &+ \int (d5) \int (d4) \mathcal{G}_0(1,4) \Sigma(4,5) \frac{\delta \mathcal{G}(1,4)}{\delta W(\mathbf{r}_3,t_3)} . \end{aligned} \tag{3.67}$$

Making use of Eqs. (3.60) and (3.64) in Eq. (3.67), we obtain

$$\begin{aligned}\frac{\delta \mathcal{G}(1,2)}{\delta W(\mathbf{r}_3,t_3)} &= \sum_{\sigma_3} \mathcal{G}_0(1,3) \mathcal{G}(3,2) \\ &+ \int (d5) \int (d4) \mathcal{G}_0(1,4) \frac{\delta \Sigma(4,5)}{\delta W(\mathbf{r}_3,t_3)} \mathcal{G}(5,2) \\ &+ \int (d5) \int (d4) \mathcal{G}_0(1,4) \Sigma(4,5) \frac{\delta \mathcal{G}(1,4)}{\delta W(\mathbf{r}_3,t_3)} . \end{aligned} \tag{3.68}$$

By iterating Eq. (3.68), we conclude that it can be expressed as

$$\begin{aligned}\frac{\delta \mathcal{G}(1,2)}{\delta W(\mathbf{r}_3,t_3)} &= \sum_{\sigma_3} \mathcal{G}(1,3) \mathcal{G}(3,2) \\ &+ \int (d5) \int (d4) \mathcal{G}(1,4) \frac{\delta \Sigma(4,5)}{\delta W(\mathbf{r}_3,t_3)} \mathcal{G}(5,2) , \end{aligned} \tag{3.69}$$

thus eliminating $\mathcal{G}_0$. The functional derivative of $\mathcal{G}$ on the right-hand side of Eq. (3.62) is given by Eq. (3.69) in terms of the renormalized Green's function $\mathcal{G}$.

Substituting Eq. (3.55) into the equation of motion of $\mathcal{G}$ given in Eq. (3.36) and then making use of Eq. (3.69), we have

$$\begin{aligned} &\left[i\hbar \frac{\partial}{\partial t_1} + \frac{\hbar^2 \nabla_1^2}{2m^*} + \mu - W(\mathbf{r}_1,t_1) \right] \mathcal{G}(1,2) \\ &= \delta(1-2) \pm i \int (d3) v(1,3) \mathcal{G}\left(3,3^+\right) \mathcal{G}(1,2) \\ &+ i \int d^3 r_3 \int_0^{-i\beta\hbar} dt_3 v(1,3) \sum_{\sigma_3} \mathcal{G}(1,3) \mathcal{G}(3,2) \\ &+ i \int d^3 r_3 \int_0^{-i\beta\hbar} dt_3 v(1,3) \int (d5) \int (d4) \mathcal{G}(1,4) \\ &\times \frac{\delta \Sigma(4,5)}{\delta W(\mathbf{r}_3,t_3)} \mathcal{G}(5,2) . \end{aligned} \tag{3.70}$$

However, upon applying $[i\hbar\partial/\partial t_1 + \hbar^2\nabla_1^2/2m^* + \mu - W(\mathbf{r}_1, t_1)]$ to Eq. (3.64) for $\mathcal{G}(1,2)$ and using Eq. (3.37) for $\mathcal{G}_0(1,2)$, we obtain

$$\begin{aligned}
&\int (d3)\Sigma(1,3)\mathcal{G}(3,2) \pm i \int (d3)v(1,3)\mathcal{G}(3,3^+)\mathcal{G}(1,2) \\
&\quad + i \int d^3 r_3 \int_0^{-i\beta\hbar} dt_3 v(1,3) \sum_{\sigma_3} \mathcal{G}(1,3)\mathcal{G}(3,2) \\
&\quad + i \int d^3 r_3 \int_0^{-i\beta\hbar} dt_3 v(1,3) \int (d5) \int (d4)\mathcal{G}(1,4) \\
&\quad \times \frac{\delta\Sigma(4,5)}{\delta W(\mathbf{r}_3, t_3)}\mathcal{G}(5,2) \,.
\end{aligned} \tag{3.71}$$

We rewrite this equation as

$$\begin{aligned}
&\int (d5)\mathcal{G}(5,2) \Bigg[\Sigma(1,5) \mp i \int (d3)v(1,3)\mathcal{G}(3,3^+)\delta(1-5) - iv(1,5)\mathcal{G}(1,5) \\
&\qquad - i \int d^3 r_3 \int_0^{-i\beta\hbar} dt_3 v(1,3) \int (d4)\mathcal{G}(1,4)\frac{\delta\Sigma(4,5)}{\delta W(\mathbf{r}_3, t_3)} \Bigg] = 0 \,,
\end{aligned} \tag{3.72}$$

which can be true if and only if

$$\begin{aligned}
\Sigma(1,2) = &\pm i \int (d3)v(1,3)\mathcal{G}(3,3^+)\delta(1-2) + iv(1,2)\mathcal{G}(1,2) \\
&+ i \int d^3 r_3 \int_0^{-i\beta\hbar} dt_3 v(1,3) \int (d4)\mathcal{G}(1,4)\frac{\delta\Sigma(4,2)}{\delta W(\mathbf{r}_3, t_3)} \,.
\end{aligned} \tag{3.73}$$

The first two terms on the right-hand side of Eq. (3.73) are just the Hartree–Fock terms. Iteration of this equation combined with Eq. (3.69) yields a propagator renormalized expansion for Σ, that is, there are no $\mathcal{G}_0$ lines in the diagrams, only $\mathcal{G}$ lines.

We have

$$\Sigma(1,2) = \pm i \int (d3) v(1,3) \mathcal{G}(3,3^+) \delta(1-2) + i v(1,2) \mathcal{G}(1,2)$$
$$+ i \int d^3 \boldsymbol{r}_3 \int_0^{-i\beta\hbar} dt_3 v(1,3) \int (d4) \mathcal{G}(1,4)$$
$$\times \frac{\delta}{\delta W(\boldsymbol{r}_3, t_3)} \Bigg[\pm i \int (d5) v(4,5) \mathcal{G}(5,5^+) \delta(4,2)$$
$$+ i v(4,2) \mathcal{G}(4,2) + i \int d^3 \boldsymbol{r}_5 \int_0^{-i\beta\hbar} dt_5 v(4,5)$$
$$\int (d6) \mathcal{G}(4,6) \frac{\delta \Sigma(6,2)}{\delta W(\boldsymbol{r}_5, t_5)} \Bigg]$$
$$= \pm \int (d3) [i v(1,3)] \mathcal{G}(3,3^+) \delta(1-2) + [i v(1,2)] \mathcal{G}(1,2)$$
$$\pm \int d^3 \boldsymbol{r}_3 \int_0^{-i\beta\hbar} dt_3 [i v(1,3)] \int (d4) \mathcal{G}(1,4)$$
$$\int (d5) [i v(4,5)] \delta(4-2) \frac{\delta \mathcal{G}(5,5^+)}{\delta W(\boldsymbol{r}_3, t_3)}$$
$$+ \int d^3 \boldsymbol{r}_3 \int_0^{-i\beta\hbar} dt_3 [i v(1,3)] \int (d4) \mathcal{G}(1,4) [i v(4,2)] \frac{\delta \mathcal{G}(4,2)}{\delta W(\boldsymbol{r}_3, t_3)}$$
$$+ \text{terms explicitly of order } \mathcal{O}(v^3) \,. \tag{3.74}$$

If we now make use of Eq. (3.69) in Eq. (3.74), this gives

$$\Sigma(1,2) = \pm i \int (d3) v(1,3) \mathcal{G}(3,3^+) \delta(1-2) + i v(1,2) \mathcal{G}(1,2)$$
$$\pm \int (d3) [i v(1,3)] \int (d5) \mathcal{G}(1,2) [i v(2,5)] \mathcal{G}(5,3) \mathcal{G}(3,5^+)$$
$$+ \int (d3) [i v(1,3)] \int (d4) \mathcal{G}(1,4) [i v(4,2)] \mathcal{G}(4,3) \mathcal{G}(3,2) \ldots \,, \tag{3.75}$$

where the first and second terms in Eq. (3.75) are the Hartree–Fock terms. The right-hand side of Eq. (3.75) is given diagrammatically in Figure 3.6.

The third-order Feynman self-energy diagrams can also be calculated. They are given in Figure 3.7

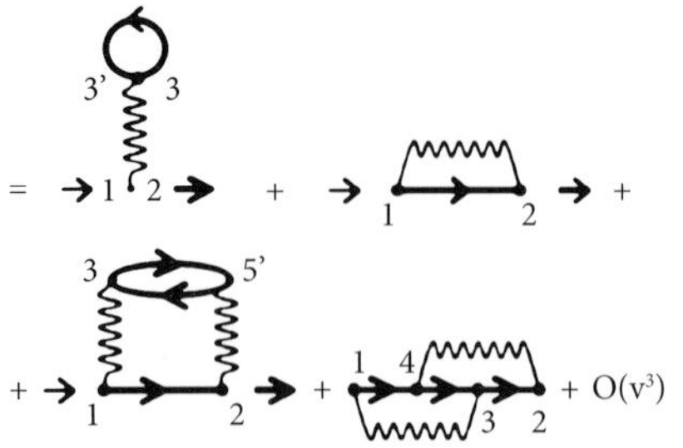

Figure 3.6 The Feynman graphs corresponding to the right-hand side of Eq. (3.75).

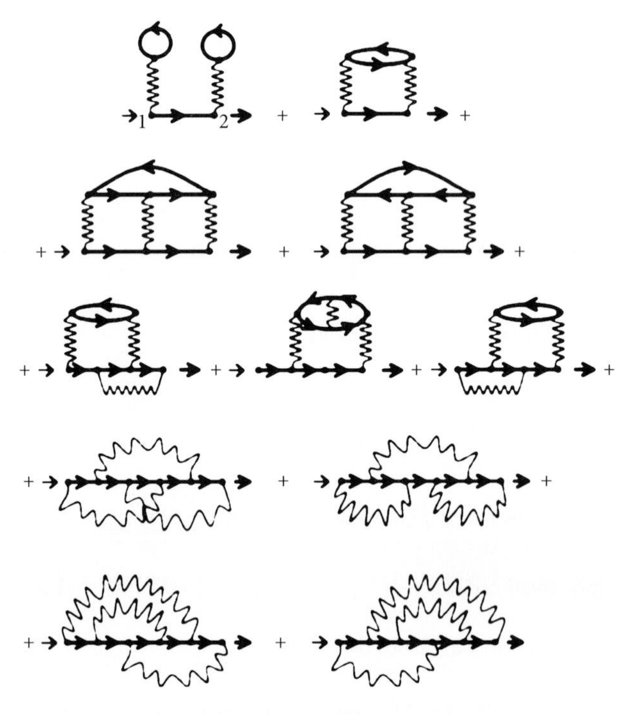

Figure 3.7 The third-order Feynman self-energy diagrams.

3.5 The Schrödinger Equation in the Hartree–Fock Approximation

At arbitrary temperature, the frequency-dependent Green's function has been calculated for a system of non-interacting electrons. In Hartree–Fock theory, we take this form as the Green's function for interacting electrons, that is,

$$G\left(\boldsymbol{x}, \boldsymbol{x}'; \omega\right) = \sum_j \varphi_j(\boldsymbol{x})\varphi_j^*\left(\boldsymbol{x}'\right)\left[\frac{f_0\left(\varepsilon_j - E_\mathrm{F}\right)}{\hbar\omega - \varepsilon_j + i\eta} + \frac{f_0(E_\mathrm{F} - \varepsilon_j)}{\hbar\omega - \varepsilon_j - i\eta}\right]. \tag{3.76}$$

Here, E_F is the Fermi energy at 0 K, ε_j the eigenenergies with eigenfunctions $\varphi_j(\boldsymbol{x})$ and $\eta = 0^+$.

Substituting Eq. (3.76) into Eq. (3.66), at $T = 0$ K we obtain

$$\Sigma_{\mathrm{HF}}\left(\boldsymbol{x},\boldsymbol{x}'\right) = \delta\left(\boldsymbol{x}-\boldsymbol{x}'\right)\int d^3\boldsymbol{x}_2 v\left(\boldsymbol{x}-\boldsymbol{x}_2\right)\sum_j\left|\varphi_j(\boldsymbol{x}_2)\right|^2\theta\left(E_F-\varepsilon_j\right)$$
$$- v\left(\boldsymbol{x}-\boldsymbol{x}'\right)\sum_j\varphi_j(\boldsymbol{x})\varphi_j^*(\boldsymbol{x}')\theta\left(E_F-\varepsilon_j\right), \tag{3.77}$$

where $\theta(x)$ is the unit step function. However, making use of Eq. (3.64) as well as

$$\hat{\mathcal{L}}_1\mathcal{G}_0\left(\boldsymbol{x}_1,\boldsymbol{x}_1';\omega\right) \equiv \left[\hbar\omega + \frac{\hbar^2\nabla_1^2}{2m^*} + \mu - V_{\mathrm{ext}}(\boldsymbol{x}_1)\right]\mathcal{G}_0\left(\boldsymbol{x}_1,\boldsymbol{x}_1';\omega\right)$$
$$= \delta\left(\boldsymbol{x}_1-\boldsymbol{x}_1'\right) \tag{3.78}$$

for a time-independent external potential $V_{\mathrm{ext}}(\boldsymbol{x})$, we have, upon applying $\hat{\mathcal{L}}_1$,

$$\hat{\mathcal{L}}_1\mathcal{G}\left(\boldsymbol{x}_1,\boldsymbol{x}_1';\omega\right) = \delta\left(\boldsymbol{x}_1-\boldsymbol{x}_1'\right) + \int d^3\boldsymbol{x}_2\Sigma_{\mathrm{HF}}(\boldsymbol{x}_1,\boldsymbol{x}_2)\mathcal{G}\left(\boldsymbol{x}_2,\boldsymbol{x}_1';\omega\right). \tag{3.79}$$

Substituting Eq. (3.76) into Eq. (3.79), multiplying by $\varphi_k(\boldsymbol{x}_1')$ and integrating over $\boldsymbol{x}_1'$, we obtain

$$\left[\hbar\omega + \frac{\hbar^2\nabla_1^2}{2m^*} + \mu - V_{\mathrm{ext}}(\boldsymbol{x}_1)\right]\frac{\varphi_k(\boldsymbol{x}_1)}{\hbar\omega-\varepsilon_k} - \int d^3\boldsymbol{x}_2\Sigma_{\mathrm{HF}}(\boldsymbol{x}_1,\boldsymbol{x}_2)\frac{\varphi_k(\boldsymbol{x}_2)}{\hbar\omega-\varepsilon_k}$$
$$= \varphi_k(\boldsymbol{x}_1). \tag{3.80}$$

That is,

$$\left[-\frac{\hbar^2\nabla_1^2}{2m^*} + V_{\mathrm{ext}}(\boldsymbol{x}_1) - \mu\right]\varphi_k(\boldsymbol{x}_1) + \int d^3\boldsymbol{x}_2\Sigma_{\mathrm{HF}}\left(\boldsymbol{x}_1,\boldsymbol{x}_2\right)\varphi_k(\boldsymbol{x}_2)$$
$$= \varepsilon_k\varphi_k\left(\boldsymbol{x}_1\right), \tag{3.81}$$

where $\Sigma_{\mathrm{HF}}(\boldsymbol{x}_1,\boldsymbol{x}_2)$ is given by Eq. (3.66).

3.6 Screened External Potential

Let us rewrite Eq. (3.38) for fermions as

$$\left\{[\mathcal{G}_0]^{-1}(1) - W(1)\right\}\mathcal{G}(\boldsymbol{r}_1,t_1;\boldsymbol{r}_1',t_1') + i\int d^3\boldsymbol{r}_2\, v(\boldsymbol{r}_1-\boldsymbol{r}_2)$$
$$\times\mathcal{G}_2(\boldsymbol{r}_1t_1,\boldsymbol{r}_2t_1;\boldsymbol{r}_1't_1',\boldsymbol{r}_2t_1^+) = \delta(\boldsymbol{r}_1-\boldsymbol{r}_1')\,\delta(t_1-t_1'), \tag{3.82}$$

where

$$[\mathcal{G}_0]^{-1}(1) = \left[i\hbar\frac{\partial}{\partial t_1} + \frac{\hbar^2\nabla_1^2}{2m^*} + \mu\right]. \tag{3.83}$$

Noting the result above in Eq. (3.55) for $\delta\mathcal{G}/\delta W$, the $\mathcal{G}_2$ term in Eq. (3.82) may be rewritten to produce

$$\left\{[\mathcal{G}_0]^{-1}(1) - W(1) + i\int d^3 r_2 v(r_1 - r_2)\,\mathcal{G}(r_2, t_1; r_2, t_1^+) - i\int d^3 r_2 \right.$$
$$\left. \times\, v(r_1 - r_2)\frac{\delta}{\delta W(r_2, t_1^+)}\right\}\mathcal{G}(r_1, t_1; r_1', t_1') = \delta\left(r_1 - r_1'\right). \tag{3.84}$$

However, in this equation, in addition to the external potential W, there is a term $\sim \mathcal{G}\mathcal{G}$ which enters in the form of an inter-particle interaction, thereby contributing to an effective potential. This is a time-dependent Hartree-like term, that is, $\int v\mathcal{G}\mathcal{G}$. We thus identify the effective time-dependent potential as

$$V(1) = W(1) - i\int d^3 r_2 v(r_1 - r_2)\,\mathcal{G}(r_2, t_1; r_2, t_1^+). \tag{3.85}$$

One can thus define an inverse dielectric function as

$$K(1,2) = \frac{\delta V(1)}{\delta W(2)} = \delta(1-2) - i\int(d3)v(1-3)\frac{\delta\mathcal{G}(3,3^+)}{\delta W(2)}, \tag{3.86}$$

which is equivalent to the following integral equation

$$K(1,2) = \delta(1-2) - i\int(d3)\int(d4)v(1-3)\frac{\delta\mathcal{G}(3,3^+)}{\delta V(4)}K(4,2). \tag{3.87}$$

The time-dependent Hartree approximation is the simplest way to solve this equation. In this approximation, the single-particle Green's function is replaced by its value for non-interacting electrons. Equation (3.60) then allows us to generate a the following integral equation for $K(1,2)$ as

$$K(1,2) = \delta(1-2) + \int(d3)\int(d4)v(1-3)\chi^0(3,4)K(4,2), \tag{3.88}$$

where

$$\chi^0(3,4) = -i\sum_{\sigma_4}\mathcal{G}_0(3,4)\mathcal{G}_0(4,3) \tag{3.89}$$

is the polarization function for non-interacting electrons and is given by a ring diagram. Equation (3.88) is the random-phase approximation (RPA) for the inverse dielectric function. It will be solved in this book for specific problems and will yield the collective excitations of a system since the poles of $K(1,2)$ correspond to the normal modes.

3.7
Retarded Polarization Function

We now obtain an explicit form for the polarization function in terms of the single-particle energy eigenvalues and eigenfunctions of a spin-degenerate fermion system. The retarded and advanced Green's functions are

$$\mathcal{G}_{\gtrless}(1,2) = -i \sum_{\nu} e^{-i(\epsilon_\nu - \mu)(t_1 - t_2)} \phi_\nu^*(\mathbf{r}_2)\phi_\nu(\mathbf{r}_1) \begin{cases} 1 - f_0(\epsilon_\nu - \mu) \\ f_0(\epsilon_\nu - \mu) \end{cases} , \tag{3.90}$$

where $\phi_\nu(\mathbf{r})$ is an eigenfunction. The retarded density–density response function is then

$$\chi^0(1,2) = 2i\theta(t_1 - t_2) \sum_{\nu,\nu'} \left[f_0(\epsilon_{\nu'} - \mu) - f_0(\epsilon_\nu - \mu) \right] \times \phi_\nu^*(\mathbf{r}_2)\,\phi_\nu(\mathbf{r}_1)\,\phi_{\nu'}^*(\mathbf{r}_1)\,\phi_{\nu'}(\mathbf{r}_2)\, e^{-i(\epsilon_\nu - \epsilon_{\nu'})(t_1 - t_2)} . \tag{3.91}$$

From this, we obtain the Fourier transform of Eq. (3.91) with time.

3.8 RPA for the Polarization Function

A self-consistent field theory for the polarization function can be obtained from the solution of the density matrix equation $i\hbar\partial\hat{\rho}/\partial t = [\hat{H}, \hat{\rho}]_-$. First, consider a *non-interacting* system subject to an external perturbation $-e\varphi_{\mathrm{ext}}(\mathbf{r}, t) = -e\tilde{\varphi}_{\mathrm{ext}}(\mathbf{r}, \omega)e^{i\omega t}$, we have $\hat{H} = \hat{H}_0 - e\varphi_{\mathrm{ext}}(\mathbf{r}, t)$ and $\hat{\rho} = \hat{\rho}_0 + \delta\hat{\rho}$, where $\delta\hat{\rho}$ is the perturbed density matrix. Denoting the eigenstates by $|\nu\rangle$, $\langle\nu|\hat{H}_0|\nu'\rangle = \varepsilon_\nu\delta_{\nu,\nu'}$, $\langle\nu|\hat{\rho}_0|\nu'\rangle = 2f_0(\varepsilon_\nu)\delta_{\nu,\nu'}$, we obtain in the lowest order of perturbation theory

$$\langle\nu|\delta\hat{\rho}|\nu'\rangle = 2e\frac{f_0(\varepsilon_\nu) - f_0(\varepsilon_{\nu'})}{\hbar\omega - \varepsilon_\nu + \varepsilon_{\nu'}}\langle\nu|\tilde{\varphi}_{\mathrm{ext}}(\mathbf{r}, \omega)|\nu'\rangle . \tag{3.92}$$

This gives rise to an induced density

$$\begin{aligned} \delta n(\mathbf{r}, \omega) &= \mathrm{Tr}\left[\delta\hat{\rho}\hat{\psi}^\dagger(\mathbf{r})\,\hat{\psi}(\mathbf{r})\right] = \sum_\nu \langle\nu|\delta\hat{\rho}\hat{\psi}^\dagger(\mathbf{r})\,\hat{\psi}(\mathbf{r})|\nu\rangle \\ &= \sum_{\nu,\nu'}\langle\nu|\delta\hat{\rho}|\nu'\rangle\langle\nu'|\hat{\psi}^\dagger(\mathbf{r})\,\hat{\psi}(\mathbf{r})|\nu\rangle \\ &= \sum_{\nu,\nu'}\langle\nu|\delta\hat{\rho}|\nu'\rangle\phi_{\nu'}^*(\mathbf{r})\,\phi_\nu(\mathbf{r}) , \end{aligned} \tag{3.93}$$

which is given in terms of the eigenfunctions $\langle\mathbf{r}|\nu\rangle = \phi_\nu(\mathbf{r})$. Making use of Eq. (3.92) in Eq. (3.93), after a little manipulation we obtain

$$\delta n(\mathbf{r}, \omega) = e\int d^3\mathbf{r}'\chi^0(\mathbf{r}, \mathbf{r}'; \omega)\,\tilde{\varphi}_{\mathrm{ext}}(\mathbf{r}', \omega) , \tag{3.94}$$

where the density–density response function $\chi^0(\mathbf{r}, \mathbf{r}'; \omega)$ is the time Fourier transform of Eq. (3.91) and is given by

$$\chi^0(\mathbf{r}, \mathbf{r}'; \omega) = 2\sum_{\nu,\nu'}\frac{f_0(\varepsilon_\nu) - f_0(\varepsilon_{\nu'})}{\hbar\omega - \varepsilon_\nu + \varepsilon_{\nu'}}\phi_{\nu'}^*(\mathbf{r})\,\phi_\nu(\mathbf{r})\,\phi_{\nu'}(\mathbf{r}')\,\phi_\nu^*(\mathbf{r}') . \tag{3.95}$$

For an interacting system of electrons, the induced density is given in linear response theory by an equation like (3.94), though in terms of a response function χ which takes account of the electron–electron interactions, that is,

$$\delta n(\mathbf{r}, \omega) = e \int d^3\mathbf{r}' \chi\left(\mathbf{r}, \mathbf{r}'; \omega\right) \tilde{\varphi}_{\text{ext}}\left(\mathbf{r}', \omega\right) . \tag{3.96}$$

For a system of interacting electrons, the induced density fluctuation results in an induced potential which is a solution of Poisson's equation

$$\nabla^2 \tilde{\varphi}_{\text{ind}}\left(\mathbf{r}, \omega\right) = \frac{4\pi e}{\epsilon_{\text{s}}} \delta n\left(\mathbf{r}, \omega\right) , \tag{3.97}$$

which has a solution of the form

$$\tilde{\varphi}_{\text{ind}}\left(\mathbf{r}, \omega\right) = \int d^3\mathbf{r}' v\left(\mathbf{r} - \mathbf{r}'\right) \delta n\left(\mathbf{r}', \omega\right) , \tag{3.98}$$

where $v(\mathbf{r} - \mathbf{r}')$ is the Coulomb potential. In the self-consistent field approximation, we determine the response function χ by assuming that

$$\delta n\left(\mathbf{r}, \omega\right) = e \int d^3\mathbf{r}' \chi^0\left(\mathbf{r}, \mathbf{r}'; \omega\right) \tilde{\Phi}_{\text{tot}}\left(\mathbf{r}', \omega\right) , \tag{3.99}$$

where $\tilde{\Phi}_{\text{tot}}\left(\mathbf{r}, \omega\right) = \tilde{\varphi}_{\text{ext}}\left(\mathbf{r}, \omega\right) + \tilde{\varphi}_{\text{ind}}\left(\mathbf{r}, \omega\right)$ must be determined self-consistently. Combining Eqs. (3.96), (3.98), and (3.99), we obtain an integral equation for $\chi\left(\mathbf{r}, \mathbf{r}'; \omega\right)$ in the RPA, that is,

$$\begin{aligned} \chi\left(\mathbf{r}, \mathbf{r}'; \omega\right) = \chi^0\left(\mathbf{r}, \mathbf{r}'; \omega\right) &+ \int d^3\mathbf{r}'' \int d^3\mathbf{r}''' \chi^0\left(\mathbf{r}, \mathbf{r}''; \omega\right) \\ &\times v\left(\mathbf{r}'' - \mathbf{r}'''\right) \chi^0\left(\mathbf{r}''', \mathbf{r}'; \omega\right) . \end{aligned} \tag{3.100}$$

3.9 Problems

1. Show explicitly that

$$\begin{aligned} &\frac{1}{3!} \int_0^t dt_1 \int_0^t dt_2 \int_0^t dt_3 \hat{T}\left[V(t_1) V(t_2) V(t_3)\right] \\ &= \int_0^t dt_1 \int_0^{t_1} dt_2 \int_0^{t_2} dt_3 V(t_1) V(t_2) V(t_3) . \end{aligned} \tag{3.101}$$

2. A useful approximation for the dielectric function for a semiconductor is

$$\epsilon_{\mathrm{r}}(q) = 1 + \frac{\epsilon(0) - 1}{\frac{1+q^2\epsilon(0)}{\gamma^2}}, \tag{3.102}$$

where

$$\gamma^2 = \lambda_{\mathrm{TF}}^2 \frac{\epsilon(0)}{\epsilon(0) - 1} \tag{3.103}$$

and $\epsilon(0)$ is the dielectric constant.

a) Show that the corresponding Thomas–Fermi equation to this is

$$\left(\nabla^2 - \gamma^2\right) V_{\mathrm{scr}} = \left[\nabla^2 - \frac{\gamma^2}{\epsilon(0)}\right] V_{\mathrm{ext}}\,. \tag{3.104}$$

b) Show that the screened potential of a point charge Q is given by

$$V_{\mathrm{scr}} = \frac{Q}{\epsilon(0) r}\left\{1 + [\epsilon(0) - 1]e^{-\gamma z}\right\}. \tag{3.105}$$

3. Generalize the material in the text to show that for an inhomogeneous electron gas, the density–density response function in the self-consistent field theory is given as a solution of

$$\chi\left(\boldsymbol{r}, \boldsymbol{r}'; \omega\right) = \chi^0\left(\boldsymbol{r}, \boldsymbol{r}'; \omega\right) + \int_V d^3 r'' \int_V d^3 r''' \times \chi^0\left(\boldsymbol{r}, \boldsymbol{r}''; \omega\right) v\left(\boldsymbol{r}'' - \boldsymbol{r}'''\right) \chi\left(\boldsymbol{r}''', \boldsymbol{r}'; \omega\right), \tag{3.106}$$

where $\chi^0(\boldsymbol{r}, \boldsymbol{r}';\ \omega)$ is the single-particle density–density response function given in the text in terms of the Fermi distribution function as well as the wave functions and eigenvalues of independent (i.e., non-interacting) electrons. In addition, v_{c} is the Coulomb potential.

4. The transverse spin susceptibility is defined by

$$\chi^{-+}\left(\boldsymbol{r}, t\right) = \sum_{\boldsymbol{p},\boldsymbol{q}} e^{i\boldsymbol{q}\cdot\boldsymbol{r}} \chi^{-+}\left(\boldsymbol{p}, \boldsymbol{q}; t\right) \tag{3.107}$$

where

$$\chi^{-+}\left(\boldsymbol{p}, \boldsymbol{q}; t\right) = i\theta(t)\left\langle\left[a^{\dagger}_{\boldsymbol{p}+\boldsymbol{q}\downarrow}(t) a_{\boldsymbol{p}\uparrow}(t), \sigma^{+}(0,0)\right]_{-}\right\rangle, \tag{3.108}$$

with

$$\sigma^{+}(0,0) = \sum_{\boldsymbol{p},\boldsymbol{q}} a^{\dagger}_{\boldsymbol{p}+\boldsymbol{q}\uparrow} a_{\boldsymbol{p}\downarrow}\,. \tag{3.109}$$

In this notation, $\theta(t)$ is the unit step function and $a^{\dagger}_{\boldsymbol{p}\uparrow}$ is the spin-↑ creation operator in the Heisenberg representation for an electron with wave vector $\boldsymbol{p}$.

a) Derive the equation of motion for $\chi^{-+}(\boldsymbol{p},\boldsymbol{q};\,t)$ for the following Hamiltonian $H = H_0 + H_1$ of N atoms:

$$H_0 = \sum_{p,\sigma} \varepsilon_p a^\dagger_{p\sigma} a_{p\sigma} \tag{3.110}$$

$$H_1 = U \sum_{i=1}^{N} n_{i\uparrow} n_{i\downarrow} = \frac{U}{N} \sum_{p,p',q} a^\dagger_{p+q\uparrow} a_{p\uparrow} a^\dagger_{p'-q\downarrow} a_{p'\downarrow}\,. \tag{3.111}$$

b) Use the generalized Hartree–Fock approximation (i.e., RPA) which consists of (a) replacing all pairs of the type $a^\dagger a$ (in the interaction term of the equation of motion) by their expectation value and taking the sum over all such averages, and (b) assuming that

$$\left\langle a^\dagger_{p\alpha} a_{p'\beta} \right\rangle = \delta_{p,p'} \delta_{\alpha\beta} f_{p\alpha}\,,$$

where $f_{p\alpha}$ is the Fermi–Dirac distribution function in order to show that

$$\chi^{-+}(\boldsymbol{q},\omega) = \sum_p \chi^{-+}(\boldsymbol{p},\boldsymbol{q};\omega) \tag{3.112}$$

is given by

$$\chi^{-+}(\boldsymbol{q},\omega) = \frac{\Gamma^{-+}(\boldsymbol{q},\omega)}{1 - U\Gamma^{-+}(\boldsymbol{q},\omega)}\,. \tag{3.113}$$

Here,

$$\Gamma^{-+}(\boldsymbol{q},\omega) = \frac{1}{N} \sum_p \frac{f_{p\uparrow} - f_{p+q\downarrow}}{\hbar\omega - \left(\tilde{\varepsilon}_{p\downarrow} - \tilde{\varepsilon}_{p+q\uparrow}\right) + i\eta} \tag{3.114}$$

is the particle–hole propagator and

$$\tilde{\varepsilon}_{p\sigma} = \varepsilon_p - \frac{U}{N} \sum_{p'} f_{p',-\sigma}$$

is the one-particle energy modified by the exchange self-energy.

c) For a certain critical value of the interaction strength U, the system can spontaneously ($\omega = 0$) acquire a uniform ferromagnetic spin density in the long wavelength limit ($q \to 0$). Show that at $T = 0\,\mathrm{K}$, this instability is given by

$$U\rho\left(\varepsilon_{\mathrm{F}}\right) = 1\,,$$

where $\rho(\varepsilon_{\mathrm{F}})$ is the single-particle density-of-states at the Fermi energy.

5. Evaluate the real and imaginary parts of the single-particle density–density response function

$$\chi^0(\boldsymbol{q},\omega) = \frac{1}{\mathcal{V}} \sum_k \frac{f_k - f_{k+q}}{\hbar\omega - \left(\varepsilon_{k+q} - \varepsilon_k\right) + i\eta} \tag{3.115}$$

for a three-dimensional (3D) electron gas (EG) – the Lindhard function. Do the same sort of calculation for a two-dimensional electron gas (2DEG) and explicitly give (a) the real and (b) the imaginary parts at $T = 0\,\mathrm{K}$ for a parabolic band, that is, $\varepsilon_k = \hbar^2 k^2/2m^*$. Also, obtain an expression for the plasmon excitations in the long wavelength limit and discuss any difference in your result and the plasmon spectrum for the 3DEG.

References

1 Mahan, G.D. (2000) *Many-Particle Physics*, 3rd edn, Plenum, New York.

2 Kadanoff, L.P. and Baym, G. (1962) *Quantum Statistical Mechanics*, W.A. Benjamin, Westview Press, Boulder.

3 Fetter, A.L. and Walecka, J.D. (2003) *Quantum Theory of Many-Particle Systems*, McGraw-Hill, New York.

4 Rickayzen, G. (1980) *Green's Functions and Condensed Matter*, Academic Press, New York.

5 Inkson, J.C. (1984) *Many-Body Theory of Solids: An Introduction*, Plenum, New York.

6 Giuliani, G.F. and Vignale, G. (2005) *Quantum Theory of the Electron Liquid*, Cambridge, UK.

7 Doniach, S. and Sondheimer, E.H. (1998) *Green's Functions for Solid State Physicists*, Benjamin.

8 Kubo, R. (1957) Statistical Mechanical Theory of Irreversible Processes. *J. Phys. Soc. Jpn*, **12**, 570–586.

9 Martin, P.C. and Schwinger, J. (1959) Theory of Many-Particle Systems. *Phys. Rev.*, **115**, 1342.

10 Lee, P.A., Stone, A.D., and Fukuyama, H. (1987) Universal Conductance Fluctuations in Metals, Effects of finite temperature, interactions, and magnetic field. *Phys. Rev. B*, **35**, 1039–1070.

11 Kubo, R., Miyake, S.J., and Hashitsume, N. (1962) in *Solid State Physics* (eds Seitz and D. Turnbull) **17**, Academic Press, New York.

12 Bastin, A., Lewiner, C., Betbeder-Matibet, O., and Noziéres, P. (1971) Quantum Oscillations of the Hall Effect of a Fermion Gas with Random Impurity Scattering. *J. Phys. Chem. Solids*, **37**, 1811–1824.

13 Baranger, H. and Stone, A.D. (1989) Electrical linear-response theory in an arbitrary magnetic field: A new Fermi-surface formation. *Phys. Rev. B*, **40**, 8169.

14 Stone, A.D. and Szafer, A. (1988) What is measured when you measure a resistance? – The Landauer Formula Revisited. *IBM J. Res. Dev.*, **32**(3), 384.

4
Plasmon Excitations in Mesoscopic Structures

4.1
Linear Response Theory and Collective Excitations

We now describe in detail an approach for the calculation of the dispersion relation of the plasmon excitations in a two-dimensional electron gas (2DEG). The technique employed in this section can be extended to many other mesoscopic structures, for example, quantum rings, wires, dots, antidots, multi-layered structures and nanotubes [1–7]. An array of carbon nanotubes is shown in Figure 4.1 [8, 9]. In the following sections, we will apply the established formalism to the calculations of plasmon excitations in a linear array of cylindrical nanotubes, quantum wires and in coupled half-plane superlattices.

Let us write the density matrix $\hat{\rho}$ of the system as a sum of two terms $\hat{\rho} = \hat{\rho}^{(0)} + \delta\hat{\rho}$, where $\hat{\rho}^{(0)}$ is the time-independent equilibrium density operator and $\delta\hat{\rho}$ represents the non-equilibrium deviation to $\hat{\rho}^{(0)}$. The total Hamiltonian operator is $\hat{H} = \hat{H}^{(0)} + \hat{H}_1$, where $\hat{H}^{(0)}$ is the Hamiltonian operator for the unperturbed system and $\hat{H}_1$ stands for the external perturbation to the system. An approximate result for the equation of motion $i\hbar\partial\hat{\rho}/\partial t = [\hat{H}, \hat{\rho}]_-$ to the lowest order in the external perturbation is

$$i\hbar\frac{\partial\delta\hat{\rho}}{\partial t} = \left[\hat{H}^{(0)}, \delta\hat{\rho}\right]_- + \left[\hat{H}_1, \hat{\rho}^{(0)}\right]_- . \tag{4.1}$$

We further denote the eigenstates of the unperturbed Hamiltonian by $|j\rangle$ with energy eigenvalues ε_j so that we have $\hat{H}^{(0)}|j\rangle = \varepsilon_j|j\rangle$ and $\hat{\rho}^{(0)}|j\rangle = 2f_0(\varepsilon_j)|j\rangle$, where the factor "2" comes from the spin degeneracy of electrons and $f_0(\varepsilon_j)$ is the equilibrium Fermi–Dirac function. Then, from Eq. (4.1), we have

$$\begin{aligned} i\hbar\langle j|\frac{\partial\delta\hat{\rho}}{\partial t}|j'\rangle &= \left(\varepsilon_j - \varepsilon_{j'}\right)\langle j|\delta\hat{\rho}|j'\rangle \\ &\quad - 2\left[f_0(\varepsilon_j) - f_0\left(\varepsilon_{j'}\right)\right]\langle j|\hat{H}_1|j'\rangle . \end{aligned} \tag{4.2}$$

Properties of Interacting Low-Dimensional Systems, First Edition. G. Gumbs and D. Huang.

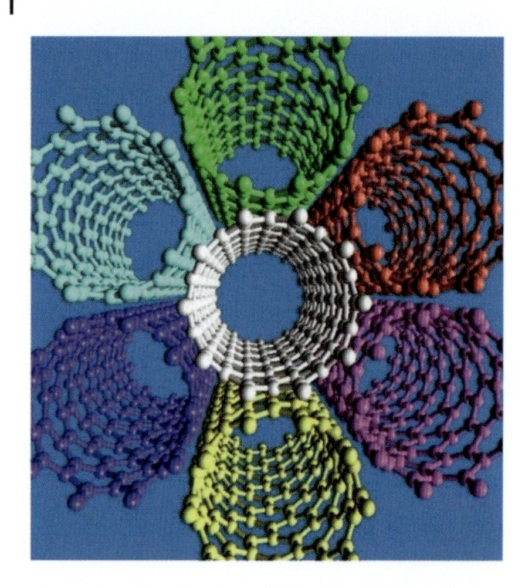

Figure 4.1 3D illustration for the carbon nanotube structure.

Taking the Fourier transform with respect to time t for $\delta\hat{\rho}(t) = \delta\hat{\rho}(0)e^{i\omega t}$ in Eq. (4.2), we obtain

$$\langle j|\delta\hat{\rho}|j'\rangle = 2\left[\frac{f_0\left(\varepsilon_j\right) - f_0\left(\varepsilon_{j'}\right)}{\hbar\omega - \left(\varepsilon_{j'} - \varepsilon_j\right)}\right]\langle j|\hat{H}_1|j'\rangle$$
$$\equiv \Pi^{(0)}_{j,j'}(\omega)\langle j|\hat{H}_1|j'\rangle. \tag{4.3}$$

The induced density fluctuation in the system is given by

$$\delta n_{\text{ind}}(\boldsymbol{r},\omega) = \text{Tr}\left\{\delta\hat{\rho}\left(\boldsymbol{r},\omega\right)\hat{\psi}^{\dagger}\left(\boldsymbol{r}\right)\hat{\psi}\left(\boldsymbol{r}\right)\right\}$$
$$= \sum_{j,j'}\langle j|\delta\hat{\rho}\left(\boldsymbol{r},\omega\right)|j'\rangle\langle j'|\hat{\psi}^{\dagger}\left(\boldsymbol{r}\right)\hat{\psi}\left(\boldsymbol{r}\right)|j\rangle = \sum_{j,j'}\langle j|\delta\hat{\rho}|j'\rangle\varphi^*_{j'}\left(\boldsymbol{r}\right)\varphi_j\left(\boldsymbol{r}\right)$$
$$= \sum_{j,j'}\Pi^{(0)}_{j,j'}(\omega)\varphi^*_{j'}\left(\boldsymbol{r}\right)\varphi_j\left(\boldsymbol{r}\right)\langle j|\hat{H}_1|j'\rangle, \tag{4.4}$$

where $\varphi_j(\boldsymbol{r}) = \langle\boldsymbol{r}|\hat{\psi}(\boldsymbol{r})|j\rangle$ is the eigenfunction and $\hat{\psi}(\boldsymbol{r})$ is a destruction operator. For a perturbation $\hat{H}_1 = \Phi(\boldsymbol{r},\omega)$, where $\Phi(\boldsymbol{r},\omega)$ is the total potential energy including the induced one, the perturbing Hamiltonian has matrix elements

$$\langle j|\hat{H}_1|j'\rangle = \int d\boldsymbol{r}'\langle j|\hat{\psi}^{\dagger}\left(\boldsymbol{r}'\right)\hat{\psi}\left(\boldsymbol{r}'\right)|j'\rangle\Phi\left(\boldsymbol{r}',\omega\right)$$
$$= \int d\boldsymbol{r}'\varphi^*_j\left(\boldsymbol{r}'\right)\Phi\left(\boldsymbol{r}',\omega\right)\varphi_{j'}\left(\boldsymbol{r}'\right). \tag{4.5}$$

Substituting Eq. (4.5) into Eq. (4.4), we obtain

$$\delta n_{\text{ind}}\left(\boldsymbol{r},\omega\right) = \int d\boldsymbol{r}'\chi^{(0)}\left(\boldsymbol{r},\boldsymbol{r}';\omega\right)\Phi\left(\boldsymbol{r}',\omega\right), \tag{4.6}$$

where

$$\chi^{(0)}\left(\boldsymbol{r},\boldsymbol{r}';\omega\right)=2\sum_{j,j'}\left[\frac{f_0\left(\varepsilon_j\right)-f_0\left(\varepsilon_{j'}\right)}{\hbar\omega-\left(\varepsilon_{j'}-\varepsilon_j\right)}\right]\varphi^*_{j'}\left(\boldsymbol{r}\right)\varphi_j\left(\boldsymbol{r}\right)\varphi^*_j\left(\boldsymbol{r}'\right)\varphi_{j'}\left(\boldsymbol{r}'\right) \tag{4.7}$$

is the density–density response function for the non-interacting electrons.

4.1.1 Screening and the Self-Consistent Field Approximation

For a 2D interacting electron system, we treat the total potential energy in the self-consistent field theory [10] and write

$$\Phi\left(\boldsymbol{r}_{||},\omega\right)=\Phi_{\text{ext}}\left(\boldsymbol{r}_{||}\right)+\Phi_{\text{ind}}\left(\boldsymbol{r}_{||},\omega\right)\,, \tag{4.8}$$

where $\Phi_{\text{ext}}(\boldsymbol{r}_{||})$ is the externally applied potential energy. In the self-consistent field approximation, the induced potential energy $\Phi_{\text{ind}}(\boldsymbol{r}_{||},\omega)$ is related to the density fluctuation $\delta n_{\text{ind}}(\boldsymbol{r}_{||},\omega)$ through Poisson's equation

$$\Phi_{\text{ind}}\left(\boldsymbol{r}_{||},\omega\right)=\frac{e^2}{4\pi\epsilon_0\epsilon_{\text{b}}}\int d\boldsymbol{r}'_{||}\frac{\delta n_{\text{ind}}\left(\boldsymbol{r}'_{||},\omega\right)}{\left|\boldsymbol{r}_{||}-\boldsymbol{r}'_{||}\right|}\,, \tag{4.9}$$

where ϵ_{b} is the background dielectric constant. Equations (4.6), (4.8), and (4.9) combine to give

$$\delta n_{\text{ind}}\left(\boldsymbol{r}_{||},\omega\right)=\int d\boldsymbol{r}'_{||}\chi\left(\boldsymbol{r}_{||},\boldsymbol{r}'_{||};\omega\right)\Phi_{\text{ext}}\left(\boldsymbol{r}'_{||}\right), \tag{4.10}$$

where the density–density response function for the interacting electrons is $\chi(\boldsymbol{r}_{||},\boldsymbol{r}'_{||};\omega)$ and is obtained by solving the following integral equation

$$\begin{aligned}\chi\left(\boldsymbol{r}_{||},\boldsymbol{r}'_{||};\omega\right)=\chi^{(0)}\left(\boldsymbol{r}_{||},\boldsymbol{r}'_{||};\omega\right)+\int d\boldsymbol{r}''_{||}\int d\boldsymbol{r}'''_{||}\\ \times\chi^{(0)}\left(\boldsymbol{r}_{||},\boldsymbol{r}''_{||};\omega\right)V_{\text{c}}\left(|\boldsymbol{r}''_{||}-\boldsymbol{r}'''_{||}|\right)\chi\left(\boldsymbol{r}'''_{||},\boldsymbol{r}'_{||};\omega\right),\end{aligned} \tag{4.11}$$

where $V_{\text{c}}(|\boldsymbol{r}_{||}-\boldsymbol{r}'_{||}|)=e^2/4\pi\epsilon_0\epsilon_{\text{b}}|\boldsymbol{r}_{||}-\boldsymbol{r}'_{||}|$ is the Coulomb potential energy.

For the homogeneous 2DEG with an area $\mathcal{A}$ and periodic boundary conditions, the electron eigenfunctions are $\varphi_{\boldsymbol{k}}(\boldsymbol{r}_{||})=e^{i\boldsymbol{k}_{||}\cdot\boldsymbol{r}_{||}}/\sqrt{\mathcal{A}}$ and the corresponding energy eigenvalues are $\varepsilon_{k_{||}}=\hbar^2k^2_{||}/2m^*$, where $\boldsymbol{k}_{||}$ and m^* are the in-plane wave vector and the effective mass of electrons, respectively. When these eigenfunctions are substituted into Eq. (4.7), it can be shown that $\chi^{(0)}(\boldsymbol{r}_{||},\boldsymbol{r}'_{||};\omega)$ only depends on the difference $(\boldsymbol{r}_{||}-\boldsymbol{r}'_{||})$. Using the Fourier transforming Eq. (4.11) with respect to the spatial variables, we obtain

$$\chi\left(q_{||},\omega\right)=\frac{\chi^{(0)}\left(q_{||},\omega\right)}{1-v_{\text{c}}\left(q_{||}\right)\chi^{(0)}\left(q_{||},\omega\right)}\,, \tag{4.12}$$

where $q_{||} = |\boldsymbol{k}'_{||} - \boldsymbol{k}_{||}|$, the Fourier transform of the Coulomb potential energy $V_c(|\boldsymbol{r}_{||} - \boldsymbol{r}'_{||}|)$ is $v_c(q_{||}) = e^2/2\epsilon_0\epsilon_b q_{||}$, and

$$\chi^{(0)}\left(q_{||}, \omega\right) = \frac{2}{A} \sum_{\boldsymbol{k}_{||}} \frac{f_0\left(\varepsilon_{k_{||}}\right) - f_0\left(\varepsilon_{|\boldsymbol{k}_{||}+\boldsymbol{q}_{||}|}\right)}{\hbar\omega - \left(\varepsilon_{|\boldsymbol{k}_{||}+\boldsymbol{q}_{||}|} - \varepsilon_{k_{||}}\right)} . \tag{4.13}$$

The plasmon excitations correspond to the zeros of the dielectric function $\epsilon(q_{||},\omega) \equiv 1 - v_c(q_{||})\chi^{(0)}(q_{||}, \omega)$. In the long wavelength limit, that is, $q_{||} \to 0$, we have

$$\epsilon\left(q_{||}, \omega\right) \approx 1 - \left(\frac{e^2}{2\epsilon_0\epsilon_b q_{||}}\right) \frac{2}{A} \sum_{\boldsymbol{k}_{||}} f_0\left(\varepsilon_{k_{||}}\right) \frac{\hbar^2 q_{||}^2/m^*}{(\hbar\omega)^2} . \tag{4.14}$$

Therefore, the 2D plasmon mode has the frequency $\omega_{q_{||}}$ as

$$\omega = \omega_{q_{||}} \equiv \sqrt{\frac{n_{2D} e^2 q_{||}}{2\epsilon_0\epsilon_b m^*}} , \tag{4.15}$$

where n_{2D} is the electron sheet density.

4.2
A Linear Array of Nanotubes

In this section, the self-consistent field theory will be applied to calculate the dispersions of plasmon excitations in a linear array of carbon nanotubes. However, we will first present some basic properties of the energy band structure which we will then use to justify the simple model employed below.

The single-wall carbon nanotube can be constructed by wrapping up a monolayer of graphene in such a way that two equivalent sites of the hexagonal lattice coincide. The wrapping vector **C** which defines the relative location of the two sites is specified by a pair of integers (n, m) which decompose **C** into two unit vectors $\hat{\boldsymbol{a}}_1$ and $\hat{\boldsymbol{a}}_2$ (i.e., $\boldsymbol{C} = n\hat{\boldsymbol{a}}_1 + m\hat{\boldsymbol{a}}_2$, where n and m are integers). The nanotube is called "armchair" if $n = m$, whereas if $m = 0$, such a carbon nanotube possesses "zigzag" chirality. All the other nanotubes belong to the "chiral" type and have a finite wrapping angle ϕ with $0° < \phi < 30°$ [11].

In a calculation, we considered a nanotube with a radius large enough not to consider the effects of the nanotube curvature on each hexagon. If the curvature is considerable, it may lead to different t-values (hopping coefficient) and some much more complicated effects.

Once we know the structure of carbon nanotubes, the electronic structure is derived by simple tight-binding calculation for the π-electrons of carbon atoms It follows that a carbon nanotube may be either metallic or semiconducting depending on its diameter and chirality. The energy gap for a semiconductor nanotube is inversely proportional to its diameter. To obtain explicit expressions for the dispersion relations, the simplest cases to consider are nanotubes with the highest symmetry.

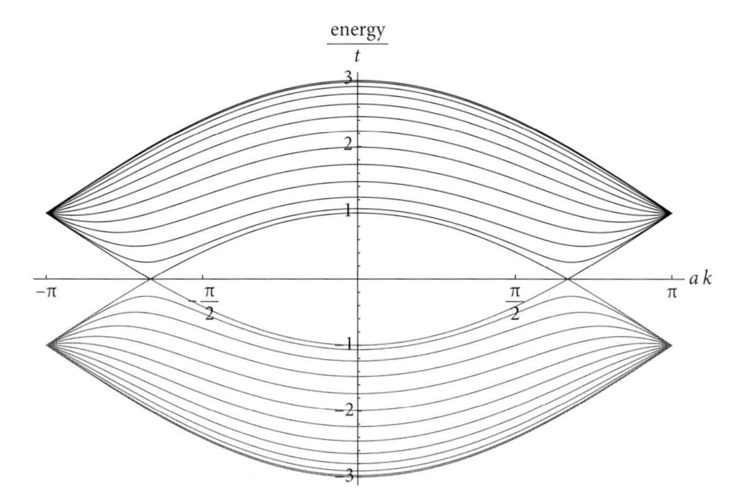

Figure 4.2 Energy dispersion for an armchair nanotube. The subbands for the valence band are plotted in gray and the conduction band subbands are plotted in black.

The appropriate periodic boundary conditions used to obtain the energy eigenvalues for the $\mathbf{C} = \{n, n\}$ defines a finite number of allowed wave vectors $k_{y,l}$ in the circumferential direction,

$$n\sqrt{3}k_{y,l}a = 2\pi l\,. \tag{4.16}$$

Here, $l = 1, \ldots, 2n$ is simply an integer in the range from 1 to n – the number of carbon atoms around the nanotube with lattice constant a. Substitution of the discrete allowed values for $k_{y,l}$ into the general dispersion formula yields an **analytic expression for the energy dispersion** for the armchair nanotube

$$E_l^{\text{armchair}} = \pm\sqrt{1 \pm 4\cos\left(\frac{l\pi}{n}\right)\cos\left(\frac{ka}{2}\right) + 4\cos^2\left(\frac{ka}{2}\right)}\,. \tag{4.17}$$

The resulting dispersion relation are plotted in Figure 4.2. Close to the separation between the valence and conduction bands, the energy dispersion is linear in the wave vector, indicating massless fermions at the K point. However, if the nanotube is doped so the Fermi energy lies within the conduction band, the electrons acquire a mass near the Fermi level and we may approximate the energy bands as parabolic. In our calculations below for plasma excitations, we use a 2D model in which a sheet of conduction electrons are wrapped around a cylinder.

4.2.1
Tight-Binding Model

Let us consider a system composed of a linear array of nanotubes, shown schematically in Figure 4.3, with their axes parallel to the z-direction. The axis of each nanotube is at $x = na$ $(n = 0, \pm1, \pm2, \ldots)$ on the x-axis.

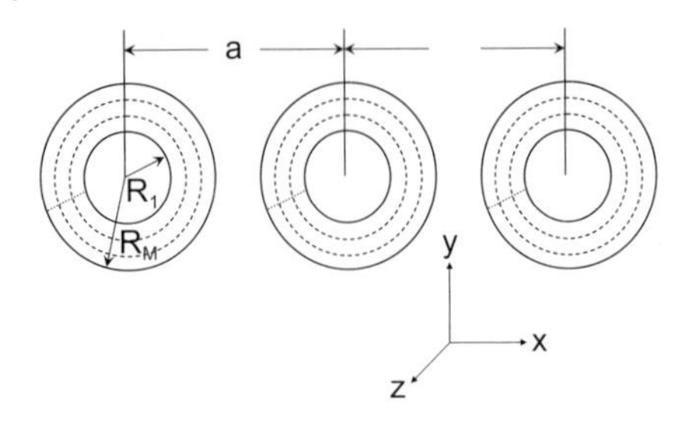

Figure 4.3 Schematic representation of the cross-sections of a linear array of tubules whose axes, pointing along the *z*-direction, are equally spaced along the *x*-axis. Here, R_j $(j = 1, 2, \ldots, M)$ are the radii of the tubules, and a is the period of the lattice.

Each nanotube consists of M concentric cylindrical tubules with radii $R_1 < R_2 < \ldots < R_M$, where $a > 2R_M$. For simplicity, we assume that each cylindrical tubule is infinitesimally thin. We will construct the electron wave functions in the form of Bloch combinations as described by Huang and Gumbs for an array of rings [3]. In the absence of tunneling between the tubules, the single-particle Bloch wave functions for the nanotube array with the periodicity of the lattice are given by

$$|j\nu\rangle = \frac{e^{ik_z z}}{\sqrt{L_z}} \frac{1}{\sqrt{N_x}} \sum_{n=-N_x/2}^{N_x/2} e^{ik_x na} \psi_{j\ell}(\boldsymbol{\rho} - na\hat{\mathbf{x}}) ,$$

$$\psi_{j\ell}(\boldsymbol{\rho}) = \frac{e^{i\ell\phi}}{\sqrt{2\pi\rho}} \varphi_j(\rho) , \tag{4.18}$$

where $\hat{\mathbf{x}}$ is the unit vector in the *x*-direction, $j = 1, 2, \ldots, M$ labels the nanotubes, $\nu = \{k_z, k_x, \ell\}$ is a composite index for the electron eigenstates, $\psi_{j\ell}(\boldsymbol{\rho})e^{ik_z z}/\sqrt{L_z}$ is the wave function for an electron in the *j*th tubule with wave vector k_z in the axial direction and angular momentum quantum number $\ell = 0, \pm 1, \pm 2, \ldots$, $|\varphi_j(\rho)|^2 = \delta(\rho - R_j)$, and $k_x = (2\pi/L_x)n$ with $n = 0, \pm 1, \pm 2, \ldots, \pm N_x/2$. Here, $N_x = L_x/a$ is the number of nanotubes in the array with periodic boundary conditions. Electron motion in the azimuthal direction around the tubule is quantized and characterized by the angular momentum quantum number ℓ, whereas motion in the axial *z*-direction is free. Thus, the electron energy spectrum in each tubule consists of 1D subbands with ℓ serving as a subband index. The energy spectrum does not depend on k_x and has the form

$$\varepsilon_{j\nu} = \frac{\hbar^2 k_z^2}{2m^*} + \frac{\hbar^2 \ell^2}{2m^* R_j^2} . \tag{4.19}$$

Plasmons can be obtained [12] from the solution of the density matrix equation in Eq. (2.12). For the self-sustaining density oscillations with $\hat{H}_1 = \Phi_{\text{ind}}$, $\langle j\nu|\hat{H}^{(0)}|j'\nu'\rangle = \varepsilon_{j\nu}\delta_{\nu,\nu'}\delta_{j,j'}$, and $\langle j\nu|\hat{\rho}^{(0)}|j'\nu'\rangle = 2f_0(\varepsilon_{j\nu})\delta_{\nu,\nu'}\delta_{j,j'}$, from Eq. (4.3), we obtain

$$\langle j\nu|\delta\hat{\rho}|j'\nu'\rangle = 2\left[\frac{f_0(\varepsilon_{j\nu}) - f_0(\varepsilon_{j'\nu'})}{\hbar\omega - (\varepsilon_{j'\nu'} - \varepsilon_{j\nu})}\right] \langle j\nu|\Phi_{\text{ind}}(\mathbf{r}, \omega)|j'\nu'\rangle, \tag{4.20}$$

where $\Phi_{\text{ind}}(\boldsymbol{r}, \omega)$ is the induced potential. The potential $\Phi_{\text{ind}}(\boldsymbol{r}, \omega)$ satisfies Poisson's equation

$$\nabla^2 \Phi_{\text{ind}}(\boldsymbol{r}, \omega) = \left(\frac{e^2}{\epsilon_0 \epsilon_b}\right) \delta n_{\text{ind}}(\boldsymbol{r}, \omega), \tag{4.21}$$

where $\delta n_{\text{ind}}(\boldsymbol{r}, \omega)$ is the fluctuation of electron density.

Making use of the relation described by Eq. (4.4)

$$\delta n_{\text{ind}}(\boldsymbol{r}, \omega) = \sum_{j,j'} \sum_{\nu,\nu'} \langle \boldsymbol{r} | j\nu \rangle \langle j\nu | \delta\hat{\rho} | j'\nu' \rangle \langle j'\nu' | \boldsymbol{r} \rangle \tag{4.22}$$

and Eq. (4.20), we can write in Fourier representation

$$\begin{aligned} \delta n_{\text{ind}}(\boldsymbol{q}, \omega) = {} & \frac{2}{\mathcal{V}} \sum_{j,j'} \sum_{\nu,\nu'} \left[\frac{f_0(\varepsilon_{j\nu}) - f_0(\varepsilon_{j'\nu'})}{\hbar\omega - (\varepsilon_{j'\nu'} - \varepsilon_{j\nu})} \right] \langle j'\nu' | e^{-i\boldsymbol{q}\cdot\boldsymbol{r}} | j\nu \rangle \\ & \times \sum_{\boldsymbol{q}'} \phi_{\text{ind}}(\boldsymbol{q}', \omega) \langle j\nu | e^{i\boldsymbol{q}'\cdot\boldsymbol{r}} | j'\nu' \rangle, \end{aligned} \tag{4.23}$$

where $\delta n_{\text{ind}}(\boldsymbol{q}, \omega)$ and $\phi_{\text{ind}}(\boldsymbol{q}, \omega)$ are 3D Fourier transforms of $\delta n_{\text{ind}}(\boldsymbol{r}, \omega)$ and $\Phi_{\text{ind}}(\boldsymbol{r}, \omega)$, respectively, $\mathcal{V}$ is the sample volume, and $\boldsymbol{q} = (q_x, q_y, q_z) \equiv (\boldsymbol{q}_\perp, q_z)$. The matrix elements $\langle j\nu | e^{i\boldsymbol{q}\cdot\boldsymbol{r}} | j'\nu' \rangle$ with wave functions $|j\nu\rangle$ given in Eq. (4.18) can be evaluated as follows

$$\langle j\nu | e^{i\boldsymbol{q}\cdot\boldsymbol{r}} | j'\nu' \rangle = \delta_{j,j'} \delta_{k_z - k'_z, q_z} \delta_{k_x - k'_x, q_x + sG} e^{-im\theta} (i)^m J_m(q_\perp R_j), \tag{4.24}$$

where $G = 2\pi/a$, $s = 0, \pm 1, \pm 2, \ldots$, $m = \ell - \ell' = 0, \pm 1, \pm 2, \ldots$, θ is the angle between $\boldsymbol{q}_\perp$ and $\hat{\boldsymbol{x}}$ $(0 \le \theta < 2\pi)$, and $J_m(x)$ is the mth order Bessel function of the first kind. Substituting Eq. (4.24) into Eq. (4.23), after some straightforward algebra, we obtain

$$\begin{aligned} \delta n_{\text{ind}}(\boldsymbol{q}, \omega) = {} & \frac{1}{\pi a L_z} \sum_{k_z} \sum_{j=1}^{M} \sum_{\ell,m} \frac{f_0(\varepsilon_{j,k_z,\ell}) - f_0(\varepsilon_{j,k_z-q_z,\ell-m})}{\hbar\omega - (\varepsilon_{j,k_z,\ell} - \varepsilon_{j,k_z-q_z,\ell-m})} e^{im\theta} \\ & \times J_m(q_\perp R_i) \sum_{s=-\infty}^{\infty} \int dq'_y \phi_{\text{ind}}(q_x + sG, q'_y, q_z; \omega) \\ & \times J_m\left(R_j \sqrt{(q_x + sG)^2 + q'^2_y}\right) \left[\frac{q_x + sG - iq'_y}{\sqrt{(q_x + sG)^2 + q'^2_y}} \right]^m . \end{aligned} \tag{4.25}$$

The potential $\phi_{\text{ind}}(\boldsymbol{q}; \omega)$ can be written in terms of $\delta n_{\text{ind}}(\boldsymbol{q}, \omega)$ as $\phi_{\text{ind}}(\boldsymbol{q}, \omega) = e^2 \delta n_{\text{ind}}(\boldsymbol{q}; \omega)/\epsilon_0 \epsilon_b q^2$. Using this relation in Eq. (4.25), we obtain

$$\delta n_{\text{ind}}(\boldsymbol{q}, \omega) = \frac{e^2}{4\pi^2 \epsilon_0 \epsilon_b a} \sum_{j,m} \chi^{(0)}_{j,m}(q_z, \omega) e^{im\theta} J_m(q_\perp R_j) U_{j,m}(q_x, q_z; \omega), \tag{4.26}$$

where

$$\chi^{(0)}_{j,m}(q_z,\omega) = 2\sum_{\ell=-\infty}^{\infty}\int dk_z \frac{f_0(\varepsilon_{j,k_z,\ell}) - f_0(\varepsilon_{j,k_z-q_z,\ell-m})}{\hbar\omega - (\varepsilon_{j,k_z,\ell} - \varepsilon_{j,k_z-q_z,\ell-m})} \tag{4.27}$$

is the density–density response function in a single cylindrical tubule of radius R_j and

$$U_{j,m}(q_x,q_z;\omega) = \sum_{s=-\infty}^{\infty}\int dq'_y \frac{\delta n_{\text{ind}}(q_x+sG,q'_y,q_z;\omega)}{(q_x+sG)^2+q'^2_y+q_z^2} \times J_m\left(R_j\sqrt{(q_x+sG)^2+q'^2_y}\right)\left[\frac{q_x+sG-iq'_y}{\sqrt{(q_x+sG)^2+q'^2_y}}\right]^m . \tag{4.28}$$

Substituting the expression for $\delta n_{\text{ind}}(\boldsymbol{q},\omega)$ given in Eq. (4.26) into Eq. (4.28), we obtain

$$U_{j,m}(q_x,q_z;\omega) - \left(\frac{e^2}{4\pi^2\epsilon_0\epsilon_b a}\right)\sum_{j'=1}^{M}\sum_{m'=0,\pm1,\dots}\chi^{(0)}_{j',m'}(q_z,\omega) \times \sum_{s=-\infty}^{\infty}\int dq_y \frac{J_{m'}\left(R_{j'}\sqrt{(q_x+sG)^2+q_y^2}\right)J_m\left(R_j\sqrt{(q_x+sG)^2+q_y^2}\right)}{(q_x+sG)^2+q_y^2+q_z^2} \times \left[\frac{q_x+sG+iq_y}{\sqrt{(q_x+sG)^2+q_y^2}}\right]^{m'-m} U_{j',m'}(q_x,q_z;\omega) = 0 . \tag{4.29}$$

This set of linear equations has nontrivial solutions provided that the following determinant is zero, that is,

$$\text{Det}\left\{\delta_{m,m'}\delta_{j,j'} - \left(\frac{e^2}{4\pi^2\epsilon_0\epsilon_b a}\right)\chi^{(0)}_{j'm'}(q_z,\omega) \times \sum_{s=-\infty}^{\infty}\int_{-\infty}^{\infty} dq_y \frac{J_{m'}\left(R_{j'}\sqrt{(q_x+sG)^2+q_y^2}\right)J_m\left(R_j\sqrt{(q_x+sG)^2+q_y^2}\right)}{(q_x+sG)^2+q_y^2+q_z^2} \times \left[\frac{q_x+sG+iq_y}{\sqrt{(q_x+sG)^2+q_y^2}}\right]^{m'-m}\right\} = 0 \tag{4.30}$$

with $m, m' = 0, \pm1, \pm2, \dots$ and $j, j' = 1, 2, \dots, M$. Equation (4.30) determines the dispersion equation for the plasmon collective excitations. At $T = 0\,\text{K}$, it is a straightforward matter to evaluate the density–density response function $\chi^{(0)}_{j,m}(q_z, \omega)$ in Eq. (4.27) with the following result for the real part

$$\mathrm{Re}\left[\chi^{(0)}_{j,m}(q_z,\omega)\right] = \frac{2m^*}{\hbar^2 q_z} \sum_{\ell=-\ell^j_M}^{\ell^j_M} \ln\left[\frac{\omega^2 - \Omega^2_-\left(\ell, m, q_z, k^{j,\ell}_{\mathrm{F}}\right)}{\omega^2 - \Omega^2_+\left(\ell, m, q_z, k^{j,\ell}_{\mathrm{F}}\right)}\right], \tag{4.31}$$

where ℓ^j_M is the maximum value of $|\ell|$ among the subbands occupied by electrons in the jth cylindrical tubule, $\hbar k^{j,\ell}_{\mathrm{F}} = \sqrt{2m^* E_{\mathrm{F}} - \hbar^2\ell^2/R^2_j}$ is the Fermi momentum in the z-direction for the subband with given ℓ, E_{F} is the Fermi energy of electrons, and

$$\Omega_\pm\left(\ell, m, q_z, k^{j,\ell}_{\mathrm{F}}\right) = \frac{\hbar\left(q_z^2 \pm 2k^{j,\ell}_{\mathrm{F}} q_z\right)}{2m^*} + \frac{\hbar\left(2m\ell + m^2\right)}{2m^* R^2_j}. \tag{4.32}$$

Equation (4.30) shows that the symmetry of the lattice is maintained in the dispersion equation and that the plasmon excitations depend on the wave vector q_x in the x-direction with period $G = 2\pi/a$ as well as the wave vector q_z.

In the limit of $a \to \infty$, the summation over the reciprocal lattice vectors in Eq. (4.30) can be transformed into an integral, and the determinantal matrix in Eq. (4.30) becomes diagonal in the indices m and m'. Using

$$\begin{aligned}\overline{V}_m\left(q_z, R_j, R_{j'}\right) &\equiv \int_0^\infty q_\perp dq_\perp \frac{J_m\left(q_\perp R_j\right) J_m\left(q_\perp R_{j'}\right)}{q_\perp^2 + q_z^2} \\ &= \begin{cases} I_m\left(q_z R_j\right) K_m\left(q_z R_{j'}\right), & R_j < R_{j'} \\ I_m\left(q_z R_{j'}\right) K_m\left(q_z R_j\right), & R_{j'} < R_j \end{cases},\end{aligned} \tag{4.33}$$

where $I_m(x)$ and $K_m(x)$ are the modified Bessel functions of the first and second kind, respectively, we obtain the following simplified dispersion equation

$$\prod_{m=0,\pm1,\ldots} \mathrm{Det}\left[\delta_{j,j'} - \frac{e^2}{4\pi^2\epsilon_0\epsilon_{\mathrm{b}}} \overline{V}_m\left(q_z, R_j, R_{j'}\right) \chi^{(0)}_{j,m}(q_z,\omega)\right] = 0. \tag{4.34}$$

Equation (4.34) is the dispersion equation for plasmons in a single nanotube consisting of M coaxial cylindrical tubules and agrees with [13]. It follows from Eq. (4.34) that the plasmon modes in a single coaxial nanotube can be labeled by the quantum numbers m and q_z where $m = \ell - \ell'$ is an angular momentum transfer in the electron intersubband transitions ($\ell \rightleftarrows \ell'$) contributing to the given plasmon mode. When a is finite, the non-diagonal in m, m' elements of the matrix in Eq. (4.30) are not equal to zero and modes with different values of m are generally coupled to each other. As shown below, this coupling modifies the plasmon spectrum.

4.2.2
Numerical Results and Discussion

For simplicity, we consider the situation when each nanotube only contains one cylindrical tubule (single-wall nanotube), that is, $j = j' = 1$ in Eq. (4.30). To closely simulate the graphene tubule, we took $\epsilon_b = 2.4$, $m^* = 0.25 m_0$ with m_0 being the bare electron mass, $R_1 = 11$ Å, $E_F = 0.6$ eV and $a = 35$ Å. The effective Bohr radius is $a_B^* \equiv 4\pi\hbar^2\epsilon_0\epsilon_b/m^*e^2 = 1.26$ Å. All calculations were carried out at zero temperature. We included the transitions $m = 0, \pm 1$ only in these calculations so that we have a 3×3 matrix for Eq. (4.30).

For the values of the parameters chosen, in each tubule there are only five subbands occupied by electrons corresponding to $\ell = 0, \pm 1, \pm 2$. For single-wall cylindrical nanotubes, Lin and Shung [13] used the same values for ϵ_b, m^*, R_1 and E_F in calculating the plasmon excitation spectrum. It was shown that there are three quasi-acoustic plasmon branches associated with intrasubband electron excitations with angular momentum transfer $m = 0$ and five optical plasmon branches associated with intersubband electron transitions with angular momentum transfer $m = \pm 1$.

In Figure 4.4, we present our results for plasmon dispersion as a function of q_z with $q_x = 0$. As it follows from Eq. (4.30), when $q_x = 0$, the elements of the determinantal matrix with $m = 0$, $m' = \pm 1$ are zero because of symmetry, thereby decoupling the intrasubband and intersubband excitations. The dispersion for intrasubband plasmons ($m = 0$) is shown in Figure 4.4a. These modes are very similar to the intrasubband plasmons in a single tubule [13]. Only undamped portions of the plasmon spectrum are shown in Figure 4.4a and, for the sake of clarity, we have omitted the boundaries of the particle–hole continuum in Figure 4.4a.

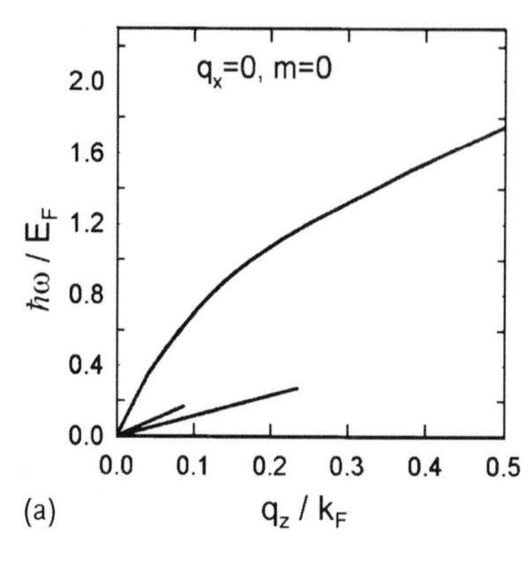

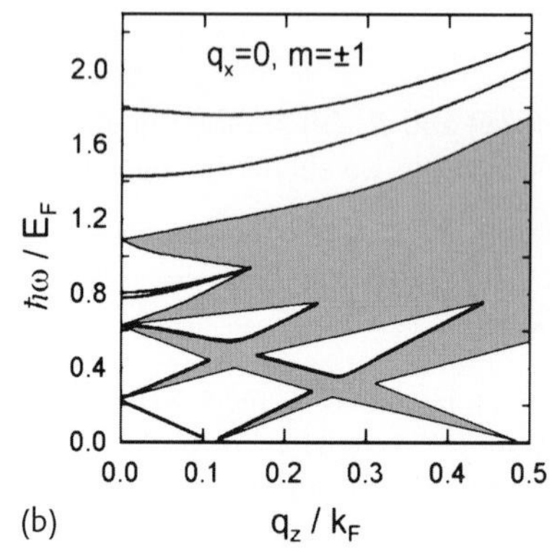

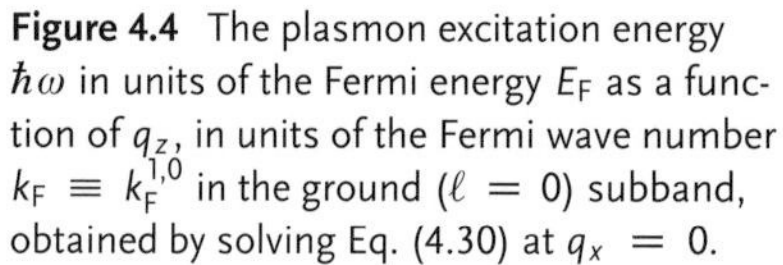
Figure 4.4 The plasmon excitation energy $\hbar\omega$ in units of the Fermi energy E_F as a function of q_z, in units of the Fermi wave number $k_F \equiv k_F^{1,0}$ in the ground ($\ell = 0$) subband, obtained by solving Eq. (4.30) at $q_x = 0$. (a) Intrasubband plasmons ($m = 0$) and (b) intersubband plasmons ($m = \pm 1$). The shaded region in (b) corresponds to the e–h continuum.

Intersubband plasmon excitations associated with electron transitions for $m = \pm 1$ are shown in Figure 4.4b along with the particle–hole continuum. The main difference between these modes in a tubule array and in a single tubule may be described as follows. In a single tubule, plasmon modes with $m = +1$ and -1 are degenerate [13]. The ordered positioning of the tubules to form a one-dimensional lattice lowers the axial symmetry of the system. The Coulomb interaction between the tubules lifts the degeneracy and splits each single tubule intersubband plasmon mode into two branches increasing the total number of modes to ten. This splitting is larger for the high-frequency modes compared with the low-frequency ones, as it is demonstrated in Figure 4.4b where the splitting of the low-frequency modes is too small to be resolved on the scale used in Figure 4.4b. When a increases, the separation between the split modes decreases, reducing to zero in the limit $a \to \infty$.

The Coulomb interaction between the tubules in the array does not qualitatively alter the dependence of the plasmon frequency on q_z but its effect is to continuously increase the plasmon frequency as the separation a between the tubules decreases. The increase in the plasmon frequency when a decreases is more pronounced for the high-frequency plasmon modes compared with the low-frequency ones where this effect is small.

When $q_x \neq 0$, the plasmon excitation spectrum becomes more complicated. Now, the modes with $m = 0$ and $m = \pm 1$ are coupled to each other [see Eq. (4.30)]. Strictly speaking, the collective excitations can not be classified as intrasubband and intersubband plasmons anymore. By solving Eq. (4.30), the excitation energies are found to be a periodic dependence on q_x with period $2\pi/a$. This dependence reflects the translational symmetry of the lattice. It is found that the modulation of the plasmon spectrum with q_x depends on the plasmon frequency. It is large for the high-frequency modes and decreases with the mode frequency.

4.3 A Linear Array of Quantum Wires

In this section, the electrodynamic model in the long-wavelength limit will be employed to calculate the dispersion relations of plasmon excitations in a linear array of quantum wires.

Here, let us consider a linear array of quantum wires, schematically shown in Figure 4.5, in which quantum wires point in the z-direction and are equally spaced along the y-axis by sitting at the positions $y = jd$ with $j = 0, \pm 1, \pm 2, \ldots$ The width of the wires in the y-direction is $2a$, while the thickness of the wires in the x-direction is zero. Here, $a/d \ll 1$ is assumed for this system. The whole array is embedded in a background with a dielectric constant ϵ_b.

The Maxwell equation for the z-component $E_z(\mathbf{r}, t)$ of a longitudinal electric field in a region without charges can be written as

$$\left[\frac{\partial^2}{\partial x^2} + \frac{\partial^2}{\partial y^2} - \beta^2\right] E_z(\mathbf{r}, t) = 0 , \tag{4.35}$$

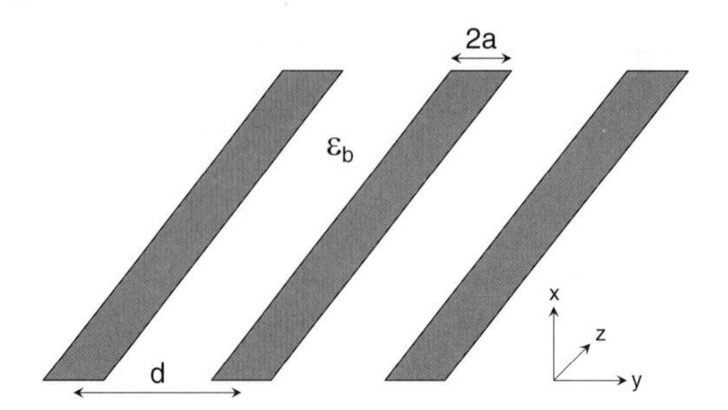

Figure 4.5 Schematic representation of a linear array of quantum wires (gray bars) pointing in the z-direction. They are equally placed along the y-axis. Here, $y = jd$ with $j = 0, \pm 1, \pm 2, \ldots$ is the position for the wires, d is the period of the lattice and ϵ_b is the background dielectric constant.

where $\beta^2 = q_z^2 - (\omega/c)^2\epsilon_b$, q_z is the field wave number in the z-direction and ω is the field angular frequency. The solution to Eq. (4.35) is found to be

$$E_z(\mathbf{r}, t) = E_0 e^{iq_z z - i\omega t} \sum_{j=-\infty}^{\infty} e^{ik_y jd} K_0(\beta\rho_j), \tag{4.36}$$

where E_0 is a constant field amplitude, k_y is the field wave number in the y-direction, $\rho_j = \sqrt{x^2 + (y - jd)^2}$, and $K_0(x)$ is the zeroth-order modified Bessel function of the second kind. Here, the Bloch periodic condition is employed in Eq. (4.36), and then, $E_z(\mathbf{r}, t)$ becomes a periodic function with respect to the transformation $y \to y \pm d$.

Using the longitudinal-field condition, that is, $\nabla \times \mathbf{E} = 0$, and Eq. (4.36), we obtain the other two components of the longitudinal electric field

$$\begin{bmatrix} E_x(\mathbf{r}, t) \\ E_y(\mathbf{r}, t) \end{bmatrix} = \left(\frac{i\beta}{q_z}\right) E_0 e^{iq_z z - i\omega t} \sum_{j=-\infty}^{\infty} e^{ik_y jd} \times \frac{1}{\rho_j} \begin{bmatrix} x \\ y - jd \end{bmatrix} K_1(\beta\rho_j), \tag{4.37}$$

where $K_1(x)$ is the first-order modified Bessel function of the second kind.

For the electrodynamic model, we must employ a relation connecting the quantum-mechanical polarizability $\chi(q_z, \omega)$ to the classical Maxwell theory [14]. At $T = 0$ K, we calculate the polarizability as [15]

$$\mathrm{Re}[\chi(q_z, \omega)] = \frac{m^* e^2}{2\pi^2\hbar^2 q_z^3 \epsilon_0 a} \ln\left[\frac{\omega^2 - \Omega_-^2(q_z)}{\omega^2 - \Omega_-^2(q_z)}\right], \tag{4.38}$$

$$\mathrm{Im}[\chi(q_z, \omega)] = \frac{m^* e^2}{2\pi^2\hbar^2 q_z^3 \epsilon_0 a} \quad \text{for} \quad \Omega_-(q_z) < \omega < \Omega_+(q_z), \tag{4.39}$$

and

$$\Omega_\pm(q_z) = \left|\frac{\hbar q_z^2}{2m^*} \pm \frac{\hbar k_F q_z}{m^*}\right|, \tag{4.40}$$

where m^* is the effective mass of electrons, $k_F = \pi n_{1D}/2$ is the Fermi wave number, and n_{1D} is the linear density of electrons in a quantum wire.

For the interface at $x = 0$, the boundary conditions for $y = \pm a$ require that the discontinuity of $D_y = \epsilon_0 \epsilon_b E_y$ be equal to $-\nabla \cdot \boldsymbol{P}_s = -i q_z \epsilon_0 \chi E_z$. This leads to the following equation:

$$-\frac{2i\beta}{q_z} K_1(\beta a) = -\frac{2iq_z}{\epsilon_b} \chi(q_z, \omega) \times \left[K_0(\beta a) + 2\sum_{j=1}^{\infty} \cos(j k_y d) K_0(j \beta d) \right] . \quad (4.41)$$

Here, the parameter a plays the role of the characteristic cut-off length. In the non-retardation limit, that is, $c \to \infty$, we have $\beta = q_z$. In addition, in the limit of $q_z a \ll 1$, we get from Eq. (4.41) the plasmon dispersion relation for the linear array of quantum wires

$$\frac{q_z^2 a}{\epsilon_b} \mathrm{Re}[\chi(q_z, \omega)] f(q_z, k_y) = 1 , \quad (4.42)$$

where the structure factor is

$$f(q_z, k_y) = K_0(q_z a) + 2\sum_{j=1}^{\infty} \cos(j k_y d) K_0(j q_z d) . \quad (4.43)$$

In Eq. (4.43), $k_y d = 0$ and π define two edges of a band for the plasmon excitation. In the long-wave length limit, that is, $q_z \to 0$, it can be proved by using Eq. (4.38) that

$$\mathrm{Re}[\chi(q_z, \omega)] \approx \frac{n_{1D} e^2}{2\pi \epsilon_0 a m^* \omega^2} . \quad (4.44)$$

Substituting Eq. (4.44) into Eq. (4.42) leads to

$$\omega^2 = \Omega_p^2 (q_z a)^2 f(q_z, k_y) , \quad (4.45)$$

where $\Omega_p = \sqrt{n_{1D} e^2 / 2\pi \epsilon_0 \epsilon_b m^* a^2}$. Equation (4.45) indicates that the plasmon dispersion is somewhat between the 2D plasmon, $\omega \propto \sqrt{q_z}$, and the acoustic-like plasmon, $\omega \propto q_z$.

4.4
Coupled Half-Plane Superlattices

In this section, the hydrodynamic model in the long-wavelength limit will be employed to calculate the dispersion relation of edge magnetoplasmon excitation in a coupled half-plane superlattices.

4.4.1
Hydrodynamic Model

For simplicity, we use a model in which arrays of infinite-thin 2DEG layers are stacked along the z-direction with a period d, as schematically illustrated in Figure 4.6, and the half-plane 2DEG layers are located in the spaces $x < 0$ (region 1) and $x > a$ (region 2) of a distance a apart (region 3 for a ditch) and embedded in a semiconductor background of dielectric constant ϵ_b. A magnetic field $\boldsymbol{B}$ lies in the z-direction perpendicular to the half-planes.

Consider a rigid positive background with charge density en_0 and a compressible electron fluid with number density $n_0 + n$. Let $n_j(\boldsymbol{r}, t)$ and $\boldsymbol{v}_j(\boldsymbol{r}, t) = (v_{jx}, v_{jy})$ denote, respectively, the small fluctuations in the electron density and the electron velocity field in the plane of the jth layer located at $z = z_j$. These amplitudes satisfy the equation of continuity, Euler's equations and Poisson's equation [16, 17]:

$$-i\omega n_j + n_0 \left(\frac{\partial v_{jx}}{\partial x} + ikv_{jy} \right) = 0 \,, \tag{4.46}$$

$$-i\omega v_{jx} + \left(\frac{s^2}{n_0} \right) \frac{\partial n_j}{\partial x} - \left(\frac{e}{m^*} \right) \frac{\partial \phi}{\partial x} + \omega_c v_{jy} = 0 \,, \tag{4.47}$$

$$-i\omega v_{jy} + iks^2 \left(\frac{n_j}{n_0} \right) - ik \left(\frac{e}{m^*} \right) \phi - \omega_c v_{jx} = 0 \,, \tag{4.48}$$

$$\left[\frac{\partial^2}{\partial x^2} + \frac{\partial^2}{\partial z^2} - k^2 \right] \phi(x, z) = \frac{e}{\epsilon_0 \epsilon_b} \sum_{j=-\infty}^{\infty} n_j(x) \delta(z - z_j) \left[\theta(-x) + \theta(x - a) \right] , \tag{4.49}$$

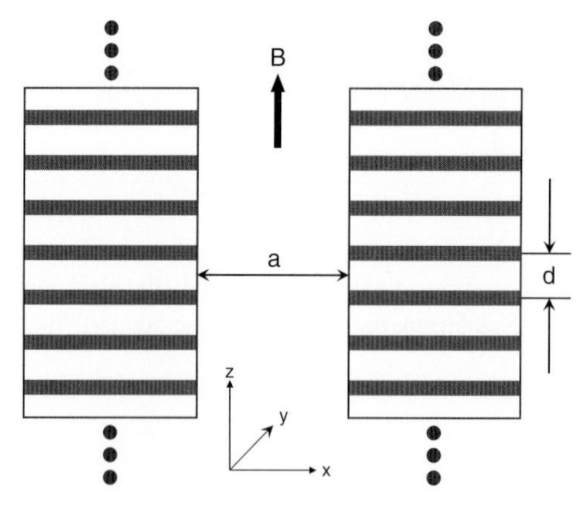

Figure 4.6 Schematic illustration of the cross-sections of coupled half-plane superlattices at a distance a apart. The 2DEG layers are staked along the z-direction with a period d. An external magnetic field $\boldsymbol{B}$ is applied in the direction perpendicular to the 2DEG layers.

where m^* is the effective mass of electrons, ω is the frequency of the self-consistent oscillation of charge-compensated 2DEGs in the system, ϕ is the electrostatic potential, and $\omega_c = eB/m^*$ is the cyclotron frequency. In addition, $\theta(x)$ is the unit step function, and s is an effective compressional wave speed. Here, since the system is translationally invariant along the y-direction, the solution may be taken as a plane wave of the form $\propto e^{iky-i\omega t}$ with amplitudes which depend on x and z. It is convenient to consider the wave number k to be positive, while ω_c can take either sign. A Fourier transform in x of Eq. (4.49) gives the ordinary differential equation

$$\left[\frac{d^2}{dz^2} - \left(k^2 + k_x^2\right)\right]\phi\,(k_x, z) = \frac{e}{\epsilon_0\epsilon_b}\sum_{j=-\infty}^{\infty} n_j(k_x)\delta(z - z_j)\,, \tag{4.50}$$

where $n_j(k_x)$ is the Fourier transform of $n_j(x)[\theta(-x) + \theta(x-a)]$. The solution of Eq. (4.50) can be written as

$$\phi\,(k_x, z) + \frac{e}{2\epsilon_0\epsilon_b}\sum_{j=-\infty}^{\infty} n_j(k_x)\left[\frac{\exp\left(-k'|z - z_j|\right)}{k}\right] = 0\,, \tag{4.51}$$

where $k' = \sqrt{k^2 + k_x^2}$. The inverse Fourier transform then gives a nonlocal integral relation between the electrostatic potential in the ℓth layer and the corresponding charge density

$$\phi\,(x, z_\ell) + \frac{e}{\epsilon_0\epsilon_b}\sum_{j=-\infty}^{\infty}\int dx' L_j\left(x - x'\right) n_j\left(x'\right)\left[\theta\left(-x'\right) + \theta\left(x' - a\right)\right] = 0\,, \tag{4.52}$$

where

$$L_j(x) = \int dk_x e^{ik_x x}\left[\frac{e^{-k'|z_\ell - z_j|}}{2k'}\right]. \tag{4.53}$$

In principle, such an integral equation can be solved by using the so-called Wiener–Hopf technique [18]. Using the Bloch condition in regions one and two, that is,

$$n_j\left(x'\right) = A\left(x'\right)e^{iq_z jd}\,, \tag{4.54}$$

we have, after the summation over j in Eq. (4.52),

$$\phi\,(x, z_\ell) + \frac{e}{2\epsilon_0\epsilon_b}\int dk_x e^{ik_x x}\left[\frac{A\,(k_x)\,S\,(k_x, k, q_z)}{k'}\right] = 0\,, \tag{4.55}$$

where

$$S\,(k_x, k, q_z) = \frac{\sinh\left(k'd\right)}{\cosh\left(k'd\right) - \cos\left(q_z d\right)}\,, \tag{4.56}$$

$$A\,(k_x) = \int dx' e^{-ik_x x'} A\left(x'\right)\,. \tag{4.57}$$

Equation (4.55) is independent of the layer label. A direct comparison of Eq. (4.52) with Eq. (4.55) yields the Fourier component of the exact kernel in Eq. (4.52):

$$L(k_x) = \frac{S\,(k_x, k, q_z)}{2\sqrt{k_x^2 + k^2}} \,. \tag{4.58}$$

Here, we introduce the expansion method [19] to Eq. (4.58), then we obtain the approximate kernel

$$L_0(k_x) = \frac{k\,f\,(k, q_z)}{2k^2 + k_x^2 g\,(k, q_z)} \,, \tag{4.59}$$

where

$$g\,(k, q_z) = 1 - \frac{k}{f\,(k, q_z)} \left[\frac{\partial f\,(k, q_z)}{\partial k}\right]. \tag{4.60}$$

Function $g(k, q_z)$ characterizes the screening correction, and $f(k, q_z) = S(k_x = 0, k, q_z)$. Here, $L(k_x)$ and $L_0(k_x)$ have the same first two terms in a power series about $k_x^2 = 0$. The inverse Fourier transform of Eq. (4.59) gives

$$L_0(x) = \frac{f(k, q_z)}{2\sqrt{2g\,(k, q_z)}} \exp\left[-k|x|\sqrt{\frac{2}{g\,(k, q_z)}}\right]. \tag{4.61}$$

As a result, the problem can be reduced to a pair of effective localized Poisson's equations

$$\left[\frac{d^2}{dx^2} - \frac{2k^2}{g}\right]\phi_{1,2}\,(x, z_i) = \frac{ek}{\epsilon_0\epsilon_\mathrm{b}}\left(\frac{f}{g}\right)\sum_{j=-\infty}^{\infty} n_j(x)\,,\, x < 0 \quad \text{or} \quad x > a,$$

$$\left[\frac{d^2}{dx^2} - \frac{2k^2}{g}\right]\phi_3\,(x, z_i) = 0\,, \quad 0 < x < a\,. \tag{4.62}$$

When Eqs. (4.46)–(4.48), and (4.62) are combined with the boundary condition that ϕ and $\partial\phi/\partial x$ continuous and that v_x vanishes there, together with the suitable boundary behavior for $|x| \to \infty$, this procedure gives the dispersion relation

$$\begin{aligned}
&D^4\omega^2\left[2\sqrt{\frac{2}{g}}C\sinh\left(ka\sqrt{\frac{2}{g}}\right) + C^2\sinh\left(ka\sqrt{\frac{2}{g}}\right) + \frac{2}{g}\sinh\left(ka\sqrt{2g}\right)\right] \\
&- 4\sqrt{\frac{2}{g}}D^2\omega_k^2\omega^2\left(\frac{f}{g}\right)\cosh\left(ka\sqrt{\frac{2}{g}}\right)\left(C + \sqrt{\frac{2}{g}}\right) \\
&+ 4\omega_k^2\left(\frac{f}{g^2}\right)\sinh\left(ka\sqrt{\frac{2}{g}}\right)\left(2\omega^2 - g\omega_\mathrm{c}^2\right) = 0\,,
\end{aligned} \tag{4.63}$$

where $\omega_k^2 = n_0 e^2 k/2\epsilon_0\epsilon_\mathrm{b} m^*$ and $\Omega_\mathrm{p}^2 = 2\omega_k^2/kd = n_0 e^2/\epsilon_0\epsilon_\mathrm{b} m^* d$ are the 2D and 3D plasma frequencies, respectively, and

$$D^2(\omega) = 2\omega_k^2\left(\frac{f}{g}\right) + \left(\omega_\mathrm{c}^2 - \omega^2\right), \tag{4.64}$$

$$C^2(\omega) = \frac{2}{D^2}\left[\omega_k^2\left(\frac{f}{g}\right) + \frac{1}{g}\left(\omega_c^2 - \omega^2\right)\right]. \tag{4.65}$$

An additional set of roots are given by $\omega^2 = \omega_c^2$ (spurious result of the approximation method) and $\omega^2 = 2\omega_k^2(f/g) + \omega_c^2$ (corresponding to the bulk continuum when $a \to 0$).

4.4.2 Numerical Results and Discussion

Equation (4.63) can be solved by the numerical method to give the edge plasmon dispersion relation and magnetic field dependence of the frequency of edge magnetoplasmons. In general, we have two branches of coupled modes, as shown in Figure 4.7. (The other two band edges corresponding to $q_z d = \pi$ have not been shown for $\omega_c \neq 0$.) When $\omega_c = 0$, there are two branches of modes due to coupling (a is finite). In the absence of a magnetic field, it should be pointed out that for the strong screening $kd \ll 1$, the frequency of the anomalous edge mode rapidly decreases when a becomes small; this is called the "softened" edge plasmon mode.

The "softened" plasmon mode can be attributed to the dramatic enhancement of complete Coulomb screening due to the strong coupling between different layers as a decreases. In the presence of magnetic field $\boldsymbol{B}$, the symmetry with respect to the $+y$- and $-y$-directions is broken, and then the edge plasmon mode will be split. However, when there are two coupled half-plane superlattices, the possible combinations of the two directions are $(+y, +y)$, $(-y, -y)$, $(+y, -y)$ and $(-y, +y)$. Of these, $(+y, +y)$ and $(-y, -y)$ are equivalent, and so are $(+y, -y)$ and $(-y, +y)$.

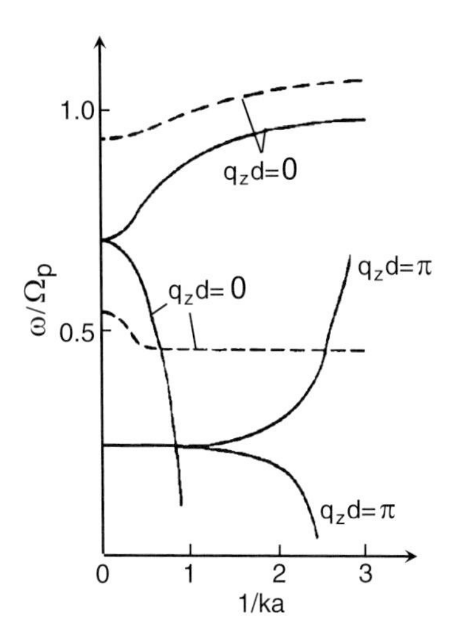

Figure 4.7 The coupling strength $1/ka$ dependence of the coupled edge mode in units of Ω_p for $kd = 0.5$ and different external magnetic fields: solid curve, $\omega_c = 0$; dashed curve, $\omega_c/\Omega_p = 0.4$ and $q_z d = 0$. The other two band edges corresponding to $q_z d = \pi$ are not shown for the broken curves.

These two kinds of combination correspond to two modes of edge magnetoplasmons. The existence of a magnetic field reduces the softening of anomalous edge mode. Also, the tops and bottoms of the two branches are interchanged.

Figure 4.8 presents the B dependence of the edge magnetoplasmon modes for different coupling strengths $1/ka$. When $\omega_c = 0$, the existence of coupling leads to the cancellation of the degeneracy of the magnetoplasmon mode. As the coupling strength increases, the splitting also increases. The normal edge mode increases with increasing magnetic field B as expected, while the dispersion of the anomalous edge mode is proportional to $1/B$ in the large-field limit. The other two band edges corresponding to $q_z d = \pi$ are not shown here.

It is very interesting to study the following special case. If a is finite, but $d \to 0$, the dispersion relation of the coupled surface modes of two half-bulks is given by

$$4R^4\omega^2 \sinh(ka) - 4R^2\omega^2\Omega_p^2 \cosh(ka) + \Omega_p^4 (\omega^2 - \omega_c^2) \sinh(ka) = 0 \,, \quad (4.66)$$

$$R^2(\omega) = \Omega_p^2 + (\omega^2 - \omega_c^2) \,, \quad (4.67)$$

which is presented in Figure 4.9. In this case, the screening becomes stronger, and the splitting of the two branches becomes smaller. The band width is zero in this case. From this, we can predict the existence of new coupled surface modes in such a system. Similar to the result in Figure 4.7, when $\omega_c = 0$, softening of the anomalous surface mode may occur. The enhancement of complete Coulomb screening weakens the interaction between electrons localized at two surfaces so that the frequency of the anomalous surface mode is decreased. The existence of a magnetic field also reduces the softening of the anomalous surface mode.

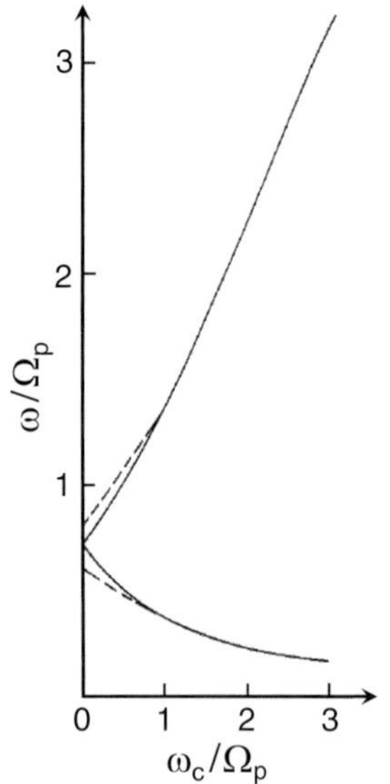

Figure 4.8 Magnetic field B dependence of the edge magnetoplasmon modes in units of Ω_p for $q_z d = 0$, $kd = 0.5$ and different coupling strengths: solid curve, a is infinite; dashed curve, weak coupling, $ka = 2$. The other two band edges corresponding to $q_z d = \pi$ are not shown for the solid and dashed curves.

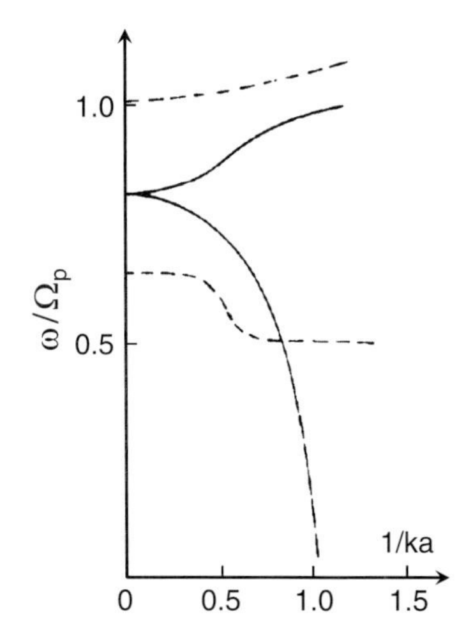

Figure 4.9 The coupling strength $1/ka$ dependence of the coupled edge modes in two coupled half-bulks ($d \to 0$) in units of Ω_p for different external magnetic fields: solid curve, $\omega_c = 0$; dashed curve, $\omega_c/\Omega_p = 0.4$.

4.5 Problems

1. Verify Eq. (4.31) for the real part of the density–density response function of a single cylindrical nanotube at $T = 0$ K.
   a) Obtain the highest-lying plasmon frequency for intra-subband transitions ($m = 0$) in the long wavelength limit.
   b) Calculate the corresponding imaginary part of the density–density response function. (c) Sketch the boundaries of single-particle excitations for $m = 0$ and $m = 1$.

2. Derive an expression which generalizes Eq. (4.31) for the density–density response function of a single nanotube in a magnetic field parallel to the axis of the nanotube.

3. When $a \to \infty$ but d is finite, show the following dispersion relations for a half-plane superlattice by making use of the general result in Eq. (4.63)

$$\omega_+ = \frac{1}{\sqrt{2}(2+g)}\mathrm{sgn}(\omega_c)\left[\sqrt{(2+g)f\omega_k^2 + g\omega_c^2} + \sqrt{g}\,|\omega_c|\right],$$

$$\omega_- = \frac{1}{\sqrt{2}(2+g)}\mathrm{sgn}\,(\omega_c)\left[\sqrt{(2+g)f\omega_k^2 + g\omega_c^2} - \sqrt{g}|\omega_c|\right],$$

   where $\mathrm{sgn}(x) = \pm 1$ is the sign function.

4. Using Eq. (4.38) for the polarizability at $T = 0$ K to verify the result in Eq. (4.44) in the long-wavelength limit, that is, $q_z \to 0$.

5. The spin–orbit interaction (SOI) Hamiltonian can be obtained from the Dirac equation in an external electromagnetic field described by a vector potential $\boldsymbol{A}$ and a scalar potential Φ by taking the non-relativistic limit up to terms quadratic in v/c inclusive (v is the velocity of electrons). This limit can be obtained in two different ways: either by a direct expansion of the Dirac equation in powers of v/c or by using the asymptotically exact Foldy–Wouthuysen formula. In either case, show that the spin–orbit Hamiltonian arising from the electrostatic confinement is given by

$$H_{\text{SO}} = \frac{\hbar}{4m^{*2}c^2}\left[\nabla V(\boldsymbol{r}) \times \hat{\boldsymbol{p}}\right] \cdot \hat{\overleftrightarrow{\sigma}} , \tag{4.68}$$

where

m^* is the effective mass of electrons,

$$\hat{\boldsymbol{p}} = -i\hbar\nabla , \quad V(\boldsymbol{r}) = -e\,\Phi(\boldsymbol{r}) \quad \text{and} \quad \hat{\overleftrightarrow{\sigma}} = (\hat{\sigma}_x, \hat{\sigma}_y, \hat{\sigma}_z)$$

is the vector of Pauli spin matrices. This Hamiltonian includes mechanisms arising from both the electric dipole moment and Thomas precession. In general, Eq. (4.68) consists of three terms arising from the spatial confinement. The z-component leads to the Rashba term for the quantum well. If, in addition, there is lateral confinement, then this may lead to additional terms in the calculation. Taking into account the electric field within the quantum well as an average $\overline{\boldsymbol{E}}$ whose direction is perpendicular to the interface of the heterojunction, the spin–orbit Hamiltonian in Eq. (4.68) can be rewritten for the Rashba coupling as

$$H_{\text{SO}} = \frac{\Delta_{\text{R}}}{\hbar}\left(\hat{\overleftrightarrow{\sigma}} \times \hat{\boldsymbol{p}}\right)_z , \tag{4.69}$$

where the z-component of the momentum does not contribute to Eq. (4.69) since in the stationary state, there is no transfer of electrons across the interface. The constant Δ_{R} contains all the universal constants from Eq. (4.68) and it is proportional to the interface electric field. Therefore, the contribution to the total electron Hamiltonian from the Rashba SOI is controlled by the value of Δ_{R}. For different systems, Δ_{R} takes on values in the range $10 \lesssim \Delta_{\text{R}} \lesssim 150$ meV Å.

a) The total Hamiltonian for free electrons in the 2DEG is a sum of the kinetic energy and H_{SO}. Since it is independent of coordinates, the wave function may be sought in the form of plane waves (for simplicity, one can denote the in-plane wave vector simply as $\boldsymbol{k}_{||}$) $\Psi_{\boldsymbol{k}_{||}}(\boldsymbol{r}_{||}) = \chi(\boldsymbol{k}_{||})e^{i\boldsymbol{k}_{||}\cdot\boldsymbol{r}_{||}}/\sqrt{\mathcal{A}}$. Here, $\boldsymbol{r}_{||}$ is the in-plane spatial coordinate, $\mathcal{A}$ is a normalization area, the spinor $\chi(\boldsymbol{k}_{||})$ satisfies the equation $H_{k_{||}}\chi(\boldsymbol{k}_{||}) = \epsilon_{k_{||}}\chi(\boldsymbol{k}_{||})$ and the Hamiltonian has the following explicit representation in spin space:

$$H_{k_{||}} = \begin{bmatrix} \frac{\hbar^2 k_{||}^2}{2m^*} & i\Delta_{\text{R}} k_{||} e^{-i\phi(k_{||})} \\ -i\Delta_{\text{R}} k_{||} e^{i\phi(k_{||})} & \frac{\hbar^2 k_{||}^2}{2m^*} \end{bmatrix} . \tag{4.70}$$

Here, $\phi(\boldsymbol{k}_{||})$ is the polar angle of the wave vector $\boldsymbol{k}_{||}$. By diagonalizing the matrix in Eq. (4.70), show that the energy eigenvalues

$$\epsilon_{k_{||}}^{\pm} = \frac{\hbar^2 k_{||}^2}{2m^*} \pm \Delta_{\mathrm{R}} k_{||} \tag{4.71}$$

with eigenspinors

$$\chi^{\pm}(\boldsymbol{k}_{||}) = \frac{1}{\sqrt{2}} \begin{bmatrix} 1 \\ \pm e^{i\phi(\boldsymbol{k}_{||})} \end{bmatrix} . \tag{4.72}$$

b) Give an explanation of the dependence of the spinor in Eq. (4.72) on the polar angle $\phi(\boldsymbol{k}_{||})$.

c) These results show that the effect of the Rashba SOI manifests itself through a mutual shift of the spin branches, resulting in an energy gap between the up "+" and down "−" spin branches. Calculate the susceptibility of a 2DEG within the RPA in the presence of this Rashba spin splitting.

6. a) Consider the Rashba SOI Hamiltonian system described above. For a total areal electron density $n_{2\mathrm{D}}$, there will be n_+ up-spins and n_- down-spins with $n_{2\mathrm{D}} = n_- + n_+$. At $T = 0\,\mathrm{K}$, show that these are determined by

$$\frac{n_s}{n_{2\mathrm{D}}} - \frac{1}{2} + sA_{\mathrm{R}} \left(\sqrt{\frac{n_s}{n_{2\mathrm{D}}}} + \sqrt{1 - \frac{n_s}{n_{2\mathrm{D}}}} \right) = 0 \,,$$

where $s = \pm$, $A_{\mathrm{R}} = k_{\mathrm{R}}/k_{\mathrm{F}}$ with $k_{\mathrm{R}} = m^* \Delta_{\mathrm{R}}/\sqrt{2}\hbar^2$ and $k_{\mathrm{F}} = \sqrt{2\pi n_{2\mathrm{D}}}$. For $A_{\mathrm{R}} < 1/2$, both bands are occupied. When $A_{\mathrm{R}} \geq 1/2$, $n_+ = 0$ and the spins are polarized in the down-spin (−) band.

b) Define the density of states (DOS) for each subband by

$$\rho_s(\varepsilon) = \sum_{\boldsymbol{k}_{||}} \delta\left(\varepsilon - \varepsilon_{k_{||},s}\right) .$$

Show that the DOS for the energy in Eq. (4.71) is given by

$$\rho_+(\varepsilon) = \theta(\varepsilon) \left(\frac{m^*}{2\pi\hbar^2} \right) \left\{ 1 - \sqrt{\frac{E_\Delta}{\varepsilon + E_\Delta}} \right\} ,$$

$$\rho_-(\varepsilon) = \left(\frac{m^*}{2\pi\hbar^2} \right) \times \left[\theta(\epsilon) \left(1 + \sqrt{\frac{E_\Delta}{\varepsilon + E_\Delta}} \right) + 2\theta(-\varepsilon)\theta(\varepsilon + E_\Delta) \sqrt{\frac{E_\Delta}{\varepsilon + E_\Delta}} \right] ,$$

where $\theta(x)$ is the unit step function and $E_\Delta = k_{\mathrm{R}} \Delta_{\mathrm{R}}/\sqrt{2}$ is a measure of the spin gap in the DOS.

c) Plot (or sketch) $\rho_+(\varepsilon), \rho_-(\varepsilon)$ and the total DOS $[\rho_+(\varepsilon) + \rho_-(\varepsilon)]$ in units of $(m^*/2\pi\hbar^2)$ as functions of ε/E_Δ for $-E_\Delta < \varepsilon < \infty$.

7. Consider a 2D periodic array of cylindrical nanotubes with their axes parallel to the z-axis and embedded in a medium with background dielectric constant ϵ_b. The periods in the x- and y-directions are d_x and d_y, respectively. Each nanotube consists of M concentric tubules of radius R_j ($j = 1, 2, \ldots, M$). Treat the electrons on the tubule as forming an electron gas. Assuming that there is no tunneling between the tubules, the single-particle eigenfunctions for the 2D periodic array are

$$\psi_{j\nu\ell}(\boldsymbol{\rho}, z) = \frac{1}{\sqrt{L_z N_x N_y}} e^{ik_z z} \sum_{n_x=-\frac{N_x}{2}}^{\frac{N_x}{2}} \sum_{n_y=-\frac{N_y}{2}}^{\frac{N_y}{2}} e^{i(k_x n_x d_x + k_y n_y d_y)} \times \Psi_{j\ell}(\boldsymbol{\rho} - n_x d_x \hat{\mathbf{x}} - n_y d_y \hat{\mathbf{y}}),$$

$$\Psi_{j\ell}(\boldsymbol{\rho}) = \frac{1}{\sqrt{2\pi}} e^{i\ell\phi} \frac{1}{\sqrt{\rho}} \Phi_j(\rho),$$

where $j = 1, 2, \ldots, M$ labels the tubules in the nanotube, $\nu = \{k_x, k_y, k_z\}$ is a composite index for the electron eigenstates, $\Psi_{j\ell}(\boldsymbol{\rho}) e^{ik_z z}/\sqrt{L_z}$ is the wave function for an electron in the jth tubule, with wave vector k_z in the axial direction and angular momentum quantum number $\ell = 0, \pm 1, \pm 2, \ldots$, $|\Phi_j(\rho)|^2 = \delta(\rho - R_j)$, $k_x = (2\pi/L_x) n_x$ and $k_y = (2\pi/L_y) n_y$ with $n_x = 0, \pm 1, \pm 2, \ldots, \pm N_x/2$ and $n_y = 0, \pm 1, \pm 2, \ldots, \pm N_y/2$. Here, $N_x = L_x/d_x$ and $N_y = L_y/d_y$ are the numbers of nanotubes in the x- and y-directions in the array with periodic boundary conditions. Electron motion in the azimuthal direction around the tubule is quantized and characterized by the angular momentum quantum number ℓ, whereas motion in the axial z-direction is free. Thus, the electron spectrum in each tubule consists of 1D subbands with ℓ serving as a subband index. The spectrum does not depend on k_x or k_y and has the form

$$\varepsilon_{jk_z\ell} = \frac{\hbar^2 k_z^2}{2m^*} + \frac{\hbar^2 \ell^2}{2m^* R_j^2}.$$

a) Making use of these results in conjunction with the methods previously employed for the linear array of nanotubes, show that the dispersion formula for plasma excitations in a 2D array of nanotubes is determined from

$$\mathrm{Det}\left[\delta_{m,m'}\delta_{j,j'} + \frac{e^2}{2\pi\epsilon_0\epsilon_b d_x d_y} \chi_{j'm'}(q_z, \omega) \times \sum_{M_1=-\infty}^{\infty} \sum_{M_2=-\infty}^{\infty} J_{m'}\left(R_{j'}\sqrt{(q_x + G_1)^2 + (q_y + G_2)^2}\right)\right.$$

$$\times \frac{J_m\left(R_j\sqrt{(q_x+G_1)^2+\left(q_y+G_2\right)^2}\right)}{(q_x+G_1)^2+\left(q_y+G_2\right)^2+q_z^2}$$

$$\times \left.\left(\frac{q_x+G_1+i\left(q_y+G_2\right)}{\sqrt{(q_x+G_1)^2+\left(q_y+G_2\right)^2}}\right)^{m'-m}\right] = 0\,, \qquad (4.73)$$

where $G_1 = 2\pi M_1/d_x$ and $G_2 = 2\pi M_2/d_y$.

b) Take the limit $d_y \to \infty$ in Eq. (4.73) and verify that the result agrees with the dispersion equation for a linear array of nanotubes.

8. a) Consider an electron in a perpendicular magnetic field B in a 2D parabolic confinement $U(r_{||}) = m^*\omega_0^2 r_{||}^2/2$ where $\boldsymbol{r}_{||}$ is the 2D coordinate vector. Show that the wave function is given by

$$\Phi_{n,m}(\rho,\phi) = \frac{1}{\sqrt{2\pi}}\sqrt{\frac{2m^*\Omega\, n!}{\hbar(|m|+n)!}}\, e^{-im\phi-\rho^2/2}\rho^{|m|}L_n^{|m|}(\rho^2)\,, \qquad (4.74)$$

where n and m are integers, m^* is the effective electron mass, $\rho = r_{||}\sqrt{m^*\Omega/\hbar}$, $\Omega = \sqrt{\omega_0^2+\omega_c^2/4}$, $L_n^{|m|}(x)$ are associated Laguerre polynomials, and $\omega_c = eB/m^*$ is the cyclotron frequency.

b) Show that the corresponding energy eigenvalues are given by

$$\varepsilon_{n,m} = (2n+|m|+1)\hbar\Omega - \frac{m}{2}\hbar\omega_c\,. \qquad (4.75)$$

c) Show that the eigenvalue problem of an electron in a parabolic confining potential under a magnetic field can be mapped onto an electron–hole pair with no Coulomb interaction between them, but in an effective magnetic field $\boldsymbol{B}_{\text{eff}}$ under some conditions you must specify.

d) Show that a conserved quantity for an isolated electron–hole pair in a magnetic field B is the exciton magnetic momentum $\hat{\boldsymbol{P}}$ defined by

$$\hat{\boldsymbol{P}} = -i\hbar\nabla_{\text{e}} - i\hbar\nabla_{\text{h}} + e\left(\boldsymbol{A}_{\text{e}} - \boldsymbol{A}_{\text{h}}\right) - e\left[\boldsymbol{B}_{\text{eff}} \times (\boldsymbol{r}_{\text{e}} - \boldsymbol{r}_{\text{h}})\right],$$

where $\boldsymbol{r}_{\text{e}}$ and $\boldsymbol{r}_{\text{h}}$ are 2D coordinate vectors of an electron and a hole. Also, $\boldsymbol{A}_{\text{e}}$ and $\boldsymbol{A}_{\text{h}}$ are the vector potentials of an electron and a hole, respectively. Use the cylindrical gauge for the vector potential, that is, $\boldsymbol{A}_{\text{e,h}} = \left(\boldsymbol{B}_{\text{eff}} \times \boldsymbol{r}_{\text{e,h}}\right)/2$.

e) Show that the eigenfunction of an electron–hole pair in a magnetic field has the form [20–23]

$$\Psi_P(\boldsymbol{R},\boldsymbol{r}) = \exp\left\{\frac{i}{\hbar}\boldsymbol{R}\cdot\left[\boldsymbol{P} + e\left(\boldsymbol{B}_{\text{eff}} \times \boldsymbol{r}\right)\right]\right\}\exp\left(\frac{i\gamma}{2\hbar}\boldsymbol{r}\cdot\boldsymbol{P}\right)\Phi\left(\boldsymbol{r}-\boldsymbol{\rho}_0\right),$$

where $\boldsymbol{R} = (m_{\text{e}}\boldsymbol{r}_{\text{e}} + m_{\text{h}}\boldsymbol{r}_{\text{h}})/(m_{\text{e}} + m_{\text{h}})$ is the coordinate of the center of mass for an electron–hole pair (m_e and m_h are the masses of electrons and holes) and $\Phi(\boldsymbol{r}-\boldsymbol{\rho}_0)$ is the wave function for the electron–hole relative motion with $\boldsymbol{r} = \boldsymbol{r}_{\rm e} - \boldsymbol{r}_{\rm h}$. Here, $\boldsymbol{\rho}_0 = (\boldsymbol{B}_{\rm eff} \times \boldsymbol{P})\,\ell_B^2/B$, $\ell_B = \sqrt{\hbar/eB}$ is the magnetic length, and

$$\Phi_{n,m}(\rho,\phi) = \frac{1}{\sqrt{2\pi}}\sqrt{\frac{2\mu\tilde{\omega}_{\rm c} n!}{\hbar(|m|+n)!}}\, e^{-im\phi-\rho^2/2}\rho^{|m|} L_n^{|m|}\left(\rho^2\right), \tag{4.76}$$

where $\mu = m_{\rm e}\, m_{\rm h}/(m_{\rm e} + m_{\rm h})$ is the effective electron mass, $\gamma = (m_{\rm h} - m_{\rm e})/(m_{\rm e} + m_{\rm h})$, $\tilde{\omega}_{\rm c} = eB_{\rm eff}/\mu$ is the cyclotron frequency, and $\rho = r\sqrt{\mu\tilde{\omega}_{\rm c}/\hbar}$. Note that in Eq. (4.76), $\mu\tilde{\omega}_c = eB_{\rm eff}$.

f) Show that the energy levels of the electron–hole pair in the magnetic field are given by

$$\mathcal{E}_{n,m} = (2n + |m| + 1)\frac{\hbar\tilde{\omega}_{\rm c}}{2} - \frac{m\gamma}{2}\hbar\tilde{\omega}_{\rm c}\,, \tag{4.77}$$

where $n = \min\{n_{\rm e}, n_{\rm h}\}$, $m = n_{\rm e} - n_{\rm h}$, and $n_{\rm e}$ and $n_{\rm h}$ denote the radial quantum numbers of an electron and a hole, respectively.

g) Use the electron–hole eigenfunction in Eq. (4.76) and the energy eigenvalues in Eq. (4.77), which are the same as the eigenfunction in Eq. (4.74) and the eigenvalues in Eq. (4.75) for an electron in a quantum dot under a magnetic field if the magnetic momentum is fixed for relative-motion of electron–hole coordinate $\boldsymbol{r} = \boldsymbol{r}_{\rm e} - \boldsymbol{r}_{\rm h}$, that is, $\boldsymbol{P} = -e(\boldsymbol{B}_{\rm eff} \times \boldsymbol{r})$, to calculate the dispersion equation for a quantum dot under the magnetic field within the RPA.

9. **Plasmons in a single graphene layer:** The spectrum of plasmon excitations in a single graphene layer, as shown in Figure 4.10a, immersed in a material with effective dielectric constant $\epsilon_{\rm b}$ without magnetic field present can be calculated in the RPA. Two bands having approximately a linear dispersion cross the Fermi level at K and K' points of the first Brillouin zone. The wave vectors of these points are given by $K = (2\pi/a)(1/3.1/\sqrt{3})$ and $K' = (2\pi/a)(2/3.0)$, with a denoting the lattice constant. The effective mass Hamiltonian in the absence of scatterers in a magnetic field applied perpendicular to the xy-plane is given by [24]

$$\mathcal{H}_0 = v_{\rm F}\begin{pmatrix} 0 & p_x^{(\rm e)} + ip_y^{(\rm e)} & 0 & 0 \\ p_x^{(\rm e)} - ip_y^{(\rm e)} & 0 & 0 & 0 \\ 0 & 0 & 0 & p_x^{(\rm h)} - ip_y^{(\rm h)} \\ 0 & 0 & p_x^{(\rm h)} + ip_y^{(\rm h)} & 0 \end{pmatrix}, \tag{4.78}$$

where $v_{\rm F} \approx 10^6$ m/s is the Fermi velocity and

$$\boldsymbol{p}^{(\rm e)} = -i\hbar\nabla_{\rm e} + e\boldsymbol{A}_{\rm e}\,, \quad \boldsymbol{p}^{(\rm h)} = -i\hbar\nabla_{\rm h} - e\boldsymbol{A}_{\rm h} \tag{4.79}$$

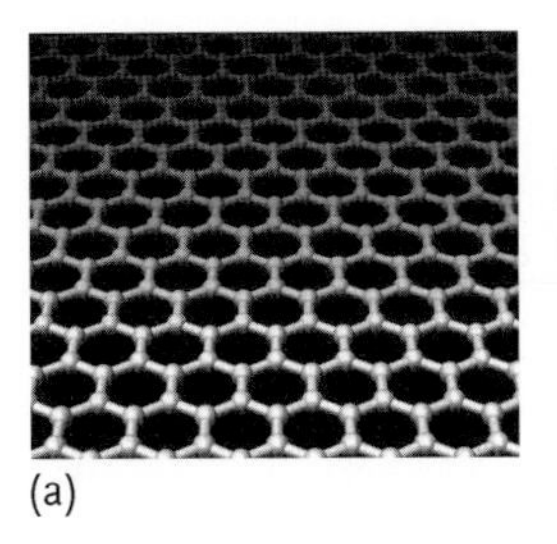
(a)

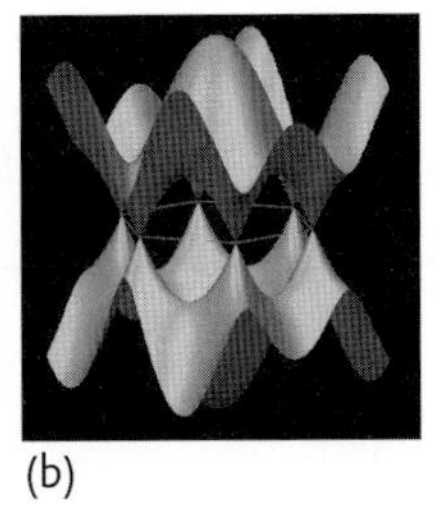
(b)

Figure 4.10 Hexagonal arrangement of carbon atoms in an ideal graphene sheet (a) and band structure of graphene sheet calculated using a simple linear combination of atomic orbitals approximation. The Fermi level (gray curve) can be tuned about the conical apices with an external gate voltage (b).

with $\boldsymbol{r}_{||}^{\mathrm{e}}$ and $\boldsymbol{r}_{||}^{\mathrm{h}}$ being the 2D position vectors of an electron and a hole, respectively, and $\boldsymbol{A}_{\mathrm{e}}$ and $\boldsymbol{A}_{\mathrm{h}}$ the vector potential of an electron and a hole.

a) In the absence of a magnetic field, show that the eigenfunction of $\mathcal{H}_0$ is given by

$$\psi\left(s, \boldsymbol{k}_{||}; \boldsymbol{r}_{||}\right) = e^{i\boldsymbol{k}_{||}\cdot\boldsymbol{r}_{||}} \frac{1}{\sqrt{2}L} \begin{pmatrix} s \\ e^{i\theta} \\ 0 \\ 0 \end{pmatrix},$$

and

$$\psi\left(s, \boldsymbol{k}_{||}; \boldsymbol{r}_{||}\right) = e^{i\boldsymbol{k}_{||}\cdot\boldsymbol{r}_{||}} \frac{1}{\sqrt{2}L} \begin{pmatrix} 0 \\ 0 \\ e^{i\theta} \\ s \end{pmatrix},$$

where $\boldsymbol{r}_{||} = (x, y)$ is 2D position vector, L^2 is a normalization area, s denotes the bands with $s = +1$ for the conduction band and $s = -1$ for the valence band. Also, the 2D wave vector $\boldsymbol{k}_{||} = (k_x, k_y)$ with $\theta = \tan^{-1}(k_y/k_x)$ is the polar angle. For this, show that the corresponding energy, as shown in Figure 4.10b, is given by

$$\varepsilon_{\mathrm{s}}\left(\boldsymbol{k}_{||}\right) = s\hbar v_{\mathrm{F}}|\boldsymbol{k}_{||}|\,.$$

b) Show that the dielectric function is given by $\epsilon(q_{||}, \omega) = 1 - V_{\mathrm{c}}(q_{||})\Pi^{(0)}(q_{||}, \omega)$, where $V_{\mathrm{c}}(q_{||}) = e^2/2\epsilon_0\epsilon_{\mathrm{b}}q_{||}$ is the 2D Coulomb interaction and $\Pi^{(0)}(q_{||}, \omega)$ is the 2D polarization function given by [25]

$$\Pi^{(0)}\left(q_{||}, \omega\right) = \frac{g_{\mathrm{s}}g_{\mathrm{v}}}{(2\pi)^2} \sum_{s,s'} \int d^2\boldsymbol{k}_{||} \left[\frac{f_{\mathrm{s}}\left(\boldsymbol{k}_{||}\right) - f_{\mathrm{s}'}\left(\boldsymbol{k}'_{||}\right)}{\hbar\omega + \varepsilon_{\mathrm{s}}\left(\boldsymbol{k}_{||}\right) - \varepsilon_{\mathrm{s}'}\left(\boldsymbol{k}'_{||}\right) + i0+}\right] \times F_{ss'}\left(\boldsymbol{k}_{||}, \boldsymbol{k}'_{||}\right),$$

where $g_s = g_v = 2$ are the spin (s) and graphene valley (v) degeneracies, $\boldsymbol{k}'_{||} = \boldsymbol{k}_{||} + \boldsymbol{q}_{||}$, $s, s' = \pm 1$ denote the band indices for the conduction $(+1)$ and the valence (-1) bands, $f_s(\boldsymbol{k}_{||}) = 1/(\exp[\varepsilon_s(\boldsymbol{k}_{||})/k_B T]+1)$ is the Fermi–Dirac function, T is the system temperature and $F_{ss'}(\boldsymbol{k}_{||}, \boldsymbol{k}'_{||})$ is the overlap of states given by

$$F_{ss'}\left(\boldsymbol{k}_{||}, \boldsymbol{k}'_{||}\right) = \frac{1 + ss' \cos\phi}{2} .$$

Here, ϕ is the angle between $\boldsymbol{k}_{||}$ and $\boldsymbol{k}'_{||}$. After performing the summation over s and s', we can rewrite the polarization function as [25]

$$\Pi^{(0)}(q_{||}, \omega) = \Pi^{+}\left(q_{||}, \omega\right) + \Pi^{-}\left(q_{||}, \omega\right), \tag{4.80}$$

where

$$\Pi^{+}\left(q_{||}, \omega\right) = \frac{g_s g_v}{2(2\pi)^2} \int d^2\boldsymbol{k}_{||} \left\{ \frac{\left[f_+\left(\boldsymbol{k}_{||}\right) - f_+\left(\boldsymbol{k}'_{||}\right)\right](1 + \cos\phi)}{\hbar\omega + \varepsilon_+\left(\boldsymbol{k}_{||}\right) - \varepsilon_+\left(\boldsymbol{k}'_{||}\right) + i0^+} \right.$$
$$\left. + \frac{f_+\left(\boldsymbol{k}_{||}\right)(1 - \cos\phi)}{\hbar\omega + \varepsilon_+\left(\boldsymbol{k}_{||}\right) - \varepsilon_-\left(\boldsymbol{k}'_{||}\right) + i0^+} - \frac{f_+\left(\boldsymbol{k}'_{||}\right)(1 - \cos\phi)}{\hbar\omega + \varepsilon_-\left(\boldsymbol{k}_{||}\right) - \varepsilon_+\left(\boldsymbol{k}'_{||}\right) + i0^+} \right\}, \tag{4.81}$$

and

$$\Pi^{-}\left(q_{||}, \omega\right) = \frac{g_s g_v}{2(2\pi)^2} \int d^2\boldsymbol{k}_{||} \left\{ \frac{\left[f_-\left(\boldsymbol{k}_{||}\right) - f_-\left(\boldsymbol{k}'_{||}\right)\right](1 + \cos\phi)}{\hbar\omega + \varepsilon_-\left(\boldsymbol{k}_{||}\right) - \varepsilon_-\left(\boldsymbol{k}'_{||}\right) + i0^+} \right.$$
$$\left. + \frac{f_-\left(\boldsymbol{k}_{||}\right)(1 - \cos\phi)}{\hbar\omega + \varepsilon_-\left(\boldsymbol{k}_{||}\right) - \varepsilon_+\left(\boldsymbol{k}'_{||}\right) + i0^+} - \frac{f_-\left(\boldsymbol{k}'_{||}\right)(1 - \cos\phi)}{\hbar\omega + \varepsilon_+\left(\boldsymbol{k}_{||}\right) - \varepsilon_-\left(\boldsymbol{k}'_{||}\right) + i0^+} \right\}. \tag{4.82}$$

We note that for intrinsic (i.e., undoped or ungated graphene), the Fermi energy $E_F = 0$, implying that the conduction band is empty while the valence band is fully occupied. In this case, if we count energy from the Fermi energy E_F, the occupancies at $T = 0\,\mathrm{K}$ are given by $f_+(\boldsymbol{k}_{||}) = 0$ and $f_-(\boldsymbol{k}_{||}) = 1$. As a result, at $T = 0\,\mathrm{K}$ the following relations are valid: $\Pi^{+}(q_{||}, \omega) = 0$ and $\Pi^{(0)}(q_{||}, \omega) = \Pi^{-}(q_{||}, \omega)$. The plasmon mode dispersion can be calculated by looking for zeros of the dynamical dielectric function, that is, $\epsilon(q_{||}, \omega) = 0$, with the polarization function $\Pi^{(0)}(q_{||}, \omega)$ defined by Eqs. (4.80)–(4.82). The spectrum of plasmons $\omega = \omega(q_{||})$ in graphene without magnetic field can be calculated as a solution of the following equation [25]:

$$V_c\left(q_{||}\right) \Pi^{(0)}\left(q_{||}, \omega\right) = 1 . \tag{4.83}$$

The physical realization of the system close to an experiment is the graphene layer sits on top of a dielectric (described by the dielectric constant

$\epsilon_b = \epsilon_s$) with another medium (vacuum) (described by the dielectric constant $\epsilon_b = 1$) above. In this case, the dielectric function and the spectrum of plasmons in the 2D electron gas in graphene is determined by Eq. (4.83) by substituting $V_c(q_{||}) = e^2/2\epsilon_0\bar{\epsilon}q_{||}$ as the 2D Coulomb interaction with the average dielectric constant $\bar{\epsilon} = (\epsilon_s + 1)/2$. The latter expression for the average dielectric function can be obtained analogously to the procedure presented in [26, 27].

10. In the preceding problem for graphene, consider the case when there is an applied perpendicular magnetic field [28, 29].
   a) Show that a conserved quantity for an isolated electron–hole pair in magnetic field B is the exciton magnetic momentum given by

$$\hat{\boldsymbol{P}} = -i\hbar\nabla_e - i\hbar\nabla_h + e(\boldsymbol{A}_e - \boldsymbol{A}_h) - e\left[\boldsymbol{B} \times \left(\boldsymbol{r}_{||}^e - \boldsymbol{r}_{||}^h\right)\right], \tag{4.84}$$

   where $\boldsymbol{A}_e$ and $\boldsymbol{A}_h$ are the vector potential of an electron and a hole, respectively. In the cylindrical gauge for vector potential, $\boldsymbol{A}_{e(h)} = \frac{1}{2}[\boldsymbol{B} \times \boldsymbol{r}_{e(h)}]$.
   b) Show that the eigenfunction $\psi(s, \boldsymbol{k}_{||}; \boldsymbol{r}_{||})$ of the Hamiltonian in Eq. (4.78) of the 2D electron–hole pair in perpendicular magnetic field B which is also the eigenfunction of the magnetic momentum $\hat{\boldsymbol{P}}$ has the form

$$\psi_{\boldsymbol{P}}(\boldsymbol{R}, \boldsymbol{r}) = \exp\left[i\left(\boldsymbol{P} + \frac{e}{2}[\boldsymbol{B} \times \boldsymbol{r}]\right)\frac{\boldsymbol{R}}{\hbar}\right]\tilde{\Phi}(\boldsymbol{r} - \boldsymbol{\rho}_0), \tag{4.85}$$

   where $\boldsymbol{R} = (\boldsymbol{r}_e + \boldsymbol{r}_h)/2$, $\boldsymbol{r} = \boldsymbol{r}_e - \boldsymbol{r}_h$, and $\boldsymbol{\rho}_0 = \hbar[\boldsymbol{B} \times \boldsymbol{P}]/(eB^2)$.
   c) Show that the wave function of the relative motion can be expressed in terms of the 2D coordinate $\boldsymbol{r}$ as harmonic oscillator eigenfunctions $\Phi_{n_1,n_2}(\boldsymbol{r})$. For an electron in Landau level n_+ and a hole in level n_-, show that the four-component wave functions for the relative coordinate are

$$\tilde{\Phi}_{n_+,n_-}(\boldsymbol{r}) = \left(\sqrt{2}L\right)^{\delta_{n_+,0}+\delta_{n_-,0}-2}\begin{pmatrix} s_+s_-\Phi_{|n_+|-1,|n_-|-1}(\boldsymbol{r}) \\ s_+\Phi_{|n_+|-1,|n_-|}(\boldsymbol{r}) \\ s_-\Phi_{|n_+|,|n_-|-1}(\boldsymbol{r}) \\ \Phi_{|n_+|,|n_-|}(\boldsymbol{r}) \end{pmatrix}, \tag{4.86}$$

   where L^2 is a normalization area, $s_\pm = \text{sgn}(n_\pm)$. The corresponding energy of the electron–hole pair $E^{(0)}_{n_+,n_-}$ is given by

$$E^{(0)}_{n_+,n_-} = \frac{\hbar v_F}{r_B}\sqrt{2}\left[\text{sgn}(n_+)\sqrt{|n_+|} - \text{sgn}(n_-)\sqrt{|n_-|}\right], \tag{4.87}$$

   where $r_B = \sqrt{\hbar/(eB)}$ is the magnetic length and the 2D harmonic oscillator wave eigenfunctions $\Phi_{n_1,n_2}(\boldsymbol{r})$ are given by

$$\Phi_{n_1,n_2}(\boldsymbol{r}) = (2\pi)^{-1/2}2^{-|m|/2}\frac{\tilde{n}!}{\sqrt{n_1!n_2!}}\frac{1}{r_B}\text{sgn}[(m)^m]\frac{r^{|m|}}{r_B^{|m|}} \times \exp\left[-im\phi - \frac{r^2}{4r_B^2}\right]L_{\tilde{n}}^{|m|}\left(\frac{r^2}{2r_B^2}\right), \tag{4.88}$$

where $L_{\tilde{n}}^{|m|}(x)$ denotes the Laguerre polynomials, $m = n_1 - n_2$ and $\tilde{n} = \min(n_1, n_2)$, and $\mathrm{sgn}[(m)^m] = 1$ for $m = 0$.

d) Making use of these results, derive an expression for the polarization function of graphene in a perpendicular magnetic field.

11. Calculate the self-energy for graphene using a screened Coulomb interaction where the dielectric function is given in the RPA. Use the cone-like approximation for the electron band structure in the problem above. Consequently, obtain an expression for the tunneling density-of-states, defined as the imaginary part of the interacting one-particle Green's function [30].

12. Consider a system of electrons (charge −e) and ions (charge +e) which are interacting via Coulomb's law. Thus, we have electron–electron, ion–ion and electron–ion interactions. Assume that a weak external electric scalar potential of the form

$$\phi_{\text{ext}}(\boldsymbol{r}, t) = \frac{\phi_0(Q, \omega)}{\mathcal{V}} e^{i(Q\cdot\boldsymbol{r}-\omega t)} e^{\eta t} \tag{4.89}$$

is applied, having been turned on at $t = -\infty$. Using the RPA to the equation of motion and working to lowest order in ϕ_0, show that the induced ion charge density is given by

$$e\bar{\rho}_+(Q, t) = \frac{\chi_+^0(Q, \omega)}{D(Q, \omega)} \phi_0(Q, \omega) e^{-i\omega t}, \tag{4.90}$$

while the induced electron charge density is given by

$$-e\bar{\rho}_-(Q, t) = \frac{\chi_-^0(Q, \omega)}{D(Q, \omega)} \phi_0(Q, \omega) e^{-i\omega t}, \tag{4.91}$$

where the common denominator is

$$D(Q, \omega) = 1 - v(Q)\left[\chi_-^0(Q, \omega) + \chi_+^0(Q, \omega)\right] \tag{4.92}$$

and $v(Q)$ is the Fourier transform of the electron–electron interaction. For small enough Q, we have

$$\chi_\pm^0(Q, \omega) = -\int d^3p \frac{\boldsymbol{Q}\cdot\nabla_p f_0(\boldsymbol{p})}{\left(\hbar\omega - \hbar^2 \boldsymbol{Q}\cdot\frac{\boldsymbol{v}}{m_\pm}\right)}. \tag{4.93}$$

Here, $m_\pm$ are the masses of electrons and ions and we used

$$f_0(\boldsymbol{k}+\boldsymbol{q}) = f_0(\boldsymbol{k}) + \boldsymbol{q}\cdot\nabla_k f_0(\boldsymbol{k}) + \dots. \tag{4.94}$$

13. Assuming that the results of the previous question are correct, it is clear that the collective modes of the electron–ion system are given by the solutions of $D(Q, \omega) = 0$.

a) Show that if ω and Q are such that

$$\omega^2 \ll v_F^2 Q^2$$
$$\omega^2 \gg \langle v^2 \rangle Q^2 \,,$$

then, an approximate solution is $\omega = sQ$, where

$$s^2 = \left(\frac{m_-}{3m_+} \right) v_F^2 \,.$$

Here, $\langle v^2 \rangle$ is the mean square ion velocity and v_F is the Fermi velocity of the electrons.

b) Show that when this phonon-like mode $\omega = sQ$ is excited in the electron–ion system, there is no net charge fluctuation associated with it, that is,

$$-e\overline{\rho}_-(Q, t) + e\overline{\rho}_+(Q, t) = 0 \,. \tag{4.95}$$

You may use results from the preceding question.

14. Using the equation of motion procedure, find the density response function $\chi^0(Q, \omega)$ for a non-interacting gas of spinless bosons (He^4 atoms, for example). In this case, the creation and destruction operators satisfy commutation relations

$$\left[\hat{a}_k, \hat{a}^\dagger_{k'}\right]_- = \delta_{k,k'} \,, \quad [\hat{a}_k, \hat{a}_{k'}]_- = 0 \,, \quad \left[\hat{a}^\dagger_k, \hat{a}^\dagger_{k'}\right]_- = 0 \,.$$

Your answer should be expressed in terms of $\langle \hat{a}^\dagger_k \hat{a}_k \rangle = n_0(\varepsilon_k)$ and $\varepsilon_k = \hbar^2 k^2 / 2m^*$, just as in the case of fermions. Secondly, show that at $T = 0$ K, your result simplifies to

$$\chi^0_{\text{Bose}}(Q, \omega) = \frac{\overline{n}\hbar^2 \frac{Q^2}{m^*}}{(\hbar\omega)^2 - \varepsilon_Q^2} \,, \quad \overline{n} = \frac{N}{\mathcal{V}} \,.$$

Hint: Remember that at $T = 0$ K, all the atoms are in the condensate, that is, they have zero momentum as a result of complete Bose–Einstein condensation.

References

1 Huang, D.H. and Antoniewicz, P.R. (1991) Coupled tunneling plasmon excitations in a planar array of quantum dots. *Phys. Rev. B*, **43**, 2169.

2 Que, W., Kirczenow, G., and Castaño, E. (1991) Nonlocal theory of collective excitations in quantum-dot arrays. *Phys. Rev. B*, **43**, 14079.

3 Huang, D.H. and Gumbs, G. (1992) Nonlocal perimeter magnetoplasmons in a planar array of narrow quantum rings. *Phys. Rev. B*, **46**, 4147.

4 Huang, D.H. and Gumbs, G. (1993) Magnetoplasmon excitations in a two-dimensional square array of antidots. *Phys. Rev. B*, **47**, 9597.

5 Giuliani, G.F. and Quinn, J.J. (1983) Charge-density excitations at the surface of a semiconductor superlattice: A new type of surface polariton. *Phys. Rev. Lett.*, **51**, 919.
6 Jain, J.K. and Allen, P.B. (1985) Plasmons in layered films. *Phys. Rev. Lett.*, **54**, 2437.
7 Jain, J.K. and Allen, P.B. (1985) Dielectric response of a semi-infinite layered electron gas and Raman scattering from its bulk and surface. *Phys. Rev. B*, **32**, 997.
8 Saito, R., Dresselhaus, G., and Dresselhaus, M.S. (1998) *Physical Properties of Carbon Nanotubes*, Imperial College Press, p. 110.
9 Saito, R., Fujita, M., Dresselhaus, G., and Dresselhaus, M.S. (1992) Electronic structure of graphene tubules based on C_{60}. *Phys. Rev. B*, **46**, 1804.
10 Ehrenreich, H. and Cohen, M.H. (1959) Self-consistent field approach to the many-electron problem. *Phys. Rev.*, **115**, 786.
11 Dresselhaus, M.S., Dresselhaus, G., and Eklund, P.C. (1996) *Science of Fullerenes and Carbon Nanotubes*, Academic Press, San Diego.
12 Gumbs, G. and Azin, G.R. (2002) Collective excitations in a linear periodic array of cylindrical nanotubes. *Phys. Rev. B*, **65**, 195407.
13 Lin, M.F. and Kenneth Shung, W.-K. (1993) Elementary excitations in cylindrical tubules. *Phys. Rev. B*, **47**, 6617.
14 Zhu, Y. and Zhou, S.X. (1988) Intrasubband collective modes in a quasi-$(1 + 1)$-dimensional semiconductor superlattice. *J. Phys. C: Solid State Phys.*, **21**, 3063.
15 Stern, F. (1967) Polarizability of a two-dimensional electron gas. *Phys. Rev. Lett.*, **18**, 546.
16 Huang, D.H., Zhu, Y., and Zhou, S.X. (1989) The softening of edge plasmons on lateral surfaces of coupled half-plane semiconductor superlattices. *J. Phys.: Condens. Matter*, **1**, 7627.
17 Huang, D.H. (1995) Coupling between surface-polariton and edge-plasmon excitations in coupled finite half-plane superlattices. *Phys. Rev. B*, **52**, 2020.
18 Carrier, G.F., Krook, M., and Pearson, C.E. (1996) *Functions of a complex variable*, McGraw-Hill, New York.
19 Fetter, A.L. (1985) Edge magnetoplasmons in a bounded two-dimensional electron fluid. *Phys. Rev. B*, **32**, 7676.
20 Gorkov, L.P. and Dzyaloshinskii, I.E. (1968) Contribution to the theory of the Mott exciton in a strong magnetic field. *JETP*, **26**, 449.
21 Lerner, I.V. and Lozovik, Y.E. (1980) Two dimensional electron–hole system in a strong magnetic field as an almost ideal exciton gas. *JETP*, **51**, 588.
22 Kallin, C. and Halperin, B.I. (1984) Excitations from a filled Landau level in the two-dimensional electron gas. *Phys. Rev. B*, **30**, 5655.
23 Kallin, C. and Halperin, B.I. (1985) Many-body effects on the cyclotron resonance in a two-dimensional electron gas. *Phys. Rev. B*, **31**, 3635.
24 Zhang, Y. and Ando, T. (2002) Hall Conductivity of a two-dimensional graphite system. *Phys. Rev. B*, **65**, 245420.
25 Hwang, E.H. and Das Sarma, S. (2007) Dielectric function, screening, and plasmons in two-dimensional graphene. *Phys. Rev. B*, **75**, 205418.
26 Eguiluz, A., Lee, T.K., Quinn, J.J., and Chiu, K.W. (1975) Interface excitations in metal-insulator-semiconductor structures. *Phys. Rev. B*, **11**, 4989.
27 Persson, B.N.J. (1984) Inelastic electron scattering from thin metal films. *Solid State Commun.*, **52**, 811.
28 Iyengar, A., Wang, J., Fertig, H.A., and Brey, L. (2007) Excitations from filled Landau levels in graphene. *Phys. Rev. B*, **75**, 125430.
29 Berman, O.L., Lozovik, Y.E., and Gumbs, G. (2008) Bose–Einstein condensation and Superfluidity of magnetoexcitons in Bilayer Graphene. *Phys. Rev. B*, **77**, 155433.
30 Gumbs, G. and Kogan, E. (2007) Effect of electron–electron interaction and plasmon excitation on the density-of-states for a two-dimensional electron liquid. *Phys. Status Solidi (b)*, **244**, 3695.

5
The Surface Response Function, Energy Loss and Plasma Instability

5.1
Surface Response Function

The subject of electron energy loss has received a considerable amount of attention over the years, with several review articles and textbooks written in the last few years [1, 2]. Unfortunately, there is no review article dealing with the formalism as it applies to nanotubes or layered 2D structures. Here, we give a formalism in terms of the surface response function. Let us first consider a structure with a planar surface and assume that the medium occupies the half-space $z > 0$. Consider a point charge $Z^* e$ moving along a prescribed trajectory $\boldsymbol{R}(t)$ outside the medium. The external potential ϕ_{ext} due to this point charge satisfies Poisson's equation

$$\nabla^2 \phi_{\text{ext}}(\boldsymbol{r}, t) = -\frac{Z^* e}{\epsilon_0} \delta\left[\boldsymbol{r} - \boldsymbol{R}(t)\right], \tag{5.1}$$

where $\boldsymbol{R}(t) = [\boldsymbol{R}_{||}(t), Z(t)]$ and $\boldsymbol{r} = (\boldsymbol{r}_{||}, z)$. Equation (5.1) has the solution

$$\phi_{\text{ext}}(\boldsymbol{r}, t) = \int \frac{d^2 q_{||}}{(2\pi)^2} \int_{-\infty}^{\infty} \frac{d\omega}{2\pi} \tilde{\phi}_{\text{ext}}(\boldsymbol{q}_{||}, \omega) e^{i \boldsymbol{q}_{||} \cdot \boldsymbol{r}_{||} - i\omega t} e^{-q_{||} z} \tag{5.2}$$

for $Z(t) < z < 0$, where

$$\tilde{\phi}_{\text{ext}}(\boldsymbol{q}_{||}, \omega) = -\frac{Z^* e}{2\epsilon_0 q_{||}} \mathcal{F}(\boldsymbol{q}_{||}, \omega) \tag{5.3}$$

with

$$\mathcal{F}(\boldsymbol{q}_{||}, \omega) \equiv \int_{-\infty}^{\infty} dt e^{q_{||} Z(t)} e^{i\omega t - i \boldsymbol{q}_{||} \cdot \boldsymbol{R}_{||}(t)} . \tag{5.4}$$

Here, $\boldsymbol{q}_{||} = (q_x, q_y)$ is a 2D wave vector in the xy-plane parallel to the surface which is in the $z = 0$ plane.

The external potential $\phi_{\text{ext}}(\boldsymbol{r}, t)$ gives rise to an induced potential. Using linear response theory to relate the induced potential to the charge density fluctuation along

Properties of Interacting Low-Dimensional Systems, First Edition. G. Gumbs and D. Huang.

with Poisson's equation, it follows that the induced potential outside the medium has the form

$$\phi_{\text{ind}}(\boldsymbol{r}, t) = -\int \frac{d^2\boldsymbol{q}_{||}}{(2\pi)^2} \int_{-\infty}^{\infty} \frac{d\omega}{2\pi} \tilde{\phi}_{\text{ext}}(\boldsymbol{q}_{||}, \omega) e^{i\boldsymbol{q}_{||}\cdot\boldsymbol{r}_{||} - i\omega t} g(q_{||}, \omega) e^{q_{||}z} \tag{5.5}$$

for $z < 0$. In this notation, $g(q_{||}, \omega)$ defines the surface response function. It has been implicitly assumed that the external potential ϕ_{ext} is so weak that the medium responds linearly to it. The function $g(q_{||}, \omega)$ is itself related to the density–density response function $\chi(z, z'; q_{||}, \omega)$ of the system of interacting particles by [3]

$$\begin{aligned} g(q_{||}, \omega) &= \frac{e^2}{2\epsilon_0 q_{||}} \int_0^{\infty} dz \int_0^{\infty} dz' e^{-q_{||}(z+z')} \chi(z, z'; q_{||}, \omega) \\ &\equiv -\int_{-\infty}^{0} dz\, e^{q_{||}z} \rho_{\text{ind}}(z; q_{||}, \omega) , \end{aligned} \tag{5.6}$$

where the second equality defines the induced surface charge density.

5.1.1
The Image Potential

Consider a stationary external point charge e located at $\boldsymbol{r}_0 = (0, 0, z_0)$ on the polar z-axis near a surface at $z = 0$. The external potential due to the presence of this point charge is obtained by solving Poisson's equation

$$\nabla^2 \phi_{\text{ext}}(\boldsymbol{r}) = -\frac{e}{\epsilon_0} \delta(\boldsymbol{r} - \boldsymbol{r}_0) . \tag{5.7}$$

For $z < z_0$, we obtain

$$\phi_{\text{ext}}(\boldsymbol{r}) = -\int \frac{d^2\boldsymbol{q}_{||}}{(2\pi)^2} e^{-q_{||}(z_0 - z)} e^{i\boldsymbol{q}_{||}\cdot\boldsymbol{r}_{||}} \frac{e}{2\epsilon_0 q_{||}} . \tag{5.8}$$

Also, for $z \geq 0$, the induced potential is given by

$$\phi_{\text{ind}}(\boldsymbol{r}) = \int \frac{d^2\boldsymbol{q}_{||}}{(2\pi)^2} e^{-q_{||}(z_0 + z)} e^{i\boldsymbol{q}_{||}\cdot\boldsymbol{r}_{||}} g(q_{||}, 0) \frac{e}{2\epsilon_0 q_{||}} , \tag{5.9}$$

where $g(q_{||}, 0) \equiv g(q_{||}, z = 0)$. Therefore, the force exerted on the external charge due to the induced charge in the medium is

$$\begin{aligned} \boldsymbol{F}_{\text{ind}} &= e \frac{\partial}{\partial z} \phi_{\text{ind}}(\boldsymbol{r}) \hat{z} \Big|_{z=z_0, \boldsymbol{r}_{||}=0} \\ &= -\frac{e^2}{2\epsilon_0} \int \frac{d^2\boldsymbol{q}_{||}}{(2\pi)^2} e^{-q_{||}(z_0+z)} e^{i\boldsymbol{q}_{||}\cdot\boldsymbol{r}_{||}} g(q_{||}, 0) \hat{z} \Big|_{z=z_0, \boldsymbol{r}_{||}=0} \\ &\equiv -\frac{\partial}{\partial z_0} \mathcal{U}_{\text{im}}(z_0) \hat{z} , \end{aligned} \tag{5.10}$$

where the image potential is defined by

$$\mathcal{U}_{\text{im}}(z_0) = -\frac{e^2}{8\pi\epsilon_0} \int_0^\infty dq_{||} e^{-2q_{||}z_0} g(q_{||}, 0) \,. \tag{5.11}$$

If we use $g(q_{||}, 0) = (\epsilon_b - 1)/(\epsilon_b + 1)$ from Eq. (5.26) below, where ϵ_b is the dielectric constant of the medium, then

$$\mathcal{U}_{\text{im}}(z_0) = -\frac{e^2}{16\pi\epsilon_0 z_0} \left(\frac{\epsilon_b - 1}{\epsilon_b + 1}\right) . \tag{5.12}$$

Therefore, the external electron polarizes the surface charge and becomes attracted to its "image charge" residing below the surface, giving rise to a spatially extended state. Because of its $1/z_0$ dependence, this potential supports an infinite number of image states having the well-known Rydberg series form $E_n = -(13.6/n^2)[(\epsilon_b - 1)/(\epsilon_b + 1)]^2$ eV, where $n = 1, 2, \ldots$ is the principal quantum number. Since $E_n \sim 1/n^2$, the states with higher n have weaker binding energies. Recently, Höfer, *et al.* [4] applied two-photon photoemission techniques to populate the coherent wave packets in image states close to a Cu(100) and Cu(111) surface. The states observed in these experiments had $n \leq 6$ and binding energies of 15–40 meV. These surface states collapsed onto the Cu surface with lifetimes of a few femtoseconds. The states with larger n have longer lifetimes [5]. For example, for the Cu(111) surface, the lifetimes of the image potential states for $n = 1, 2, 3$ are $\tau_1 \approx 40$ fs, $\tau_2 \approx 110$ fs and $\tau_3 \approx 300$ fs, respectively [4].

5.1.2
A Bi-Layer System

Let us consider the following arrangement in Figure 5.1. Here, a 2DEG layer is located at $z = 0$ and $z = a$ with a medium of dielectric constant ϵ between them. We take the total potential consisting of the external potential from an impinging beam of charge, for example, and the induced electrostatic potential in the "vicinity" of

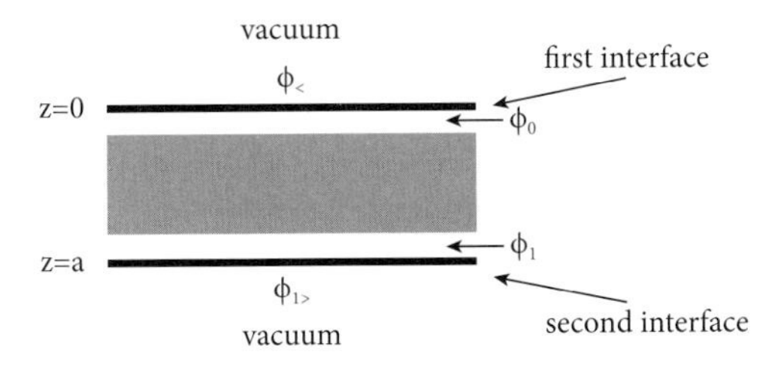

Figure 5.1 A pair of two-dimensional electron gas layers separated by a distance a with a material of dielectric constant ϵ between them.

the $z = 0$ layer (between the external probe and the surface) to be

$$\begin{aligned} \phi_<(z) &= e^{-q_{||}z} - g(q_{||}, \omega) e^{q_{||}z} , \\ \phi_0(z) &= a_0 e^{q_{||}z} + b_0 e^{-q_{||}z} , \\ \phi_{0>}(z) &= t_0 e^{-q_{||}z} + r_0 e^{q_{||}z} , \end{aligned} \tag{5.13}$$

where all the coefficients are independent of the z coordinate. We use the boundary condition, that is, $\phi_<(0) = \phi_0(0)$ at the vacuum–surface interface and

$$\frac{d\phi_0(0)}{dz} - \frac{d\phi_<(0)}{dz} = -\frac{\sigma_0}{\epsilon_0} , \tag{5.14}$$

where σ_0 is the induced surface charge density in the 2DEG layer at $z = 0$. For the same reason, we have $\phi_0(0) = \phi_{0>}(0)$ and

$$\epsilon \frac{d\phi_{0>}(0)}{dz} - \frac{d\phi_0(0)}{dz} = 0 . \tag{5.15}$$

By substituting Eq. (5.13) into Eqs. (5.14) and (5.15), we obtain

$$a_0 = -g - \frac{\sigma_0}{2\epsilon_0 q_{||}} , \tag{5.16}$$

$$b_0 = 1 + \frac{\sigma_0}{2\epsilon_0 q_{||}} , \tag{5.17}$$

$$r_0 = \frac{1}{2}(1 - g) - \frac{1}{2\epsilon}\left(1 + g + \frac{\sigma_0}{\epsilon_0 q_{||}}\right) , \tag{5.18}$$

$$t_0 = \frac{1}{2}(1 - g) + \frac{1}{2\epsilon}\left(1 + g + \frac{\sigma_0}{\epsilon_0 q_{||}}\right) . \tag{5.19}$$

At the interface $z = a$, the induced potential is given by

$$\begin{aligned} \phi_1(z) &= a_1 e^{q_{||}z} + b_1 e^{-q_{||}z} , \\ \phi_{1>}(z) &= t_1 e^{q_{||}z} , \end{aligned} \tag{5.20}$$

where the coefficients a_1, b_1 and t_1 are also determined from the boundary condition at $z = a$ with the induced surface charge density σ_1. However, in general, we have in linear response theory $\sigma = \chi^0 \phi$, where χ^0 is the susceptibility for non-interacting particles. Therefore, $\sigma_0 = \chi^0 \phi_0 = (1 - g)\chi^0$ and $\sigma_1 = \chi^0 \phi(z = a)$. After some algebra, we obtain

$$g(q_{||}, \omega) = 1 + 2 \frac{\left(1 + \epsilon - \frac{\chi^0}{\epsilon_0 q_{||}}\right) - \left(1 - \epsilon - \frac{\chi^0}{\epsilon_0 q_{||}}\right) e^{-2q_{||}a}}{\left(1 - \epsilon - \frac{\chi^0}{\epsilon_0 q_{||}}\right)^2 e^{-2q_{||}a} - \left(1 + \epsilon - \frac{\chi^0}{\epsilon_0 q_{||}}\right)^2} . \tag{5.21}$$

The normal modes are obtained by setting the denominator equal to zero, that is,

$$1 - \frac{\chi^0}{\epsilon_0 q_{||}} = \epsilon \left[\frac{e^{-q_{||}a} \pm 1}{e^{-q_{||}a} \mp 1}\right] . \tag{5.22}$$

In the limit $q_{||}a \to \infty$, we obtain the surface response from Eq. (5.21) for a single plane as

$$g_{sp}(q_{||}, \omega) = 1 - \frac{2}{1 + \epsilon - \frac{\chi^0}{\epsilon_0 q_{||}}}, \tag{5.23}$$

which was first derived in the paper by Persson [6]. The poles of the surface response function are solutions of a dispersion equation which agrees with that obtained by Eguiluz, Lee, Quinn, and Chiu [7].

If we set $\chi^0 = 0$ in Eq. (5.21), we obtain the surface response function for a slab of dielectric material as

$$\begin{aligned} g_{slab}(q_{||}, \omega) &= \frac{2\sinh(q_{||}a)}{[(1-\epsilon)/(1+\epsilon)]\, e^{-q_{||}a} - [(1+\epsilon)/(1-\epsilon)]\, e^{q_{||}a}} \\ &= 2\beta\left(\frac{\epsilon - 1}{\epsilon + 1}\right) e^{-q_{||}a}\sinh(q_{||}a), \end{aligned} \tag{5.24}$$

where

$$\beta \equiv \left[1 - \left(\frac{\epsilon - 1}{\epsilon + 1}\right)^2 e^{-2q_{||}a}\right]^{-1}. \tag{5.25}$$

In the limit $a \to \infty$, Eq. (5.24) becomes

$$g_{slab}(q_{||}, \omega) \to \frac{\epsilon - 1}{\epsilon + 1}. \tag{5.26}$$

Let us now use the long-wavelength approximation

$$\chi^0(q_{||}, \omega) \approx \frac{n_{2D} e^2 q_{||}^2}{m^* \omega^2} \tag{5.27}$$

for the 2DEG with electron effective mass m^* and sheet density n_{2D} in Eq. (5.22). We obtain the dispersion relation for plasmons in the long-wavelength limit as

$$\omega_+^2 = \frac{n_{2D} e^2 q_{||}}{2\epsilon_0 \epsilon\, m^*}\left[1 + e^{-q_{||}a}\right], \tag{5.28}$$

$$\omega_-^2 = \frac{n_{2D} e^2 q_{||}}{2\epsilon_0 \epsilon\, m^*}\left[1 - e^{-q_{||}a}\right]. \tag{5.29}$$

5.1.3 A Dielectric Slab

We now use the method of images to determine the Coulomb interaction $V_c(\boldsymbol{r}, \boldsymbol{r}')$ for a dielectric slab, whose surfaces are planar and perpendicular to the z-axis. In the case of a semi-infinite geometry, the potential is the sum of a "bulk" term and an "image" term depending on the distance between $\boldsymbol{r}$ and the image of $\boldsymbol{r}'$. In the case of a film, infinitely many image points $\mathbf{r}'_\lambda = (\mathbf{r}'_{||}, -z')$ and $(\mathbf{r}'_{||}, \pm z' + 2ja)$, where $j = \pm 1, \pm 2, \ldots$ and a is the thickness of the film, must be introduced to satisfy the boundary conditions. Consequently, the Coulomb interaction $V_{\text{film}}(z, z'; q_{||})$ for a film can be written as a sum of image contributions depending on $\mathbf{r} - \mathbf{r}'_\lambda$. For this case, we have

$$V_{\text{film}}(z, z'; q_{||}) = \frac{e^2}{2\epsilon_0 \epsilon q_{||}} \left(\Phi_1 + \Phi_2^< + \Phi_2^>\right), \tag{5.30}$$

where

$$\begin{aligned} \Phi_1 &= \sum_{j=-\infty}^{\infty} \alpha^{2|j|} e^{-q_{||}|z-z'-2ja|} \\ &= \beta \left[e^{-q_{||}|z-z'|} + \alpha^2 e^{-2q_{||}a} e^{q_{||}|z-z'|} \right], \end{aligned} \tag{5.31}$$

$\alpha \equiv (\epsilon - 1)/(\epsilon + 1)$, and β was defined above in Eq. (5.25). Also,

$$\Phi_2^< = \sum_{j=0}^{\infty} \alpha^{2j+1} e^{-q_{||}|z+z'+2ja|} = \beta \alpha e^{-q_{||}|z+z'|}, \tag{5.32}$$

$$\Phi_2^> = \sum_{j=1}^{\infty} \alpha^{2j-1} e^{-q_{||}|z+z'-2ja|} = \beta \alpha e^{-2q_{||}a} e^{q_{||}|z+z'|}. \tag{5.33}$$

Therefore, we eventually obtain

$$\begin{aligned} V_{\text{film}}(z, z'; q_{||}) = \frac{\beta e^2}{2\epsilon_0 \epsilon q_{||}} \Big[& e^{-q_{||}|z-z'|} + \alpha e^{-q_{||}|z+z'|} \\ & + \alpha^2 e^{-2q_{||}a} e^{q_{||}|z-z'|} + \alpha e^{-2q_{||}a} e^{q_{||}|z+z'|} \Big]. \end{aligned} \tag{5.34}$$

5.1.4
A Layered 2DEG System

We now turn to calculating the surface response function for a layered 2DEG structure with separation a between the layers. The electron–electron interaction $V_c(z, z'; q_{||})$ is given by Eq. (5.34). The total potential at a point z is the sum of the external and induced potentials, where the induced potential is

$$\phi_{\text{ind}}^>(z) = \int dz' \left[\epsilon^{-1}(z, z') - \delta(z - z')\right] \phi_0(z'). \tag{5.35}$$

Here, $\phi_0(z)$ is the external electrostatic potential for a slab of dielectric material and is obtained from Eqs. (5.16), (5.17), and (5.24). The inverse dielectric function in Eq. (5.35) is given by

$$\epsilon^{-1}(z, z') = \delta(z - z') + \int dz'' V_c(z, z'') \chi(z'', z') \tag{5.36}$$

with the polarization function obeying

$$\chi(z,z') = \sum_{j,j'} \delta(z - ja)\tilde{\chi}(ja, j'a)\delta\left(z' - j'a\right) . \tag{5.37}$$

In the RPA, $\tilde{\chi}(ja, j'a)$ is a solution of

$$\begin{aligned}\tilde{\chi}(ja, j'a) &= \tilde{\chi}^{(0)}(ja, ja)\delta_{j,j'} + \tilde{\chi}^{(0)}(ja, ja) \\ &\quad \times \sum_{j''} V_c(ja, j''a)\tilde{\chi}(j''a, j'),\end{aligned} \tag{5.38}$$

where $\tilde{\chi}^{(0)}(ja, j'a)$ is the single-particle polarization function.

Equations (5.35)–(5.38) show that

$$\phi^{>}_{\text{ind}}(z) = \sum_{j,j'} V_c(z, ja)\tilde{\chi}(ja, j'a)\phi_0(j'a) . \tag{5.39}$$

However, $\phi_{<}(0) = [1 - g_{\text{LEG}}(q_{||}, \omega)]$ in Eq. (5.13) must match $[1 - g_{\text{slab}}(q_{||}, \omega)] + \phi^{>}_{\text{ind}}(0)$. We then obtain the surface response function for the layered EG to be

$$g_{\text{LEG}}(q_{||}, \omega) = g_{\text{slab}}(q_{||}, \omega) - \sum_{j,j'} V_c(0, ja)\tilde{\chi}(ja, j'a)\phi_0(j'a) . \tag{5.40}$$

This closed-form analytic solution must be used in conjunction with Eq. (5.38) for $\tilde{\chi}(ja, j'a)$ to obtain the surface response function. For example, for a biplane, we have $j, j' = 0, 1$ in Eq. (5.38) and we must solve for $\tilde{\chi}(a, a)$, $\tilde{\chi}(0, 0)$, $\tilde{\chi}(a, 0)$ and $\tilde{\chi}(0, a)$. In the symmetric case, $\tilde{\chi}(a, a) = \tilde{\chi}(0, 0)$ and $\tilde{\chi}(a, 0) = \tilde{\chi}(0, a)$. A straightforward calculation shows that

$$\tilde{\chi}(a, a) = \left[\frac{\tilde{\chi}^{(0)}(q_{||}, \omega)}{D(q_{||}, \omega)}\right]\left[1 - V_{00}(q_{||})\tilde{\chi}^{(0)}(q_{||}, \omega)\right], \tag{5.41}$$

$$\tilde{\chi}(a, 0) = \left[\frac{\tilde{\chi}^{(0)}(q_{||}, \omega)}{D(q_{||}, \omega)}\right] V_{01}(q_{||})\tilde{\chi}^{(0)}(q_{||}, \omega) , \tag{5.42}$$

where $\tilde{\chi}^{(0)} = \tilde{\chi}^{(0)}(0, 0)$ is the single-particle response function on either plane, $V_{00} = V_c(0, 0)$ and $V_{01} = V_c(0, a)$. Also, we have introduced the dispersion function

$$D(q_{||}, \omega) = \left[1 - V_{00}(q_{||})\tilde{\chi}^{(0)}(q_{||}, \omega)\right]^2 - \left[V_{01}(q_{||})\tilde{\chi}^{(0)}(q_{||}, \omega)\right]^2 . \tag{5.43}$$

5.2 Electron Energy Loss for a Planar Surface

The imaginary part of the surface response function Im(g) can be identified with the power absorption in the semiconductor due to the electronic excitations induced by the evanescent external potential. The power absorption is obtained by integrating the Poynting vector over the surface and over time [8]. This gives

$$\Delta\mathcal{E} = \epsilon_0 \int d^2 \mathbf{r}_{||} \int_{-\infty}^{\infty} dt \left[\frac{\partial\phi^*(\mathbf{r}, t)}{\partial t}\frac{\partial\phi(\mathbf{r}, t)}{\partial z}\right]_{z=0} , \tag{5.44}$$

where $\phi = \phi_{\text{ind}} + \phi_{\text{ext}}$ from Eq. (5.13) is the total potential with

$$\phi(\mathbf{r}, t) = \int \frac{d^2 \mathbf{q}_{||}}{(2\pi)^2} \int_{-\infty}^{\infty} \frac{d\omega}{2\pi} \left[e^{-q_{||} z} - g(q_{||}, \omega) e^{q_{||} z} \right] \times e^{i \mathbf{q}_{||} \cdot \mathbf{r}_{||} - i\omega t} \tilde{\phi}_{\text{ext}}(\mathbf{q}_{||}, \omega) . \quad (5.45)$$

Substituting ϕ into the expression for $\Delta\mathcal{E}$, after some algebra, we obtain

$$\Delta\mathcal{E} = \frac{(Z^* e)^2}{2\epsilon_0} \int \frac{d^2 \mathbf{q}_{||}}{(2\pi)^2} \int_{-\infty}^{\infty} \frac{d\omega}{2\pi} \frac{\omega}{q_{||}} \left| \mathcal{F}(\mathbf{q}_{||}, \omega) \right|^2 \text{Im} \left[g(q_{||}, \omega) \right] , \quad (5.46)$$

where $\mathcal{F}(\mathbf{q}_{||}, \omega)$ is given by Eq. (5.4) and $\text{Im}\left[g(q_{||}, \omega)\right]$ is called the loss function.

5.2.1
Transfer-Matrix Method

We now generalize the method of calculation for the surface response function to a multi-layer 2DEG superlattice with arbitrary separation between the layers. The 2DEG is embedded in a material with background dielectric constant ϵ_{b}. Suppose the layers are located at $z = z_j$ where $j = 0, 1, 2, \ldots, L$. Generalizing the notation used in Section 5.1.2, we denote the potential on both sides of the interface of the layer at $z = z_j$ for $1 \leq j \leq L - 1$

$$\begin{aligned} \phi_{j<} &= A_j e^{-q_{||}(z - z_j)} + B_j e^{q_{||}(z - z_j)} , \\ \phi_j &= a_j e^{-q_{||}(z - z_j)} + b_j e^{q_{||}(z - z_j)} , \\ \phi_{j>} &= t_j e^{-q_{||}(z - z_j)} + r_j e^{q_{||}(z - z_j)} . \end{aligned} \quad (5.47)$$

By making use of standard electromagnetic theory boundary condition at $z = z_j$, we relate the pairs of coefficients (A_j, B_j) and (A_{j+1}, B_{j+1}) by

$$\begin{pmatrix} A_{j+1} \\ B_{j+1} \end{pmatrix} = \overleftrightarrow{T}_j \begin{pmatrix} A_j \\ B_j \end{pmatrix} , \quad (5.48)$$

where the transfer-matrix $\overleftrightarrow{T}_j$ satisfies

$$\begin{aligned} \frac{1}{2}\text{Tr}\left(\overleftrightarrow{T}_j\right) &= \cosh(q_{||} d_j) - \alpha \sinh(q_{||} d_j) , \\ \text{Det}\left(\overleftrightarrow{T}_j\right) &= \frac{1}{8}\left(\epsilon_{\text{b}}^2 + 7\right) \equiv \eta(\epsilon_{\text{b}}) , \end{aligned} \quad (5.49)$$

$d_j = z_{j+1} - z_j$, and $\alpha = e^2 \chi^{(0)} / 2\epsilon_0 \epsilon_{\text{b}} q_{||}$. At the first layer $z = z_0$, we have

$$\begin{pmatrix} A_1 \\ B_1 \end{pmatrix} = \overleftrightarrow{T}_0 \begin{pmatrix} 1 \\ -g \end{pmatrix} . \quad (5.50)$$

Combining these equations, we obtain

$$\begin{pmatrix} A_{j+1} \\ B_{j+1} \end{pmatrix} = \overleftrightarrow{T}_j \otimes \overleftrightarrow{T}_{j-1} \otimes \ldots \otimes \overleftrightarrow{T}_1 \otimes \overleftrightarrow{T}_0 \begin{pmatrix} 1 \\ -g \end{pmatrix} \equiv \overleftrightarrow{\mathcal{M}} \begin{pmatrix} 1 \\ -g \end{pmatrix} , \tag{5.51}$$

which defines the 2×2 total transfer-matrix $\overleftrightarrow{\mathcal{M}}$. Here, $\otimes$ represents the product of two matrices. The potential at the last layer $z = z_L$, that is,

$$\phi_{L>} = t_L e^{q_{||}(z-z_L)} , \tag{5.52}$$

yields

$$\begin{pmatrix} A_L \\ B_L \end{pmatrix} = \frac{1}{2} \begin{pmatrix} 1 & 1 \\ 1 & -1 \end{pmatrix} \begin{pmatrix} 1 \\ 1 - 2\epsilon_b \alpha \end{pmatrix} t_L . \tag{5.53}$$

Eliminating the coefficient t_L by making use of Eqs. (5.51) and (5.53), we obtain the surface response function for the superlattice as

$$g(q_{||}, \omega) = \frac{\epsilon_b \alpha \mathcal{M}_{11} - (1 - \epsilon_b \alpha)\mathcal{M}_{21}}{\epsilon_b \alpha \mathcal{M}_{12} - (1 - \epsilon_b \alpha)\mathcal{M}_{22}} . \tag{5.54}$$

Let us now apply these results to the Fibonacci superlattice with lattice spacing in a sequence $abaab\ldots$ [9, 10]. This gives

$$\begin{aligned}
&\overleftrightarrow{\mathcal{M}}_1 = \overleftrightarrow{T}_a , \\
&\overleftrightarrow{\mathcal{M}}_2 = \overleftrightarrow{T}_b \otimes \overleftrightarrow{T}_a , \\
&\overleftrightarrow{\mathcal{M}}_3 = \overleftrightarrow{T}_a \otimes \overleftrightarrow{T}_b \otimes \overleftrightarrow{T}_a = \overleftrightarrow{\mathcal{M}}_1 \otimes \overleftrightarrow{\mathcal{M}}_2 , \\
&\overleftrightarrow{\mathcal{M}}_4 = \overleftrightarrow{T}_b \otimes \overleftrightarrow{T}_a \otimes \overleftrightarrow{T}_a \otimes \overleftrightarrow{T}_b \otimes \overleftrightarrow{T}_a = \overleftrightarrow{\mathcal{M}}_2 \otimes \overleftrightarrow{\mathcal{M}}_3 , \\
&\quad \vdots \\
&\overleftrightarrow{\mathcal{M}}_{j+1} = \overleftrightarrow{\mathcal{M}}_{j-1} \otimes \overleftrightarrow{\mathcal{M}}_j .
\end{aligned} \tag{5.55}$$

Now, introduce the notation

$$\text{Det}\left(\overleftrightarrow{\mathcal{M}}_j\right) = G_j > 0 , \quad x_j = \frac{\text{Tr}\left(\overleftrightarrow{\mathcal{M}}_j\right)}{2\sqrt{G_j}} . \tag{5.56}$$

Then, making use of the identity

$$\overleftrightarrow{\mathcal{M}}_{j+1} + G_j \overleftrightarrow{\mathcal{M}}_{j-2}^{-1} = \overleftrightarrow{\mathcal{M}}_{j-1} \otimes \overleftrightarrow{\mathcal{M}}_j + G_j \overleftrightarrow{\mathcal{M}}_{j-1} \otimes \overleftrightarrow{\mathcal{M}}_j^{-1} , \tag{5.57}$$

we obtain the recursion relation [11]

$$x_{j+1} = 2x_j x_{j-1} - x_{j-2} . \tag{5.58}$$

5.2.2 Motion Parallel to the Surface

In this case, we have

$$\mathcal{F}\left(\boldsymbol{q}_{||},\omega\right)=\int\limits_{-\Delta T/2}^{\Delta T/2} dt e^{q_{||}Z_0} e^{i(\omega-\boldsymbol{q}_{||}\cdot\boldsymbol{v}_{||})t}\,, \tag{5.59}$$

where $Z(t)=Z_0<0$, $\boldsymbol{R}_{||}(t)=\boldsymbol{v}_{||}t$ and $\Delta T\to\infty$. We then have

$$\left|\mathcal{F}(\boldsymbol{q}_{||},\omega)\right|^2=2\pi\Delta T e^{-2q_{||}|Z_0|}\delta(\omega-\boldsymbol{q}_{||}\cdot\boldsymbol{v}_{||})\,, \tag{5.60}$$

which leads to

$$\frac{\Delta\mathcal{E}}{\Delta T}=\frac{(Z^*e)^2}{2\epsilon_0}\int\frac{d^2\boldsymbol{q}_{||}}{(2\pi)^2}e^{-2q_{||}|Z_0|}\left(\frac{\boldsymbol{q}_{||}\cdot\boldsymbol{v}_{||}}{q_{||}}\right)\text{Im}\left[g(\boldsymbol{q}_{||},\boldsymbol{q}_{||}\cdot\boldsymbol{v}_{||})\right]. \tag{5.61}$$

Furthermore, let us define a friction parameter $\overleftrightarrow{\eta}$ for a force $\boldsymbol{f}$ in a general way by

$$\boldsymbol{f}=-\overleftrightarrow{\eta}[\boldsymbol{R}(t)]\cdot\boldsymbol{v}(t)\,. \tag{5.62}$$

Therefore, the energy loss is

$$\Delta\mathcal{E}=\int\limits_{-\Delta T/2}^{\Delta T/2}dt\,\boldsymbol{f}\cdot\boldsymbol{v}(t)=-\int\limits_{-\Delta T/2}^{\Delta T/2}dt\boldsymbol{v}(t)\cdot\overleftrightarrow{\eta}[\boldsymbol{R}(t)]\cdot\boldsymbol{v}(t)\,. \tag{5.63}$$

In the fast particle approximation, $\boldsymbol{v}$ is independent of time and the friction parameter is weakly dependent on time and becomes a scalar. In this case, we then have

$$\frac{\Delta\mathcal{E}}{\Delta T}=-\eta v^2\,, \tag{5.64}$$

that is,

$$\eta=-\frac{1}{v^2}\frac{\Delta\mathcal{E}}{\Delta T}\,. \tag{5.65}$$

5.2.3 Motion Perpendicular to the Surface

Assume that the charged particle recedes from the surface so that $Z(t)=Z_0-v_\perp t$ with $Z_0<0$ and $\boldsymbol{R}_{||}(t)=0$ for $0\le t\le t_0$. Then, we have

$$\begin{aligned}\mathcal{F}(\boldsymbol{q}_{||},\omega)&=\int\limits_0^{t_0}dt e^{-q_{||}(|Z_0|+v_\perp t)}e^{i\omega t}\\&=\frac{e^{-q_{||}|Z_0|}}{v_\perp q_{||}-i\omega}\left[1-e^{-(v_\perp q_{||}-i\omega)t_0}\right].\end{aligned} \tag{5.66}$$

If we let $t_0 \to \infty$, then we get $|\mathcal{F}(\boldsymbol{q}_{||}, \omega)|^2 = e^{-2q_{||}|Z_0|}/(\omega^2 + v_\perp^2 q_{||}^2)$, which yields

$$\Delta\mathcal{E} = \frac{(Z^* e)^2}{2\epsilon_0} \int \frac{d^2 \boldsymbol{q}_{||}}{(2\pi)^2} \int_{-\infty}^{\infty} \frac{d\omega}{2\pi} \left(\frac{\omega}{q_{||}}\right) \frac{e^{-2q_{||}|Z_0|}}{\omega^2 + v_\perp^2 q_{||}^2} \operatorname{Im}\left[g(q_{||}, \omega)\right] . \quad (5.67)$$

We note that $g(q_{||}, \omega)$ has contributions from both particle–hole and collective excitations.

5.2.4
The Inverse Dielectric Function Formalism

Let us consider a particle with charge $Z^* e$ moving with a parallel velocity $\boldsymbol{v}_{||}$ near a medium with a planar surface located in the $z = 0$ plane. The particle is initially located at $\boldsymbol{r}_0 = (0, 0, z_0)$ on the z-axis. The screened potential at the space-time point $1 = (\boldsymbol{r}_1, t_1)$ due to an external potential U is given by

$$V(1) = \int (d2) \epsilon^{-1}(1, 2)\, U(2) = \int dz_2 \int \frac{d^2 \boldsymbol{q}_{||}}{(2\pi)^2} \int \frac{d\omega}{2\pi} \times e^{i\boldsymbol{q}_{||}\cdot \boldsymbol{r}_1 - i\omega t_1} \epsilon^{-1}(z_1, z_2; \boldsymbol{q}_{||}, \omega)\, V_c(z_2, q_{||}; \omega) \,, \quad (5.68)$$

where

$$U(\boldsymbol{r}, t) = \frac{Z^* e}{4\pi\epsilon_0\epsilon_b |\boldsymbol{r} - \boldsymbol{v}_{||} t - z_0 \hat{\boldsymbol{z}}|} \,, \quad (5.69)$$

$$V_c(z, q_{||}; \omega) = \frac{\pi Z^* e}{\epsilon_0 \epsilon_b q_{||}} e^{-q_{||}|z - z_0|} \delta\left(\omega - \boldsymbol{q}_{||} \cdot \boldsymbol{v}_{||}\right) , \quad (5.70)$$

and ϵ_b is the background dielectric constant for the medium. The induced dynamic charge density is

$$\rho(1) = -\epsilon_0\epsilon_b \nabla_1^2 \int dz_4 \int \frac{d^2 \boldsymbol{q}'_{||}}{(2\pi)^2} \int \frac{d\omega'}{2\pi} e^{i\boldsymbol{q}'_{||}\cdot \boldsymbol{r}_1 - i\omega' t_1} \times \left[\epsilon^{-1}(z_1, z_4; \boldsymbol{q}'_{||}, \omega') - \delta(z_1 - z_4)\right] V_c\left(z_4, q'_{||}; \omega'\right) . \quad (5.71)$$

Therefore, the force acting on the moving electron parallel to the surface is

$$\begin{aligned} \boldsymbol{F}_{||} &= \int d^3 \boldsymbol{r}_1 \rho(1) \frac{\partial}{\partial \boldsymbol{r}_{1||}} V(1) \\ &= -i\epsilon_0\epsilon_b \int dz_1 \int dz_2 \int dz_4 e^{-q_{||}(|z_2 - z_0| + |z_4 - z_0|)} \int \frac{d^2 \boldsymbol{q}_{||}}{(2\pi)^4} \\ &\quad \times \boldsymbol{q}_{||} \epsilon^{-1}(z_1, z_2; \boldsymbol{q}_{||}, \boldsymbol{q}_{||} \cdot \boldsymbol{v}_{||}) \left(\frac{\pi Z^* e}{\epsilon_0 \epsilon_b q_{||}}\right)^2 \\ &\quad \times \left(\frac{\partial^2}{\partial z_1^2} - q_{||}^2\right) \left[\epsilon^{-1}(z_1, z_4; -\boldsymbol{q}_{||}, -\boldsymbol{q}_{||} \cdot \boldsymbol{v}_{||}) - \delta(z_1 - z_4)\right] . \end{aligned} \quad (5.72)$$

The double ϵ^{-1} term vanishes and we have

$$\begin{aligned} \boldsymbol{F}_{||} = i\epsilon_0\epsilon_b \int dz_1 \int dz_2 \int dz_4 e^{-q_{||}(|z_2-z_0|+|z_4-z_0|)} \\ \int \frac{d^2\boldsymbol{q}_{||}}{(2\pi)^4}\epsilon^{-1}(z_1, z_2; \boldsymbol{q}_{||}, \boldsymbol{q}_{||}\cdot\boldsymbol{v}_{||}) \\ \times \boldsymbol{q}_{||}\left(\frac{\pi Z^* e}{\epsilon_0\epsilon_b q_{||}}\right)^2 \left(\frac{\partial^2}{\partial z_1^2} - q_{||}^2\right)\delta(z_1 - z_4)\,. \end{aligned} \tag{5.73}$$

However, using

$$\begin{aligned} \frac{\partial^2}{\partial z_1^2}\left[\int dz_4 \delta(z_1 - z_4) e^{-q_{||}|z_4 - z_0|}\right] \\ = -2q_{||}\delta(z_1 - z_0)e^{-q_{||}|z_1-z_0|} + q_{||}^2 e^{-q_{||}|z_1 - z_0|}\,, \end{aligned} \tag{5.74}$$

we obtain

$$\begin{aligned} \boldsymbol{F}_{||} &= -i\frac{(Z^*e)^2}{2\epsilon_0\epsilon_b}\int dz_2 \int \frac{d^2\boldsymbol{q}_{||}}{(2\pi)^2}\left(\frac{\boldsymbol{q}_{||}}{q_{||}}\right) e^{-q_{||}|z_2-z_0|}\epsilon^{-1}(z_0, z_2; \boldsymbol{q}_{||}, \boldsymbol{q}_{||}\cdot\boldsymbol{v}_{||}) \\ &= \frac{(Z^*e)^2}{2\epsilon_0\epsilon_b}\int dz_2 \int \frac{d^2\boldsymbol{q}_{||}}{(2\pi)^2}\left(\frac{\boldsymbol{q}_{||}}{q_{||}}\right) e^{-q_{||}|z_2-z_0|} \\ &\quad \times \operatorname{Im}\left[\epsilon^{-1}(z_0, z_2; \boldsymbol{q}_{||}, \boldsymbol{q}_{||}\cdot\boldsymbol{v}_{||})\right], \end{aligned} \tag{5.75}$$

which gives the rate of loss of energy to the moving particle [12] as

$$\begin{aligned} P = \boldsymbol{F}_{||}\cdot\boldsymbol{v}_{||} = \frac{(Z^*e)^2}{2\epsilon_0\epsilon_b}\int dz_2 \int \frac{d^2\boldsymbol{q}_{||}}{(2\pi)^2}\left(\frac{\boldsymbol{q}_{||}\cdot\boldsymbol{v}_{||}}{q_{||}}\right) e^{-q_{||}|z_2-z_0|} \\ \times \operatorname{Im}\left[\epsilon^{-1}(z_0, z_2; \boldsymbol{q}_{||}, \boldsymbol{q}_{||}\cdot\boldsymbol{v}_{||})\right]. \end{aligned} \tag{5.76}$$

The inverse dielectric function for a 2DEG located at $z = 0$ is

$$\epsilon^{-1}(z_1, z_2; \boldsymbol{q}_{||}, \omega) = \delta(z_1 - z_2) - \delta(z_2)\left[\frac{e^{-q_{||}|z_1|}}{1 + \alpha_{2D}(q_{||}, \omega)}\right], \tag{5.77}$$

where $\alpha_{2D}(q_{||}, \omega)$ is the polarization function of the 2DEG and its real and imaginary parts at $T = 0$ K are given by

$$\begin{aligned} \operatorname{Im}\left[\alpha_{2D}(q_{||}, \omega)\right] = \frac{m^* e^2}{\pi\epsilon_0\epsilon_b\hbar^3 q_{||}}\sqrt{\frac{m^*}{2q_{||}^2}}\sum_{\pm}(\pm 1)\theta\left[E_F - \frac{m^*}{2q_{||}^2}\left(\omega \mp \frac{\hbar q_{||}^2}{2m^*}\right)^2\right] \\ \times \sqrt{E_F - \frac{m^*}{2q_{||}^2}\left(\omega \mp \frac{\hbar q_{||}^2}{2m^*}\right)^2}\,, \end{aligned} \tag{5.78}$$

$$\begin{aligned} \operatorname{Re}\left[\alpha_{2D}(q_{||}, \omega)\right] = \frac{m^* e^2}{2\pi\epsilon_0\epsilon_b\hbar^3 q_{||}^3}\left\{\frac{\hbar q_{||}^2}{2} + m^*\sqrt{1 - \frac{2q_{||}^2 E_F}{m^*\left(\frac{\hbar q_{||}^2}{2m^*} - \omega\right)^2}}\right. \\ \left.\times\left(\omega - \frac{\hbar q_{||}^2}{2m^*}\right)\theta\left[\frac{m^*}{2q_{||}^2}\left(\frac{\hbar q_{||}^2}{2m^*} - \omega\right)^2 - E_F\right] + (\omega \to -\omega)\right\}, \end{aligned} \tag{5.79}$$

where $(\omega \to -\omega)$ represents the term with the replacement of ω to $-\omega$ in the first term and $\theta(x)$ is the unit step function.

In the limit of a *slowly moving* particle, that is, small $v_{||}$, the plasmon excitations are not important and

$$\mathrm{Im}\left[\frac{1}{1+\alpha_{2D}(q_{||},\omega)}\right] \approx -\frac{\mathrm{Im}\left[\alpha_{2D}(q_{||},\omega)\right]}{\left\{1+\mathrm{Re}\left[\alpha_{2D}(q_{||},0)\right]\right\}^2}\,, \tag{5.80}$$

so that the rate of loss of energy in this case is given approximately by

$$\begin{aligned} P = {} & \frac{(Z^*e)^2 v_{||}}{4\pi\epsilon_0\epsilon_b}\int_0^{\sqrt{8m^*E_F}/\hbar}\frac{q_{||}dq_{||}}{2\pi}\frac{e^{-2q_{||}|z_0|}}{\left[1+\frac{m^*e^2}{2\pi\epsilon_0\epsilon_b\hbar^2 q_{||}}\right]^2}\int_0^{2\pi}d\phi\cos\phi \\ & \times\left(\frac{m^*e^2}{\pi\epsilon_0\epsilon_b\hbar^3 q_{||}}\right)\sqrt{\frac{m^*}{2q_{||}^2}}\sum_{\pm}(\pm 1)\theta\left[E_F-\frac{m^*}{2q_{||}^2}\left(q_{||}v_{||}\cos\phi\mp\frac{\hbar q_{||}^2}{2m^*}\right)^2\right] \\ & \times\sqrt{E_F-\frac{m^*}{2q_{||}^2}\left(q_{||}v_{||}\cos\phi\mp\frac{\hbar q_{||}^2}{2m^*}\right)^2}\,. \end{aligned} \tag{5.81}$$

5.3 Plasma Instability for a Planar Surface

When a charged particle current passes by in the vicinity of a surface, it serves to excite plasma modes which may subsequently become unstable. The drift-induced instability has been studied theoretically for semiconductor superlattices [13, 14], the 2DEG [15] and high-T_c superconductors [16] and nanotubes [17]. This was carried out by solving the plasma dispersion formula in the complex frequency domain. That region of plasma frequency with an imaginary part has a finite lifetime due to the instability of the collective-mode excitation.

In the RPA, the dynamic dielectric function for a 2DEG is given in the preceding chapter as $\epsilon(q_{||},\omega) = 1 - v_c(q_{||})\chi(q_{||},\omega)$ where $v_c(q_{||})$ is the 2D Coulomb interaction and $\chi(q_{||},\omega)$ is the 2D polarization function [18]. It may be shown that the dispersion equation for plasmons in a superlattice of period a is obtained by solving

$$1 - v_c(q_{||})\chi^{(0)}(q_{||},\omega)S(q_{||},k_z) = 0\,, \tag{5.82}$$

where $\chi^{(0)}(q_{||},\omega)$ is the polarization function for a single layer of 2DEG. Also, $S(q_{||},k_z)$ is the structure factor determining the phase coherence of the collective excitation in different layers given by

$$S(q_{||},k_z) = \sum_{\ell'} e^{-q_{||}|\ell-\ell'|a - ik_z(\ell-\ell')a} = \frac{\sinh(q_{||}a)}{\cosh(q_{||}a)-\cos(k_z a)}\,, \tag{5.83}$$

where $\ell, \ell' = 0, \pm 1, \pm 2, \ldots, \pm\infty$ are the possible indices of the layers in the superlattice. Note that the periodicity ensures that $S(q_{||},k_z)$ is independent of the layer index ℓ. It has been shown that there is an enhanced plasmon instability in superlattices. This is due to oscillations occurring in-phase in different layers. The enhancement is demonstrated in a larger imaginary part for the frequency of plasmon excitation.

The plasmon instability may be shown to exist in an alternative calculation. Let us consider a system shown in Figure 5.2. For $z < 0$, the half-space is filled with air. The half-space $z > 0$ is occupied by a semi-infinite doped semiconductor. The interface at $z = 0$ consists of a current-driven conducting sheet and a metal grating on top of it. Optically, this sheet and the grating can be considered to be in one plane. They can, however, still be electrically separated from each other by an energy barrier. Light incident from $z < 0$ is diffracted by the grating in both the reflection ($z < 0$) and the transmission ($z > 0$) regions. The diffraction (Bragg) modes of electromagnetic fields (EMFs) are produced by the induced optical polarization at the interface from both the conducting sheet and grating. In addition, all the Bragg modes of EMFs are mixed nonlocally with each other by the grating.

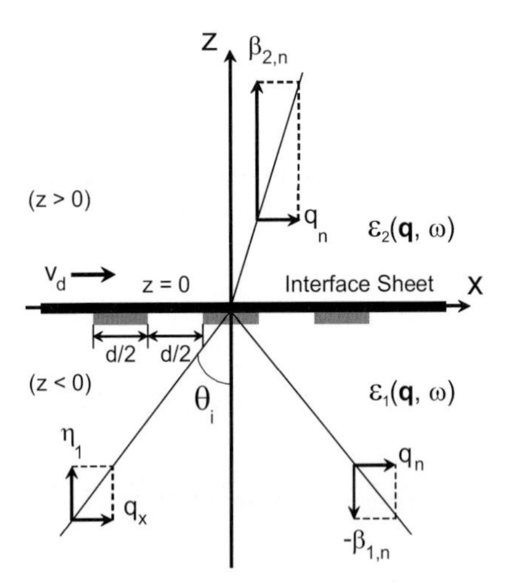

Figure 5.2 Schematic of the half-space of air ($z < 0$, region-1) and a semi-infinite conducting bulk ($z > 0$, region-2). The surface at $z = 0$ is covered by a conducting sheet (black) as well as a conductive grating (gray) with period d. The incident angle is θ_i on the air side and the bulk dielectric functions are $\epsilon_1(q, \omega) = 1$ for $z < 0$ and $\epsilon_2(q, \omega)$ for $z > 0$. The wave vector of the incident field is $(q_x, 0, \eta_1)$. The wave vector of the reflected field is $(q_n, 0, -\beta_{1,n})$, and the wave vector of the transmitted field is $(q_n, 0, \beta_{2,n})$. A DC current (indicated by an arrow) is driven in the conducting sheet along the x-direction with a drift velocity v_d.

The grating is periodic in the x-direction. Setting $q_y = 0$ and denoting the electric field below the interface by $\boldsymbol{E}_1$ and above by $\boldsymbol{E}_2$, we obtain [19, 20]

$$\boldsymbol{E}_1(q_n, \omega; z) = e^{i\eta_1^{\mathrm{T}} z} \begin{bmatrix} A_x^{\mathrm{T}}(q_x; \omega) \\ A_y^{\mathrm{T}}(q_x; \omega) \\ -\frac{q_x}{\eta_1^{\mathrm{T}}} A_x^{\mathrm{T}}(q_x; \omega) \end{bmatrix} \delta_{n,0} + \sum_{n'=-\infty}^{\infty} e^{-i\beta_{1,n+n'}^{\mathrm{T}} z} \begin{bmatrix} B_x^{\mathrm{T}}(q_{n+n'}; \omega) \\ B_y^{\mathrm{T}}(q_{n+n'}; \omega) \\ \frac{q_{n+n'}}{\beta_{1,n+n'}^{\mathrm{T}}} B_x^{\mathrm{T}}(q_{n+n'}; \omega) \end{bmatrix}, \quad \text{for } z < 0 \,, \tag{5.84}$$

$$\boldsymbol{E}_2(q_n, \omega; z) = \sum_{n'=-\infty}^{\infty} e^{i\beta_{2,n+n'}^{\mathrm{T}} z} \begin{bmatrix} C_x^{\mathrm{T}}(q_{n+n'}; \omega) \\ C_y^{\mathrm{T}}(q_{n+n'}; \omega) \\ -\frac{q_{n+n'}}{\beta_{2,n+n'}^{\mathrm{T}}} C_x^{\mathrm{T}}(q_{n+n'}; \omega) \end{bmatrix} + \sum_{n'=-\infty}^{\infty} e^{i\beta_{2,n+n'}^{L} z} \begin{bmatrix} C_x^{L}(q_{n+n'}; \omega) \\ 0 \\ \frac{\beta_{2,n+n'}^{L}}{q_{n+n'}} C_x^{L}(q_{n+n'}; \omega) \end{bmatrix}, \quad \text{for } z > 0 \,. \tag{5.85}$$

In this notation, n and n' are integers, the angular frequency of the incident electromagnetic field (EMF) is ω, the wave vector $q_x = (\omega/c)\sin\theta_i$, $\eta_1^{\mathrm{T}} = (\omega/c)\cos\theta_i$, θ_i is the incident angle, and $q_n = q_x + n(2\pi/d)$ where d is the period of the grating. In addition, for the transverse (T) EMF, we obtain $\beta_{1,n}^{\mathrm{T}}$ and $\beta_{2,n}^{\mathrm{T}}$ from the dispersion relation

$$\begin{bmatrix} \left(\beta_{1,n}^{\mathrm{T}}\right)^2 \\ \left(\beta_{2,n}^{\mathrm{T}}\right)^2 \end{bmatrix} = \left(\frac{\omega}{c}\right)^2 \begin{bmatrix} 1 \\ \epsilon_2^{\mathrm{T}}(q_n, \beta_{2,n}^{\mathrm{T}}; \omega) \end{bmatrix} - q_n^2 \,, \tag{5.86}$$

where the region-1 ($z < 0$) is filled with air. On the other hand, for the longitudinal (L) EMF, we obtain $\beta_{2,n}^{L}$ as a root of the equation

$$\epsilon_2^{L}(q_n, \beta_{2,n}^{L}; \omega) = 0 \,. \tag{5.87}$$

Moreover, the summation over n' in Eqs. (5.84) and (5.85) represents the contributions from all Bragg modes ($n' \neq 0$) of the reflected and transmitted electric fields in the presence of the conductive grating.

The magnetic field components can be simply obtained through the relation $\boldsymbol{H} = (-i/\omega\mu_0)\nabla \times \boldsymbol{E}^{\mathrm{T}}$. When $q_y = 0$, from Eq. (5.84), we get the magnetic field $\boldsymbol{H}_1$ below the interface

$$H_1^{x}(q_n, \omega; z) = \left(\frac{-i}{\omega\mu_0}\right) \Bigg[-i\eta_1^{\mathrm{T}} e^{i\eta_1^{\mathrm{T}} z} A_y^{\mathrm{T}}(q_x; \omega)\delta_{n,0} + \sum_{n'=-\infty}^{\infty} i\beta_{1,n+n'}^{\mathrm{T}} e^{-i\beta_{1,n+n'}^{\mathrm{T}} z} B_y^{\mathrm{T}}(q_{n+n'}; \omega) \Bigg], \tag{5.88}$$

$$H_1^y(q_n,\omega;z) = \left(\frac{-i}{\omega\mu_0}\right)\left(\frac{\omega^2}{c^2}\right)\left[\left(\frac{i}{\eta_1^{\rm T}}\right)e^{i\eta_1^{\rm T}z}A_x^{\rm T}(q_x;\omega)\delta_{n,0} + \sum_{n'=-\infty}^{\infty}\left(\frac{-i}{\beta_{1,n+n'}^{\rm T}}\right)e^{-i\beta_{1,n+n'}^{\rm T}z}B_x^{\rm T}(q_{n+n'};\omega)\right], \tag{5.89}$$

$$H_1^z(q_n,\omega;z) = \left(\frac{-i}{\omega\mu_0}\right)\left[iq_x e^{i\eta_1^{\rm T}z}A_y^{\rm T}(q_x;\omega)\delta_{n,0} + \sum_{n'=-\infty}^{\infty} iq_{n+n'}e^{-i\beta_{1,n+n'}^{\rm T}z}B_y^{\rm T}(q_{n+n'};\omega)\right]. \tag{5.90}$$

In addition, the magnetic field $\boldsymbol{H}_2$ above the interface can be calculated by means of Eq. (5.85)

$$H_2^x(q_n,\omega;z) = -\left(\frac{-i}{\omega\mu_0}\right)\sum_{n'=-\infty}^{\infty} i\beta_{2,n+n'}^{\rm T}e^{i\beta_{2,n+n'}^{\rm T}z}C_y^{\rm T}(q_{n+n'};\omega)\ , \tag{5.91}$$

$$H_2^y(q_n,\omega;z) = \left(\frac{-i}{\omega\mu_0}\right)\left(\frac{\omega^2}{c^2}\right)\sum_{n'=-\infty}^{\infty}\left(\frac{i}{\beta_{2,n+n'}^{\rm T}}\right)\epsilon_2^{\rm T}(q_{n+n'},\beta_{2,n+n'}^{\rm T};\omega) \times e^{i\beta_{2,n+n'}^{\rm T}z}C_x^{\rm T}(q_{n+n'};\omega)\ , \tag{5.92}$$

$$H_2^z(q_n,\omega;z) = \left(\frac{-i}{\omega\mu_0}\right)\sum_{n'=-\infty}^{\infty} iq_{n+n'}e^{i\beta_{2,n+n'}^{\rm T}z}C_y^{\rm T}(q_{n+n'};\omega)\,. \tag{5.93}$$

For an incident field with an *s*-polarization, we have $A_x^{\rm T}(q_x;\omega) = 0$ and $A_y^{\rm T}(q_x;\omega) = E_0$, where E_0 is the amplitude of the incident electric field. On the other hand, for an incident field with a *p*-polarization, we have $A_y^{\rm T}(q_x;\omega) = 0$ and $A_x^{\rm T}(q_x;\omega) = \eta_1^{\rm T}H_0/(\omega\epsilon_0)$, where H_0 is the amplitude of the incident magnetic field.

For a conducting bulk medium, we use the RPA to obtain the transverse dielectric function in the long-wavelength limit as

$$\epsilon_2^{\rm T}(q_x,q_z;\omega) = \epsilon_{\rm b}\left[1 - \frac{n_{\rm 3D}e^2}{\epsilon_0\epsilon_{\rm b}m^*\omega(\omega+i\gamma_0)}\right], \tag{5.94}$$

where $\epsilon_{\rm b}$ is the background dielectric constant of the bulk, $n_{\rm 3D}$ is the electron concentration, m^* is the effective mass of an electron in the bulk, and γ_0 is the homogeneous broadening describing the finite lifetime of excited electrons. In addition, by using the hydrodynamic model, the nonlocal longitudinal dielectric function in the long-wavelength limit can be expressed as

$$\epsilon_2^L(q_x,q_z;\omega) = \epsilon_{\rm b}\left\{1 - \frac{n_{\rm 3D}e^2}{\epsilon_0\epsilon_{\rm b}m^*\left[\omega(\omega+i\gamma_0) - \xi\left(q_x^2+q_z^2\right)\right]}\right\}, \tag{5.95}$$

where $\xi = 3v_{\rm F}^2/5$ and $v_{\rm F} = \hbar(3\pi^2 n_{\rm 3D})^{1/3}/m^*$ is the Fermi velocity of electrons in the bulk.

For the specular ($n = 0$) mode, the boundary conditions for the EMFs at the interface require

$$B_x^{\mathrm{T}}(q_x;\omega) - C_x^{\mathrm{T}}(q_x;\omega) - C_x^{L}(q_x;\omega) = -A_x^{\mathrm{T}}(q_x;\omega)\,, \tag{5.96}$$

$$B_y^{\mathrm{T}}(q_x;\omega) - C_y^{\mathrm{T}}(q_x;\omega) = -A_y^{\mathrm{T}}(q_x;\omega)\,, \tag{5.97}$$

$$\begin{aligned}
&-\frac{i\beta_1^{\mathrm{T}}c^2}{\omega^2}B_y^{\mathrm{T}}(q_x;\omega) - \frac{i\beta_2^{\mathrm{T}}c^2}{\omega^2}C_y^{\mathrm{T}}(q_x;\omega)\\
&-\sum_{n'=-\infty}^{\infty}\left[\bar{\chi}_s(q_{n'};\omega) + \frac{1}{2}\bar{\chi}_g(q_{n'};\omega)\right]B_y^{\mathrm{T}}(q_{n'};\omega)\\
&=\left[\bar{\chi}_s(q_x;\omega) + \frac{1}{2}\bar{\chi}_g(q_x;\omega) - \frac{i\eta_1^{\mathrm{T}}c^2}{\omega^2}\right]A_y^{\mathrm{T}}(q_x;\omega)\,,
\end{aligned} \tag{5.98}$$

$$\begin{aligned}
&-\frac{i}{\beta_1^{\mathrm{T}}}B_x^{\mathrm{T}}(q_x;\omega) - \frac{i}{\beta_2^{\mathrm{T}}}\epsilon_2^{\mathrm{T}}(q_x,\beta_2^{\mathrm{T}};\omega)C_x^{\mathrm{T}}(q_x;\omega)\\
&-\sum_{n'=-\infty}^{\infty}\left[\bar{\chi}_s(q_{n'};\omega) + \frac{1}{2}\bar{\chi}_g(q_{n'};\omega)\right]B_x^{\mathrm{T}}(q_{n'};\omega)\\
&=\left[\bar{\chi}_s(q_x;\omega) + \frac{1}{2}\bar{\chi}_g(q_x;\omega) - \frac{i}{\eta_1^{\mathrm{T}}}\right]A_x^{\mathrm{T}}(q_x;\omega)\,,
\end{aligned} \tag{5.99}$$

$$\begin{aligned}
&\frac{q_x}{\beta_2^{\mathrm{T}}}\left[\epsilon_2^{\mathrm{T}}(q_x,\beta_2^{\mathrm{T}};\omega) - \epsilon_b\right]C_x^{\mathrm{T}}(q_x;\omega) + \frac{\beta_2^{L}}{q_x}\epsilon_b C_x^{L}(q_x;\omega)\\
&+ i q_x\sum_{n'=-\infty}^{\infty}\left[\bar{\chi}_s(q_{n'};\omega) + \frac{1}{2}\bar{\chi}_g(q_{n'};\omega)\right]\\
&\times\left[C_x^{\mathrm{T}}(q_{n'};\omega) + C_x^{L}(q_{n'};\omega)\right] = 0\,.
\end{aligned} \tag{5.100}$$

The summation over n' in Eqs. (5.98)–(5.100) describes the excitation of diffracted EMFs through the specular mode due to an incident EMF by interacting with both the conductive grating ($\bar{\chi}_g$) and the current-driven conducting sheet ($\bar{\chi}_s$).

For the Bragg modes with even integers $n \neq 0$, the boundary conditions for the EMFs lead to

$$B_x^{\mathrm{T}}(q_n;\omega) - C_x^{\mathrm{T}}(q_n;\omega) - C_x^{L}(q_n;\omega) = 0\,, \tag{5.101}$$

$$B_y^{\mathrm{T}}(q_n;\omega) - C_y^{\mathrm{T}}(q_n;\omega) = 0\,, \tag{5.102}$$

$$\beta_{1,n}^{\mathrm{T}}B_y^{\mathrm{T}}(q_n;\omega) + \beta_{2,n}^{\mathrm{T}}C_y^{\mathrm{T}}(q_n;\omega) = 0\,, \tag{5.103}$$

$$\frac{1}{\beta_{1,n}^{\mathrm{T}}}B_x^{\mathrm{T}}(q_n;\omega) + \frac{1}{\beta_{2,n}^{\mathrm{T}}}\epsilon_2^{\mathrm{T}}(q_n,\beta_{2,n}^{\mathrm{T}};\omega)C_x^{\mathrm{T}}(q_n;\omega) = 0\,, \tag{5.104}$$

$$\frac{q_n}{\beta_{2,n}^{\mathrm{T}}}\left[\epsilon_2^{\mathrm{T}}(q_n,\beta_{2,n}^{\mathrm{T}};\omega) - \epsilon_b\right]C_x^{\mathrm{T}}(q_n;\omega) + \frac{\beta_{2,n}^{L}}{q_n}\epsilon_b C_x^{L}(q_n;\omega) = 0\,. \tag{5.105}$$

For the Bragg modes with odd integers n, the boundary conditions for the EMFs take the form of

$$B_x^{\mathrm{T}}(q_n;\omega) - C_x^{\mathrm{T}}(q_n;\omega) - C_x^L(q_n;\omega) = 0\,, \tag{5.106}$$

$$B_y^{\mathrm{T}}(q_n;\omega) - C_y^{\mathrm{T}}(q_n;\omega) = 0\,, \tag{5.107}$$

$$\begin{aligned}&-\frac{i\beta_{1,n}^{\mathrm{T}}c^2}{\omega^2}B_y^{\mathrm{T}}(q_n;\omega) - \frac{i\beta_{2,n}^{\mathrm{T}}c^2}{\omega^2}C_y^{\mathrm{T}}(q_n;\omega)\\ &-\frac{(-1)^{\frac{|n|-1}{2}}}{|n|\pi}\sum_{n'=-\infty}^{\infty}\bar{\chi}_g(q_{n+n'};\omega)B_y^{\mathrm{T}}(q_{n+n'};\omega) = 0\,,\end{aligned} \tag{5.108}$$

$$\begin{aligned}&-\frac{i}{\beta_{1,n}^{\mathrm{T}}}B_x^{\mathrm{T}}(q_n;\omega) - \frac{i}{\beta_{2,n}^{\mathrm{T}}}\epsilon_2^{\mathrm{T}}\left(q_n,\beta_{2,n}^{\mathrm{T}};\omega\right)C_x^{\mathrm{T}}(q_n;\omega)\\ &-\frac{(-1)^{\frac{|n|-1}{2}}}{|n|\pi}\sum_{n'=-\infty}^{\infty}\bar{\chi}_g(q_{n+n'};\omega)B_x^{\mathrm{T}}(q_{n+n'};\omega) = 0\,,\end{aligned} \tag{5.109}$$

$$\begin{aligned}&\frac{q_n}{\beta_{2,n}^{\mathrm{T}}}\left[\epsilon_2^{\mathrm{T}}(q_n,\beta_{2,n}^{\mathrm{T}};\omega) - \epsilon_{\mathrm{b}}\right]C_x^{\mathrm{T}}(q_n;\omega) + \frac{\beta_{2,n}^L}{q_n}\epsilon_{\mathrm{b}}C_x^L(q_n;\omega)\\ &+\frac{i(-1)^{\frac{|n|-1}{2}}}{|n|\pi}q_n\sum_{n'=-\infty}^{\infty}\bar{\chi}_g(q_{n+n'};\omega)\left[C_x^{\mathrm{T}}(q_{n+n'};\omega) + C_x^L(q_{n+n'};\omega)\right] = 0\,.\end{aligned} \tag{5.110}$$

The summation over n' in Eqs. (5.108)–(5.110) represents the nonlocal mixing among the Bragg modes ($n' \neq -n$) and the specular mode ($n' = -n$) by interacting with the conductive grating through induced sheet currents.

Equations (5.96)–(5.110) constitute a complete set of linear equations (or a linear matrix equation with a coefficient matrix and a source-term vector) with respect to the independent variables $B_x^{\mathrm{T}}(q_n,\omega)$, $B_y^{\mathrm{T}}(q_n,\omega)$, $C_x^{\mathrm{T}}(q_n,\omega)$, $C_y^{\mathrm{T}}(q_n,\omega)$, and $C_x^L(q_n,\omega)$ for $n = 0, \pm1, \pm2, \ldots$. The components of the source-term vector contain $A_x^{\mathrm{T}}(q_x,\omega)$ for the p-polarization or $A_y^{\mathrm{T}}(q_x,\omega)$ for the s-polarization.

For the conducting sheet and grating, their dielectric functions are given by $\bar{\chi}_s(q_x;\omega) = \chi_e(q_x;\omega - q_x v_{\mathrm{d}})$ and $\bar{\chi}_g(q_x;\omega) = \chi_e(q_x;\omega)$. Here, v_{d} is the drift velocity of electrons under a bias field and the electron sheet polarizability in the RPA is given by [21]

$$\begin{aligned}\chi_e(q_x;\omega) = \frac{2n_{2\mathrm{D}}e^2m_s^*}{\epsilon_0\hbar^2k_F|q_x|^3}&\left\{\left[2z - C_-\sqrt{(z-u)^2-1} - C_+\sqrt{(z+u)^2-1}\right]\right.\\ &\left.+\,i\left[D_-\sqrt{1-(z-u)^2} - D_+\sqrt{1-(z+u)^2}\right]\right\},\end{aligned} \tag{5.111}$$

where $n_{2\mathrm{D}}$ is the electron density, $k_{\mathrm{F}} = \sqrt{2\pi n_{2\mathrm{D}}}$ is the Fermi wave number of electrons in the sheet, and m_s^* is the effective mass of electrons in the sheet. Moreover, we have defined the notations in Eq. (5.111): $u = m_s^*\omega/\hbar k_{\mathrm{F}}|q_x|$, $z = |q_x|/2k_{\mathrm{F}}$, $C_+ = (z+u)/|z+u|$ and $D_+ = 0$ ($C_+ = 0$ and $D_+ = 1$) for $|z+u| > 1$

($|z+u|<1$), and $C_- = (z-u)/|z-u|$ and $D_- = 0$ ($C_- = 0$ and $D_- = 1$) for $|z-u|>1$ ($|z-u|<1$).

Equations (5.96)–(5.110) all together can be casted into the following matrix equation:

$$\overleftrightarrow{M}(q_x;\omega)\begin{bmatrix} B_x^{\mathrm{T}}(q_{-n_{\max}};\omega) \\ \vdots \\ B_x^{\mathrm{T}}(q_{n_{\max}};\omega) \\ B_y^{\mathrm{T}}(q_{-n_{\max}};\omega) \\ \vdots \\ B_y^{\mathrm{T}}(q_{n_{\max}};\omega) \\ C_x^{\mathrm{T}}(q_{-n_{\max}};\omega) \\ \vdots \\ C_x^{\mathrm{T}}(q_{n_{\max}};\omega) \\ C_y^{\mathrm{T}}(q_{-n_{\max}};\omega) \\ \vdots \\ C_y^{\mathrm{T}}(q_{n_{\max}};\omega) \\ C_x^{L}(q_{-n_{\max}};\omega) \\ \vdots \\ C_x^{L}(q_{n_{\max}};\omega) \end{bmatrix} = \begin{bmatrix} -A_x^{\mathrm{T}}(q_x;\omega) \\ -A_y^{\mathrm{T}}(q_x;\omega) \\ \frac{\omega}{c}\left[\bar{\chi}_s(q_x;\omega)+\frac{\bar{\chi}_g(q_x;\omega)}{2}-\frac{i\eta_1^{\mathrm{T}}c^2}{\omega^2}\right]A_y^{\mathrm{T}}(q_x;\omega) \\ \frac{\omega}{c}\left[\bar{\chi}_s(q_x;\omega)+\frac{\bar{\chi}_g(q_x;\omega)}{2}-\frac{i}{\eta_1^{\mathrm{T}}}\right]A_x^{\mathrm{T}}(q_x;\omega) \\ 0 \\ \vdots \\ 0 \end{bmatrix}, \tag{5.112}$$

where the cutoff $|n| \le n_{\max}$ is taken for integer n, $q_n = q_x + n(2\pi/d)$, and $\overleftrightarrow{M}(q_x;\omega)$ is a $(10n_{\max}+5)\times(10n_{\max}+5)$ coefficient matrix. When $A_x^{\mathrm{T}}(q_x;\omega) = A_y^{\mathrm{T}}(q_x;\omega) = 0$ for the null external source on the right-hand side of Eq. (5.112), the non-zero solution of the matrix equation requires the condition $\mathrm{Det}[\overleftrightarrow{M}(q_x;\omega)] = 0$ for the self-sustaining oscillations in the system. By setting $\mathrm{Re}\{\mathrm{Det}[\overleftrightarrow{M}(q_x;\omega)]\} = 0$ and treating real ω and q_x independently, we are able to get the real roots $\omega = \omega_j(q_x)$ for the dispersions of the single-particle and plasmon excitations in the system with $j = 1, 2, \ldots$ corresponding to the index of the jth root. The plasmon excitations of the system are subjected to an additional condition, that is, $\mathrm{Im}\{\mathrm{Det}[\overleftrightarrow{M}(q_x;\omega)]\} = 0$. On the other hand, we can also set $\mathrm{Det}[\overleftrightarrow{M}(q_x;\tilde{\omega})] = 0$ and treat complex $\tilde{\omega}$ and real q_x independently for a complex equation. As a result, the plasmon excitation will have a finite lifetime if $\mathrm{Im}(\tilde{\omega}) < 0$. In addition, if $\mathrm{Im}(\tilde{\omega}) > 0$, the amplitude of the plasma wave will increase with time, that is, the plasma instability [15] occurs in the system.

5.4 Energy Transfer in Nanotubes

Over the years, there have been several papers devoted to the theory of electron energy loss spectroscopy (EELS) [22–32]. These works were concerned with EELS for films, the two-dimensional electron gas (2DEG) and nanowires. Most recently, there have been several papers dealing with the theory of EELS for cylindrical nanotubes and cylindrical cavities. However, the published work so far [33–39] does not fully compare the contributions from plasmons and single-particle excitations for nanotubes. This scattering problem deserves some attention and will be the subject of this section. We will present a model and theory within the RPA for EELS in cylindrical nanotubes. The solid metallic cylinder is a special case which is obtained when the polarization function is set equal to infinity. However, this latter model does not allow the contributions to energy loss from individual subbands to be analyzed. Our model provides a way of separating and calculating the plasmon and particle–hole contributions.

5.4.1 Energy Loss on a Single Wall Nanotube

We assume that a particle of mass m^* and charge Q moves with velocity $\boldsymbol{v} = v\hat{\boldsymbol{z}}$ parallel to the axis of the nanotube (of radius R) with impact parameter $\rho_0 > R$, that is, we take the particle to move outside the cylinder. The position vector of the particle at any time t is given by $\boldsymbol{r}_0 = (\rho_0, \phi_0, z_0 = vt)$.

The total electrostatic potential Φ at any point of space for any time t satisfies Poisson's equation

$$\nabla^2 \Phi_1(\rho, \phi, z, t) = 0 , \quad \rho < R \tag{5.113}$$

$$\nabla^2 \Phi_2(\rho, \phi, z, t) = -\frac{4\pi Q}{\epsilon_2}\delta(\boldsymbol{r} - \boldsymbol{r}_0) , \quad \rho > R , \tag{5.114}$$

along with the boundary conditions

$$\Phi_1(\rho = R, \phi, z, \omega) = \Phi_2(\rho = R, \phi, z, \omega) , \tag{5.115}$$

$$(\boldsymbol{D}_2 - \boldsymbol{D}_1) \cdot \hat{\boldsymbol{n}}_{21}|_{\rho=R} = 4\pi\sigma(R, \phi, z, \omega) , \tag{5.116}$$

where σ is the induced surface charge density and $\hat{\boldsymbol{n}}_{21} = \hat{\boldsymbol{\rho}}$. The solution Φ_1 of Eq. (5.113) can be Fourier expanded as

$$\Phi_1(\rho,\phi,z,t) = \frac{Q}{\pi\epsilon_1}\sum_{L=-\infty}^{\infty} e^{iL(\phi-\phi_0)} \times \int_{-\infty}^{\infty} dq_z e^{iq_z(z-vt)} C_L^{<}(q_z) I_L(q_z\rho) , \tag{5.117}$$

and from this we find

$$\Phi_1(\rho,\phi,z,\omega) = \frac{2Q}{\epsilon_1}\sum_{L=-\infty}^{\infty} e^{iL(\phi-\phi_0)} \times \int_{-\infty}^{\infty} dq_z C_L^{<}(q_z) I_L(q_z\rho)\delta(\omega - q_z v) . \tag{5.118}$$

The potential Φ_2 can be written using superposition as

$$\Phi_2(\rho,\phi,z,t) = \frac{Q}{\epsilon_2|\boldsymbol{r}-\boldsymbol{r}_0(t)|} + \Phi_{\text{ind}}^{>}(\rho,\phi,z,t) , \tag{5.119}$$

where $(Q/\epsilon_2|\boldsymbol{r}-\boldsymbol{r}_0(t)|)$ is the potential of the moving charged particle and $\Phi_{\text{ind}}^{>}$ is the potential due to the induced surface charge density (a solution of Laplace equation for $\rho > R$). We can write these electric potentials in cylindrical coordinates as

$$\frac{Q}{|\boldsymbol{r}-\boldsymbol{r}_0(t)|} = \frac{Q}{\pi}\sum_{L=-\infty}^{\infty} e^{iL(\phi-\phi_0)} \times \int_{-\infty}^{\infty} dq_z e^{iq_z(z-vt)} I_L(q_z\rho_<) K_L(q_z\rho_>) , \tag{5.120}$$

$$\Phi_{\text{ind}}^{>}(\rho,\phi,z,t) = \frac{Q}{\pi\epsilon_2}\sum_{L=-\infty}^{\infty} e^{iL(\phi-\phi_0)} \times \int_{-\infty}^{\infty} dq_z e^{iq_z(z-vt)} C_L^{>}(q_z) K_L(q_z\rho) , \tag{5.121}$$

where $\rho_<$ ($\rho_>$) is the smaller (larger) of ρ and ρ_0. Therefore, Eq. (5.119) can be written as

$$\Phi_2(\rho,\phi,z,t) = \frac{Q}{\pi\epsilon_2}\sum_{L=-\infty}^{\infty} e^{iL(\phi-\phi_0)} \int_{-\infty}^{\infty} dq_z e^{iq_z(z-vt)} \times \left[I_L(q_z\rho_<) K_L(q_z\rho_>) + C_L^{>}(q_z) K_L(q_z\rho) \right] , \tag{5.122}$$

or

$$\Phi_2(\rho, \phi, z, \omega) = \frac{2Q}{\epsilon_2} \sum_{L=-\infty}^{\infty} e^{iL(\phi-\phi_0)} \int_{-\infty}^{\infty} dq_z e^{iq_z z} \left[I_L(q_z \rho_<) K_L(q_z \rho_>) \right.$$
$$\left. + C_L^>(q_z) K_L(q_z \rho) \right] \delta(\omega - q_z v) . \tag{5.123}$$

The induced charge density can be found in this case by substituting Eqs. (5.118), (5.122), and (5.123) into the boundary conditions in Eqs. (5.115) and (5.116). In this way, we find that

$$\rho(\mathbf{r}, \omega) = -\frac{Qe^2}{\pi \epsilon_1 R} \delta(\rho - R) \sum_{L=-\infty}^{\infty} e^{iL(\phi-\phi_0)} \int_{-\infty}^{\infty} dq_z e^{iq_z z}$$
$$\times \chi_L(q_z, \omega) C_L^<(q_z) I_L(q_z R) \delta(\omega - q_z v) , \tag{5.124}$$

and that the induced surface charge density is given by

$$\sigma(R, \phi, z, \omega) = -\frac{Qe^2}{\pi \epsilon_1 R} \sum_{L=-\infty}^{\infty} e^{iL(\phi-\phi_0)} \int_{-\infty}^{\infty} dq_z e^{iq_z z} \chi_L(q_z, \omega)$$
$$\times C_L^<(q_z) I_L(q_z R) \delta(\omega - q_z v) . \tag{5.125}$$

The boundary conditions Eqs. (5.115) and (5.116) can now be written using the expressions found for Φ_1, Φ_2 and σ as

$$\epsilon_2 C_L^<(q_z) I_L(q_z R) = \epsilon_1 \left[I_L(q_z R) K_L(q_z \rho_0) + C_L^>(q_z) K_L(q_z R) \right] , \tag{5.126}$$

$$C_L^<(q_z) I_L'(q_z R) - C_L^>(q_z) K_L'(q_z R) = K_L(q_z \rho_0) I_L'(q_z R)$$
$$- \frac{2e^2}{\epsilon_1 q_z R} \chi_L(q_z, \omega = q_z v) C_L^<(q_z) I_L(q_z R) . \tag{5.127}$$

The solutions of the set of Eqs. (5.126) and (5.127) are given by the expressions

$$C_L^<(q_z) = \frac{\epsilon_1 K_L(q_z \rho_0)}{D_L(q_z, \omega = q_z v)} , \tag{5.128}$$

$$C_L^>(q_z) = -\left[2e^2 I_L(q_z R) \chi_L(q_z, \omega = q_z v) + q_z R(\epsilon_1 - \epsilon_2) I_L'(q_z R) \right]$$
$$\times \frac{K_L(q_z \rho_0) I_L(q_z R)}{D_L(q_z, \omega = q_z v)} , \tag{5.129}$$

with

$$D_L(q_z, \omega) = \epsilon_1 + (\epsilon_1 - \epsilon_2) q_z R I_L(q_z R) K_L'(q_z R)$$
$$+ 2e^2 \chi_L(q_z, \omega) I_L(q_z R) K_L(q_z R) , \tag{5.130}$$

or

$$D_L(q_z,\omega) = \left[\epsilon_1 + (\epsilon_1-\epsilon_2)q_z R I_L(q_z R) K_L'(q_z R)\right] \times \left[1+\alpha_L(q_z,\omega) K_L(q_z R)\right], \tag{5.131}$$

where

$$\alpha_L(q_z,\omega) = \frac{2e^2 I_L(q_z R) K_L(q_z R)}{\epsilon_1 + (\epsilon_1-\epsilon_2)q_z R I_L(q_z R) K_L'(q_z R)} \chi_L(q_z,\omega) \tag{5.132}$$

is the polarizability function of the electron gas, and $\epsilon_L(q_z,\omega) = 1+\alpha_L(q_z,\omega)$ is the dielectric function. We can now find the induced potential because of the charge density fluctuations on the surface of the cylinder. For $\rho > R$, using Eqs. (5.121) and (5.129), we find

$$\Phi^{>}_{\text{ind}}(\boldsymbol{r},t) = -\frac{Q}{\pi\epsilon_2}\sum_{L=-\infty}^{\infty} e^{iL(\phi-\phi_0)} \int_{-\infty}^{\infty} dq_z e^{iq_z(z-vt)} K_L(q_z\rho) K_L(q_z\rho_0) I_L(q_z R) \times \left[\frac{1}{K_L(q_z R)}\frac{\alpha_L(q_z,\omega)}{1+\alpha_L(q_z,\omega)} + q_z R(\epsilon_1-\epsilon_2)\frac{I_L'(q_z R)}{D_L(q_z,\omega)}\right]_{\omega=q_z v}, \tag{5.133}$$

while for $\rho < R$, using Eqs. (5.117) and (5.128), we obtain [since $\Phi_1(\boldsymbol{r},t) = \Phi^{<}_{\text{ind}}(\boldsymbol{r},t)$]

$$\Phi^{<}_{\text{ind}}(\boldsymbol{r},t) = \frac{Q}{\pi}\sum_{L=-\infty}^{\infty} e^{iL(\phi-\phi_0)} \int_{-\infty}^{\infty} dq_z e^{iq_z(z-vt)} \frac{I_L(q_z\rho) K_L(q_z\rho_0)}{D_L(q_z,\omega=q_z v)}. \tag{5.134}$$

The force applied to the moving charged particle from the induced charge density on the surface of the nanotube is therefore given by

$$\boldsymbol{F} = -Q\nabla\Phi^{>}_{\text{ind}}(\boldsymbol{r},t)\Big|_{\boldsymbol{r}=(\rho_0,\phi_0,z_0=vt)}, \tag{5.135}$$

and the rate of energy loss can be calculated as

$$\frac{dW}{dt} = \boldsymbol{F}\cdot\boldsymbol{v} = -Qv\frac{\partial\Phi^{>}_{\text{ind}}(\rho,\phi,z,t)}{\partial z}\Bigg|_{\boldsymbol{r}=(\rho_0,\phi_0,z_0=vt)}. \tag{5.136}$$

We then obtain

$$\frac{dW}{dt} = \frac{2Q^2}{\pi}v\sum_{L=-\infty}^{\infty}\int_0^{\infty} dq_z q_z K_L^2(q_z\rho_0)\frac{I_L(q_z R)}{K_L(q_z R)}\,\text{Im}\left[\frac{1}{\epsilon_L(q_z,\omega=q_z v)}\right] \times \frac{1}{\epsilon_1 + q_z R(\epsilon_1-\epsilon_2) I_L(q_z R) K_L'(q_z R)}. \tag{5.137}$$

This result generalizes that obtained by Arista and Fuentes [40] for a cylindrical cavity which corresponds to neglecting the response function $\chi_L(q_z, \omega)$ for the electron gas on the surface of the cylinder. Equation (5.137) includes contributions to the total energy loss from all possible linear momentum transfers q_z along the axis of the nanotube and all transitions (angular momentum transfers) L within and between different subbands. However, only excitations of frequencies $\omega = q_z v$ can contribute to the stopping power. The imaginary part of the Fourier transform of the dielectric function $\varepsilon_L(q_z, \omega)$ enters the energy loss formula. This means that we can separate the contributions to Eq. (5.137) from plasmons and particle–hole excitations in a similar way to that done by Horing, Tso and Gumbs [32] in calculations of the stopping power of a 2D sheet of electron gas if we express

$$\epsilon_L(q_z, \omega) = \epsilon_{1,L}(q_z, \omega) + i\epsilon_{2,L}(q_z, \omega) , \tag{5.138}$$

$$\frac{1}{\epsilon_L(q_z, \omega)} = \frac{-\epsilon_{2,L}(q_z, \omega)}{\epsilon_{1,L}^2(q_z, \omega) + \epsilon_{2,L}^2(q_z, \omega)} , \tag{5.139}$$

where $\epsilon_{1,L}(q_z, \omega)$ and $\epsilon_{2,L}(q_z, \omega)$ are the real and imaginary parts respectively of $\epsilon_L(q_z, \omega)$. The imaginary part function in Eq. (5.137) is multiplied by a kinematical factor depending on the impact parameter, momentum transfer and velocity v which can be adjusted experimentally. There is a contribution from the integrand whenever either

$$\text{(a) } \epsilon_{2,L}(q_z, \omega = q_z v) = \text{Im}\left[\alpha_L(q_z, \omega = q_z v)\right] \neq 0 \quad \text{or}$$
$$\text{(b) } \epsilon_{1,L}(q_z, \omega = q_z v) = 0 \quad \text{and} \quad \epsilon_{2,L}(q_z, \omega = q_z v) = 0 .$$

When case (a) is applied, we have Landau damping and the particle–hole region (see Figure 5.6) contributes to the energy loss. In case (b), however, the dispersion equation for plasmon excitations is satisfied on the surface of the cylinder and the plasmons make a contribution. In this case, we use Dirac's identity

$$\lim_{\epsilon \to 0^+} \frac{\epsilon}{x^2 + \epsilon^2} = \pi \delta(x) \tag{5.140}$$

and we find that Eq. (5.137) becomes

$$\left.\frac{dW}{dt}\right|_{\text{plasmons}} = -2Q^2 v \sum_{L=-\infty}^{\infty} \int_0^{\infty} dq_z \frac{q_z \delta(\omega - q_z v)}{\left|\frac{d\epsilon_{1,L}(q_z,\omega)}{d\omega}\right|_{\omega=\omega_L(q_z)}} \times K_L^2(q_z \rho_0) \frac{I_L(q_z R)}{K_L(q_z R)} \frac{1}{\epsilon_1 + q_z R(\epsilon_1 - \epsilon_2) I_L(q_z R) K_L'(q_z R)} , \tag{5.141}$$

where $\omega_L(q_z)$ is the solution of $\epsilon_{1,L}(q_z, \omega) = 0$, that is, it represents the plasmon dispersion curves. In the case that the charged particle moves parallel to the axis

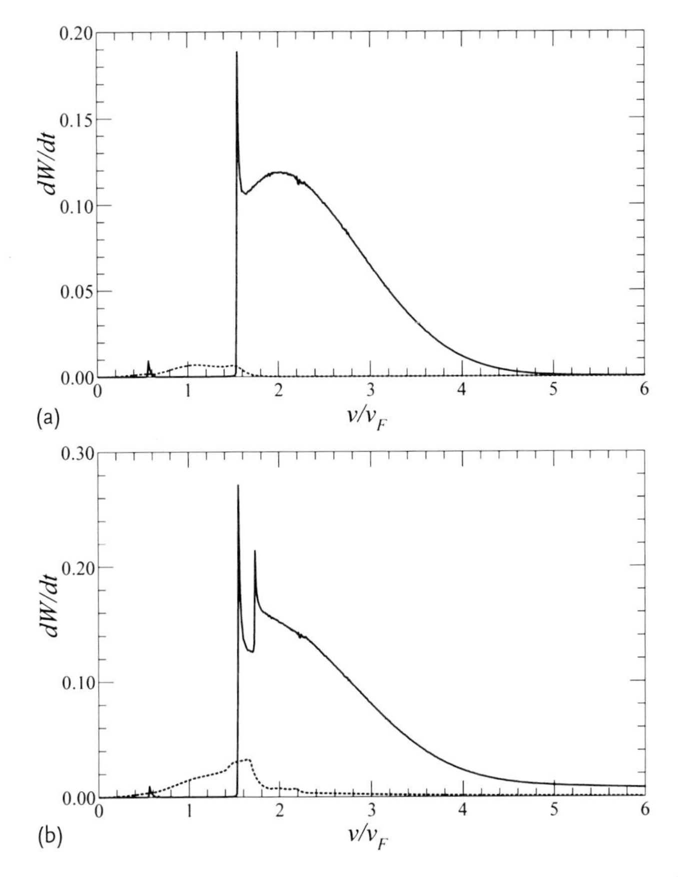

Figure 5.3 The energy loss for plasmon (solid lines) and particle–hole modes (dotted lines) as a function of the charged particle velocity parallel to the axis of the cylinder when (a) $\rho_0 = 0$ Å and (b) $\rho_0 = 5$ Å. The energy-loss rate is expressed in units of $e^2 k_F^2 v_F$.

of the tubule at distance ρ_0 with $\rho_0 < R$, the induced potential, charge density and stopping power can be obtained in a similar way.

We have calculated the total rate of loss of energy from all subband transitions at $T = 0$ K, as described by Eq. (5.137). We simulated a graphene tubule by choosing $\epsilon_1 = \epsilon_2 = 2.4$, $m^* = 0.25 m_e$, $R = 11$ Å and $E_F = 0.6$ eV. We included all the transitions with $|L| \leq 10$ in calculating dW/dt. There are only five subbands occupied by electrons corresponding to $\ell = 0, \pm 1, \pm 2$. Our results are presented in Figures 5.3 and 5.4 as functions of v/v_F, where the Fermi velocity $v_F = 918.88$ km/s. In Figure 5.3a, the charged particle travels along the axis of the tubule ($\rho_0 = 0$). Here, the plasmon contribution to the rate of loss of energy is larger than that from single-particle excitations except in the low-velocity limit. For fast moving particles, only plasmons play a role. For both curves, dW/dt initially increases with v, but then decreases after reaching a maximum. Therefore, the dominant contribution to the energy loss comes from those excitations whose phase velocities lie close to

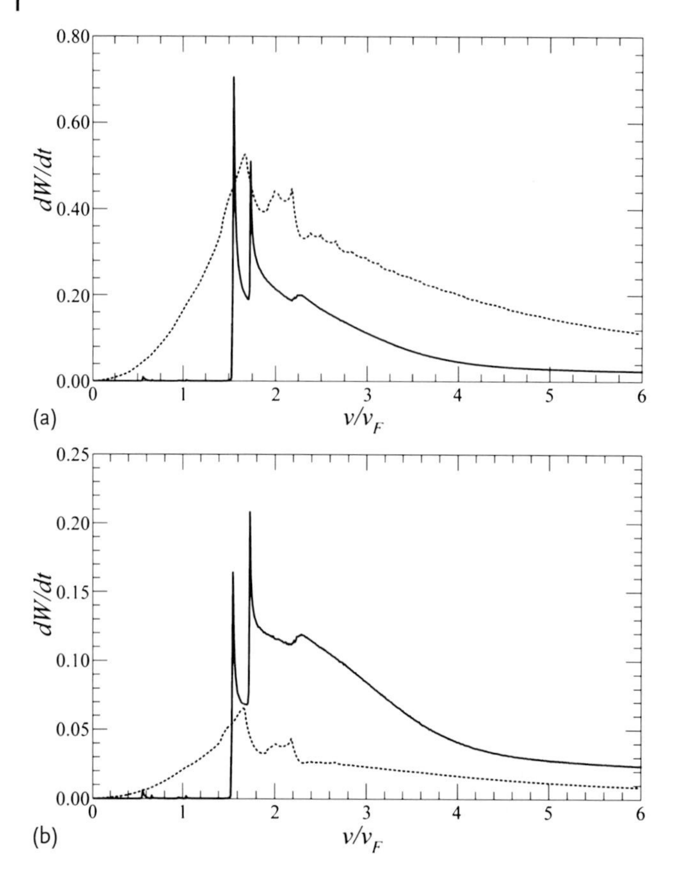

Figure 5.4 The energy loss for plasmon (solid lines) and particle–hole modes (dotted lines) as a function of the charged particle velocity parallel to the axis of the cylinder when (a) $\rho_0 = 10\,\text{Å}$ and (b) $\rho_0 = 15\,\text{Å}$. The energy-loss rate is expressed in units of $e^2 k_F^2 v_F$.

where the maximum occurs. We examined where this peak occurs for each term in the sum over L. Our calculations show that for fixed ρ_0, the rate of loss of energy has a peak for each value of L contributing to the sum in Eq. (5.137). When the charged particle moves along the axis of the nanotube, only the $L = 0$ transitions contribute. This is explained by the behavior of the Bessel function $I_L(q_z \rho_0)$. However, when the particle trajectory is not along the axis of the cylinder, terms with $L \neq 0$ contribute. The peak position is not the same for each value of L. This accounts for the multiple peaks in the total energy loss dW/dt and is elaborated on below for both plasmon and particle–hole modes for a chosen ρ_0. As a matter of fact, the lowest subband transitions corresponding to $L = 0, 1, 2$ mainly contribute to the total stopping power when $\rho_0 \neq 0$.

In Figure 5.3b, we choose $\rho_0 = 5\,\text{Å}$ so that the charged particle trajectory is almost halfway between the axis and the surface of the cylinder. Comparing these results with those in Figure 5.3a, the particle–hole mode contribution increases for

$v \ll v_F$. In Figure 5.4a, we choose $\rho_0 = 10$ Å. For this case when the charged particle trajectory is close to the surface of the cylinder, the energy loss due to particle–hole excitations surpasses that from plasmons for all values of the charged particle velocity, except when $v \approx 1.8 v_F$. Both the plasmon and single-particle contributions are increased when the charged particle trajectory is set closer to the cylinder surface. In Figure 5.4b, it is shown that as ρ_0 is increased with the particle trajectory outside the cylinder, the plasmon contribution to the energy loss is larger than that of the single-particle. Our calculations have shown that when the charged particle trajectory is at the same distance from the surface either inside or outside the cylinder, there is a small difference in dW/dt. The small difference is due to the asymmetry of the induced potential with respect to the cylindrical surface.

In Figure 5.5, we plot the contributions to the stopping power from the lowest occupied subbands $L = 0, \pm 1, \pm 2$ when the charged particle distance from the axis of the cylinder is $\rho_0 = 15$ Å. Figure 5.5a is the contribution from plasmon excita-

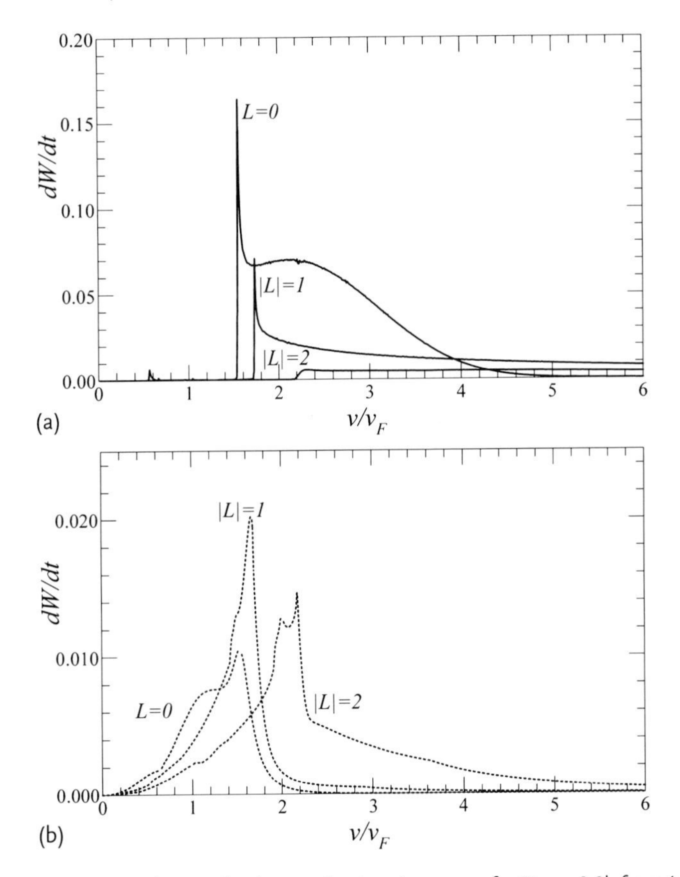

Figure 5.5 The contributions to the stopping power for Figure 3.2b from (a) plasmon and (b) particle–hole excitations for intersubband transitions with L. dW/dt is expressed in the same unit as in Figures 5.3 and 5.4.

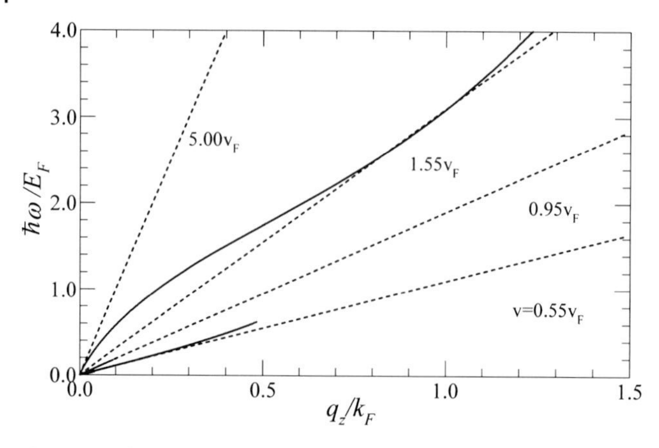

Figure 5.6 The $L = 0$ plasmon excitation energies for a single wall nanotube of $R = 11$ Å. The dashed lines are $\omega = q_z v$, that is, $\hbar\omega/E_F = 2(q_z/k_F)(v/v_F)$.

tions, whereas Figure 5.5b shows the stopping power of particle–hole modes. In Figure 5.5a, the transition with $L = 0$ subband is larger than any other intersubband transitions producing plasmons excitations. In fact, the contribution from plasmons excitations decreases with increasing value of L. However, Figure 5.5b shows that the $|L| = 1$ and $|L| = 2$ subband contributions to the stopping power from particle–hole excitations exceed that from the $L = 0$ subband. The contributions from higher subbands decreases with increasing L. Each curve in Figure 5.5 has a peak at a value of velocity which shifts to the right as L is increased. Similar results were obtained when the ρ_0 value was changed and the charged particle trajectory was either inside or outside the nanotube.

The numerical results in Figure 5.3 show that there is a sudden increase in the stopping power of plasmon excitations for specific values of charged particle velocity. The highest threshold value is at $v = 1.55 v_F$ regardless of the chosen impact parameter ρ_0. To explain these characteristics, we have plotted in Figure 5.6 the three acoustic plasmon modes obtained by solving the dispersion equation. These plasmons have Landau damping when they enter the particle–hole mode regions which are not shown in Figure 5.6. This explains why each mode only exists over a range of frequency and wave vector. Only when the plasmon frequency satisfies $\omega = q_z v$ can we find that there is a contribution to the energy loss. The slope of this straight line increases with v, demonstrating that there is a range of values of charged particle speed below which the stopping power is negligible. The small value of dW/dt for $v \ll v_F$ comes from a narrow region when the excitation energy satisfies $\hbar\omega/E_F \ll 1$. Our numerical calculations clearly show that there is a threshold contribution coming from the plasmon modes whenever the straight line $\omega = q_z v$ just touches the plasmon branches. That is, the plasmon modes start contributing when the $\omega = q_z v$ straight line crosses the lowest plasmon branch. The threshold for the lowest mode corresponds to a speed $v = 0.55 v_F$. This mode gives a small peak in our energy loss plots at that particle velocity. According to

Figure 5.6, the middle plasmon branch contributes when the speed of the charged particle is $v = 0.95v_F$. The contribution to the energy loss from this mode is small as can be deduced from Figure 5.6. Finally, when the charged particle speed lies in the range $1.55v_F < v < 5.0v_F$, the most energetic plasmon branch contributes. When this branch starts to contribute, we have the high peak in our energy loss plots at the threshold speed $v = 1.55v_F$. There is a single-particle excitation region below the lowest acoustic plasmon branch which leads to energy loss to these excitations at low charged particle velocities.

5.5 Problems

1. Generalize Eq. (5.21) to a bilayer system of parallel 2DEG layers with separation a and susceptibility χ_1 and χ_2 by showing that the surface response function is given by

$$g = 1 - 2\epsilon_r \frac{\mathcal{N}}{\mathcal{D}} \tag{5.142}$$

with

$$\mathcal{N} = \left[\epsilon_r + \frac{4\pi e^2 \chi_2}{q_{\|}} + \epsilon(\omega)\right] - \left[\epsilon_r + \frac{4\pi e^2 \chi_2}{q_{\|}} - \epsilon(\omega)\right] e^{-2q_{\|}a}, \tag{5.143}$$

$$\begin{aligned}\mathcal{D} = &\left[\epsilon_r + \frac{4\pi e^2 \chi_1}{q_{\|}} + \epsilon(\omega)\right]\left[\epsilon_r + \frac{4\pi e^2 \chi_2}{q_{\|}} + \epsilon(\omega)\right] \\ &- \left[\epsilon_r + \frac{4\pi e^2 \chi_1}{q_{\|}} - \epsilon(\omega)\right]\left[\epsilon_r + \frac{4\pi e^2 \chi_2}{q_{\|}} - \epsilon(\omega)\right] e^{-2q_{\|}a},\end{aligned} \tag{5.144}$$

where $\epsilon(\omega)$ is the dielectric constant of the medium between the layers and ϵ_r is the background dielectric constant of the material in which the layers are embedded.

2. Consider a semi-infinite jellium occupying the half-space $z < 0$ and consider the response of this system to an external potential of the form

$$\phi_{ext} = e^{q_{\|}z} e^{i q_{\|}\cdot r_{\|}} + \text{c.c.},$$

where c.c. represents the complex conjugate of the first term. Since an arbitrary external potential can always be decomposed into evanescent plane waves of this type in a region of space where there are no external charges, we have $\nabla^2 \phi_{ext} = 0$. By integrating the Poynting vector over the jellium surface, that is,

$$\mathcal{W} = \int d^2 r_{\|} \frac{1}{\hbar\omega} \hat{z} \cdot (\boldsymbol{E} \times \boldsymbol{H}),$$

where $\hat{z}$ is a unit vector perpendicular to the surface, prove that the power absorption in the material due to the electronic excitations induced by an evanescent external potential is given in terms of the transition rate $\mathcal{W}$ by the expression

$$\mathrm{Im}(g) \propto \frac{e^2 \hbar \mathcal{W}}{\epsilon_0 \mathcal{S} q_{||}} ,$$

where $\mathcal{S}$ is the surface area.

3. Treating the electrons on the surface of a cylindrical nanotube as an electron gas interacting through the Coulomb potential, obtain an expression for the image potential in terms of the radius R of the cylinder and the angular momentum quantum number ℓ of a charged particle around the axis.

4. Prove that the dispersion equation for a periodic superlattice is given by Eq. (5.82).

5. Prove the recursion relation in Eq. (5.58). Also, obtain the recursion relation for the sequence which satisfies $\overleftrightarrow{\mathcal{M}}_{j+1} = \overleftrightarrow{\mathcal{M}}_{j-1}^{2} \otimes \overleftrightarrow{\mathcal{M}}_{j}$.

6. Consider a slab of material of thickness a and background dielectric constant ϵ_b. Suppose its surfaces are located at $z = 0$ and $z = a$ and for $z < 0$, the medium has dielectric constant ϵ_L while for $z > a$, the medium has dielectric constant ϵ_R. Show that the surface response function is given by

$$g = \frac{(\epsilon_b + \epsilon_L)(\epsilon_b - \epsilon_R) e^{-q_{||} a} - (\epsilon_b + \epsilon_R)(\epsilon_b - \epsilon_L) e^{q_{||} a}}{(\epsilon_b - \epsilon_L)(\epsilon_b - \epsilon_R) e^{-q_{||} a} - (\epsilon_b + \epsilon_R)(\epsilon_b - \epsilon_L) e^{q_{||} a}} .$$

7. For a particle of charge $Z^* e$ with coordinates $(0, 0, z_0)$ moving with a parallel velocity $\boldsymbol{v}_{||}$ in the vicinity of a medium with a planar surface located at $z = 0$, the force at right angle to the surface is given in terms of the inverse dielectric function ϵ^{-1} of the medium by

$$\begin{aligned} \boldsymbol{F}_{\perp} = {} & \frac{(Z^* e)^2}{16\pi^2 \epsilon_0} \hat{\boldsymbol{z}} \int dz_1 \int dz_2 \int dz_4 \int \frac{d^2 q_{||}}{q_{||}^2} e^{-q_{||}|z_2 - z_0|} e^{-q_{||}|z_4 - z_0|} \\ & \times \frac{\partial}{\partial z_1} \epsilon^{-1}(z_1, z_2; \boldsymbol{q}_{||}, \boldsymbol{q}_{||} \cdot \boldsymbol{v}_{||}) \\ & \times \left(\frac{\partial^2}{\partial z_1^2} - q_{||}^2 \right) \left[\epsilon^{-1}(z_1, z_4; -\boldsymbol{q}_{||}, -\boldsymbol{q}_{||} \cdot \boldsymbol{v}_{||}) - \delta(z_1 - z_4) \right] , \end{aligned}$$

where $\hat{\boldsymbol{z}}$ is a unit vector perpendicular to the planar surface. Estimate the contribution to $F_{\perp}$ for slow charge particles.

8. In the absence of a grating, only the $n = 0$ specular mode exists. Using Eqs. (5.96)–(5.100) to show that the linear equations in Eq. (5.112) reduces

to

$$\overleftrightarrow{M}\begin{bmatrix} B_x^{\mathrm{T}} \\ B_y^{\mathrm{T}} \\ C_x^{\mathrm{T}} \\ C_y^{\mathrm{T}} \\ C_x^{L} \end{bmatrix}_{\mathrm{s}} = \begin{bmatrix} 0 \\ -E_0 \\ \left(\bar{\chi}_{\mathrm{s}} - \frac{i\eta_1^{\mathrm{T}} c^2}{\omega^2}\right)\frac{\omega}{c}E_0 \\ 0 \\ 0 \end{bmatrix}$$

for the *s*-polarization and

$$\overleftrightarrow{M}\begin{bmatrix} B_x^{\mathrm{T}} \\ B_y^{\mathrm{T}} \\ C_x^{\mathrm{T}} \\ C_y^{\mathrm{T}} \\ C_x^{L} \end{bmatrix}_{\mathrm{p}} = \begin{bmatrix} -\frac{\eta_1^{\mathrm{T}}}{\omega\epsilon_0}H_0 \\ 0 \\ 0 \\ \left(\bar{\chi}_{\mathrm{s}} - \frac{i}{\eta_1^{\mathrm{T}}}\right)\frac{\eta_1^{\mathrm{T}}}{\epsilon_0 c}H_0 \\ 0 \end{bmatrix}$$

for the *p*-polarization, where E_0 and H_0 are the amplitudes for the incident electric and magnetic fields and the coefficient matrix $\overleftrightarrow{M}$ is given by

$$\begin{bmatrix} 1 & 0 & -1 & 0 & -1 \\ 0 & 1 & 0 & -1 & 0 \\ 0 & -\bar{\chi}_{\mathrm{s}} - \frac{i\eta_1^{\mathrm{T}} c^2}{\omega^2} & 0 & -\frac{i\beta_2^{\mathrm{T}} c^2}{\omega^2} & 0 \\ -\bar{\chi}_{\mathrm{s}} - \frac{i}{\eta_1^{\mathrm{T}}} & 0 & -\frac{i\epsilon_2^{\mathrm{T}}}{\beta_2^{\mathrm{T}}} & 0 & 0 \\ 0 & 0 & \frac{q_x}{\beta_2^{\mathrm{T}}}(\epsilon_2^{\mathrm{T}} - \epsilon_{\mathrm{b}}) + iq_x\bar{\chi}_{\mathrm{s}} & 0 & \frac{\beta_2^{L}}{q_x}\epsilon_{\mathrm{b}} + iq_x\bar{\chi}_{\mathrm{s}} \end{bmatrix}.$$

9. Use Eq. (5.82) to calculate the behavior of the plasmon frequency as $q_{||} \to 0$.

10. Let us consider electrons in a single graphene layer in the xy-plane in the absence of magnetic field. The Hamiltonian for non-interacting electrons in one valley of a graphene layer without scatterers is given by the following equation. Here, we neglect the Zeeman splitting and assume valley energy degeneracy, describing the eigenstates by two pseudospin *s*. We have

$$\hat{H}^{(0)} = v_{\mathrm{F}}\begin{pmatrix} 0 & \hat{p}_x - i\hat{p}_y \\ \hat{p}_x + i\hat{p}_y & 0 \end{pmatrix}, \tag{5.145}$$

where $\hat{\boldsymbol{p}} = -i\hbar\nabla$, $v_{\mathrm{F}} = \sqrt{3}at/2\hbar$ is the Fermi velocity with $a = 2.566$ Å denoting the lattice constant, $t \approx 2.71$ eV the overlap integral between nearest-neighbor carbon atoms. Making use of the single-particle states, obtain an expression for the response function $\chi^{(0)}(q_{||}, \omega)$ and deduce the plasmon dispersion in the long-wavelength limit.

11. Derive an expression for the surface response function of a sheet of graphene using the band structure depicted in Figure 4.10. Use this result to obtain the image potential of graphene.

12. **Analytical Expression for the Absorption Coefficient:** Use self-consistent field theory for the infrared absorption coefficient defined as

$$\beta_{\text{abs}}(\omega) = \hbar\omega \times \frac{\text{Number of transitions per unit volume and time}}{\text{Incident Flux}} \tag{5.146}$$

where

- Energy flux = energy density × velocity of flow,
- Energy density in the medium $= \epsilon_b |\boldsymbol{E}^{\text{ext}}|^2$ where ϵ_b is the averaged optical dielectric constant of the system and $\boldsymbol{E}^{\text{ext}}$ is the uniform external electric field,
- When averaged over a cycle, $\overline{\text{Energy density}} = \frac{1}{2}\epsilon_b |\boldsymbol{E}^{\text{ext}}|^2$,
- Propagation velocity is c/n_r, where n_r is the index of refraction, show that

$$\beta_{\text{abs}}(\omega) = \frac{N_0}{n_r(\omega)\varepsilon_0 c}\omega[1 + \rho_{\text{ph}}(\omega)] \operatorname{Im} \alpha_L(\omega) , \tag{5.147}$$

in terms of the Lorentz ratio

$$\alpha_L(\omega) \equiv \frac{-e}{|\boldsymbol{E}_{\text{ext}}|^2} \frac{1}{L_z A} \int d\boldsymbol{r} \delta\mathcal{N}_{\text{ind}}(\boldsymbol{r}, \omega)\boldsymbol{r} \cdot \boldsymbol{E}_{\text{ext}} . \tag{5.148}$$

In this notation, $\delta\mathcal{N}_{\text{ind}}(\boldsymbol{r}, \omega)$ is the induced electron density, N_0 is the number of charge carriers per unit volume, $\rho_{\text{ph}}(\omega) = 1/\left[e^{\hbar\omega/k_B T} - 1\right]$ is the photon distribution function, and the refractive index is given by

$$\sqrt{2}n_r(\omega) = \left[\epsilon_b + \frac{\operatorname{Re}\alpha_L(\omega)}{\varepsilon_0} + \sqrt{\left(\epsilon_b + \frac{\operatorname{Re}\alpha_L(\omega)}{\varepsilon_0}\right)^2 + \left(\frac{\operatorname{Im}\alpha_L(\omega)}{\varepsilon_0}\right)^2}\right]^{1/2} \tag{5.149}$$

where "Re" and "Im" denote taking the real and imaginary parts, respectively.

13. Noting that the linear response of the electron density $n(\boldsymbol{r}, t)$ to an external potential φ^{ext} is the induced density

$$\begin{aligned} \delta\mathcal{N}_{\text{ind}}(\boldsymbol{r}, t) &= \int_{-\infty}^{\infty} dt' \int d\boldsymbol{r}' \left\langle \frac{1}{i\hbar}[n(\boldsymbol{r}, t), n(\boldsymbol{r}', t')] \right\rangle \theta(t - t')\varphi^{\text{ext}}(\boldsymbol{r}') , \\ &= \int_{-\infty}^{\infty} dt' \int d\boldsymbol{r}' \chi(\boldsymbol{r}, \boldsymbol{r}', t - t')\varphi^{\text{ext}}(\boldsymbol{r}') , \end{aligned} \tag{5.150}$$

where $\chi(\boldsymbol{r}, \boldsymbol{r}', t - t')$ is the density–density response function, show that the absorption coefficient discussed in the preceding problem may be expressed in terms of the surface response function for a slab with planar surfaces perpendicular to the z-axis with

$$\beta_{\rm abs}(\omega) \propto \frac{1}{n_{\rm r}(\omega)}\omega[1+\rho_{\rm ph}(\omega)]\,{\rm Im}\int\limits_{-\infty}^{\infty} dz \int\limits_{-\infty}^{\infty} dz' g(q_{||},z,z')\Phi(z)\Phi(z') \,, \tag{5.151}$$

where

$$\varphi^{\rm ext}(\boldsymbol{r} \sim e^{iq_{||}\cdot r_{||}}\Phi(z) \tag{5.152}$$

and

$$g(q_{||},z,z') \equiv \int d\boldsymbol{r}_{||}\int d\boldsymbol{r}'_{||}e^{iq_{||}\cdot r_{||}}\chi(\boldsymbol{r}_{||},\boldsymbol{r}'_{||};z,z',\omega)e^{iq_{||}\cdot r'_{||}} \tag{5.153}$$

is the surface response function.

References

1 Ibach, H. and Mills, D.L. (1982) *Electron energy loss spectroscopy and surface vibrations*, Academic, New York.

2 Echenique, P.M., Flores, F., and Ritchie, R.H. (1990) *Dynamic screening of ions in condensed matter*, Vol. **43** of Solid State Phys., (eds H. Ehrenreich and D. Turnbull), Academic Press, Boston, p. 229.

3 Persson, B.N.J. and Zaremba, E. (1985) Electron–hole pair production at metal surfaces. *Phys. Rev. B*, **31**, 1863.

4 Höfer, U., Shumay, I.L., Reuß, C., Thomann, U., Wallauer, W., and Fauster, T. (1997) Time-resolved coherent photoelectron spectroscopy of quantized electronic states on metal surfaces. *Science*, **277**, 1480.

5 Echenique, P.M. and Pendry, J.B. (1990) Theory of image states at metal surfaces. *Prog. Surf. Sci.*, **32**, 111.

6 Persson, B.N.J. (1984) Inelastic electron scattering from thin metal films. *Solid State Commun.*, **52**, 811.

7 Eguiluz, A., Lee, T.K., Quinn, J.J., and Chiu, K.W. (1975) Interface excitations in metal-insulator-semiconductor structures. *Phys. Rev. B*, **11**, 4989.

8 Gumbs, G. (1989) Inelastic electron scattering from the surfaces of semiconductor multilayers: Study of subband structure within the quantum well. *Phys. Rev. B*, **39**, 5186.

9 Gumbs, G. and Ali, M.K. (1988) Dynamical maps, Cantor spectra, and localization for Fibonacci and related quasiperiodic lattices. *Phys. Rev. Lett.*, **60**, 1081.

10 Steinhardt, P.J. and Ostlund, S. (1986) *The physics of quasicrystals*, World Scientific, Singapore.

11 Kohmoto, M., Kadanoff, L.P., and Tang, C. (1983) Localization problem in one dimension: Mapping and escape. *Phys. Rev. Lett.*, **50**, 1870.

12 Gumbs, G. (1988) Fast-particle energy loss to a layered electron gas. *Phys. Rev. B*, **37**, 10184.

13 Bakshi, P., Cen, J., and Kempa, K. (1988) Amplification of surface modes in type II semiconductor superlattices. *J. Appl. Phys.*, **64**, 2243.

14 Cen, J., Kempa, K., and Bakshi, P. (1988) Amplification of a new surface plasma mode in the type-I semiconductor superlattice. *Phys. Rev. B*, **38**, 10051.

15 Kempa, K., Bakshi, P., Cen, J., and Xie, H. (1991) Spontaneous generation of plasmons by ballistic electrons. *Phys. Rev. B*, **43**, 9273.

16 Kempa, K., Cen, J., and Bakshi, P. (1989) Current-driven plasma instabilities in superconductors. *Phys. Rev. B*, **39**, 2852.

17 Balassis, A. and Gumbs, G. (2006) Plasmon instability in energy transfer from a current of charged particles to multiwall and cylindrical nanotube arrays based on self-consistent field theory. *Phys. Rev. B*, **74** 045420(8).

18 Ando, T., Fowler, A.B., and Stern, F. (1982) Electronic properties of two-dimensional systems. *Rev. Mod. Phys.*, **54**, 437.

19 Huang, D.H., Rhodes, C., Alsing, P.M., and Cardimona, D.A. (2006) Effects of longitudinal field on transmitted near field in doped semi-infinite semiconductors with a surface conducting sheet. *J. Appl. Phys.*, **100**, 113711.

20 Gumbs, G. and Huang, D.H. (2007) Electronically modulated two-dimensional plasmons coupled to surface plasmon modes. *Phys. Rev. B*, **75**, 115314.

21 Stern, F. (1967) Polarizability of a two-dimensional electron gas. *Phys. Rev. Lett.*, **18**, 546.

22 Ritchie, R.H. (1957) Plasma Losses by Fast Electrons in Thin Films. *Phys. Rev.*, **106**, 874.

23 Garcia de Abajo, F.J. and Howie, A. (2002) Retarded field calculation of electron energy loss in inhomogeneous dielectrics. *Phys. Rev. B*, **65**, 115418.

24 Gumbs, G. and Horing, N.J.M. (1991) Plasma losses by charged particles in thin films: Effects of spatial dispersion, phonons, and magnetic field. *Phys. Rev. B*, **43**, 2119.

25 Pendry, J.B. and Moreno, L.M. (1994) Energy loss by charged particles in complex media. *Phys. Rev. B*, **50**, 5062.

26 Pitarke, J.M., Pendry, J.B., and Echenique, P.M. (1997) Electron energy loss in composite systems. *Phys. Rev. B*, **55**, 9550.

27 Echenique, P.M. and Pendry, J.B. (1975) Absorption profile at surfaces. *J. Phys. C: Solid State*, **8**, 2936.

28 Zabala, N., Ogando, E., Rivacoba, A., and Garcia de Abajo, F.J. (2001) Inelastic scattering of fast electrons in nanowires: A dielectric formalism approach. *Phys. Rev. B*, **64**, 205410.

29 Garcia-Lekue, A. and Pitarke, J.M. (2001) Energy loss of charged particles interacting with simple metal surfaces. *Phys. Rev. B*, **64**, 035423.

30 Aminov, K.L. and Pedersen, J.B. (2001) Quantum theory of high-energy electron transport in the surface region. *Phys. Rev. B*, **63**, 125412.

31 Ferrel, T.L., Echenique, P.M., and Ritchie, R.H. (1979) Friction parameter of an ion near a metal surface. *Solid State Commun.*, **32**, 419.

32 Horing, N.J.M., Tso, H.C., and Gumbs, G. (1987) Fast-particle energy loss in the vicinity of a two-dimensional plasma. *Phys. Rev. B*, **36**, 1588.

33 Bertsch, G.F., Esbensen, H., and Reed, B.W. (1998) Electron energy-loss spectrum of nanowires. *Phys. Rev. B*, **58**, 14031.

34 Gervasoni, J.L. and Arista, N.R. (2003) Plasmon excitations in cylindrical wires by external charged particles. *Phys. Rev. B*, **68**, 235302.

35 Wang, Y.N. and Miskovic, Z.L. (2002) Energy loss of charged particles moving in cylindrical tubules. *Phys. Rev. A*, **66**, 042904.

36 Stephan, O., Taverna, D., Kociak, M., Suenaga, K., Henrard, L., and Colliex, C. (2002) Dielectric response of isolated carbon nanotubes investigated by spatially resolved electron energy-loss spectroscopy: From multiwalled to single-walled nanotubes. *Phys. Rev. B*, **66**, 155422.

37 Pitarke, J.M. and Garcia-Vidal, F.J. (2001) Electronic response of aligned multishell carbon nanotubes. *Phys. Rev. B*, **63**, 073404.

38 Rivacoba, A. and Garcia de Abajo, F.J. (2003) Electron energy loss in carbon nanostructures. *Phys. Rev. B*, **67**, 085414.

39 Arista, N.R. and Fuentes, M.A. (2001) Interaction of charged particles with surface plasmons in cylindrical channels in solids. *Phys. Rev. B*, **63**, 165401.

40 Arista, N.R. (2001) Interaction of ions and molecules with surface modes in cylindrical channels in solids. *Phys. Rev. A*, **64**, 032901.

6
The Rashba Spin–Orbit Interaction in 2DEG

6.1
Introduction to Spin–Orbit Coupling

The electronic transport and photonic effects in a two-dimensional electron gas (2DEG) such as that found at a semiconductor heterojunction of GaAs/AlGaAs have been the subject of interest and discussion for many years now [1]. Related physical properties of narrow strips of 2DEG have also been the subject of experimental and theoretical investigations because of their potential for device applications in the field of nanotechnology [2, 3]. It is thus necessary to include the edge effects in a model of a narrow nanoribbon [4–6]. Here, we analyze the role played by the boundaries in a nanoribbon of 2DEG where the Rashba spin–orbit interaction (SOI) is included.

It is well established that spin–orbit coupling is an essentially relativistic effect: an electron moving in an external electric field sees a magnetic field in its rest frame. In a semiconductor, the interaction causes an electron's spin to precess as it moves through the material; this is the basis of various proposed "spintronic" devices. In nano-structures, quantum confinement can change the symmetry of the spin–orbit interaction. The relativistic motion of an electron is described by the Dirac equation. The effects form an electric dipole moment and the Thomas precession which is due to the rotational kinetic energy in the electric field [7, 8]. The two mechanisms accidentally have very close mathematical form and consequently combine in a very elegant way.

We include the effects due to edges through quasi-square-well boundary conditions. As a result, we are not able to solve the Rashba SOI model Hamiltonian to obtain analytic solutions for the eigenenergies and eigenfunctions analytically. The reason for this is due to the fact that the solution manifestly contains quantum interference effects from multiple effects off the edges. We solved the eigenvalue problem numerically, obtaining the energies as a function of the wave vector k_y parallel to the edge of the nanoribbon shown schematically in Figure 6.1. Our results show that for realistic values of the Rashba spin–orbit coupling and narrow nanoribbons, the lowest two energy levels are well separated from the higher excitation energies. Based on the fact that there is such an energy gap, we exploit this in our dispersion equation for the collective plasmon excitations [9–11].

Properties of Interacting Low-Dimensional Systems, First Edition. G. Gumbs and D. Huang.

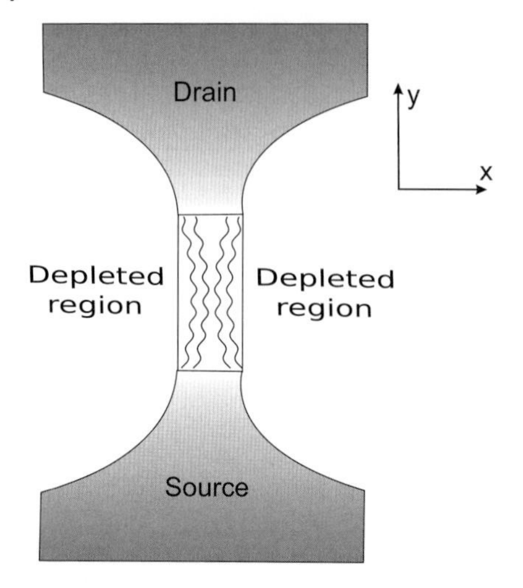

Figure 6.1 Schematic illustration of the nanowire of 2DEG between a source and drain.

6.2 Spin–Orbit Coupling in the Dirac Equation

It is now well established that spin–orbit coupling is an essentially relativistic effect. The relativistic motion of an electron is described by the Dirac equation that contains both effects (electric dipole and Thomas precession) in the spin–orbit interaction and does so in a very elegant way [12, 13]. The SOI Hamiltonian can be obtained from the Dirac equation by taking the non-relativistic limit of the Dirac equation up to terms quadratic in v/c inclusive. This limit can be attained in two different ways: either by direct expansion of the Dirac equation in powers of v/c or by the asymptotically exact Foldy–Wouthuysen transformation [14]. Here, we will only present the method using the Dirac equation.

The Dirac equation in an external electromagnetic field described by the vector potential $\boldsymbol{A}$ (magnetic field $\boldsymbol{B} = \nabla \times \boldsymbol{A}$) and the scalar potential Φ has the following form

$$(\hat{\mathcal{E}} - e\Phi)\Psi = c(\hat{\alpha} \cdot \hat{\boldsymbol{P}})\Psi + \hat{\beta} mc^2 \Psi \,, \tag{6.1}$$

where $\hat{\mathcal{E}} = i\hbar\partial/\partial t$ is the electron energy operator, $\hat{\boldsymbol{P}} = \hat{\boldsymbol{p}} - e\boldsymbol{A}$ is the electron momentum in the electromagnetic field, and $\hat{\boldsymbol{p}} = -i\hbar\nabla$. The wave function $\Psi = \Psi(x)$, where $x = (\boldsymbol{r}, ict)$ is the space-time four-dimensional vector coordinate. We write Ψ as a bi-spinor with spinor components φ and χ,

$$\Psi = \begin{bmatrix} \varphi(x) \\ \chi(x) \end{bmatrix} . \tag{6.2}$$

For a free particle, the spinor, φ, describes the spin-up and spin-down states with a positive energy, while the spinor, χ, corresponds to the spin-up and spin-down states with a negative energy. The Dirac matrices $\hat{\alpha}$ and $\hat{\beta}$ are defined as

$$\hat{\alpha} \equiv \begin{bmatrix} 0 & \overleftrightarrow{\sigma} \\ \overleftrightarrow{\sigma} & 0 \end{bmatrix} , \quad \hat{\beta} \equiv \begin{bmatrix} \hat{\sigma}_0 & 0 \\ 0 & \hat{\sigma}_0 \end{bmatrix} , \tag{6.3}$$

where $\hat{\sigma}_0$ and $\overleftrightarrow{\sigma} = \{\hat{\sigma}_x, \hat{\sigma}_y, \hat{\sigma}_z\}$ are the Pauli matrices in spin space.

A convenient way to study the non-relativistic limit of the Dirac equation (6.1) is to explicitly separate the positive and negative energy states by rewriting Eq. (6.1) in terms of the spinor components φ and χ. We obtain

$$\begin{aligned} \mathcal{T}\varphi &= c(\overleftrightarrow{\sigma} \cdot \hat{\boldsymbol{P}})\chi , \\ (\mathcal{T} + 2mc^2)\chi &= c(\overleftrightarrow{\sigma} \cdot \hat{\boldsymbol{P}})\varphi , \end{aligned} \tag{6.4}$$

where $\mathcal{T} = \mathcal{E} - mc^2 - e\Phi$ is the kinetic energy of the electron corresponding to positive energy states. The non-relativistic limit implies that the kinetic energy should be much less than the rest energy mc^2. This means that the left-hand side of the second equation in Eq. (6.4) is approximately equal to $2mc^2\chi$ to the lowest order and we obtain in a straightforward way

$$\chi = \left[\left(1 - \frac{\mathcal{T}}{2mc^2}\right)\frac{(\overleftrightarrow{\sigma} \cdot \hat{\boldsymbol{P}})}{2mc} + \mathcal{O}\left(\frac{v^4}{c^4}\right)\right]\varphi . \tag{6.5}$$

Consequently, after we substitute Eq. (6.5) into the first equation in Eq. (6.4), we obtain $\mathcal{E}'\varphi = \hat{H}\varphi$, where $\mathcal{E}' = \mathcal{E} - mc^2$ and

$$\hat{H} \equiv \frac{(\overleftrightarrow{\sigma} \cdot \hat{\boldsymbol{P}})}{2m}\left[1 - \frac{\mathcal{E}' - V(\boldsymbol{r})}{2mc^2}\right](\overleftrightarrow{\sigma} \cdot \hat{\boldsymbol{P}}) + V(\boldsymbol{r}) \tag{6.6}$$

with $V(\boldsymbol{r}) = e\Phi$. By using algebraic properties of the Pauli matrices to rewrite the momentum-dependent terms in the Hamiltonian equation (6.6) as

$$\left(1 - \frac{\mathcal{E}'}{2mc^2}\right)\frac{(\overleftrightarrow{\sigma} \cdot \hat{\boldsymbol{P}})^2}{2m} = \left(1 - \frac{\mathcal{E}'}{2mc^2}\right)\frac{\left(\hat{\boldsymbol{P}}\right)^2}{2m} - \frac{e\hbar}{2mc}\hat{\sigma} \cdot \boldsymbol{B} + \mathcal{O}\left(\frac{v^3}{c^3}\right) , \tag{6.7}$$

$$(\overleftrightarrow{\sigma} \cdot \hat{\boldsymbol{P}})\frac{V(\boldsymbol{r})}{2mc^2}(\overleftrightarrow{\sigma} \cdot \hat{\boldsymbol{P}}) = \frac{V(\boldsymbol{r})}{2mc^2}\frac{\hat{\boldsymbol{P}}^2}{2m} + \frac{\hbar}{2mc^2}\left[\nabla V(\boldsymbol{r}) \times \hat{\boldsymbol{p}}\right] \cdot \hat{\sigma} \tag{6.8}$$

$$-\frac{\hbar}{4m^2c^2}\left[\nabla V(\boldsymbol{r}) \cdot \hat{\boldsymbol{p}}\right] + \mathcal{O}\left(\frac{v^3}{c^3}\right) . \tag{6.9}$$

The final step in transforming the Hamiltonian is to choose the correct normalization for the wave functions corresponding to the approximation which employs up to quadratic terms in v/c. Using Eq. (6.5), the normalization condition for the wave function can be written asymptotically as

$$\int d^3 r \Psi^\dagger(r)\Psi(r) \approx \left[1 + \frac{\left(\overleftrightarrow{\sigma}\,\hat{P}\right)^2}{4m^2c^2}\right]\varphi^\dagger(r)\varphi(r)$$

$$\approx \left[1 + \frac{\hat{p}^2}{4m^2c^2}\right]\varphi^\dagger(r)\varphi(r) = 1\,. \tag{6.10}$$

From Eq. (6.10), it follows that the spinor φ is not a normalized eigenvector, while the normalized [to order $(v/c)^2$] function Ψ is given by

$$\Psi \approx \eta\,(\hat{p})\,\varphi \quad \text{with} \quad \eta\,(\hat{p}) \approx 1 + \frac{\hat{p}^2}{8m^2c^2}\,. \tag{6.11}$$

When we substitute the representation $\varphi = [\eta(\hat{p})]^{-1}\Psi$ from Eq. (6.11) into Eq. (6.6), we obtain the following equation in terms of Ψ, that is,

$$\tilde{\mathcal{H}}\Psi = \tilde{\mathcal{E}}\Psi \quad \text{with} \quad \tilde{\mathcal{E}} \equiv \eta\,(\hat{p})\,\mathcal{E}'\eta\,(\hat{p})^{-1}, \quad \tilde{\mathcal{H}} \equiv \eta\,(\hat{p})\,\hat{H}\eta\,(\hat{p})^{-1}. \tag{6.12}$$

Equations (6.7) and (6.9) imply that we might treat the normalization function $\eta(\hat{p})$ different from unity in the definition of the Hamiltonian $\tilde{\mathcal{H}}$ in Eq. (6.12) only in the last term of Eq. (6.6), which gives

$$\eta\,(\hat{p})\,V\,(r)\,\eta\,(\hat{p})^{-1} \approx V\,(r) - \frac{1}{8m^2c^2}\{\hbar^2\nabla^2 V\,(r)$$

$$+\, 2i\hbar\,[\nabla V\,(r)\cdot\hat{p}]\} + \mathcal{O}\left(\frac{v^2}{c^2}\right)\,. \tag{6.13}$$

Combining Eqs. (6.7), (6.9), and (6.13), we finally obtain the Hamiltonian for an electron in the quadratic $[\mathcal{O}(v^2/c^2)]$ approximation as the sum

$$\tilde{\mathcal{H}} = \hat{\mathcal{H}}_{SO} + \Delta\hat{H}\,, \tag{6.14}$$

where $\Delta\hat{H}$ is the free Hamiltonian and

$$\hat{\mathcal{H}}_{SO} = \frac{\hbar}{4m^2c^2}\,[\nabla V\,(r)\times\hat{p}]\cdot\overleftrightarrow{\sigma} \tag{6.15}$$

describes the SOI within the material and includes both contributions to the spin–orbit coupling from the electric dipole and the Thomas precession (caused by the electric field) mechanisms. This result is *general* since it was derived from the Dirac equation (6.1), an exact relativistic equation for the electron, and includes all possible relativistic effects, whatever might be their kinetic source. When an electron gas at a heterojunction is confined in xy-plane so that the electrostatic potential is uniform along the heterostructure interface and varies only along the z-axis, the Hamiltonian in Eq. (6.15) contains just the contribution arising from its confinement along the z-direction. For a quasi-one-dimensional structure, a second term must now be added to account for the extra local confinement produced by the electric field within the 2D plane.

6.3 Rashba Spin–Orbit Coupling for a Quantum Wire

For quantum wires, the width of the potential well is comparable with the spatial spread of the electron wave functions in the z-direction. Therefore, in order to determine an effective electric field acting on electrons in the potential well, one should calculate an average of the electric field $E(z)$ over the range of the z variable where the wave function is essentially finite. Consequently, one can model the averaged electric field by a potential. In principle, all potential profiles can be classified in two ways. In the first case, the average electric field is negligible, although $E(z)$ may not be zero or even small. This applies for symmetric potentials such as the square and parabolic quantum wells. However, for asymmetric quantum wells, the average electric field is non-zero in the direction perpendicular to the plane of the 2DEG and is called the interface or quantum well electric field. In experimentally achievable semiconductor heterostructures, this field can be as large as 10^7 V/cm. Therefore, from Eq. (6.15), there should be an additional (compared with the infinite 3D crystal) mechanism of spin–orbit coupling associated with this field and is usually referred to as the Rashba SOI for quantum wells [15]. When we take into account that the quantum well electric field is perpendicular to the heterojunction interface, the spin–orbit Hamiltonian has a contribution which can be written for the Rashba coupling as

$$\hat{\mathcal{H}}_{\mathrm{SO}}^{(\alpha)} = \frac{\alpha}{\hbar}\left(\overleftrightarrow{\sigma} \times \hat{\boldsymbol{p}}\right)_z \tag{6.16}$$

within the zero z-component (stationary situation, no electron transfer across the interface). The constant α includes universal constants from Eq. (6.15) and it is proportional to the interface electric field. The value of α determines the contribution of the Rashba spin–orbit coupling to the total electron Hamiltonian. This constant may have values from (1–10) meV nm.

Within the single-band effective mass approximation, [16, 17] the Hamiltonian of a quasi-one-dimensional electron system (Q1DES) can be written as

$$\hat{\mathcal{H}} = \frac{\hat{p}^2}{2m^*} + V_{\mathrm{C}}(\boldsymbol{r}) + \hat{\mathcal{H}}_{\mathrm{SO}}\,, \tag{6.17}$$

where the electron effective mass m^* incorporates both the crystal lattice and interaction effects. The form of the Hamiltonian derived from the relativistic 4×4 Dirac equation is similar to that which follows from the $(8 \times 8) - \boldsymbol{k} \cdot \boldsymbol{p}$ Hamiltonian [18]. Moroz and Barnes [4] chose the lateral confining potential $V_{\mathrm{C}}(\boldsymbol{r})$ as a parabola which would be appropriate for very narrow wires since the electrons would be concentrated at the bottom of the potential. Such narrow Q1DES are difficult to achieve experimentally. We are not aware of any experimental evidence or measurement of the features arising from the spin–orbit coupling resulting from the parabolic confining potential employed by Moroz and Barnes. Thus, in this part, we explore the effects of confinement in which the electrons are essentially free over a wide range except close to the edges where the potential rises sharply to confine them. The in-plane electric field $\boldsymbol{E}_{\mathrm{C}}(\boldsymbol{r})$ associated with $V_{\mathrm{C}}(\boldsymbol{r})$ is given by $\boldsymbol{E}_{\mathrm{C}}(\boldsymbol{r}) = -\nabla V_{\mathrm{C}}(\boldsymbol{r})$. We assume that the SOI Hamiltonian in Eq. (6.17) is formed by two contributions: $\hat{\mathcal{H}}_{\mathrm{SO}} = \hat{\mathcal{H}}_{\mathrm{SO}}^{(\alpha)} + \hat{\mathcal{H}}_{\mathrm{SO}}^{(\beta)}$. The first one, $\hat{\mathcal{H}}_{\mathrm{SO}}^{(\alpha)}$, (Eq. (6.16)) arises from the asymmetry of the quantum well, that is, from the Rashba mechanism of the spin–orbit coupling. For convenience, in what follows, we will refer to the Rashba mechanism of the spin–orbit coupling as α-coupling. If the lateral confinement is sufficiently strong for narrow and deep potentials or sharp and high potentials at the edges, then the electric field associated with it may not be negligible compared with the interface-induced (Rashba) field. We use a square well potential $V(x) = V_0[\theta(-x) + \theta(x - \mathcal{W})]$ for a conducting channel of width $\mathcal{W}$ with barrier height V_0. Also, $\theta(x)$ in the unit step function. For this potential, the Hamiltonian gives a term

$$\begin{aligned}\hat{\mathcal{H}}_{\mathrm{SO}}^{(\beta)} &= -i\beta\sigma_z\left(\frac{\mathcal{W}}{\ell_0}\right)\left\{\exp\left[-\frac{(x-\mathcal{W})^2}{2\ell_0^2}\right] - \exp\left[-\frac{x^2}{2\ell_0^2}\right]\right\}\frac{\partial}{\partial y}\\ &\equiv i\beta\mathcal{F}(x)\sigma_z\frac{\partial}{\partial y}\,,\end{aligned} \tag{6.18}$$

in which we approximate the derivative of the step function by a Gaussian of width ℓ_0 at the edges $x = 0$ and $x = \mathcal{W}$. In Eq. (6.18), $\mathcal{F}(x)$ is related to the electric field due to confinement in the x-direction. Since $\ell_0 \ll \mathcal{W}$ characterizes the steepness of the potentials at the two edges, we are at liberty to use a range of values of the ratio of these two lengths, keeping in mind that the in-plane confinement must be appreciable if the β-term is to play a role. Therefore, in most of our calculations, we only use one small value of $\ell_0/\mathcal{W}$ to illustrate the effects arising from our model on the conductance and thermoelectric power. We introduced the parameter $\beta = \hbar^2 V_0/(4\sqrt{2\pi}m^{*2}c^2\mathcal{W})$, which is expressed in terms of fundamental constants as well as V_0 and $\mathcal{W}$. The β is another Rashba parameter due to the electric confinement along the x-direction. Comparison of typical electric fields originating from the quantum well and lateral confining potentials allows one to conclude that a reasonable estimate [4] for β should be roughly 10% of α. The β-SOI term in Eq. (6.18) is asymmetric about the mid-plane $x = \mathcal{W}/2$ and varies quadratically with the displacement from either edge. In this quasi-square well potential, the electron wave functions slightly penetrate the barrier regions. However, we only need energy levels for the calculations of ballistic transport electrons, not the wave functions, if we assume electronic system is a quasi-one-dimensional one.

The eigenfunctions for the nanowire have the form

$$\varphi(\boldsymbol{r}) = \frac{e^{ik_y y}}{\sqrt{L_y}}\begin{bmatrix}\psi_A(x)\\ \psi_B(x)\end{bmatrix}. \tag{6.19}$$

Since the nanowire is translationally invariant in the y-direction with $k_y = (2\pi/L_y)n$, where L_y is a normalization length and $n = 0, \pm1, \pm2, \ldots$, we must solve for $\psi_A(x)$ and $\psi_B(x)$ in Eq. (6.19) numerically due to the presence of edges at $x = 0$ and $x = \mathcal{W}$. Substituting the wave function in Eq. (6.19) into the

Schrödinger equation, that is, $\hat{H}_{SO}\varphi(\boldsymbol{r}) = \varepsilon\varphi(\boldsymbol{r})$ with ε being the eigenenergy, we obtain the two coupled equations

$$\begin{aligned}
&-\frac{\hbar^2}{2m^*}\left(\frac{d^2}{dx^2} - k_y^2\right)\psi_A(x) + \alpha\left(\frac{d}{dx} + k_y\right)\psi_B(x)\\
&-\beta k_y \mathcal{F}(x)\psi_A(x) = \varepsilon\psi_A(x)\,,\\
&-\frac{\hbar^2}{2m^*}\left(\frac{d^2}{dx^2} - k_y^2\right)\psi_B(x) - \alpha\left(\frac{d}{dx} - k_y\right)\psi_A(x)\\
&+\beta k_y \mathcal{F}(x)\psi_B(x) = \varepsilon\psi_B(x)\,.
\end{aligned} \tag{6.20}$$

In the absence of any edges, we may set $\mathcal{F}(x) = 0$, $\psi_A(x) = \mathcal{A}e^{ik_x x}$ and $\psi_B(x) = \mathcal{B}e^{ik_x x}$, where $\mathcal{A}$ and $\mathcal{B}$ are independent of x, which then yields a pair of simultaneous algebraic equations for states A and B. However, in the case when there exist edges, we have a pair of coupled differential equations to solve for ψ_A and ψ_B which may be analyzed when only β is not zero and then when both Rashba parameters are non-zero.

Two parameters of interest are

$$\ell_\alpha = \hbar^2/2m^*\alpha\,, \quad \ell_\beta = \hbar^2/2m^*\beta\,, \tag{6.21}$$

with three ratios

$$\tau_\alpha = \mathcal{W}/\ell_\alpha\,, \quad \tau_\beta = \mathcal{W}/\ell_\beta\,, \quad \tau_0 = \mathcal{W}/\ell_0\,. \tag{6.22}$$

In our numerical calculations below, we will use these three ratios to determine how narrow the nanowire is and how strong the Rashba interaction are.

Here, the energy bands are symmetric with respect to the wave numbers $\pm k_y$. For the symmetric bands, energy dispersion ε_{j,k_y}, the Fermi function $f_0(\varepsilon_{j,k_y})$, and the group velocity v_{j,k_y} satisfy the relations: $\varepsilon_{j,k_y} = \varepsilon_{j,-k_y}$, $f_0(\varepsilon_{j,k_y}) = f_0(\varepsilon_{j,-k_y})$, and $v_{j,k_y} = -v_{j,-k_y}$. Therefore, one can write the following equation [19] in a form which includes only positive values of the wave number for the ballistic heat ($\mathcal{Q}^{(1)}$) and charge ($\mathcal{Q}^{(0)}$) currents, that is,

$$\begin{aligned}
\mathcal{Q}^{(\ell)} = {}& \frac{eV_b(-e)^{1-\ell}}{\pi}\sum_j\left(\int_{\varepsilon_{j,k_0}}^{\varepsilon_{j,k_1}} + \int_{\varepsilon_{j,k_1}}^{\varepsilon_{j,k_2}} + \ldots + \int_{\varepsilon_{j,k_N}}^{\infty}\right)\mathrm{sgn}(v_{j,k_y})\\
&\times\left(\varepsilon_{j,k_y} - \mu\right)^\ell\left[\frac{\partial f_0(\varepsilon_{j,k_y})}{\partial\varepsilon_{j,k_y}}\right]d\varepsilon_{j,k_y}\,,
\end{aligned} \tag{6.23}$$

where $\ell = 0, 1$, V_b is the bias voltage between the source and drain electrodes, $\mathrm{sgn}(x)$ is the sign function, $k_0 = 0$, and μ is the chemical potential. In Eq. (6.23), the whole energy integration performed over the range $0 \le k_y < \infty$ is divided into the sum of many sub-integrations between two successive extremum points ε_{j,k_n} for $0 \le n \le N$, and ε_{j,k_N} is the last minimum point. For each sub-integration over k_y, ε_{j,k_y} is a monotonic function. In addition, each sub-integration in Eq. (6.23) can be calculated analytically, leading to the following expression for electron-diffusion thermoelectric power

$$S_{\mathrm{d}} = \frac{Q^{(1)}}{TQ^{(0)}} = -\frac{k_{\mathrm{B}}}{eg} \sum_{j,n} C_{j,n} \times \left[\beta\left(\varepsilon_{j,k_n} - \mu\right) f_0(\varepsilon_{j,k_n}) + \ln\left(e^{\beta(\mu-\varepsilon_{j,k_n})} + 1\right)\right], \tag{6.24}$$

where T is the temperature, $\beta = 1/k_{\mathrm{B}}T$, and the dimensionless conductance g is given by

$$g = \sum_{j,n} C_{j,n} f_0(\varepsilon_{j,k_n}) . \tag{6.25}$$

Physically, the quantity g defined in Eq. (6.25) represents the number of pairs of the Fermi points at $T = 0\,\mathrm{K}$. In Eqs. (6.24) and (6.25), the summations over n are for all the energy-extremum points on each jth spin-split subband in the range $0 \le k_y < \infty$. The quantity ε_{j,k_n} is the energy at the extremum point $k_y = k_{j,n}$. For a given jth spin-split subband, $C_{j,n} = 1$ (or $C_{j,n} = -1$) for a local energy minimum (maximum) point. The physical conductance G is related to g for spin-split subbands through

$$G = \left(\frac{e^2}{h}\right) g . \tag{6.26}$$

6.4
SOI Effects on Conductance and Electron-Diffusion Thermoelectric Power

In Figure 6.2, we have displayed comparisons of modified electron density ($n_{1\mathrm{D}}$) dependence of the ballistic conductance (G) and the electron-diffusion thermoelectric power (S_{d}) by the α-term in the SOI when $T = 4\,\mathrm{K}$ and $\mathcal{W} = 568.7\,\text{Å}$. From Figure 6.2a, we find that, as $\alpha = 0$ (black curve), a number of steps in G show up as a result of successive populations of more and more spin-degenerate subbands. In addition, the observed plateau becomes wider and wider as higher and higher subbands are occupied by electrons due to increased energy-level separation, resulting from the high potential barriers at the two edges. The finite-temperature effect can easily be seen from the smoothed steps in this figure. As α is increased to 0.5 eV Å (gray curve), the steps are rightward shifted to higher electron densities due to an enhanced density-of-states from the flattened subband dispersion curves by SOI. However, the step sharpness remains constant in this case. Furthermore, there exists no "pole-like feature" [4] in this figure which can be traced back to the absence of spike-like feature in the subband dispersion curves, leading to additional local energy minimum/maximum points. The suppressed spike-like feature in the subband dispersion curves can be explained by a nonlinear x dependence near the center ($x = \mathcal{W}/2$) of a transverse symmetric potential well with a large β value

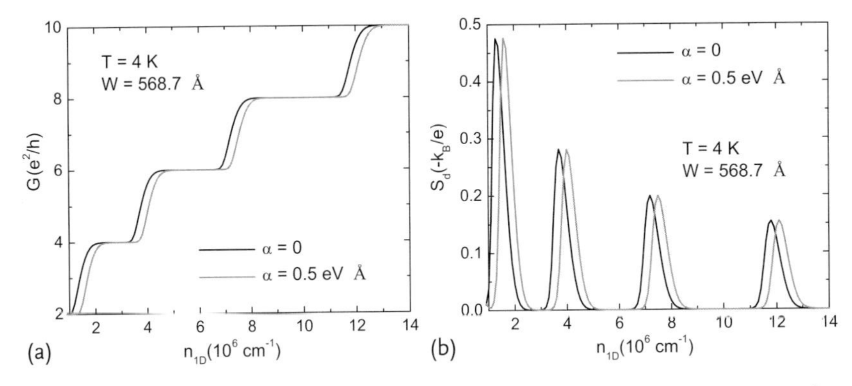

Figure 6.2 Comparisons of the conductance G (a) as well as the electron-diffusion thermoelectric power S_d (b) as a function of the electron density n_{1D} with a wire width $\mathcal{W} = 568.7$ Å and a temperature $T = 4$ K for $\alpha = 0$ (black curves) and $\alpha = 0.5$ eV Å (gray curves), respectively. Here, $\tau_\beta = 1.0$ and $\tau_0 = 10^3$ are chosen for the calculations in these two figures.

in our model for wide quantum wires, instead of a linear x dependence close to the center of the confining potential in the model proposed by Moroz and Barnes [4] for narrow quantum wires. We also see sharp peaks in S_d from Figure 6.2b as $\alpha = 0$ (black curve), corresponding to the steps in G, which again comes from successive population of spin-degenerate subbands with increased n_{1D} [20]. The center of a plateau in G aligns with the minimum of S_d between two peaks. The peaks (gray curve) are rightward shifted accordingly in electron density when a finite value of α is assumed.

In Figure 6.3, we have compared the results of G and S_d for two values of wire width $\mathcal{W}$ at $T = 4$ K and $\alpha = 0.5$ eV Å. We find from Figure 6.3a that as $\mathcal{W}$ decreases from 1137.4 Å (black curve) to 568.7 Å (gray curves), the steps in G are leftward shifted in electron density, and meanwhile, the steps become sharpened. The step

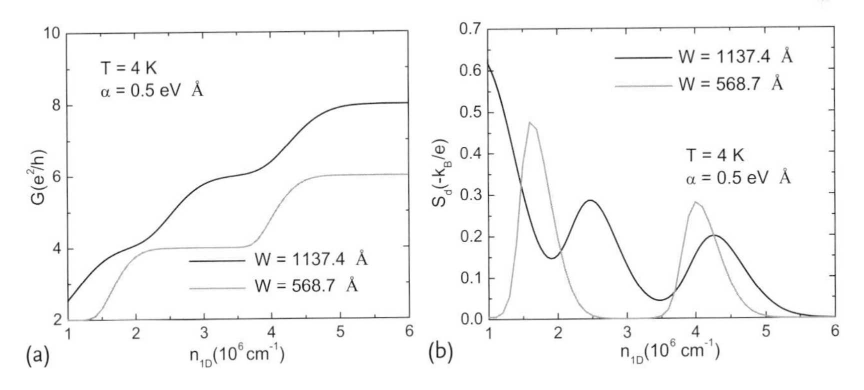

Figure 6.3 Comparisons of G (a) and S_d (b) as a function of n_{1D} at $T = 4$ K and with $\alpha = 0.5$ eV Å for $\mathcal{W} = 1137.4$ Å (black curves) and $\mathcal{W} = 568.7$ Å (gray curves), separately. Here, $\tau_\beta = 1.0$ and $\tau_0 = 10^3$ are chosen for the calculations in these two figures.

shifting is a result of the reduction of SOI effect due to a smaller value for τ_α (proportional to $\alpha\mathcal{W}$) with a fixed value of α. This leads to a leftward shift in steps for the same reason given for Figure 6.2a. The step sharpening, on the other hand, comes from the significantly increased subband separation (proportional to $1/\mathcal{W}^2$), which effectively suppresses the thermal-population effect on G for smoothing out the conductance steps. The shifting of steps in G with $\mathcal{W}$ is also reflected in S_d, as shown in Figure 6.3b. The peaks of S_d get sharpened due to the suppression of S_d in the density region corresponding to the widened plateaus of G.

6.5 Problems

1. Consider the two-dimensional (2D) Hamiltonian

$$H = -\frac{\boldsymbol{p}^2}{2m^*} + \frac{\Delta_\mathrm{R}}{\hbar}(\hat{\sigma} \times \boldsymbol{p})_z , \tag{6.27}$$

where Δ_R is the Rashba parameter for spin–orbit coupling, m^* is the electron effective mass, $\boldsymbol{p} = -i\hbar\nabla$ is the in-plane momentum and $\hat{\sigma} = (\sigma_x, \sigma_y, \sigma_z)$ is the vector of Pauli spin matrices.

a) Verify that the eigenfunctions are given by ($s = \pm 1$)

$$\psi_{\boldsymbol{k},s} = \frac{1}{\sqrt{2}} \begin{pmatrix} 1 \\ s\frac{(k_y - ik_x)}{k} \end{pmatrix} \frac{e^{i\boldsymbol{k}\cdot\boldsymbol{r}}}{\sqrt{A}} \tag{6.28}$$

with corresponding eigenvalues

$$\epsilon_{\boldsymbol{k},s} = \frac{\hbar^2 \boldsymbol{k}^2}{2m^*} + s\Delta_\mathrm{R} k , \tag{6.29}$$

where A is a normalization area and the 2D wave vector $\boldsymbol{k} = (k_x.k_y)$.

b) For a total areal electron density $n_{2\mathrm{D}}$, there will be $n_+ \uparrow$-spins and $n_- \downarrow$-spins with $n_{2\mathrm{D}} = n_- + n_+$. At $T = 0\,\mathrm{K}$, show that these are determined by

$$\frac{n_s}{n_{2\mathrm{D}}} - \frac{1}{2} + sA_\mathrm{R}\left[\left(\frac{n_s}{n_{2\mathrm{D}}}\right)^{1/2} + \left(1 - \frac{n_s}{n_{2\mathrm{D}}}\right)^{1/2}\right] = 0 , \tag{6.30}$$

where $A_\mathrm{R} = k_\mathrm{R}/k_\mathrm{F}$ with $k_\mathrm{R} = m^*\Delta_\mathrm{R}/\sqrt{2}\hbar^2$ and $k_\mathrm{F} = (2\pi n_{2\mathrm{D}})^{1/2}$. For $A_\mathrm{R} < 1/2$, both bands are occupied. When $A_\mathrm{R} \geq 1/2$, $n_+ = 0$ and the spins are polarized in the $\downarrow$ band.

2. Define the density of states (DoS) for each subband

$$\rho_s(\epsilon) = \sum_{\boldsymbol{k}} \delta(\epsilon - E_{\boldsymbol{k},s}). \tag{6.31}$$

Show that the DoS for the energy in Eq. (6.29) is given by

$$\rho_{\uparrow}(\epsilon) = \eta_{+}(\epsilon)\left(\frac{m^*}{2\pi\hbar^2}\right)\left\{1 - \sqrt{\frac{E_\Delta}{\epsilon + E_\Delta}}\right\} \tag{6.32}$$

$$\rho_{\downarrow}(\epsilon) = \left(\frac{m^*}{2\pi\hbar^2}\right)\left\{\eta_{+}(\epsilon)\left(1 + \sqrt{\frac{E_\Delta}{\epsilon + E_\Delta}}\right)\right.$$

$$\left. +2\eta_{+}(-\epsilon)\eta_{+}(\epsilon + E_\Delta)\sqrt{\frac{E_\Delta}{\epsilon + E_\Delta}}\right\}, \tag{6.33}$$

where $\eta_{+}(\epsilon)$ is the unit step function and $E_\Delta = k_{\mathrm{R}}\Delta_{\mathrm{R}}/\sqrt{2}$ is a measure of the spin gap in the DoS. Equations (6.32) and (6.33) show that the total DoS $\rho(\epsilon) = \rho_{\uparrow}(\epsilon) + \rho_{\downarrow}(\epsilon)$ is $m^*/\pi\hbar^2$ for $\epsilon \geq 0$ which is equal to the DoS for a spin degenerate 2D electron system.

Plot (or sketch) $\rho_{\uparrow}(\epsilon), \rho_{\downarrow}(\epsilon)$ and the total DoS $\rho_{\uparrow}(\epsilon), \rho_{\downarrow}(\epsilon)$ in units of $\frac{m^*}{2\pi\hbar^2}$ as functions of ϵ/E_Δ for $-E_\Delta < \epsilon < \infty$.

References

1. Davies, J.H. (1998) *The Physics of Low-Dimensional Semiconductors*, Cambridge University Press, New York.
2. Smith, L.W., Hew, W.K., Thomas, K.J., Pepper, M., Farrer, I., Anderson, D., Jones, G.A.C., and Ritchie, D.A. (2009) Row coupling in an interacting quasi-one-dimensional quantum wire investigated using transport measurements. *Phys. Rev. B*, **80**, 041306.
3. Hew, W.K., Thomas, K.J., Pepper, M., Farrer, I., Anderson, D., Jones, G.A.C., and Ritchie, D.A. (2009) Incipient formation of an electron lattice in a weakly confined quantum wire. *Phys. Rev. Lett.*, **102**, 056804.
4. Moroz, A.V. and Barnes, C.H.W. (1999) Effect of the spin–orbit interaction on the band structure and conductance of quasi-one-dimensional systems. *Phys. Rev. B*, **60**, 14272.
5. Pershin, Y.V., Nesteroff, J.A., and Privman, V. (2004) Effect of spin–orbit interaction and in-plane magnetic field on the conductance of a quasi-one-dimensional system. *Phys. Rev. B*, **69**, 121306.
6. Gumbs, G. (2004) Effect of spin–orbit interaction on the plasma excitations in a quantum wire. *Phys. Rev. B*, **70**, 235314.
7. Darwin, C.G. (1928) On the magnetic moment of the electron. *Proc. R. Soc. Lond.*, **120**, 621.
8. Fisher, G.P. (1971) The electric dipole moment of a moving magnetic dipole. *Am. J. Phys.*, **39**, 1528.
9. Kushwaha, M.S. (2007) Theory of magnetoplasmon excitations in Rashba spintronic quantum wires: Maxons, rotons, and negative-energy dispersion. *Phys. Rev. B*, **76**, 245315.
10. Gumbs, G. (2006) Effect of spontaneous spin depopulation on the ground-state energy of a two-dimensional spintronic system. *Phys. Rev. B*, **73**, 165315.
11. Hai, G.-Q. and Tavares, M.R.S. (2000) Tunneling-assisted acoustic-plasmonquasiparticle excitation resonances in coupled quasi-one-dimensional electron gases. *Phys. Rev. B*, **61**, 1704.
12. Itzykson, C. and Zuber, J.-B. (1980) *Quantum Field Theory*, McGraw-Hill, New York.

13 Thankappan, V.K. (1993) *Quantum Mechanics*, John Wiley & Sons, Inc., New York.

14 Foldy, L.L. and Wouthuysen, S.A. (1950) On the Dirac theory of spin 1/2 particles and its non-relativistic limit. *Phys. Rev.*, **78**, 29.

15 Bychkov, Y.A. and Rashba, E.I. (1984) Oscillatory effects and the magnetic susceptibility of carriers in inversion layers. *J. Phys. C*, **17**, 6039.

16 Li, M.F. (1994) *Modern Semiconductor Quantum Physics*, World Scientific, Singapore.

17 Ridley, B.K. (1993) *Quantum Processes in Semiconductors*, Clarendon Press, Oxford.

18 Darnhofer, T. and Rössler, U. (1993) Effects of band structure and spin in quantum dots. *Phys. Rev. B*, **47**, 16020.

19 Lyo, S.K. and Huang, D.H. (2004) Quantized magneto-thermopower in tunnel-coupled ballistic channels: Sign reversal and oscillations. *J. Phys.: Condens. Matter*, **16**, 3379.

20 Lyo, S.K. and Huang, D.H. (2002) Magnetoquantum oscillations of thermoelectric power in multisublevel quantum wires. *Phys. Rev. B*, **66**, 155307.

7
Electrical Conductivity: the Kubo and Landauer–Büttiker Formulas

In this chapter, we present a general formalism for obtaining the electrical linear response for a conductor with multiple leads in an arbitrary magnetic field and for a given impurity configuration [1–14]. The theory is convenient for analyzing and predicting transport behavior in mesoscopic conductors. The current I_m through the mth lead is derived in terms of the voltages V_n applied at the nth lead, that is, $I_m = \sum_n g_{mn} V_n$ where the conductance coefficients g_{mn} are expressed in terms of Green's functions. Furthermore, we obtain results for the longitudinal and Hall resistances in terms of Green's functions. With the use of scattering theory, we show that g_{mn} can be interpreted as the sum of all transmission coefficients between leads m and n, as was first shown using phenomenological arguments by Büttiker [1]. We show how to use this formalism to determine the conditions satisfied by the scattering matrix for the occurrence of the quantum Hall effect.

We note that a considerable amount of work has been done over the years dealing with the quantum mechanical linear response theory for the total current flowing in and out of the system in response to voltages applied at its boundaries. This started with the seminal work of Kubo and Greenwood [2–4] and was built on and extended by many authors over the years. Our presentation closely follows the work of Baranger and Stone [5]. We shall first derive an expression for the spatially varying, nonlocal frequency-dependent conductivity response function $\sigma(x, x'; \omega)$ which describes the current density response to an electric field. Using this result, we shall obtain the conductance coefficients described above.

7.1
Quantum Mechanical Current

The current operator for an electron of effective mass m^* and with charge $-e$ is

$$j = -\frac{e}{2m^*}\left[\hat{\psi}^\dagger(\boldsymbol{P}\hat{\psi}) + (\boldsymbol{P}\hat{\psi})^\dagger\hat{\psi}\right] \tag{7.1}$$

where $\hat{\psi}^\dagger, \hat{\psi}$ are creation and destruction operators and $\boldsymbol{P} = -i\hbar\nabla + e\boldsymbol{A}(\boldsymbol{x})$ with $\boldsymbol{A}(\boldsymbol{x})$ denoting the vector potential. We denote the total Hamiltonian as $H = H_0 + H_1$ where $H_0 = \boldsymbol{P}^2/2m^* + U(\boldsymbol{x})$ is the unperturbed Hamiltonian with

Properties of Interacting Low-Dimensional Systems, First Edition. G. Gumbs and D. Huang.

eigenfunctions $\phi_\alpha(x)$ and eigenenergies ε_α and $H_1 = -eV(x,t)$ where $V(x,t)$ is the perturbing voltage. Since H_0 commutes with P, for the matrix element of the current operator, we have

$$\begin{aligned} j_{\beta\alpha}(x) = \langle\phi_\beta|j|\phi_\alpha\rangle &= -\frac{e}{2m^*}\left[\phi_\beta(x)^*(P\phi_\alpha(x)) + (P\phi_\beta(x))^*\phi_\alpha(x)\right] \\ &= \frac{ie\hbar}{2m^*}\left[\phi_\beta(x)^* D\phi_\alpha(x) - \left(D^*\phi_\beta^*(x)\right)\phi_\alpha(x)\right], \end{aligned} \tag{7.2}$$

where $D = \nabla + ieA/\hbar$. Equation (7.2) is conveniently rewritten as

$$j_{\beta\alpha}(x) = \frac{ie\hbar}{2m^*}\left[\phi_\beta(x)^* \overleftrightarrow{D}\phi_\alpha(x)\right] \equiv \frac{ie\hbar}{2m^*}W_{\beta\alpha}(x), \tag{7.3}$$

where, for two arbitrary functions f and g, we have defined the double-sided derivative operator by

$$\left[f\overleftrightarrow{D}g\right] = f(x)Dg(x) - g(x)D^*f(x). \tag{7.4}$$

7.2
The Statistical Current

Setting $\hat{\rho} = \hat{\rho}_0 + \hat{\rho}_1(t)$ in the equation of motion for the density matrix $i\hbar d\hat{\rho}/dt = [H,\hat{\rho}]$, we obtain

$$i\hbar\frac{d}{dt}(\hat{\rho}_1)_{\alpha\beta} = \varepsilon_{\alpha\beta}(\hat{\rho}_1)_{\alpha\beta} + ef_{\alpha\beta}V_{\alpha\beta}F(t), \tag{7.5}$$

where $\varepsilon_{\alpha\beta} = \varepsilon_\alpha - \varepsilon_\beta$, $f_{\alpha\beta} = f_0(\varepsilon_\alpha) - f_0(\varepsilon_\beta)$, $V_{\alpha\beta} = \int dx\phi_\alpha^*(x)V(x)\phi_\beta(x)$ and we took the external perturbation as separate functions of space and time, that is, $V(x,t) = V(x)F(t)$. Taking $F(t) = \cos(\omega t)e^{-\delta|t|}$ for $t < 0$ where δ is small and positive, Eq. (7.5) is easily solved and gives the result

$$(\hat{\rho}_1)_{\alpha\beta}(t<0) = \frac{e}{2}f_{\alpha\beta}V_{\alpha\beta}e^{\delta t}\left[\frac{e^{-i\omega t}}{\hbar\omega - \varepsilon_{\alpha\beta} + i\hbar\delta} - \frac{e^{i\omega t}}{\hbar\omega + \varepsilon_{\alpha\beta} - i\hbar\delta}\right]. \tag{7.6}$$

We now make use of this result to obtain the statistical current $J(x,t) = J_0(x) + J_1(x,t)$ where

$$J_0(x) = \sum_{\alpha,\beta}(\hat{\rho}_0)_{\alpha\beta}J_{\beta\alpha}(x) = \frac{ie\hbar}{2m^*}\sum_\alpha f_0(\varepsilon_\alpha)W_{\alpha\alpha}(x), \tag{7.7}$$

$$J_1(x,t) = \sum_{\alpha,\beta}\left[\hat{\rho}_1(t)\right]_{\alpha\beta}J_{\beta\alpha}(x). \tag{7.8}$$

However, it can be shown in a straightforward way that

$$\nabla \cdot \boldsymbol{W}_{\alpha\beta}(\boldsymbol{x}) = \frac{2m^*}{\hbar^2} \varepsilon_{\alpha\beta} \phi_\alpha^*(\boldsymbol{x}) \phi_\beta(\boldsymbol{x}), \tag{7.9}$$

which means that $\nabla \cdot \boldsymbol{j}_0(\boldsymbol{x}) = 0$ since $\nabla \cdot \boldsymbol{W}_{\alpha\alpha}(\boldsymbol{x}) = 0$. This result implies that there is no net current flowing in or out of the equilibrium system in response to an external field. Therefore, we only deal with $\boldsymbol{j}_1(\boldsymbol{x}, t)$.

By substituting Eqs. (7.3) and (7.6) into Eq. (7.8), we obtain in the limit as $\delta \to 0^+$

$$\begin{aligned} \boldsymbol{j}_1(\boldsymbol{x}, t) = {} & \frac{ie^2\hbar^3}{8m^{*2}} \sum_{\alpha,\beta} \boldsymbol{W}_{\beta\alpha}(\boldsymbol{x}) \frac{f_{\beta\alpha}}{\varepsilon_{\beta\alpha}} \int_{\mathcal{A}} d\boldsymbol{x}' \, \boldsymbol{W}_{\alpha\beta}(\boldsymbol{x}') \cdot \boldsymbol{E}(\boldsymbol{x}') \, e^{i\omega t} \\ & \times \left[\frac{\mathcal{P}}{\varepsilon_{\beta\alpha} - \hbar\omega} + \frac{\mathcal{P}}{\varepsilon_{\beta\alpha} + \hbar\omega} - i\pi\delta(\varepsilon_{\beta\alpha} - \hbar\omega) - i\pi\delta(\varepsilon_{\beta\alpha} + \hbar\omega) \right], \end{aligned} \tag{7.10}$$

where $\mathcal{A}$ is the cross-sectional area of the leads and $\mathcal{P}$ denotes the principal part. Equation (7.10) immediately yields the dynamical conductivity $\sigma(\boldsymbol{x}, \boldsymbol{x}'; \omega)$ defined by

$$\boldsymbol{j}_1(\boldsymbol{x}, t) = \int d\boldsymbol{x}' \sigma(\boldsymbol{x}, \boldsymbol{x}'; \omega) \cdot \boldsymbol{E}(\boldsymbol{x}', t), \tag{7.11}$$

where $\boldsymbol{E}(\boldsymbol{x}, t) = -\nabla V(\boldsymbol{x}) e^{i\omega t}$ is the external electric field. This formalism thus gives the current and nonlocal response function in terms of the basis of eigenstates for H_0. In a straightforward way, we have

$$\begin{aligned} \nabla \cdot \boldsymbol{j}_1(\boldsymbol{x}, \omega) = {} & -\frac{\pi e^2}{2} \omega \\ & \times \sum_{\alpha,\beta} f_{\beta\alpha} V_{\alpha\beta} \left\{ \left[\delta(\varepsilon_{\beta\alpha} - \hbar\omega) - \delta(\varepsilon_{\beta\alpha} + \hbar\omega) \right] \right. \\ & \left. + \left(\frac{i\varepsilon_{\beta\alpha}}{\pi\hbar\omega} \right) \left[\frac{\mathcal{P}}{\varepsilon_{\beta\alpha} - \hbar\omega} + \frac{\mathcal{P}}{\varepsilon_{\beta\alpha} + \hbar\omega} \right] \right\} \phi_\beta^*(\boldsymbol{x}) \phi_\alpha(\boldsymbol{x}) \end{aligned} \tag{7.12}$$

so that $\nabla \cdot \boldsymbol{j}_1(\boldsymbol{x}, \omega) \sim \omega$ for high frequencies.

7.3 A Green's Function Formalism

It is convenient to express $\sigma(\boldsymbol{x}, \boldsymbol{x}', \omega)$ in terms of retarded and advanced Green's functions

$$G_\varepsilon^\pm(\boldsymbol{x}, \boldsymbol{x}') = \sum_\alpha \frac{\phi_\alpha(\boldsymbol{x}) \phi_\alpha(\boldsymbol{x}')}{\varepsilon - \varepsilon_\alpha + i0+}. \tag{7.13}$$

We separate $\sigma(\boldsymbol{x}, \boldsymbol{x}', \omega)$ obtained from Eq. (7.10) into two parts according to its δ-function and principal part contributions, that is, $\sigma = \sigma_s + \sigma_{as}$ with

$$\sigma_s(\boldsymbol{x}, \boldsymbol{x}', \omega) = \frac{e^2\hbar^3\pi}{8m^{*2}} \sum_{\alpha,\beta} \frac{f_{\beta\alpha}}{\varepsilon_{\beta\alpha}} \left[\delta(\varepsilon_{\beta\alpha} - \hbar\omega) + \delta(\varepsilon_{\beta\alpha} + \hbar\omega)\right] \times \boldsymbol{W}_{\beta\alpha}(\boldsymbol{x})\, \boldsymbol{W}_{\alpha\beta}(\boldsymbol{x}')\,, \tag{7.14}$$

$$\sigma_{as}(\boldsymbol{x}, \boldsymbol{x}', \omega) = \frac{e^2\hbar^3\pi}{8m^{*2}} \sum_{\alpha,\beta} \frac{f_{\beta\alpha}}{\varepsilon_{\beta\alpha}} \left[\frac{\mathcal{P}}{\varepsilon_{\beta\alpha} - \hbar\omega} + \frac{\mathcal{P}}{\varepsilon_{\beta\alpha} + \hbar\omega}\right] \times \boldsymbol{W}_{\beta\alpha}(\boldsymbol{x})\, \boldsymbol{W}_{\alpha\beta}(\boldsymbol{x}')\,. \tag{7.15}$$

Making use of the result in Eq. (7.3) to express $\boldsymbol{W}_{\beta\alpha}(\boldsymbol{x})\,\boldsymbol{W}_{\alpha\beta}(\boldsymbol{x}')$ in terms of the eigenfunctions of the unperturbed Hamiltonian H_0 and subsequently in terms of the retarded and advanced Green's functions, we obtain

$$\begin{aligned}\sigma_s(\boldsymbol{x}, \boldsymbol{x}', \omega) = &-\frac{e^2\hbar^2}{32\pi m^{*2}\omega} \int_{-\infty}^{\infty} d\varepsilon\, f_0(\varepsilon) \\ &\times \Big\{\Delta G_\varepsilon(\boldsymbol{x}, \boldsymbol{x}')\, \overset{\leftrightarrow}{\boldsymbol{D}}{}^{*} \overset{\leftrightarrow}{\boldsymbol{D}}{}' \left[\Delta G_{\varepsilon+\hbar\omega}(\boldsymbol{x}', \boldsymbol{x}) - \Delta G_{\varepsilon-\hbar\omega}(\boldsymbol{x}', \boldsymbol{x})\right] \\ &+ \left[\Delta G_{\varepsilon+\hbar\omega}(\boldsymbol{x}, \boldsymbol{x}') - \Delta G_{\varepsilon-\hbar\omega}(\boldsymbol{x}, \boldsymbol{x}')\right] \overset{\leftrightarrow}{\boldsymbol{D}}{}^{*} \overset{\leftrightarrow}{\boldsymbol{D}}{}'\, \Delta G_\varepsilon(\boldsymbol{x}', \boldsymbol{x})\Big\}\,, \end{aligned} \tag{7.16}$$

$$\begin{aligned}\sigma_{as}(\boldsymbol{x}, \boldsymbol{x}', \omega) = &\frac{e^2\hbar^2}{32\pi m^{*2}\omega} \int_{-\infty}^{\infty} d\varepsilon\, f_0(\varepsilon) \\ &\times \Big\{\left[\Sigma G_{\varepsilon-\hbar\omega}(\boldsymbol{x}, \boldsymbol{x}') - \Sigma G_{\varepsilon+\hbar\omega}(\boldsymbol{x}, \boldsymbol{x}')\right] \overset{\leftrightarrow}{\boldsymbol{D}}{}^{*} \overset{\leftrightarrow}{\boldsymbol{D}}{}'\, \Delta G_\varepsilon(\boldsymbol{x}', \boldsymbol{x}) \\ &- \Delta G_\varepsilon(\boldsymbol{x}, \boldsymbol{x}')\, \overset{\leftrightarrow}{\boldsymbol{D}}{}^{*} \overset{\leftrightarrow}{\boldsymbol{D}}{}' \left[\Sigma G^{+}_{\varepsilon-\hbar\omega}(\boldsymbol{x}', \boldsymbol{x}) - \Sigma G_{\varepsilon+\hbar\omega}(\boldsymbol{x}', \boldsymbol{x})\right]\Big\}\,, \end{aligned} \tag{7.17}$$

where $\Delta G_\varepsilon(\boldsymbol{x}, \boldsymbol{x}') = G^{+}_\varepsilon(\boldsymbol{x}, \boldsymbol{x}') - G^{-}_\varepsilon(\boldsymbol{x}, \boldsymbol{x}')$ and $\Sigma G_\varepsilon(\boldsymbol{x}, \boldsymbol{x}') = G^{+}_\varepsilon(\boldsymbol{x}, \boldsymbol{x}') + G^{-}_\varepsilon(\boldsymbol{x}, \boldsymbol{x}')$. Combining the results in Eqs. (7.16) and (7.17), we obtain the nonlocal conductivity

$$\begin{aligned}\sigma(\boldsymbol{x}, \boldsymbol{x}', \omega) = &\frac{e^2\hbar^2}{16\pi m^{*2}\omega} \int_{-\infty}^{\infty} d\varepsilon\, f_0(\varepsilon) \\ &\times \Big\{G^{+}_\varepsilon(\boldsymbol{x}, \boldsymbol{x}')\, \overset{\leftrightarrow}{\boldsymbol{D}}{}^{*} \overset{\leftrightarrow}{\boldsymbol{D}}{}' \left[G^{-}_{\varepsilon+\hbar\omega}(\boldsymbol{x}', \boldsymbol{x}) - G^{-}_{\varepsilon-\hbar\omega}(\boldsymbol{x}', \boldsymbol{x})\right] \\ &+ \left[G^{+}_{\varepsilon+\hbar\omega}(\boldsymbol{x}, \boldsymbol{x}') - G^{+}_{\varepsilon-\hbar\omega}(\boldsymbol{x}, \boldsymbol{x}')\right] \overset{\leftrightarrow}{\boldsymbol{D}}{}^{*} \overset{\leftrightarrow}{\boldsymbol{D}}{}'\, G^{-}_\varepsilon(\boldsymbol{x}', \boldsymbol{x}) \\ &- G^{-}_\varepsilon(\boldsymbol{x}, \boldsymbol{x}')\, \overset{\leftrightarrow}{\boldsymbol{D}}{}^{*} \overset{\leftrightarrow}{\boldsymbol{D}}{}' \left[G^{-}_{\varepsilon+\hbar\omega}(\boldsymbol{x}', \boldsymbol{x}) - G^{-}_{\varepsilon-\hbar\omega}(\boldsymbol{x}', \boldsymbol{x})\right] \\ &- \left[G^{+}_{\varepsilon+\hbar\omega}(\boldsymbol{x}, \boldsymbol{x}') - G^{+}_{\varepsilon-\hbar\omega}(\boldsymbol{x}, \boldsymbol{x}')\right] \overset{\leftrightarrow}{\boldsymbol{D}}{}^{*} \overset{\leftrightarrow}{\boldsymbol{D}}{}'\, G^{+}_\varepsilon(\boldsymbol{x}', \boldsymbol{x})\Big\}\,. \end{aligned}$$

(7.18)

7.4
The Static Limit

In the static DC limit where $\omega \to 0$, it follows from Eq. (7.12) that $\nabla \cdot \boldsymbol{j}_1(\boldsymbol{x}, t) = 0$. Also, from Eq. (7.18), it follows that

$$\sigma(\boldsymbol{x}, \boldsymbol{x}', \omega = 0) = \frac{e^2\hbar^3}{8\pi m^{*2}} \int_{-\infty}^{\infty} d\varepsilon \left[-\frac{d f_0(\varepsilon)}{d\varepsilon}\right] \times G_\varepsilon^+(\boldsymbol{x}, \boldsymbol{x}') \overset{\leftrightarrow}{\boldsymbol{D}}{}^* \overset{\leftrightarrow}{\boldsymbol{D}}{}' G_\varepsilon^-(\boldsymbol{x}', \boldsymbol{x}) - \frac{e^2\hbar^3}{8\pi m^{*2}} \int_{-\infty}^{\infty} d\varepsilon\, f_0(\varepsilon) \left\{ \frac{d G_\varepsilon^+(\boldsymbol{x}, \boldsymbol{x}')}{d\varepsilon} \overset{\leftrightarrow}{\boldsymbol{D}}{}^* \overset{\leftrightarrow}{\boldsymbol{D}}{}' G_\varepsilon^+(\boldsymbol{x}', \boldsymbol{x}) + G_\varepsilon^-(\boldsymbol{x}, \boldsymbol{x}') \overset{\leftrightarrow}{\boldsymbol{D}}{}^* \overset{\leftrightarrow}{\boldsymbol{D}}{}' \frac{d G_\varepsilon^-(\boldsymbol{x}', \boldsymbol{x})}{d\varepsilon} \right\} . \tag{7.19}$$

We now obtain the total DC ($\omega = 0$) transport current I_m coming out of lead m by making use of our results above. From Eq. (7.11) for the current density, we have

$$I_m = \int_{C_m} d y_m \, \boldsymbol{j}_1(x_m, y_m) \cdot \hat{\boldsymbol{x}}_m \, , \tag{7.20}$$

where for each lead a local coordinate system (x_m, y_m) is chosen so that $\hat{\boldsymbol{x}}_m$ is an outward pointing unit vector and C_m is a cross-section line in lead m. Following a straightforward calculation, it can be shown that

$$I_m = \sum_n g_{mn} V_n \, , \tag{7.21}$$

where the conductance coefficient between two leads is equal to the flux of σ into those leads and is given by

$$g_{mn} = -\int_{C_m} d y_m \int_{C_n} d y_n \hat{\boldsymbol{x}}_m \cdot \sigma(\boldsymbol{x}, \boldsymbol{x}') \cdot \hat{\boldsymbol{x}}_n \, . \tag{7.22}$$

Equation (7.21) relates the current coming out of lead m to the applied voltage at lead n.

In the zero frequency limit, the σ_s-part becomes

$$\sigma_s(\boldsymbol{x}, \boldsymbol{x}', \omega = 0) = -\frac{e^2\hbar^3}{16\pi m^{*2}} \int_{-\infty}^{\infty} d\varepsilon \left[-\frac{d f_0(\varepsilon)}{d\varepsilon}\right] \times \Delta G_\varepsilon(\boldsymbol{x}, \boldsymbol{x}') \overset{\leftrightarrow}{\boldsymbol{D}}{}^* \overset{\leftrightarrow}{\boldsymbol{D}}{}' \Delta G_\varepsilon(\boldsymbol{x}', \boldsymbol{x}) . \tag{7.23}$$

In taking the $\omega \to 0$ limit for σ_{as}, we note that the terms in square brackets in Eq. (7.17) may be written as derivatives of $\Sigma\, G^{\pm}$ and we have

$$\sigma_{as}(\boldsymbol{x}, \boldsymbol{x}', \omega = 0) = -\frac{e^2\hbar^3}{16\pi m^{*2}} \int_{-\infty}^{\infty} d\varepsilon\, f_0(\varepsilon) \left\{ \frac{d\Sigma\, G_{\varepsilon}(\boldsymbol{x}, \boldsymbol{x}')}{d\varepsilon} \overset{\leftrightarrow}{\boldsymbol{D}}{}^{*} \overset{\leftrightarrow}{\boldsymbol{D}}{}' \Delta G_{\varepsilon}(\boldsymbol{x}', \boldsymbol{x}) - \Delta G_{\varepsilon}(\boldsymbol{x}, \boldsymbol{x}') \overset{\leftrightarrow}{\boldsymbol{D}}{}^{*} \overset{\leftrightarrow}{\boldsymbol{D}}{}' \frac{d\Sigma\, G_{\varepsilon}^{+}(\boldsymbol{x}', \boldsymbol{x})}{d\varepsilon} \right\} . \tag{7.24}$$

We also note that Eqs. (7.23) and (7.24) satisfy

$$\sigma_s(\boldsymbol{x}', \boldsymbol{x}, \omega = 0) = \sigma_s(\boldsymbol{x}, \boldsymbol{x}', \omega = 0) , \tag{7.25}$$

$$\sigma_{as}(\boldsymbol{x}', \boldsymbol{x}, \omega = 0) = -\sigma_{as}(\boldsymbol{x}, \boldsymbol{x}', \omega = 0) , \tag{7.26}$$

which are the Onsager relations for conductivity.

7.5 Model and Single-Particle Eigenstates

In this section, we present a model calculation for 1D periodic modulation [6–9]. We diagonalize the Hamiltonian to obtain the eigenvalues as functions of the modulation strength and the magnetic field. The single-particle Hamiltonian for a 1D periodic potential in the xy-plane in a uniform perpendicular magnetic field $\boldsymbol{B}$ is given in the Landau gauge by

$$\mathcal{H}_0 = \frac{1}{2m^*} \left[-i\hbar\nabla + e\boldsymbol{A}(\boldsymbol{r})\right]^2 + U_{\mathrm{L}}(x) , \tag{7.27}$$

where $\boldsymbol{A}(\boldsymbol{r}) = (0, Bx, 0)$ is the vector potential in the Landau gauge. In Eq. (7.27), the 1D lattice potential $U_{\mathrm{L}}(x)$ can be taken as having the following form

$$U_{\mathrm{L}}(x) = U_0 \left[\cos\left(\frac{2\pi x}{a}\right)\right]^{2N} , \tag{7.28}$$

where U_0 is either positive or negative, a is a lattice constant of the artificially imposed periodic modulation potential, and N is the power for determining the width of the quantum wire potential. The reason for taking $2N$ in the exponent is to always obtain a positive (or negative) potential when U_0 is taken as a positive (or negative) value. The value of N describes the steepness, or degree of modulation, of the potential. When N is sufficiently large, the steep slopes make it reasonable to approximate the lattice potential by the simple form $U_{\mathrm{L}}(x) = V_0 \sum_j \delta(x - ja)$ with $V_0 = U_0 a$. In dealing with lattice scattering, the potential should be determined self-consistently, taking many-body effects into account. The dependence of the potential on the 2D electron density $n_{2\mathrm{D}}$ can be incorporated into our theory by treating U_0 as an adjustable parameter. The reason is that U_0 is expected to only be weakly dependent on magnetic field. In our model, we wish to include the effects due to tunneling by adjusting both the strength of the potential U_0 as well as its steepness.

In the absence of impurities, the single-particle eigenfunctions are determined through

$$\psi_{j,X_0}(\boldsymbol{r}) = \sum_n C_n(j, X_0)\phi^{(0)}_{n,X_0}(\boldsymbol{r}) , \tag{7.29}$$

where

$$\phi^{(0)}_{n,X_0}(\boldsymbol{r}) = \frac{\exp\left(-\frac{iX_0 y}{\ell_H^2}\right)}{\sqrt{L_y}}\sqrt{\frac{1}{\pi^{1/2}\ell_H 2^n n!}} \times \exp\left[-\frac{(x-X_0)^2}{2\ell_H^2}\right] H_n\left(\frac{x-X_0}{\ell_H}\right) . \tag{7.30}$$

In Eqs. (7.29) and (7.30), $n = 0, 1, 2, \ldots$ is a Landau-level index and $H_n(x)$ is the nth order Hermite polynomial. Also, $L_y = N_y a$ is the sample length in the y-direction, $X_0 = k_y \ell_H^2$ is the guiding center, $\ell_H = \sqrt{\hbar/eB}$ is the magnetic length and k_y is a wave vector along the y-direction. The expansion coefficients $C_n(j, X_0)$ in Eq. (7.29) are determined from the following matrix equation:

$$\sum_n \left\{ \left[E_n^{(0)} - E_j(X_0)\right]\delta_{n,n'} + \frac{U_0 a}{\ell_H \pi^{1/2}} \times \sqrt{\frac{1}{2^{n+n'} n! n'!}} B_{n,n'}(X_0) \right\} C_n(j, X_0) = 0 \tag{7.31}$$

as well as the orthonormality condition:

$$\sum_n [C_n(j, X_0)]^* C_n\left(j', X_0'\right) = \delta_{j,j'}\delta_{X_0,X_0'} . \tag{7.32}$$

From this calculation, we also obtain the secular equation which determines the energy eigenvalues $E_j(X_0)$, that is,

$$\mathrm{Det}\left\{ \left[E_n^{(0)} - E_j(X_0)\right]\delta_{n,n'} + \frac{U_0 a}{\ell_H \pi^{1/2}} \times \sqrt{\frac{1}{2^{n+n'} n! n'!}} B_{n,n'}(X_0) \right\} = 0. \tag{7.33}$$

Here, $E_n^{(0)} = (n + 1/2)\hbar\omega_c$ is the energy for the nth Landau level with eigenfunction $\phi_n^{(0)}(\boldsymbol{r})$ in the absence of scatterers and $\omega_c = eB/m^*$. The matrix element $B_{n,n'}(X_0)$ appearing in Eqs. (7.31) and (7.33) is defined as follows for the potential in Eq. (7.28),

$$B_{n,n'}(X_0) = \int_{-\infty}^{\infty} dx \exp\left[-\frac{(x-X_0)^2}{\ell_H^2}\right] \cos^{2N}\frac{2\pi x}{a} \times H_n\left(\frac{x-X_0}{\ell_H}\right) H_{n'}\left(\frac{x-X_0}{\ell_H}\right). \tag{7.34}$$

Since the coefficient matrix in Eq. (7.31) is real and symmetric, meaning that $C_n(j, X_0)$ must be real. The Fermi energy is determined from the 2D electron density

$$n_{2D} = \frac{2N_y \Phi}{\mathcal{A}} \sum_j \int_{-G\ell_H^2/2}^{G\ell_H^2/2} \frac{dX_0}{a} \int_{-\infty}^{\infty} dE \times f_0(E) D_{j,X_0}(E), \tag{7.35}$$

in terms of the Fermi–Dirac function $f_0(E)$, the sample area $\mathcal{A}$ and the partial density-of-states

$$D_{j,X_0}(E) = \frac{1}{\sqrt{2\pi}\gamma_0} \exp\left\{-\frac{[E - E_j(X_0)]^2}{2\gamma_0^2}\right\}. \tag{7.36}$$

In this notation, $\Phi = Ba^2/\phi_0$ with $\phi_0 = h/e$, $G = 2\pi/a$, γ_0 is a magnetic field-dependent parameter chosen to represent weak scattering by impurities. We shall use $\hbar\omega_c$ as an energy scale, a as a length scale, and adopt periodic boundary conditions in the y-direction so that $k_y = 2\pi L/a$ where $L = -N_y/2, \ldots, [(N_y/2)-1]$ are integers. Also, we shall express the scattering in terms of the dimensionless quantity $\bar{U}_0 \equiv m^* U_0 a^2/\sqrt{2}\pi\hbar^2$ and the number N_x of unit cells in the x-direction. In carrying out our numerical calculations, we must choose a value of $\bar{U}_0$. If the lattice period is $a = 200$ nm, then for $U_0 = 0.156$ meV, we have $\bar{U}_0 = 1.235$. For the same lattice constant, if $U_0 = 12.65$ meV, this corresponds to a value of $\bar{U}_0 = 100$. This range for scattering potential strength U_0 justifies our choice below. The available DC and frequency-dependent transport experimental data was obtained at very low temperature. Consequently, we carry out our numerical calculations in the zero temperature ($T = 0$ K) limit. Also, the modulation not only determines the quantum magneto-transport properties in the presence of the lattice, but also the electron distribution. Using the eigenstates determined from Eq. (7.33), we may calculate the distribution of charge for a chosen electron density, that is,

$$n(\mathbf{r}) = \frac{2\Phi}{N_x} \sum_j \int_{-G\ell_H^2/2}^{G\ell_H^2/2} \frac{dX_0}{a} \left|\psi_{j,X_0}(\mathbf{r})\right|^2 f_0[E_j(X_0)]. \tag{7.37}$$

The modulation makes the electron density distribution nonuniform. This is produced by having finite values for more than one of the coefficients $C_n(j, X_0)$ for the wave function in Eq. (7.29).

7.6 Averaged Conductivity

By neglecting the effects from the small current and voltage leads connected to the sample, we may take the external electric field as uniform over the whole array. This is justified since the size of a sample used in an experiment is very small compared with the scale over which the electric field varies. Upon averaging the response matrix in Eq. (7.18) over the whole array, we obtain the averaged dynamic conductivity for this system as

$$\begin{aligned}\sigma_{\mu\nu}(\omega) &= \frac{1}{\mathcal{A}}\int d\mathbf{r}\int d\mathbf{r}'\sigma_{\mu\nu}(\mathbf{r},\mathbf{r}',\omega)\\ &= -\frac{ie^2}{2\omega\mathcal{A}}\int_{-\infty}^{\infty} d\varepsilon\, f_0(\varepsilon)\mathrm{Tr}\Big[v_\mu\delta(\varepsilon-H)v_\nu\left(\hat{G}^-_{\varepsilon+\hbar\omega}-\hat{G}^-_{\varepsilon-\hbar\omega}\right)\\ &\quad - v_\mu\left(\hat{G}^+_{\varepsilon+\hbar\omega}-\hat{G}^+_{\varepsilon-\hbar\omega}\right)v_\nu\delta(\varepsilon-H)\Big].\end{aligned} \tag{7.38}$$

This is a generalized form for the Kubo formula for the conductivity at finite frequencies. In this notation, the velocity operator is defined as $\boldsymbol{v} = -(i\hbar\mathbf{D})/m^*$ and $\hat{G}_\varepsilon = (\varepsilon - H)^{-1}$.

We now approximate the Dyson equation $\hat{G} = \hat{G}^0 + \hat{G}^0\hat{\Sigma}\hat{G}$ by $\hat{G}_\alpha = \hat{G}^0_\alpha + \hat{\Sigma}\left(\hat{G}^0_\alpha\right)^2$, where $\hat{G}^0$ is the single-particle Green's function and $\hat{\Sigma}$ is the self-energy. This approximation is reasonable for a weak impurity scattering potential. In this approximation, we may separate the contributions to $\sigma_{\mu\nu}(\omega)$ from its "band" part arising from the subbands and a term arising from electron–impurity scattering, that is, $\sigma_{\mu\nu}(\omega) \approx \sigma^{(0)}_{\mu\nu}(\omega) + \sigma^{(1)}_{\mu\nu}(\omega)$, where

$$\begin{aligned}\sigma^{(0)}_{\mu\nu}(\omega) &= -\frac{ie^2}{2\omega\mathcal{A}}\sum_{m,n} v^{mn}_\mu v^{nm}_\nu\int_{-\infty}^{\infty} d\varepsilon\, f_0(\varepsilon)\\ &\quad\times\left[\frac{\delta(\varepsilon-\varepsilon_m)}{\varepsilon-\varepsilon_n+\hbar\omega-i\eta}-\frac{\delta(\varepsilon-\varepsilon_m)}{\varepsilon-\varepsilon_n-\hbar\omega-i\eta}\right.\\ &\quad\left.-\frac{\delta(\varepsilon-\varepsilon_n)}{\varepsilon-\varepsilon_m+\hbar\omega+i\eta}+\frac{\delta(\varepsilon-\varepsilon_n)}{\varepsilon-\varepsilon_m-\hbar\omega+i\eta}\right]\\ &= -\frac{ie^2}{2\omega\mathcal{A}}\sum_{m,n} v^{mn}_\mu v^{nm}_\nu\Bigg\{[f_0(\varepsilon_m)-f_0(\varepsilon_n)]\\ &\quad\times\left[\mathcal{P}\left(\frac{1}{\varepsilon_m-\varepsilon_n+\hbar\omega}\right)-\mathcal{P}\left(\frac{1}{\varepsilon_m-\varepsilon_n-\hbar\omega}\right)\right]\\ &\quad+i\pi\left[f_0(\varepsilon_n-\hbar\omega)-f_0(\varepsilon_n)\right]\delta(\varepsilon_m-\varepsilon_n+\hbar\omega)\\ &\quad-i\pi\left[f_0(\varepsilon_n+\hbar\omega)-f_0(\varepsilon_n)\right]\delta(\varepsilon_m-\varepsilon_n-\hbar\omega)\Bigg\},\end{aligned} \tag{7.39}$$

$$\sigma^{(1)}_{\mu\nu}(\omega) = -\frac{e^2}{4\pi\omega\mathcal{A}}\sum_{m,n} v_\mu^{mn} v_\nu^{nm} \int_{-\infty}^{\infty} d\varepsilon\, f_0(\varepsilon)$$
$$\times \Bigg\{\Delta G^0_{\varepsilon,n}\left[\left(G^{0(+)}_{\varepsilon+\hbar\omega,m}\right)^2 \Sigma^{(+)}(\varepsilon+\hbar\omega) - \left(G^{0(+)}_{\varepsilon-\hbar\omega,m}\right)^2 \Sigma^{(+)}(\varepsilon-\hbar\omega)\right]$$
$$+\left[\Delta G^0_{\varepsilon+\hbar\omega,n} - \Delta G^0_{\varepsilon-\hbar\omega,n}\right]\left(G^{0(+)}_{\varepsilon,m}\right)^2 \Sigma^{(+)}(\varepsilon)$$
$$-\Delta G^0_{\varepsilon,m}\left[\left(G^{0(-)}_{\varepsilon+\hbar\omega,n}\right)^2 \Sigma^{(-)}(\varepsilon+\hbar\omega) - \left(G^{0(-)}_{\varepsilon-\hbar\omega,n}\right)^2 \Sigma^{(-)}(\varepsilon-\hbar\omega)\right]$$
$$-\left[\Delta G^0_{\varepsilon+\hbar\omega,m} - \Delta G^0_{\varepsilon-\hbar\omega,m}\right]\left(G^{0(-)}_{\varepsilon,n}\right)^2 \Sigma^{(-)}(\varepsilon)\Bigg\}. \tag{7.40}$$

Here, $G^{0(\pm)}_{\varepsilon,m} = 1/(\varepsilon - \varepsilon_m \pm i\eta)$, $\Sigma^{(\pm)}(\varepsilon) = \Sigma(\varepsilon \pm i\eta)$ and $v_\mu^{mn} = \langle m|v_\mu|n\rangle$ is a velocity matrix element for eigenstates $|m\rangle$ and $|n\rangle$. In the zero-frequency limit, the results in Eqs. (7.39) and (7.40) become

$$\sigma^{(0)}_{\mu\nu}(0) = -\frac{ie^2\hbar}{\mathcal{A}}\sum_{m,n} v_\mu^{mn} v_\nu^{nm}\left[\frac{f_0(\varepsilon_m) - f_0(\varepsilon_n)}{\varepsilon_m - \varepsilon_n}\right]\mathcal{P}\left(\frac{1}{\varepsilon_m - \varepsilon_n}\right)$$
$$+\frac{\pi e^2\hbar}{\mathcal{A}}\sum_{m,n} v_\mu^{mn} v_\nu^{nm}\left[-\frac{d f_0(\varepsilon_m)}{d\varepsilon_m}\right]\delta(\varepsilon_m - \varepsilon_n)\,, \tag{7.41}$$

$$\sigma^{(1)}_{\mu\nu}(0) = -\frac{ie^2\hbar}{\mathcal{A}}\int_{-\infty}^{\infty} d\varepsilon\left[-\frac{d f_0(\varepsilon)}{d\varepsilon}\right]\sum_{m,n} v_\mu^{mn} v_\nu^{nm}$$
$$\times\left[\frac{\delta(\varepsilon-\varepsilon_m)\Sigma^{(-)}(\varepsilon)}{(\varepsilon-\varepsilon_n-i0^+)^2} - \frac{\delta(\varepsilon-\varepsilon_n)\Sigma^{(+)}(\varepsilon)}{(\varepsilon-\varepsilon_m+i0^+)^2}\right]. \tag{7.42}$$

Clearly, the $[f_0(\varepsilon_m) - f_0(\varepsilon_n)]$ terms in $\sigma^{(0)}_{xx}(0)$ and $\sigma^{(0)}_{xx}(\omega)$ are zero by symmetry and, therefore, only the second terms contribute. In our calculations, we assume that the impurity distribution is not dense and the impurities are not correlated. In this case, the scattering potential is short-ranged due to screening by the electrons, and the self-energy is independent of all quantum numbers [10]. In this limit, we can include impurity scattering effects and obtain the self-consistent equation for the self-energy as

$$\Sigma(\varepsilon) = \frac{2\pi\hbar^2 f n_I}{m^*}\left\{1 - \frac{2\pi\hbar^2 f N_y \Phi}{m^*\mathcal{A}}\sum_j \int_{-G\ell_H^2/2}^{G\ell_H^2/2} \frac{dX_0}{a}\left[\frac{1}{\varepsilon - E_j(X_0) - \Sigma(\varepsilon)}\right]\right\}^{-1}, \tag{7.43}$$

where $n_{\rm I}$ is the impurity density and f is an impurity scattering amplitude. In our numerical calculations, we chose $f = 0.62$ and $n_{\rm I} = 0.01 n_{\rm 2D}$. Here, we have included multiple scattering from a single scatterer in the self-consistent t-matrix approximation. From this self-energy, we can further calculate the density-of-states

$$D(\varepsilon) = \frac{2N_y \Phi}{\pi A} \sum_j \int_{-G\ell_H^2/2}^{G\ell_H^2/2} \frac{dX_0}{a} \times \frac{\mathrm{Im}\{\Sigma^{(-)}(\varepsilon)\}}{\left[\varepsilon - E_j(X_0) - \mathrm{Re}\{\Sigma^{(-)}(\varepsilon)\}\right]^2 + \left[\mathrm{Im}\{\Sigma^{(-)}(\varepsilon)\}\right]^2} . \tag{7.44}$$

The real and imaginary parts of the self-energy determine the peak shift and the broadening of the peaks in the density-of-states from the impurity scattering, respectively. In the absence of impurities, the single-particle Green's functions are determined by the eigenfunctions.

Moreover, for the case without scatterers, we have $\sigma_{xx}(\omega) = 0$ and

$$\sigma_{yx}(\omega) = \frac{e^2}{h}\left(\frac{1}{\Phi}\right)\frac{n_{2D}a^2}{1-\left(\frac{\omega}{\omega_c}\right)^2} . \tag{7.45}$$

It can be shown that the current vertex correction due to impurity averaging for the single-particle Green's function vanishes when it is assumed that the self-energy is independent of all quantum numbers. Using Eqs. (7.39) and (7.40), we obtain closed-form analytic expressions for the longitudinal and transverse conductivities. We will numerically calculate both the band and impurity parts of the conductivity, although the impurity scattering is weak compared with the lattice scattering. In our previous work, we did not give the conductivity due to electron–impurity effects which we include to leading order only. Straightforward calculations show that

$$\sigma_{xx}^{(0)}(\omega) = \frac{\pi e^2 \hbar^2 \Phi N_y}{\omega m^{*2} A} \sum_{j,j'} \int_{-G\ell_H^2/2}^{G\ell_H^2/2} \frac{dX_0}{a} \left(F^{(2)}_{j,X_0;j',X_0}\right)^2 \times \int_{-\infty}^{\infty} d\varepsilon\, D_{j',X_0}(\varepsilon) \left\{\left[f_0(\varepsilon - \hbar\omega) - f_0(\varepsilon)\right] D_{j,X_0}(\varepsilon - \hbar\omega) - \left[f_0(\varepsilon + \hbar\omega) - f_0(\varepsilon)\right] D_{j,X_0}(\varepsilon + \hbar\omega)\right\} . \tag{7.46}$$

By replacing $F^{(2)}_{j,X_0;j',X_0}$ with $F^{(3)}_{j,X_0;j',X_0} - (ieB/\hbar)F^{(1)}_{j,X_0;j',X_0}$, defined below in Eqs. (7.50)–(7.52), respectively, we can obtain $\sigma_{yy}^{(0)}(\omega)$. In the zero-frequency limit, we obtain

$$\sigma_{xx}^{(0)}(0) = \frac{2\pi e^2 \hbar^3 \Phi N_y}{m^{*2} A} \sum_{j,j'} \int_{-G\ell_H^2/2}^{G\ell_H^2/2} \frac{dX_0}{a} \left(F^{(2)}_{j,X_0;j',X_0}\right)^2 \times \int_{-\infty}^{\infty} d\varepsilon\, D_{j,X_0}(\varepsilon) D_{j',X_0}(\varepsilon) \left[-\frac{d f_0(\varepsilon)}{d\varepsilon}\right] . \tag{7.47}$$

For the Hall conductivity, we also obtain the band part as

$$\sigma_{yx}^{(0)}(\omega) = \frac{4e^2\hbar^3\Phi N_y}{m^{*2}\mathcal{A}} \sum_{j,j'}{}' \int_{-G\ell_H^2/2}^{G\ell_H^2/2} \frac{dX_0}{a} \frac{1}{\left[\varepsilon_j(X_0) - \varepsilon_{j'}(X_0)\right]^2 - (\hbar\omega)^2}$$
$$\times \left[\frac{1}{\ell_H^2} \operatorname{Re}\left\{F^{(1)}_{j,X_0;j',X_0} F^{(2)}_{j',X_0;j,X_0}\right\} + \operatorname{Im}\left\{F^{(3)}_{j,X_0;j',X_0} F^{(2)}_{j',X_0;j,X_0}\right\}\right]$$
$$\times \int_{-\infty}^{\infty} d\varepsilon\, D_{j,X_0}(\varepsilon)\, f_0(\varepsilon) . \tag{7.48}$$

In the above, the summation with prime means that all terms with $|E_j(X_0) - E_{j'}(X_0)| = \hbar\omega$ must be excluded. From the inverse of the conductivity matrix, we obtain the resistivities

$$\rho_{xx}(\omega) = \frac{\sigma_{yy}(\omega)}{\mathcal{L}} , \quad \rho_{yy}(\omega) = \frac{\sigma_{xx}(\omega)}{\mathcal{L}} ,$$
$$\rho_{xy}(\omega) = -\rho_{yx}(\omega) = \frac{\sigma_{yx}(\omega)}{\mathcal{L}} , \tag{7.49}$$

where $\mathcal{L} = \sigma_{xx}(\omega)\sigma_{yy}(\omega) + [\sigma_{yx}(\omega)]^2$. Here, the structure factors determining the group velocities are

$$F^{(1)}_{j,X_0;j',X_0} = F^{(1)}_{j',X_0;j,X_0} = \int_{\mathcal{A}} d^2\boldsymbol{r}\,\psi^*_{j,X_0}(\boldsymbol{r})\, x\, \psi_{j',X_0}(\boldsymbol{r}) = \frac{\ell_H}{\sqrt{2}}$$
$$\times \sum_n C_n(j, X_0)\left[\sqrt{n+1}\,C_{n+1}(j', X_0) + \sqrt{n}\,C_{n-1}(j', X_0)\right] , \tag{7.50}$$

$$F^{(2)}_{j,X_0;j',X_0} = -F^{(2)}_{j',X_0;j,X_0} = \int_{\mathcal{A}} d^2\boldsymbol{r}\,\psi^*_{j,X_0}(\boldsymbol{r}) \frac{\partial}{\partial x} \psi_{j',X_0}(\boldsymbol{r}) = \frac{1}{\sqrt{2}\ell_H}$$
$$\times \sum_n C_n(j, X_0)\left[\sqrt{n+1}\,C_{n+1}(j', X_0) - \sqrt{n}\,C_{n-1}(j', X_0)\right] , \tag{7.51}$$

$$F^{(3)}_{j,X_0;j',X_0} = F^{(3)}_{j',X_0;j,X_0} = -i \int_{\mathcal{A}} d^2\boldsymbol{r}\,\psi^*_{j,X_0}(\boldsymbol{r}) \frac{\partial}{\partial y} \psi_{j',X_0}(\boldsymbol{r})$$
$$= -\frac{X_0}{\ell_H^2}\delta_{j,j'} , \tag{7.52}$$

where the coefficients $C_n(j', X_0)$ are all real.

7.7
Applications to One-Dimensional Density Modulated 2DEG

In this section, we present numerical results for the Fermi energy, energy eigenvalues, the conductivities and the resistivities. In Figure 7.1a, we plot the energy eigenvalues as a function of the wave vector k_y. Here, we chose $\Phi = 1.1$, $\bar{U}_0 = 10.235$, $N_x = N_y = 20$ and $N = 10$ in the modulation potential. Figure 7.1a shows that the Landau levels are shifted upward by the scattering potential. The lowest Landau level is weakly dependent on k_y. There is mixing of the Landau orbits for the higher levels. Consequently, the higher energy eigenvalues vary with k_y and this dependence is stronger for the larger energies. The variation of the energy eigenvalues with magnetic flux Φ is shown in Figure 7.1b for chosen $\bar{U}_0$. The mixing of the Landau orbits increases with magnetic flux. When the lattice potential is increased, the energy levels are shifted upward. As the lattice potential is further increased, the Landau levels overlap at low magnetic fields. The kinks appearing in the eigenvalue spectrum correspond to a rapid increase in the density of states at that magnetic field and the Fermi energy is pinned there. This is the reason for the staircase structure in E_F.

In Figure 7.2, we plot E_F as a function of Φ for $n_{2D}a^2 = 1$. We chose two values for the interaction potential. In each case, we took $N = 10$ in the scattering potential. There are small kinks corresponding to pinning between sub-Landau levels. The Fermi energy depends on the strength of the scattering potential. As the magnetic field increases, the Fermi energy decreases. Our results show that when the lattice scattering is increased, the Fermi energy is shifted upward to higher value. This is a result of the increased scattering of the electrons by the lattice. Above a threshold value of magnetic flux, the cyclotron radius becomes much smaller than the modulation period and the system behaves like a homogeneous 2DES, resulting in a constant value for E_F.

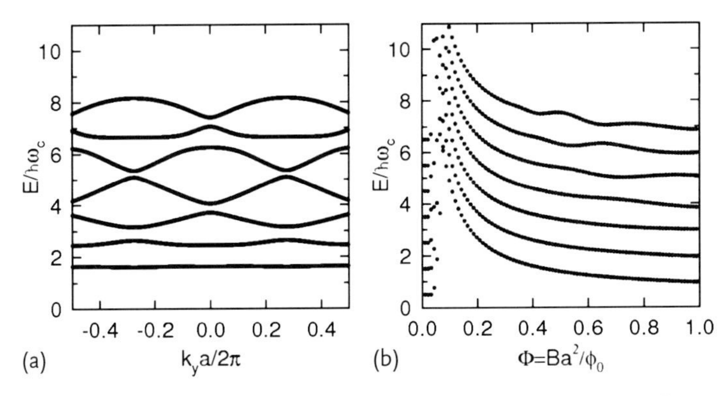

Figure 7.1 (a) The energy eigenvalues as a function of $k_y a$ for $\Phi = 1.1$, $\bar{U}_0 = 10.235$. (b) For $k_y a = \pi$, the eigenvalues are plotted as a function of Φ when $\bar{U}_0 = 1.235$. The parameters used in the calculation are given in the text.

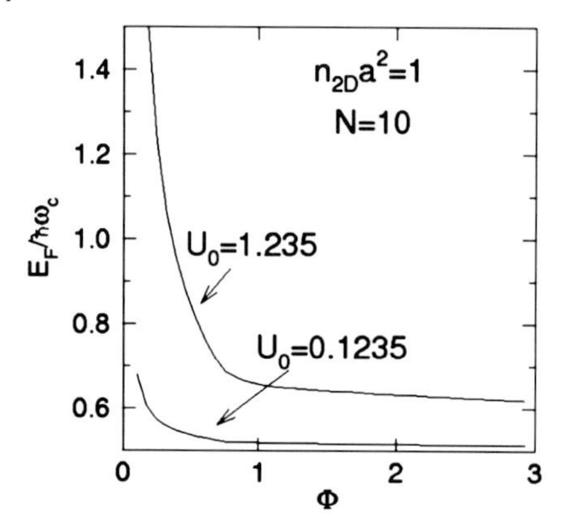

Figure 7.2 Plot of the Fermi energy E_F as a function of magnetic flux Φ for a chosen density and two values of the scattering potential $\bar{U}_0$. The parameters used in the calculation are given in the text.

Figure 7.3 shows the longitudinal and transverse conductivity coefficients $\sigma_{xx}(\omega)$ and $\sigma_{yx}(\omega)$ as functions of the flux Φ. The results compare the depen-

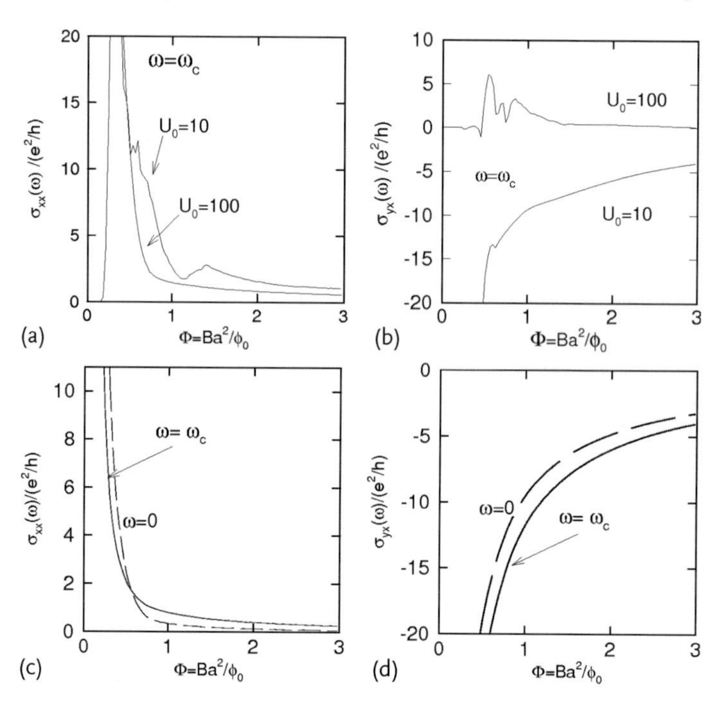

Figure 7.3 Plots of σ_{xx} and σ_{yx} as functions of the flux Φ in units of $\phi_0 = h/e$. We set $n_{2D}a^2 = 1$, $\sqrt{2}\gamma_0 = E_F/1.5$, and $N = 10$. In (a) and (b), we chose $\omega = \omega_c$ and two values of $\bar{U}_0 = 10$ and 100. In (c) and (d) $\bar{U}_0 = 1.235$ for $\omega = \omega_c$ (solid line) and $\omega = 0$ (dashed line). The impurity density is $n_I = 0.01 n_{2D}$ and the impurity scattering amplitude is $f = 0.62$.

dence on frequency and scattering potential. At zero frequency, $\sigma_{xx}(\omega)$ decreases as the lattice potential is increased at low and high magnetic fields. This behavior is unlike what was found for a two-dimensional array of scatterers. There, it was shown that the lattice enhances the forward-scattering of electrons at low magnetic fields but this conductivity is suppressed at large magnetic fields at $\omega = 0$. In Figure 7.3a,b, we demonstrate the effect of finite frequency and large scattering amplitude on both $\sigma_{xx}(\omega)$ and $\sigma_{yx}(\omega)$. Here, there exist oscillations arising from the commensurability effects between the cyclotron orbits and the widths of the wires confining the electrons. These commensurability Weiss oscillations [11, 12] occur at finite and zero frequency only when the electrons do not diffuse away from their confinement regions. This is substantiated by our calculations for smaller values of $\bar{U}_0$ in Figure 7.3c and d where there are no Weiss oscillations. That is, the nature of the oscillations for the magnetoconductivity is similar to the well-known Weiss oscillations which were observed for the magnetoresistivity of a modulated 2DES. Our model allows us to give a semiquantitative analysis of the magnetic field effect as the modulation strength is varied. At low magnetic fields, there is a range of magnetic flux where the conductivity is quenched when the scattering amplitude is large, as it is in Figure 7.3a,b. The results are shown when the impinging frequency is the same as the cyclotron frequency. At large magnetic field, there are no Weiss oscillations since the cyclotron radius is much smaller than the width of the wire. It is in the intermediate region of the magnetic field that the commensurability effects are obtained. Our results in Figure 7.3c and d show that in the weak modulation limit, the transverse conductivity behaves like the unmodulated 2DES (see Eq. (7.29)). As Φ is increased, finite ω enhances the backward scattering significantly in the longitudinal conductivity $\sigma_{xx}(\omega)$. We showed previously that the effect due to finite frequency for a 2D array of scatters may serve to enhance the forward scattering and lead to an increased conductivity. In Figure 7.3d, the transverse conductivity $\sigma_{yx}(\omega)$ is plotted as a function of Φ at zero and finite frequency for fixed $\bar{U}_0$. It is found that $\sigma_{yx}(\omega)$ is affected when a finite frequency field is applied throughout the range of chosen magnetic fields. There are two main observations. The first is that for $\omega = 0$ and $\omega = \omega_c$, the coefficient $\sigma_{yx}(\omega)$ is negative for this strength of scattering potential. Increasing the value of $\bar{U}_0$ affects $\sigma_{yx}(\omega)$ dramatically, as we show in Figure 7.3d. This is an indication of the effect which backscattering of the electron orbits has on the transport. The strong modulation quenches the transverse conductivity when the orbital radius is large enough for electron scattering off the walls quantum wires to dominate. In general, the effect of finite frequency is to reduce the transverse Hall and longitudinal conductivity, at low magnetic fields. However, at higher magnetic fields, the reverse effect takes place.

In Figure 7.4, we have plotted the longitudinal resistivity ρ_{xx} and transverse Hall resistivity ρ_{xy} as functions of Φ for different values of $\bar{U}_0$ and ω with $n_{2D}a^2 = 1$. For the smaller scattering strength, that is, $\bar{U}_0 = 10$, Figure 7.4a,b show that ρ_{xy} is almost linear in Φ at high magnetic field when either $\omega = 0$ or is finite. As the magnetic flux is decreased, there is a quenching of ρ_{xy}. Our results also show that the quenching at finite frequency for ρ_{xy} in Figure 7.4b occurs over a wider

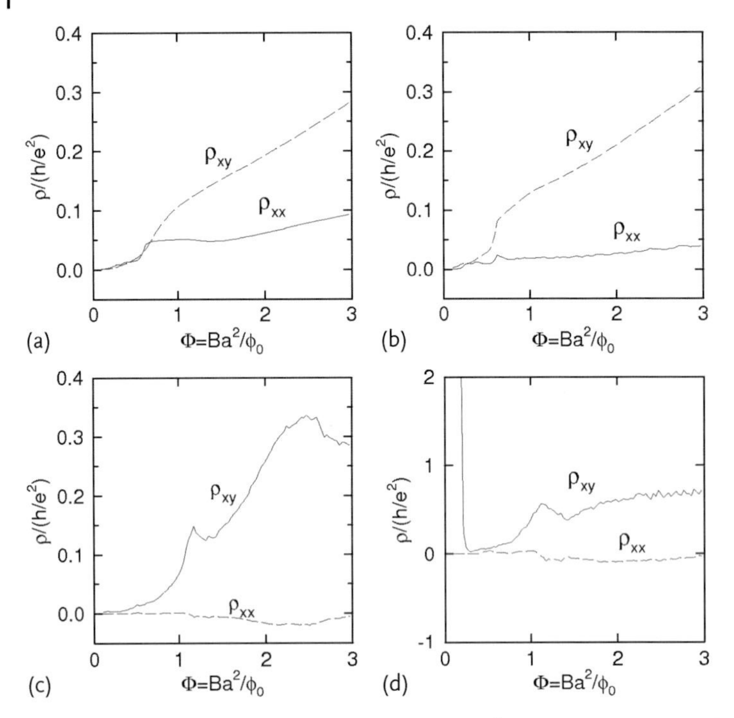

Figure 7.4 The Hall resistivity $\rho_{xy}(\omega)$ and longitudinal resistivity $\rho_{xx}(\omega)$ for electron density $n_{2D}a^2 = 1$, $\sqrt{2}\gamma_0 = E_F/1.5$ and $N = 10$. In (a) $\omega = 0$ and $\bar{U}_0 = 10$; (b) $\hbar\omega = 0.01E_F$ and $\bar{U}_0 = 10$; (c) $\omega = 0$ and $\bar{U}_0 = 100$; (d) $\hbar\omega = 0.01E_F$ and $\bar{U}_0 = 100$. The impurity density is $n_I = 0.01n_{2D}$ and the impurity scattering amplitude is $f = 0.62$.

range of magnetic field. The effect of finite frequency in Figure 7.4b serves to decrease ρ_{xy} and ρ_{xx} in Figure 7.4a but both are positive over the range of magnetic field for which the graphs are plotted. When the scattering strength is increased to $\bar{U}_0 = 100$, Weiss oscillations appear in the components of the magnetoresistivity shown in Figure 7.4a–c. These commensurate oscillations appear in the strong modulation limit at zero frequency as well as for the two frequencies chosen corresponding to $\omega = \omega_c$ and $\hbar\omega = 0.01E_F$. These oscillations are larger for ρ_{xy} than ρ_{xx} and do not appear in the low magnetic field regime since the cyclotron radius is much larger than the period of the density modulation in this case. When $\bar{U}_0$ is increased, electrons are more likely confined. So the Hall resistivities are increased with increased $\bar{U}_0$. When $\bar{U}_0 = 100$, the potential strength is large enough to produce backscattering in the direction of the density modulation along the x-direction to yield the commensurability oscillations and the negative values of ρ_{xx}. When $\bar{U}_0$ is large, Landau level mixing occurs and even the Landau levels in the high magnetic field regime are broadened and close to each other. The wave function overlap leads to tunneling between the confined regions and thus contribute more to ρ_{xx} and ρ_{xy}. However, when $\bar{U}_0 = 100$, the overlap due to guiding center mixing causes more trapping by the potential and eventually, the net force overcomes the Lorentz force, producing a negative longitudinal resistivity which is enhanced when the impinging radiation frequency is at resonance with the cyclotron frequency.

7.8 Scattering Theory Formalism

We now make use of the result for σ given in Eq. (7.19) to express the conductance coefficients in terms of the retarded and advanced Green's functions. From the definition of g_{mn} in Eq. (7.22), we obtain

$$g_{mn} = -\frac{e^2\hbar^3}{8\pi m^{*2}} \int_{-\infty}^{\infty} d\varepsilon \left[-\frac{d f_0(\varepsilon)}{d\varepsilon}\right] \int_{C_m} dy_m \int_{C_n} dy_n G_\varepsilon^+(\mathbf{x}, \mathbf{x}') \times \left[\overleftrightarrow{\mathbf{D}}^* \cdot \hat{\mathbf{x}}_m\right]\left[\overleftrightarrow{\mathbf{D}}' \cdot \hat{\mathbf{x}}_n\right] G_\varepsilon^-(\mathbf{x}', \mathbf{x}) \quad (7.53)$$

since the $G_\varepsilon^+ G_\varepsilon^+$ and the $G_\varepsilon^- G_\varepsilon^-$ terms vanish identically. Next, in order to obtain the multiprobe Landauer formula [13] derived by Büttiker, Baranger and Stone showed that

$$g_{mn} = \frac{e^2}{h} \int_{-\infty}^{\infty} d\varepsilon \left[-\frac{d f_0(\varepsilon)}{d\varepsilon}\right] \sum_{\alpha,\beta} \left|t_{mn,\alpha\beta}\right|^2 \quad \text{for } m \neq n \,, \quad (7.54)$$

where $t_{mn,\alpha\beta}$ is the transmission amplitude from mode β in lead n to mode α in lead m and the matrix $\{t_{mn,\alpha\beta}\}$ is a unitary S matrix. Also, since no current flows when V_n is the same for all n, it follows from Eq. (7.21) that $\sum_n g_{mn} = 0$ and consequently that

$$g_{mm} = -\sum_{n \neq m} g_{mn} = -\frac{e^2}{h} \int_{-\infty}^{\infty} d\varepsilon \left[-\frac{d f_0(\varepsilon)}{d\varepsilon}\right] \sum_{n \neq m} T_{mn}(\varepsilon) \,, \quad (7.55)$$

where $T_{mn}(\varepsilon) \equiv \sum_{\alpha,\beta} \left|t_{mn,\alpha\beta}\right|^2$. Assuming that all channels are identical and $\sum_n T_{mn}(\varepsilon) = N_c$ which is the number of propagating states or channels, it follows that

$$g_{mm} = -\frac{e^2}{h} \int_{-\infty}^{\infty} d\varepsilon \left[-\frac{d f_0(\varepsilon)}{d\varepsilon}\right] \left[N_c - T_{mm}(\varepsilon)\right] , \quad (7.56)$$

and therefore for $m \neq n$,

$$\begin{aligned} g_{mm} &= \frac{e^2}{h} \int_{-\infty}^{\infty} d\varepsilon \left[-\frac{d f_0(\varepsilon)}{d\varepsilon}\right] T_{mn}(\varepsilon) \quad \text{at } T \neq 0\,\mathrm{K} \\ &= \frac{e^2}{h} T_{mn}(E_F) \quad \text{at } T = 0\,\mathrm{K} \,, \end{aligned} \quad (7.57)$$

where E_F is the Fermi energy. As a result, the current coming out of lead m is related to the voltage on lead n at $T = 0\,\text{K}$ by

$$\begin{aligned} I_m &= \sum_n g_{mn} V_n \\ &= \frac{e^2}{h}\left[-N_c V_m + \sum_{n=1}^{N_L} T_{mn} V_n\right], \end{aligned} \tag{7.58}$$

which is the Landauer formula derived by Büttiker (see also, Baranger and Stone). Equation (7.58) may also be expressed in terms of the chemical potential $\mu_n = (-e)V_n$ in each lead. We observe that Eq. (7.21) can be inverted as a matrix equation to give the voltage $\{V_n\}$ in terms of the current and the inverse of the conductance matrix $\{g_{mn}\}$. This means that the voltage fluctuations can be calculated from a knowledge of the fluctuations in $\{g_{mn}\}$.

The novel features of the multiprobe Landauer formula, Eq. (7.58), has been discussed at length in the literature, including the review article by Stone and Szafer [14] that describes its historical development. Equation (7.58) treats the current and voltage probes on an equal footing and consequently involves the full S matrix which describes scattering into all the probes instead of just scattering between the current source and sink.

7.9 Quantum Hall Effect

Several authors have used the Landauer formula in Eq. (7.58) and simpler versions in which the probes are not introduced explicitly to analyze the quantum Hall effect. Equation (7.58), based on linear response theory, gives both the longitudinal and Hall resistance for mesoscopic systems. However, the question which must be addressed is what conditions must the conductance matrix $\{g_{mm}\}$ satisfy for there to be a quantized Hall resistance and also for there to be zero longitudinal resistance for a range of measurements as the magnetic field is varied. Baranger and Stone have shown that for systems with an arbitrary number of leads Eq. (7.58) possesses some general properties that are related to the quantum Hall effect. We now summarize their results which utilizes edge states.

Suppose a strong perpendicular magnetic field is applied to the sample. The carriers entering the sample are in edge states in the leads since the Lorentz force on these carriers pushes them on to the edges of the leads. As the carriers enter the sample, the edge states are mixed if there is bulk disorder and the states are no longer those of the leads. However, there are still current-carrying edge states if the energy of the disordered potential is much less than $\hbar\omega_c$ and the Fermi energy is not close to a bulk Landau level. It is reasonable to assume that carriers in the edge states will not backscatter more than a cyclotron radius so that we have $T_{m+1,m} \approx N_c$ and all other transmission coefficients as well as the reflection coefficient R_{mm} are zero. For simplicity, it is being assumed that all leads have the same number of channels. The system exhibits the integer quantum Hall effect if all resistances measured on the same side of the current path are zero, yielding zero longitudinal resistance, and all resistances measured across the current path give $R_H = (h/e^2)(1/N_c)$, the "*quantized Hall resistance*". Thus the system exhibits the complete integer Hall effect for any number of current leads.

7.10 Problems

1. By using the equation of motion for the density matrix $\hat{\rho}$, derive the result in Eq. (7.5).
2. Given $V(\boldsymbol{x}, t) = V(\boldsymbol{x})F(t)$ and $F(t) = \cos(\omega t)e^{-\delta|t|}$ for $t < 0$, prove the expression in Eq. (7.6) is the solution for the equation of motion in Eq. (7.5).
3. By substituting Eqs. (7.3) and (7.6) into Eq. (7.8), derive the result in Eq. (7.10) in the limit as $\delta \to 0$.
4. By using the expression in Eq. (7.10), derive the result in Eq. (7.12).
5. By combining the results in Eqs. (7.16) and (7.17), prove the expression in Eq. (7.18).
6. By using the result in Eq. (7.11) for $\omega = 0$, prove the nonlocal current-voltage relation in Eq. (7.21).
7. By using the expression for σ in Eq. (7.19) and that for g_{mn} in Eq. (7.22), derive the result in Eq. (7.53).

References

1 Büttiker, M. (1986) Four-terminal phase-coherent conductance. *Phys. Rev. Lett.*, **57**, 1761.
2 Kubo, R. (1957) Statistical-Mechanical Theory of Irreversible Processes. *J. Phys. Soc. Jpn.*, **12**, 570.
3 Kubo, R. (1959) *Lectures in Theoretical Physics*, (eds W.E. Brittin and L.G. Dunham), Interscience, New York, **1**, pp. 120–203.
4 Kubo, R. (1966) *Rep. Prog. Phys.*, **29**, 255.
5 Baranger, H.U. and Stone, A.D. (1989) Electrical linear-response theory in an arbitrary magnetic field: A new Fermi-surface formation. *Phys. Rev. B*, **40**, 8169.
6 Huang, D.H. and Gumbs, G. (1993) Quenching of the Hall effect in strongly modulated two-dimensional electronic systems. *Phys. Rev. B*, **48**, 2835.
7 Huang, D.H., Gumbs, G., and MacDonald, A.H. (1993) Comparison of magnetotransport in two-dimensional arrays of quantum dots and antidots. *Phys. Rev. B*, **48**, 2843.
8 Huang, D.H. and Gumbs, G. (1995) Quantum magnetotransport theory for bound-state electrons in lateral surface superlattices. *Phys. Rev. B*, **51**, 5558.
9 Gumbs, G. (2005) Dynamic resistivity of a two-dimensional electron gas with electric modulation. *Phys. Rev. B*, **72**, 125342.
10 Abrikosov, A.A., Gorkov, L.P., and Dzyaloshinski, I.E. (1963) *Methods*

of Quantum Field Theory in Statistical Physics, Prentice-Hall, Englewood Cliffs, New Jersey.

11 Weiss, D., von Klitzing, K., Ploog, K., and Weimann, G. (1989) Magnetoresistance oscillations in a two-dimensional electron gas induced by a submicrometer periodic potential. *Europhys. Lett.*, **8**, 179.

12 Weiss, D., Roukes, M.L., Menschig, A., Grambow, P., von Klitzing, K., and Weimann, G. (1991) Electron pinball and commensurate orbits in a periodic array of scatterers. *Phys. Rev. Lett.*, **66** 2790.

13 Landauera, R. (1970) Electrical resistance of disordered one-dimensional lattices. *Philos. Mag.*, **21**, 863.

14 Stone, A.D. and Szafer, A. (1988) What is measured when you measure a resistance? The Landauer formula revisited. *IBM J. Res. Dev.*, **32**, 384.

8
Nonlocal Conductivity for a Spin-Split Two-Dimensional Electron Liquid

8.1
Introduction

The transport properties of very small disordered conductors must be characterized by their conductance instead of their bulk conductivity. Several recent investigations have demonstrated the importance of statistical fluctuations arising from localization and quantum interference [1–20]. These fluctuations seem to be universal and of magnitude e^2/h for metallic samples. In the localized regime, they diverge. At low temperatures and in the metallic regime, the dependence of the conductance on the chemical potential and magnetic field shows reproducible aperiodic fluctuations of this order of magnitude. The effect of spin–orbit interaction (SOI) on these fluctuations is desirable.

The motivation for this section is as follows. An interesting aspect of the transport properties of low-dimensional structures is the nonlocal behavior of its electrical conductivity. In the metallic regime, transport phenomena are not adequately described using "local" conductivities such as the Drude form with a range of the mean-free path and where a long ranged part has been omitted [1]. When this correction is made, it has been shown that in the delocalized regime, the conductivity has a long-ranged diffusion part. However, in two dimensions (2D), noninteracting waves are localized when there is no SOI [9, 21]. The localization is primarily due to the interference of the waves scattered by the disorder. As shown in [1], the diffusion part is long-ranged in the absence of SOI. The reason for this difference is that the impurity-averaged Green's functions are short-ranged but the diffusion ladder was shown not to constrain the distance between its two end points in the absence of SOI.

So far, the response to both a static and dynamic electric field has been investigated. Of growing interest in the literature is the effect of SOI of the Rashba [22–24] or Dresselhaus [25] type on the conductance of asymmetric two-dimensional electron systems [17, 26–28]. The paper by Zhang and Ma [26] discussed how the Hall resistivity could be finite for sufficiently strong Rashba coupling in the absence of a perpendicular magnetic field. These authors also found an interesting behavior for the longitudinal resistivity which decreases as the Rashba SOI increases. In the work of Sinova, *et al.* [27], it was shown that for high-mobility systems with sub-

Properties of Interacting Low-Dimensional Systems, First Edition. G. Gumbs and D. Huang.

stantial Rashba SOI, a spin current that flows perpendicular to the charge current is intrinsic since an external magnetic field is not required to observe the transverse spin current. This phenomenon is referred to as the spin Hall effect (SHE) in 2D systems with Rashba-type SOI. The first step in our investigation is to calculate the conductivity tensor in the long wavelength limit and for low frequencies. This is carried out by averaging the nonlocal conductivity, that is, calculating $\langle \overleftrightarrow{\sigma}(\boldsymbol{r}, \boldsymbol{r}'; \omega)\rangle$ using the Kubo formula for the nonlocal conductivity. This involves the Green's functions with SOI. Our derived result contains contributions from both intraband and interband scattering. When the SOI is neglected, our result for intraband scattering reduces to the well-known Drude form. However, in the presence of SOI, both intraband and interband contributions depend on a logarithmic term which has no classical analogue. This term arises from the quantum mechanical nature of the SOI. We obtain a generalization of the nonlocal conductivity of metallic 2D systems with SOI [1]. The correction we obtain to the local conductivity is described by a propagator. This result was obtained starting with $\overleftrightarrow{\sigma}(\boldsymbol{r}, \boldsymbol{r}'; \omega)$ in coordinate space. Several steps are necessary to arrive at our result as we now describe in the outline of this chapter.

In Section 8.2, we present the frequency-dependent conductivity tensor in coordinate space and discuss the approximations we use in our calculations. In Section 8.3, we obtain the self-energy of the single-particle Green's function which we then employ to get an explicit formula for the scattering time of each subband determining the conductivity. Making use of this Green's function in Section 8.4, we calculate the frequency-dependent impurity-averaged conductivity in the long wavelength limit. The ladder diagram corrections to the local conductivity are calculated in Section 8.5. We express the result in coordinate space in order to gain some insight of its significance from a physical point of view. Some numerical results are presented in Section 8.6.

8.2
Kubo Formula for Conductivity

In this section, we consider electrons confined to move in two dimensions. If we neglect effects due to anisotropy for a 2D electron gas in the xy-plane with spin–orbit coupling, the Hamiltonian is [22–24]

$$H = \begin{pmatrix} -\frac{\hbar^2}{2m^*}\nabla^2 & \alpha_R \nabla_- \\ \alpha_R \nabla_+ & -\frac{\hbar^2}{2m^*}\nabla^2 \end{pmatrix}, \tag{8.1}$$

where m^* is the effective mass of the electron, $\nabla^2 = \partial^2/\partial x^2 + \partial^2/\partial y^2$, $\nabla_\pm = \partial_x \pm i\partial_y$. Also, α_R is the Rashba parameter for the electron system. The energy eigenvalues of the Hamiltonian H are, in terms of an in-plane wave vector $\boldsymbol{k}$, given by

$$\varepsilon_{k,\lambda} = \frac{\hbar^2 k^2}{2m^*} + \lambda \alpha_R k \tag{8.2}$$

with eigenfunctions

$$\psi_{k,\lambda}(\boldsymbol{r}) = \left[1,\ \lambda/k\left(k_y - ik_x\right)\right]\frac{e^{i\boldsymbol{k}\cdot\boldsymbol{r}}}{\sqrt{\mathcal{A}}}\,, \tag{8.3}$$

where $\lambda = \pm 1$ and $\mathcal{A}$ is a normalization area. Experiments by Nitta, *et al.* [29] demonstrated the realization of the 2D asymmetric inversion layer with spin–orbit interaction.

Using the Kubo formula for zero magnetic field and finite frequency ω in the absence of magnetic field, we have in terms of the retarded and advanced Green's functions $G_\varepsilon^\pm(\boldsymbol{r},\boldsymbol{r}')$ [20, 30, 31]

$$\begin{aligned}
\overleftrightarrow{\sigma}(\boldsymbol{r},\boldsymbol{r}';\omega) &= \left(\frac{e^2\hbar^2}{16\pi m^{*2}\omega}\right)\int_{-\infty}^{\infty} d\varepsilon\, f_0(\varepsilon) \\
&\times\Big\{\left[G_\varepsilon^+(\boldsymbol{r},\boldsymbol{r}') - G_\varepsilon^-(\boldsymbol{r},\boldsymbol{r}')\right]\overleftrightarrow{\nabla}\overleftrightarrow{\nabla}'\left[G_{\varepsilon+\hbar\omega}^-(\boldsymbol{r}',\boldsymbol{r}) - G_{\varepsilon-\hbar\omega}^-(\boldsymbol{r}',\boldsymbol{r})\right] \\
&- \left[G_{\varepsilon+\hbar\omega}^+(\boldsymbol{r},\boldsymbol{r}') - G_{\varepsilon-\hbar\omega}^+(\boldsymbol{r},\boldsymbol{r}')\right]\overleftrightarrow{\nabla}\overleftrightarrow{\nabla}'\left[G_\varepsilon^+(\boldsymbol{r}',\boldsymbol{r}) - G_\varepsilon^-(\boldsymbol{r}',\boldsymbol{r})\right]\Big\}\,,
\end{aligned} \tag{8.4}$$

where $f_0(\varepsilon)$ is the Fermi–Dirac distribution function, and the antisymmetric operator is defined by $f\overleftrightarrow{\nabla}g = f(\boldsymbol{r})\nabla g(\boldsymbol{r}) - g(\boldsymbol{r})\nabla f(\boldsymbol{r})$ with $f(\boldsymbol{r})$ and $g(\boldsymbol{r})$ being arbitrary functions of the position vector $\boldsymbol{r}$. Taking the space and time Fourier transforms of Eq. (8.4), we obtain the configuration-averaged conductivity as

$$\begin{aligned}
\langle\sigma_{\alpha\beta}(q,\omega)\rangle &= \left(\frac{e^2\hbar^3}{8m^{*2}}\right)\sum_{\lambda,\lambda'}\sum_{k,k'}\delta_{k',k+q}\left(k_\alpha + k'_\alpha\right)\left(k_\beta + k'_\beta\right) \\
&\times\int_{-\infty}^{\infty}\frac{d\varepsilon}{2\pi}\left[\frac{f_0(\varepsilon-\hbar\omega) - f_0(\varepsilon)}{\hbar\omega}\left\langle G_\lambda^+(\varepsilon,\boldsymbol{k})G_{\lambda'}^-(\varepsilon-\hbar\omega,\boldsymbol{k}+\boldsymbol{q})\right\rangle\right. \\
&+ \left.\frac{f_0(\varepsilon) - f_0(\varepsilon+\hbar\omega)}{\hbar\omega}\left\langle G_\lambda^+(\varepsilon,\boldsymbol{k})G_{\lambda'}^-(\varepsilon+\hbar\omega,\boldsymbol{k}+\boldsymbol{q})\right\rangle\right]\,,
\end{aligned} \tag{8.5}$$

where $\boldsymbol{q}$ is the wave vector and the brackets $\langle\cdots\rangle$ denote impurity averaging. If we use the free-particle Green's function

$$G_{0,\lambda}^\pm(\varepsilon,\boldsymbol{k}) = \frac{1}{\varepsilon - \varepsilon_{k,\lambda} \pm i0^+} \tag{8.6}$$

with the linear spin-split band $\varepsilon_{k,\lambda}$ given above. Then, the conductivity would be infinite since no scattering is involved.

For a finite conductivity $\overleftrightarrow{\sigma}$, we must include scattering. Impurity averaging allows the survival of only those diagrams where at least two impurity lines end at the same impurity so that if V_q denotes the impurity-scattering potential, the diagram

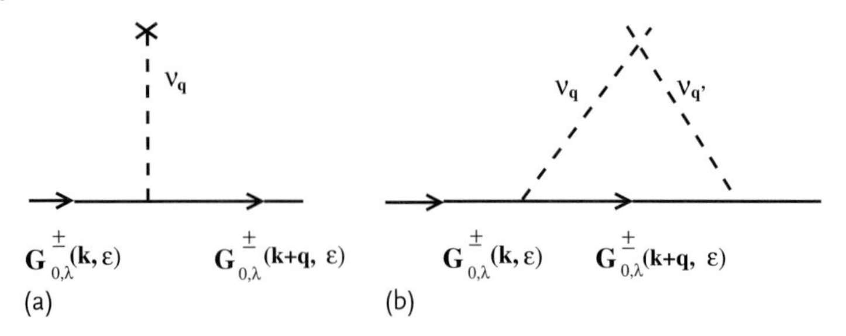

Figure 8.1 The diagram in (a) does not contribute to the conductivity in our calculations. The diagram in (b) is included. The cross represents an impurity and the dashed line with a cross denotes scattering off the impurity.

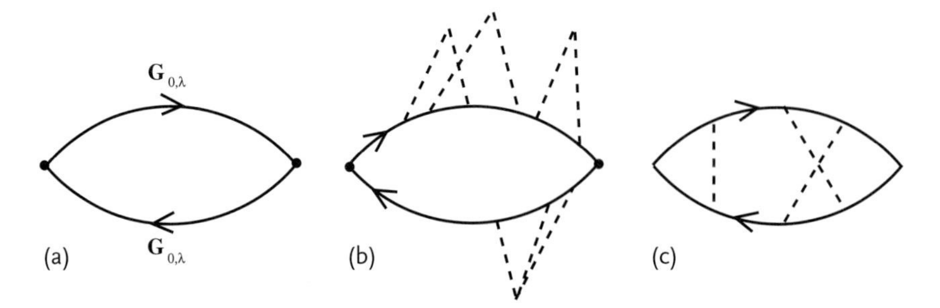

Figure 8.2 (a) The lowest-order diagram for the conductivity. The diagrams for its corrections are from the self-energy in (b) and from the vertex in (c).

in Figure 8.1a dies, whereas the diagram in Figure 8.1b survives if $\boldsymbol{q}+\boldsymbol{q}'=0$. Also, we show below that there are two types of corrections to Figure 8.2a, namely, the self-energy corrections in Figure 8.2b and the vertex corrections in Figure 8.2c. We now present the details for these calculations.

8.3
The Self-Energy and Scattering Time

The self-energy is calculated from the Dyson equation

$$G_{\lambda}^{\pm}(\varepsilon,\boldsymbol{k}) = G_{0,\lambda}^{\pm}(\varepsilon,\boldsymbol{k}) + G_{0,\lambda}^{\pm}(\varepsilon,\boldsymbol{k}) \times \left[\int \frac{d^2\boldsymbol{q}}{(2\pi)^2}|V_q|^2 G_{0,\lambda}^{\pm}(\varepsilon,\boldsymbol{k}+\boldsymbol{q})\right] G_{\lambda}^{\pm}(\varepsilon,\boldsymbol{k})\,. \tag{8.7}$$

The solution of Eq. (8.7) is straightforward. Denoting the real and imaginary parts of the self-energy by $-A_{\pm}$ and $-B_{\pm}$ so that

$$
\begin{aligned}
-\left[A_\pm(\varepsilon,\lambda) + i B_\pm(\varepsilon,\lambda)\right] &= \int \frac{d^2\boldsymbol{q}}{(2\pi)^2} |V_q|^2 G^{\pm}_{0,\lambda}(\varepsilon, \boldsymbol{k}+\boldsymbol{q}) \\
&= \int \frac{d^2\boldsymbol{q}}{(2\pi)^2} \frac{|V|^2}{\varepsilon - \frac{\hbar^2 q^2}{2m^*} - \lambda \alpha_{\mathrm{R}} q \pm i\eta} \\
&= \int \frac{d^2\boldsymbol{q}}{(2\pi)^2} \mathcal{P}\left(\frac{|V|^2}{\varepsilon - \frac{\hbar^2 q^2}{2m^*} - \lambda \alpha_{\mathrm{R}} q} \right) \\
&\quad \mp i\pi \int \frac{d^2\boldsymbol{q}}{(2\pi)^2} |V|^2 \delta\left(\varepsilon - \frac{\hbar^2 q^2}{2m^*} - \lambda \alpha_{\mathrm{R}} q \right), \qquad (8.8)
\end{aligned}
$$

where, for simplicity, we set $|V_q|^2 = |V|^2$, independent of wave vector, and we have

$$
G^{\pm}_{\lambda}(\varepsilon, \boldsymbol{k}) = \frac{1}{\varepsilon + A_\pm(\varepsilon,\lambda) - \frac{\hbar^2 k^2}{2m^*} - \lambda \alpha_{\mathrm{R}} k + i B_\pm(\varepsilon,\lambda)} \qquad (8.9)
$$

by neglecting the infinitesimal η and keeping the finite $B_\pm(\varepsilon,\lambda)$. Equation (8.9) clearly shows that the mean-free time is $\tau_\lambda(\varepsilon)$ with $\tau_\lambda^{-1}(\varepsilon) \equiv 2B_\pm/\hbar$ and

$$
\tau_\lambda^{-1}(\varepsilon) = \frac{2|V|^2}{\hbar} \int \frac{d^2\boldsymbol{q}}{(2\pi)^2} \delta\left(\varepsilon - \frac{\hbar^2 q^2}{2m^*} - \lambda \alpha_{\mathrm{R}} q \right) \equiv \frac{2\pi \mathcal{N}_\lambda(\varepsilon) |V|^2}{\hbar}, \qquad (8.10)
$$

where $\mathcal{N}_\lambda(\varepsilon)$ is the density-of-states (DOS) for the subband $\lambda = \pm 1$ at energy ε [32]. Also, since the energy is shifted by an amount independent of the wave vector, we can neglect $A_\pm$. Therefore, the total DOS is not affected by the SOI, and the sum of τ_+^{-1} and τ_-^{-1} is a constant [32].

8.4
Drude-Type Conductivity for Spin-Split Subband Model

With the finite-lifetime Green's function

$$
G^{\pm}_{\lambda}(\varepsilon, \boldsymbol{k}) = \frac{1}{\varepsilon - \frac{\hbar^2 k^2}{2m^*} - \lambda \alpha_{\mathrm{R}} k \pm \frac{i\hbar}{2\tau_\lambda}}, \qquad (8.11)
$$

we can calculate the conductivity using Eq. (8.5). We note that the use of self-energy corrections amounts to setting $\langle G^- G^+ \rangle = \langle G^- \rangle \langle G^+ \rangle$ in Eq. (8.5) [33]. The vertex corrections will go beyond this approximation. In the long wavelength limit, we then have, to the lowest order in the disorder,

$$
\begin{aligned}
\sigma_{\alpha\beta}(q=0,\omega) &= \left(\frac{e^2 \hbar^3}{2m^{*2}} \right) \sum_{\lambda,\lambda'} \int_{-\infty}^{\infty} \frac{d\varepsilon}{2\pi} \left\{ \frac{f_0(\varepsilon - \hbar\omega) - f_0(\varepsilon)}{\hbar\omega} \Lambda^{(\lambda,\lambda')}_{\alpha\beta}(\varepsilon,\omega) \right. \\
&\quad \left. + \frac{f_0(\varepsilon) - f_0(\varepsilon + \hbar\omega)}{\hbar\omega} \Lambda^{(\lambda,\lambda')}_{\alpha\beta}(\varepsilon,-\omega) \right\}, \qquad (8.12)
\end{aligned}
$$

where we must evaluate the integral over the wave vector in calculating the function defined by

$$\begin{aligned} \Lambda_{\alpha\beta}^{(\lambda,\lambda')}(\varepsilon,\omega) &\equiv \sum_{\boldsymbol{k}} k_\alpha k_\beta G_\lambda^+(\varepsilon,\boldsymbol{k})\, G_{\lambda'}^-(\varepsilon-\hbar\omega,\boldsymbol{k}) \\ &= \sum_{\boldsymbol{k}} \left[\frac{k^2}{2}\delta_{\alpha\beta} + (1-\delta_{\alpha\beta})k_\alpha k_\beta\right] G_\lambda^+(\varepsilon,\boldsymbol{k})\, G_{\lambda'}^-(\varepsilon-\hbar\omega,\boldsymbol{k}) \end{aligned} \tag{8.13}$$

with the understanding that there is a cut-off for the upper limit of the $\boldsymbol{k}$-integral for convergence. The $\delta_{\alpha\beta}$-term has a $q = 0$ contribution and all contributions $\sim q, q^2, \cdots$ can be shown to be higher order in the perturbation parameter $(k_F\ell_\lambda)^{-1}$, where ℓ_λ is the mean-free path for subband$-\lambda$. This follows from Eq. (8.11) by writing the product of Green's functions in Eq. (8.5) in partial fractions, then expanding $G_\lambda^\pm(\varepsilon,\boldsymbol{k}+\boldsymbol{q})$ in powers of q. The $(1-\delta_{\alpha\beta})$-term obviously does not have a $q = 0$-contribution (due to the asymmetry of the integrand in Eq. (8.13)). Furthermore, those contributions from finite q are also of higher order in $(k_F\ell_\lambda)^{-1}$. We also make use of this approximation in Section 8.5 when we calculate the vertex corrections. Therefore, to the lowest order in $(k_F\ell_\lambda)^{-1}$, we obtain

$$\Lambda_{\alpha\beta}^{(\lambda,\lambda')}(\varepsilon,\omega) = \frac{\delta_{\alpha\beta}}{2}\left(\frac{m^*}{2\hbar^3}\right) M_{\lambda,\lambda'}(\varepsilon,\omega)\,, \tag{8.14}$$

where

$$\begin{aligned} M_{\lambda,\lambda'}(\varepsilon,\omega) \approx{}& \left[k_\varepsilon^2 + \left(\frac{m^*\alpha_R}{\hbar^2}\right)^2\right]^{-1/2} \\ &\times \left\{ \frac{k_\lambda^3(\varepsilon)}{\frac{1}{2}\left[\tau_\lambda^{-1}(\varepsilon)+\tau_{\lambda'}^{-1}(\varepsilon)\right] - i\omega - \frac{i}{\hbar}(\lambda'-\lambda)\alpha_R k_\lambda(\varepsilon)} \right. \\ &\times \left[1 - \frac{i}{\pi}\ln\left(\frac{\frac{\lambda m^*\alpha_R}{\hbar^2} + \sqrt{k_\varepsilon^2 + \left(\frac{m^*\alpha_R}{\hbar^2}\right)^2}}{-\frac{\lambda m^*\alpha_R}{\hbar^2} + \sqrt{k_\varepsilon^2 + \left(\frac{m^*\alpha_R}{\hbar^2}\right)^2}}\right)\right] \\ &+ \frac{k_{\lambda'}^3(\varepsilon)}{\frac{1}{2}\left[\tau_\lambda^{-1}(\varepsilon)+\tau_{\lambda'}^{-1}(\varepsilon)\right] - i\omega - \frac{i}{\hbar}(\lambda'-\lambda)\alpha_R k_{\lambda'}(\varepsilon)} \\ &\times \left. \left[1 - \frac{i}{\pi}\ln\left(\frac{\frac{\lambda' m^*\alpha_R}{\hbar^2} + \sqrt{k_\varepsilon^2 + \left(\frac{m^*\alpha_R}{\hbar^2}\right)^2}}{-\frac{\lambda' m^*\alpha_R}{\hbar^2} + \sqrt{k_\varepsilon^2 + \left(\frac{m^*\alpha_R}{\hbar^2}\right)^2}}\right)\right]\right\}, \end{aligned} \tag{8.15}$$

and we assumed that $(k_F\ell_\lambda)^{-1} \ll \sqrt{1+(m^*\alpha_R/\hbar^2 k_F)^2}$. In this notation, $\ell_\lambda = (\hbar k_\lambda(E_F)/m^*)\tau_\lambda$ and

$$k_\lambda(\varepsilon) = -\lambda\frac{m^*\alpha_R}{\hbar^2} + \sqrt{k_\varepsilon^2 + \left(\frac{m^*\alpha_R}{\hbar^2}\right)^2}$$

with the definition of $k_\varepsilon = \sqrt{2m^*\varepsilon/\hbar^2}$.

In the low-frequency limit, that is, $\hbar\omega \ll E_F$, and at low temperature, the difference in Fermi functions in Eq. (8.12) yields a delta function and the Green's functions must be evaluated at the Fermi energy E_F. We then obtain the leading-order contribution to the conductivity as

$$\sigma_{\alpha\beta}(q = 0, \omega) = \frac{1}{2}\left[\sigma^{D}_{\alpha\beta}(\omega) + \sigma^{D}_{\alpha\beta}(-\omega)\right], \tag{8.16}$$

where the Drude-type result for conductivity is given by

$$\sigma^{D}_{\alpha\beta}(\omega) = \frac{\delta_{\alpha\beta}}{2\pi}\left(\frac{e^2}{m^*}\right)\sum_{\lambda,\lambda'} M_{\lambda,\lambda'}(E_F, \omega) \tag{8.17}$$

with $M_{\lambda,\lambda'}(E_F, \omega)$ defined in Eq. (8.15). One can express Eq. (8.17) in terms of the electron density in each subband using $n_\lambda = k_\lambda^2/4\pi$ for the λ subband. There are logarithmic terms in $\sigma^{D}_{\alpha\beta}(\omega)$ arising from SOI and which have no counterpart in a classical equation-of-motion approach. It is present in both intraband ($\lambda' = \lambda$) and interband ($\lambda' \neq \lambda$) scattering terms. When $\alpha_R = 0$, for intraband scattering, the contribution coincides with the well-known result for a single species of scatterers. For interband scattering, the average of the inverse scattering times $1/\tau_\pm(E_F)$ takes account of a (+) spin scattering off a (−) spin. Also, the densities of both these carriers enter this term. The frequency $\alpha_R k_\lambda(E_F)/\hbar$ which arises from the SOI also plays a role in this expression.

8.5
Vertex Corrections to the Local Conductivity

The crossover from weak localization [34, 35] to antilocalization in high-mobility GaAs/AlGaAs 2DEG systems has been realized by Miller, *et al.* [36] using tunable gates to control SOI effects. These authors presented a theory beyond the diffusion approximation to analyze their experimental data. The theory was successful in explaining the magnetotransport when the SO precession frequency becomes comparable with the inverse transport scattering time. The way in which quantum interference effects influence the conductivity was also included in the theory of Miller, *et al.* Some discussion of the relationship between weak localization and SO rotations with a phase coherence length was also given. These are all interesting concerns of the subject and may be further investigated using field theoretic techniques such as presented in this chapter. Hikami, Larkin, and Nagaoka [37] showed that the correction to the Drude conductivity in 2D is logarithmic and scales like $\ln(L/\ell)$, where L is the sample length. Lee and Ramakrishnan [34] also showed that the diffusive part of the conductivity comes from the ladder diagrams. The maximally crossed diagrams contribute to higher order in the impurity concentration.

We now deal with the corrections to the Drude formula at zero frequency. The leading-order corrections are *ladder diagrams* as shown in Figure 8.3a and are the

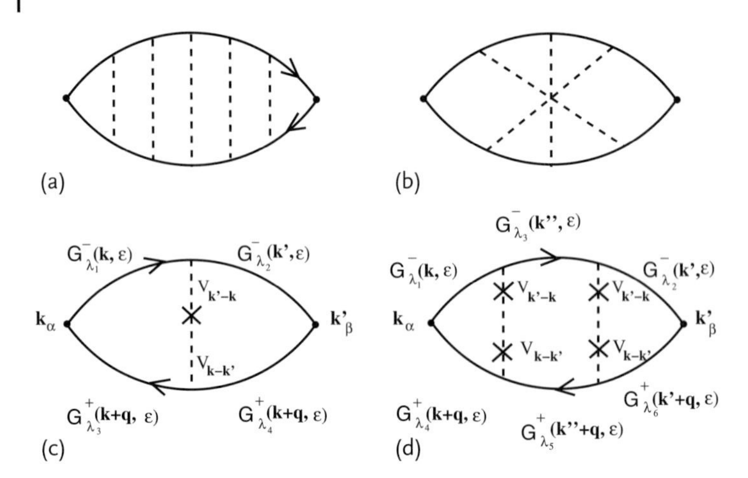

Figure 8.3 (a) Typical Ladder diagram, (b) typical maximally crossed diagram, (c) the simplest ladder diagram, and (d) the term corresponding to Eq. (8.19).

simplest to deal with. The contributions of diagrams with crosses are smaller. The most relevant ones are *maximally* crossed diagrams shown in Figure 8.3b, corresponding to weak localization. They describe the leading terms responsible for backscattering and consequently the modification of the diffusion constant [9]. The simplest *ladder* diagram is as given in Figure 8.3c [38].

The Green's functions describe the propagation of the electron and the hole. Their interference due to impurity scattering has been described diagrammatically in the paper by Lee and Ramakrishnan [34]. The contributions to the conductivity from various types of diagrams have also been analyzed in [34] and in the book by Abrikosov, Gorkov, and Dzyaloshinski [39]. In the absence of SOI, they also noted that the maximally crossed diagrams lead to interference effects of higher order in the impurity concentration and are the crucial localizing factor. Also, it was pointed out in [34] that the ladder diagram as shown in Figure 8.3 is responsible for diffusion. In the presence of SOI, as long as we are in the weak localization regime, these arguments are valid.

We note that higher order corrections to the Drude conductivity contain a scattering time that is weighted by a factor which takes account of large angle and long-range scattering processes. Mahan [31] showed that this arises from vertex corrections which can be evaluated perturbatively for low impurity concentration. Fetter and Walecka [40] have calculated in the ladder approximation the effective interaction for a dilute many-particle system interacting with singular repulsive potentials. This effective potential has a long-range contribution due to the vertex corrections.

From the Kubo formula, the ladder diagram corresponds to, assuming that the scattering is isotropic with $|V_k|^2 = |V|^2$,

$$\sum_{k,k'} \sum_{\lambda_i} k_\alpha k'_\beta |V|^2 \times G^-_{\lambda_1}(k, E_F) G^-_{\lambda_2}(k', E_F) G^+_{\lambda_3}(k+q, E_F) G^+_{\lambda_4}(k'+q, E_F) . \tag{8.18}$$

The next term is shown in Figure 8.3d and is equal to

$$\sum_{k,k',k''} \sum_{\lambda_i} k_\alpha k'_\beta |V|^2 G^-_{\lambda_1}(k, E_F) G^-_{\lambda_2}(k', E_F) G^-_{\lambda_3}(k'', E_F) \times G^+_{\lambda_4}(k+q, E_F) G^+_{\lambda_5}(k''+q, E_F) G^+_{\lambda_6}(k'+q, E_F) . \tag{8.19}$$

As long as the scattering is isotropic, it is easy to write the sum of diagrams in Figure 8.4 in compact form as

$$\mathcal{J}_\alpha(q)|V|^2 \left[1 + |V|^2 \Pi(q) + \left(|V|^2 \Pi(q)\right)^2 + \cdots\right] \mathcal{J}_\beta(q) = \mathcal{J}_\alpha(q) \mathcal{J}_\beta(q) \mathcal{D}(q) , \tag{8.20}$$

where the diffusion propagator is given by

$$\mathcal{D}(q) \equiv \frac{|V|^2}{1 - |V|^2 \Pi(q)} , \tag{8.21}$$

and

$$\mathcal{J}_\alpha(q) = \frac{1}{\mathcal{A}} \sum_k \sum_{\lambda,\lambda'} k_\alpha G^-_\lambda(k, E_F) G^+_{\lambda'}(k+q, E_F) , \tag{8.22}$$

$$\Pi(q) = \frac{1}{\mathcal{A}} \sum_k \sum_{\lambda,\lambda'} G^-_\lambda(k, E_F) G^+_{\lambda'}(k+q, E_F) \tag{8.23}$$

where $\mathcal{A}$ is the sample area.

Thus, we obtain the vertex correction to the local conductivity given by

$$\delta \langle \sigma_{\alpha\beta}(q, \omega = 0) \rangle = \left(\frac{e^2 \hbar^3}{\pi m^{*2}}\right) \mathcal{J}_\alpha(q) \mathcal{J}_\beta(q) \mathcal{D}(q) . \tag{8.24}$$

We now make use of the result in Eq. (8.24) to investigate the behavior of the conductivity to the lowest order in $(k_F \ell_\lambda)^{-1}$. After a tedious but straightforward calculation, we obtain

$$\mathcal{J}_\alpha(q) \approx -\frac{i}{2k_F^2} q_\alpha \frac{m^{*2}}{\hbar^4} \frac{1}{\sqrt{k_F^2 + \left(\frac{m^* \alpha_R}{\hbar^2}\right)^2}} \sum_{\lambda,\lambda'} \left. \frac{\partial \mathcal{F}_{\lambda,\lambda'}(X)}{\partial X} \right|_{X=1} , \tag{8.25}$$

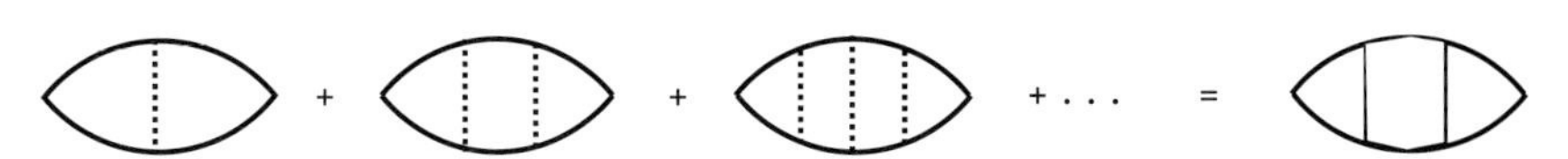

Figure 8.4 Sum of ladder diagrams given by Eq. (8.20).

where

$$
\begin{aligned}
&\mathcal{F}_{\lambda,\lambda'}(X)\\
&\approx \frac{2k_\lambda^3(E_\mathrm{F}) + \lambda' \frac{2m^*\alpha_\mathrm{R}}{\hbar^2} k_\lambda^2(E_\mathrm{F})}{(\lambda'-\lambda)\frac{2m^*\alpha_\mathrm{R}}{\hbar^2} k_\lambda(E_\mathrm{F}) - \frac{im^*}{\hbar}\left[\tau_\lambda^{-1}(E_\mathrm{F}) + \tau_{\lambda'}^{-1}(E_\mathrm{F})\right] - k_\mathrm{F}^2(X-1)}\\
&\quad\times\left[1 - \frac{i}{\pi}\ln\left(\frac{\frac{\lambda m^*\alpha_\mathrm{R}}{\hbar^2} + \sqrt{k_\mathrm{F}^2 + \left(\frac{m^*\alpha_\mathrm{R}}{\hbar^2}\right)^2}}{-\frac{\lambda m^*\alpha_\mathrm{R}}{\hbar^2} + \sqrt{k_\mathrm{F}^2 + \left(\frac{m^*\alpha_\mathrm{R}}{\hbar^2}\right)^2}}\right)\right]\\
&\quad+ \frac{2k_{\lambda'}^3(E_\mathrm{F}) + \lambda' \frac{2m^*\alpha_\mathrm{R}}{\hbar^2} k_{\lambda'}^2(E_\mathrm{F})}{(\lambda'-\lambda)\frac{2m^*\alpha_\mathrm{R}}{\hbar^2} k_{\lambda'}(E_\mathrm{F}) - \frac{im^*}{\hbar}\left[\tau_\lambda^{-1}(E_\mathrm{F}) + \tau_{\lambda'}^{-1}(E_\mathrm{F})\right] - k_\mathrm{F}^2(X-1)}\\
&\quad\times\left[1 - \frac{i}{\pi}\ln\left(\frac{\frac{\lambda' m^*\alpha_\mathrm{R}}{\hbar^2} + \sqrt{k_\mathrm{F}^2 X + \left(\frac{m^*\alpha_\mathrm{R}}{\hbar^2}\right)^2}}{-\frac{\lambda' m^*\alpha_\mathrm{R}}{\hbar^2} + \sqrt{k_\mathrm{F}^2 X + \left(\frac{m^*\alpha_\mathrm{R}}{\hbar^2}\right)^2}}\right)\right].
\end{aligned}
\tag{8.26}
$$

Our calculation also shows that

$$
\Pi(q) \approx A - Bq^2 . \tag{8.27}
$$

In this notation,

$$
\begin{aligned}
A &= \rho_\mathrm{2D}\left(\frac{\pi}{\hbar}\right)\left[k_\mathrm{F}^2 + \left(\frac{m^*\alpha_\mathrm{R}}{\hbar^2}\right)^2\right]^{-1/2}\\
&\quad\times\sum_{\lambda,\lambda'} \frac{\frac{1}{2}\left[\tau_\lambda^{-1}(E_\mathrm{F}) + \tau_{\lambda'}^{-1}(E_\mathrm{F})\right] k_\lambda(E_\mathrm{F})}{(\lambda'-\lambda)^2\left(\frac{\alpha_\mathrm{R} k_\lambda(E_\mathrm{F})}{\hbar}\right)^2 + \frac{1}{4}\left[\tau_\lambda^{-1}(E_\mathrm{F}) + \tau_{\lambda'}^{-1}(E_\mathrm{F})\right]^2}\\
&\quad\times\left[1 - \frac{i}{\pi}\ln\left(\frac{\frac{\lambda m^*\alpha_\mathrm{R}}{\hbar^2} + \sqrt{k_\mathrm{F}^2 + \left(\frac{m^*\alpha_\mathrm{R}}{\hbar^2}\right)^2}}{-\frac{\lambda m^*\alpha_\mathrm{R}}{\hbar^2} + \sqrt{k_\mathrm{F}^2 + \left(\frac{m^*\alpha_\mathrm{R}}{\hbar^2}\right)^2}}\right)\right],
\end{aligned}
\tag{8.28}
$$

where $\rho_\mathrm{2D} = m^*/\pi\hbar^2$ is the density-of-states for a homogeneous 2D electron system in the absence of SOI and

$$
B = -\rho_\mathrm{2D}\left(\frac{\pi}{\hbar}\right)\frac{1}{4k_\mathrm{F}^4}\left[k_\mathrm{F}^2 + \left(\frac{m^*\alpha_\mathrm{R}}{\hbar^2}\right)^2\right]^{-1/2} \sum_{\lambda,\lambda'} \left.\frac{\partial^2 \mathcal{S}_{\lambda,\lambda'}}{\partial X^2}\right|_{X=1}, \tag{8.29}
$$

with

$$\mathcal{S}_{\lambda,\lambda'}(X)$$

$$\approx -\frac{k_\lambda(E_F)\left[k_\lambda(E_F)+2\lambda' m^*\alpha_R/\hbar^2\right]^2}{\frac{1}{2}\left[\tau_\lambda^{-1}(E_F)+\tau_{\lambda'}^{-1}(E_F)\right]+i(\lambda'-\lambda)\alpha_R k_\lambda(E_F)/\hbar - i\hbar k_\lambda^2(E_F)(X-1)/m^*}$$

$$\times\left[1-\frac{i}{\pi}\ln\left(\frac{\frac{\lambda m^*\alpha_R}{\hbar^2}+\sqrt{k_F^2+\left(\frac{m^*\alpha_R}{\hbar^2}\right)^2}}{-\frac{\lambda m^*\alpha_R}{\hbar^2}+\sqrt{k_F^2+\left(\frac{m^*\alpha_R}{\hbar^2}\right)^2}}\right)\right]$$

$$+\frac{k_{\lambda'}(E_F)\left[k_{\lambda'}(E_F)+2\lambda' m^*\alpha_R/\hbar^2\right]^2}{\frac{1}{2}\left[\tau_\lambda^{-1}(E_F)+\tau_{\lambda'}^{-1}(E_F)\right]+i(\lambda'-\lambda)\alpha_R k_\lambda(E_F)/\hbar - i\hbar k_\lambda^2(E_F)(X-1)/m^*}$$

$$\times\left[1-\frac{i}{\pi}\ln\left(\frac{\frac{\lambda' m^*\alpha_R}{\hbar^2}+\sqrt{k_F^2 X+\left(\frac{m^*\alpha_R}{\hbar^2}\right)^2}}{-\frac{\lambda' m^*\alpha_R}{\hbar^2}+\sqrt{k_F^2 X+\left(\frac{m^*\alpha_R}{\hbar^2}\right)^2}}\right)\right]. \tag{8.30}$$

Making use of these results in Eq. (8.24), we obtain the vertex correction to the conductivity as

$$\delta\left\langle\sigma_{\alpha\beta}(q,\omega=0)\right\rangle = -\sigma_0^D\frac{q_\alpha q_\beta}{q^2}C(q)\,, \tag{8.31}$$

where the expression for $C(q)$ is given explicitly as

$$C(q) = -\left(\frac{e^2 m^{*2}}{4\pi\hbar^5 k_F^4}\right)\left(\frac{q^2}{A-Bq^2}\right)\frac{1}{k_F^2+\left(\frac{m^*\alpha_R}{\hbar^2}\right)^2}$$

$$\times\left(\sum_{\lambda,\lambda'}\left.\frac{\partial\mathcal{F}_{\lambda,\lambda'}(X)}{\partial X}\right|_{X=1}\right)^2\frac{1}{\sigma_0^D}\,. \tag{8.32}$$

Collecting these results, we obtain the conductivity in q-space to be given by

$$\left\langle\sigma_{\alpha\beta}(q,\omega=0)\right\rangle = \sigma_0^D\left[\delta_{\alpha\beta}-\frac{q_\alpha q_\beta}{q^2}C(q)\right], \tag{8.33}$$

and $\sigma_{\alpha\beta}^D(\omega=0)=\delta_{\alpha\beta}\sigma_0^D$. In the absence of the SOI, the above results reduce to the following:

$$\mathcal{F}_{\lambda,\lambda'}(X)|_{\alpha_R=0} = -\frac{4k_F^3}{\frac{2im^*}{\hbar\tau}+k_F^2(X-1)}\,, \tag{8.34}$$

$$\mathcal{S}_{\lambda,\lambda'}(X)|_{\alpha_R=0} = \frac{2k_F^3}{\frac{1}{\tau}-i\hbar k_F^2\frac{(X-1)}{m^*}}\,, \tag{8.35}$$

$$A|_{\alpha_R=0} = \frac{4\pi\tau}{\hbar}\rho_{2D}\,,\quad B|_{\alpha_R=0} = \frac{4\pi\hbar k_F^2\tau^3}{m^{*2}}\rho_{2D}\,, \tag{8.36}$$

where τ is the average time between collisions in the absence of SOI. Also, from Eq. (8.36), we obtain

$$\left.\frac{A}{Bk_{\mathrm{F}}^2}\right|_{\alpha_{\mathrm{R}}=0} = \frac{1}{(k_{\mathrm{F}}\ell)^2}\,, \tag{8.37}$$

where ℓ is the mean free path when $\alpha_{\mathrm{R}} = 0$. Thus, for $k_{\mathrm{F}}\ell \gg 1$, it is straightforward to show from Eqs. (8.34) to (8.36) that $C(q) = 1$ [4].

At this point, we note that we employed a long wavelength approximation in calculating the conductivity. This is valid in the metallic regime, also mentioned in Section 8.4. If we are to use an inverse Fourier transform to real space, it means that we are only going to capture the short-range behavior of the conductivity. Additional terms in q-space would be needed to obtain the conductivity over an extended region of coordinate space. A back-Fourier transformation to coordinate space leads to a modification of the result given by Kane, Serota and Lee [1], that is,

$$\left\langle\sigma_{\alpha\beta}(\boldsymbol{r},\boldsymbol{r}';\omega=0)\right\rangle = \sigma_0^{\mathrm{D}}\left[\delta_{\alpha\beta}\overline{\delta(\boldsymbol{r}-\boldsymbol{r}')} - \nabla_\alpha\nabla'_\beta d(\boldsymbol{r},\boldsymbol{r}')\right], \tag{8.38}$$

where $-\nabla^2 d(\boldsymbol{r},\boldsymbol{r}') = C(\boldsymbol{r}-\boldsymbol{r}')$ and $\overline{\delta(\boldsymbol{r}-\boldsymbol{r}')}$ is a short-range function. Mathematically, this is a delta-function. However, since the inclusion of small but finite wave numbers would broaden this function, we follow [1] and replace it by a short-range function. The detailed structure of $\overline{\delta(\boldsymbol{r}-\boldsymbol{r}')}$ is not important to this formalism. However, our derived result shows that unlike the case when there is no SOI, $C(q)$ is not unity but depends on the value of α_{R}. This means that $C(\boldsymbol{r}-\boldsymbol{r}')$ is not a delta function and the electrons do not diffuse through the sample. The significance of this result lies in the following. The sample-averaged nonlocal conductivity $\left\langle\sigma_{\alpha\beta}(\boldsymbol{r},\boldsymbol{r}')\right\rangle$ in [1] satisfies $\nabla_\alpha\left\langle\sigma_{\alpha\beta}(\boldsymbol{r},\boldsymbol{r}')\right\rangle = 0$, where $\sigma_0 = en_0\mu_0$ in terms of the electron density n_0 and the mobility μ_0. This is so since we have $\nabla_\alpha\langle\sigma_{\alpha\beta}(\boldsymbol{r},\boldsymbol{r}')\rangle = \sigma_0[\delta_{\alpha\beta}\nabla\overline{\delta(\boldsymbol{r}-\boldsymbol{r}')} - \nabla'_\beta\nabla_\alpha\nabla_\alpha d(\boldsymbol{r}-\boldsymbol{r}')] = 0$ when one uses $\nabla_\alpha^2 d(\boldsymbol{r}-\boldsymbol{r}') = -\overline{\delta(\boldsymbol{r}-\boldsymbol{r}')}$ for the diffusion propagator in the absence of SOI. The presence of SOI means that there is no time reversal symmetry and this condition is not satisfied. This is why the function $C(q)$ appears in Eq. (8.31).

Since the derivation of the sample-averaged conductivity is based on a diagrammatic diffusion approximation, it *should* also be possible to derive it from a simple drift-diffusion approximation for the current density. Let us first consider the case as considered in [1] when there is no SOI, that is, we let $j_\alpha(\boldsymbol{r}) = en(\boldsymbol{r})\mu(E)E_\alpha(\boldsymbol{r}) + eD(E)\nabla_\alpha n(\boldsymbol{r})$, where $n(\boldsymbol{r})$ is the electron density function, $\mu(E)$ and $D(E)$ are the mobility and diffusion constant which in general depend on the external electric field $\boldsymbol{E}$. In linear response theory, $\mu(E) \approx \mu(E=0) \equiv \mu_0$, $D(E) \approx D(E=0) \equiv D_0$ and $n(\boldsymbol{r}) \approx n_0$ in the first term of the equation which gives the current density so that

$$j_\alpha(\boldsymbol{r}) = en_0\mu_0 E_\alpha(\boldsymbol{r}) + eD_0\nabla_\alpha n(\boldsymbol{r})\,. \tag{8.39}$$

Now, in order to express j_α by a conductivity, we have to relate $\nabla_\alpha n(\boldsymbol{r})$ to $\boldsymbol{E}$. This can be done using current conservation in the steady state. Since $\nabla\cdot\boldsymbol{j}(\boldsymbol{r}) = 0$, we obtain

$$\nabla^2 n(\mathbf{r}) = -\left(\frac{n_0 \mu}{D_0}\right) \nabla \cdot E(\mathbf{r}) .$$

Making use of this equation, we obtain the nonlocal form of the conductivity in [1, Eq. (2.7)] for a sample with either insulating or conducting boundaries. In the presence of SOI, the first term in the current density in Eq. (8.39) must be replaced by $\sim \int_{\text{sample}} d\mathbf{r}' C(\mathbf{r}, \mathbf{r}') E_\beta(\mathbf{r}')$. This reflects the way in which SOI affects the conductivity and we have an explicit expression for the function $C(\mathbf{r}, \mathbf{r}')$. For further discussions of the conductivity tensor in 2D in the absence of SOI, see the paper by Goodman and Serota [4].

In the presence of SOI, the spins become delocalized as seen in the transverse resistivity ρ_{xy} of the spin Hall effect. We note that the SOI involves a $(\overleftrightarrow{\sigma} \times \mathbf{p})$ term which may be either positive or negative. Consequently, σ_{xy} may be enhanced but not σ_{xx}, and the behavior of the resulting transverse resistivity requires further analysis. We also note that the sum of the diagrams which were included is not the only contribution with scattering from several impurities. So that any disagreement with experiment may have to be addressed by including non-ladder diagrams. However, as discussed by Mahan [31], the omitted terms are not important when the impurity concentration $n_i \to 0$.

8.6 Numerical Results for Scattering Times

The scattering times τ_λ defined in Eq. (8.10) may be rewritten in terms of the 2D density-of-states ρ_{2D} as $\tau_\lambda = (\hbar^3/2m^*|V|^2)\rho_{2D}/\mathcal{N}_\lambda(E_F)$. In Figure 8.5, we plot

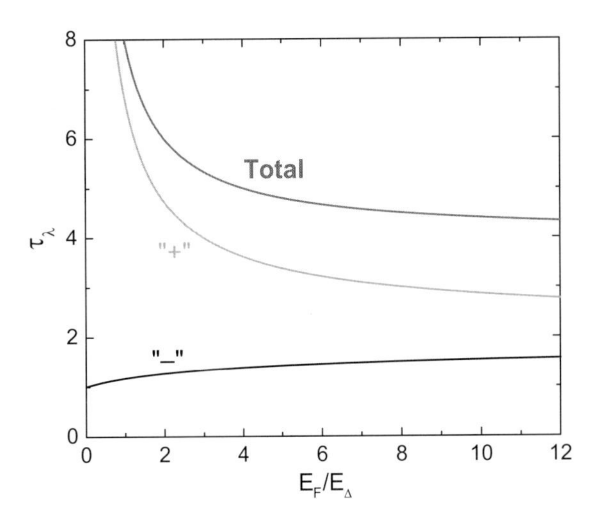

Figure 8.5 The scattering times τ_+, τ_- and their sum in units of $\hbar^3/2m^*|V|^2$ as functions of the Fermi energy E_F in units of E_Δ, defined in the text.

τ_+ and τ_- as well as their sum as functions of the Fermi energy E_F/E_Δ. Here, $E_\Delta = k_R\alpha_R/\sqrt{2}$ with $k_R = m^*\alpha_R/\sqrt{2}\hbar^2$. When $E_F < 0$, only the "-" subband is occupied and τ_- gradually increases from zero at the bottom of the band. At the bottom of the "+" subband, the DOS $\mathcal{N}_+$ for the "+" spins is zero [32]. This results in the large value of τ_+ near $E_F = 0$. Thus, there are ranges of E_F where the metallic regime of the conductivity is valid, that is, when $(k_F\ell_\lambda)^{-1} \ll 1$.

8.7
Related Results in 3D in the Absence of SOI

In this section, we summarize some related results for a three-dimensional electron gas when there is no SOI present. Our calculations are restricted to lowest order in $(k_F\ell)^{-1}$. With an electron energy $\varepsilon_k = \hbar^2k^2/2m^*$, and scattering time τ, we have

$$\mathcal{J}_\alpha(q) \equiv \frac{1}{\mathcal{V}}\sum_k k_\alpha G^-(E_F, \boldsymbol{k})G^+(E_F, \boldsymbol{k}+\boldsymbol{q}) = \left(\frac{2m^*}{\hbar^2}\right)^2 \int \frac{d^3\boldsymbol{k}}{(2\pi)^3}$$
$$\times \left[\frac{k_\alpha}{(k_F^2 - k^2) - \frac{im^*}{\hbar\tau}}\right]\left[\frac{1}{(k_F^2 - k^2 - 2\boldsymbol{k}\cdot\boldsymbol{q} - q^2) + \frac{im^*}{\hbar\tau}}\right]. \tag{8.40}$$

We note that in Eq. (8.40), the q-dependence occurs only in one factor. If $q = 0$, the integral is zero by symmetry. Denoting the second factor in Eq. (8.40) by $Q(q)$, we expand this function in powers of q so that we have

$$Q(q) = Q(0) + \sum_{\alpha'} q_{\alpha'} \left.\frac{\partial Q(q)}{\partial q_{\alpha'}}\right|_{q=0} + \cdots$$
$$= Q(0) + \sum_{\alpha'} q_{\alpha'} \frac{2k_{\alpha'}}{\left[k^2 - k_F^2 - \frac{im^*}{\hbar\tau}\right]^2} + \mathcal{O}(q^2)\,.$$

Since the $Q(0)$ term does not contribute to the integral in Eq. (8.40), after some rearrangement, we obtain

$$\mathcal{J}_\alpha(q) \approx -\frac{2}{3}q_\alpha\left(\frac{2m^*}{\hbar^2}\right)^2 \frac{1}{(2\pi)^3}\left(4\pi k_F^2\right)\int_0^\infty dk\,k^2$$
$$\times \left[\frac{1}{k^2 - k_F^2 + \frac{im^*}{\hbar\tau}}\right]\left[\frac{1}{\left(k^2 - k_F^2 - \frac{im^*}{\hbar\tau}\right)^2}\right], \tag{8.41}$$

where we assumed that $k_F^2 \gg m^*/\hbar\tau$, that is, $k_F\ell \gg 1$. If we now set $A = k_F^2 - im^*/\hbar\tau$ and $B = k_F^2 + im^*/\hbar\tau$, then the k-integral in Eq. (8.41) can be expressed as a partial derivative with respect to B of $(A - B)^{-1}\int_0^\infty dk[A/(k^2 - A) - B/(k^2 - B)]$. When this integration is carried out, consistent with the limit $k_F\ell \gg 1$, we obtain the result

$$\mathcal{J}_\alpha(q) \approx -\frac{i}{3\pi}q_\alpha\frac{\tau^2 m^{*3} v_F^3}{\hbar^5}\,. \tag{8.42}$$

We now turn to a calculation of

$$\Pi(q) \equiv \frac{1}{\mathcal{V}}\sum_{k} G^{-}(E_{\mathrm{F}}, \boldsymbol{k})G^{+}(E_{\mathrm{F}}, \boldsymbol{k}+\boldsymbol{q}) = \left(\frac{2m^*}{\hbar^2}\right)^2 \int \frac{d^3\boldsymbol{k}}{(2\pi)^3} \times \left[\frac{1}{(k_{\mathrm{F}}^2 - k^2) - \frac{im^*}{\hbar\tau}}\right]\left[\frac{1}{(k_{\mathrm{F}}^2 - k^2 - 2\boldsymbol{k}\cdot\boldsymbol{q} - q^2) + \frac{im^*}{\hbar\tau}}\right]. \tag{8.43}$$

This calculation proceeds along lines which are similar to the above, except that the second factor which we denoted by $Q(q)$, when expanded in powers of q, makes a contribution for $q = 0$. The linear term in q does not, but the term of order q^2 does contribute. We have

$$\Pi(q) \approx \left(\frac{2m^*}{\hbar^2}\right)^2 \int \frac{d^3\boldsymbol{k}}{(2\pi)^3}\left[\frac{1}{k^2 - k_{\mathrm{F}}^2 + \frac{im^*}{\hbar\tau}}\right]\left\{\frac{1}{k^2 - k_{\mathrm{F}}^2 - \frac{im^*}{\hbar\tau}} - \frac{q^2}{\left(k^2 - k_{\mathrm{F}}^2 - \frac{im^*}{\hbar\tau}\right)^2} + \frac{4k^2\frac{q^2}{3}}{\left(k^2 - k_{\mathrm{F}}^2 - \frac{im^*}{\hbar\tau}\right)^3}\right\}. \tag{8.44}$$

The first term in Eq. (8.44) is equal to $(2m^*/\hbar^2)^2(\hbar\tau k_{\mathrm{F}}/2\pi m^*) = 2\pi\tau\mathcal{N}(E_{\mathrm{F}})/\hbar$ expressed in terms of the 3D density-of-states at the Fermi level, where $\mathcal{N}(E_{\mathrm{F}}) = m^{*2}v_{\mathrm{F}}/\pi^2\hbar^3 = 3n_{3\mathrm{D}}/2E_{\mathrm{F}}$ with $n_{3\mathrm{D}}$ being the electron density. The second term in Eq. (8.44) is given by

$$-4\pi\left(\frac{2m^*}{\hbar^2}\right)^2 q^2 \frac{1}{(2\pi)^3}\frac{1}{2}\frac{\frac{i\pi}{k_{\mathrm{F}}}}{\left(\frac{-im^*}{\hbar k_{\mathrm{F}}\tau}\right)^2}.$$

The third term in Eq. (8.44) is approximately

$$\frac{4}{3}q^2(4\pi)\left(\frac{2m^*}{\hbar^2}\right)^2\frac{1}{(2\pi)^3}k_{\mathrm{F}}^2\int_0^\infty dk\left[\frac{k^2}{k^2 - k_{\mathrm{F}}^2 + \frac{im^*}{\hbar\tau}}\right] \times \left[\frac{1}{\left(k^2 - k_{\mathrm{F}}^2 - \frac{im^*}{\hbar\tau}\right)^3}\right] = i\left(\frac{2m^*}{\hbar^2}\right)^2\frac{k_{\mathrm{F}}^2 q^2}{6} \times \left\{-\frac{\frac{1}{2}k_{\mathrm{F}}^3}{\left(\frac{-im^*}{\hbar k_{\mathrm{F}}\tau}\right)^2} + \frac{\frac{1}{k_{\mathrm{F}}^2}}{\left(\frac{-im^*}{\hbar k_{\mathrm{F}}\tau}\right)^3}\right\}. \tag{8.45}$$

Collecting these results, we obtain

$$\Pi(q) \approx \frac{2\pi\tau}{\hbar}\mathcal{N}(E_{\mathrm{F}})(1 - D\tau q^2). \tag{8.46}$$

In this notation, $D = v_{\mathrm{F}}^2\tau/3$ is the diffusion constant. Combining these results, we have in three dimensions, $\mathcal{D}(q) \approx [\hbar/2\pi\tau^2\mathcal{N}(E_{\mathrm{F}})](1/Dq^2)$. Consequently, we obtain the vertex correction

$$\delta\left\langle\sigma_{\alpha\beta}(q, \omega = 0)\right\rangle = -\frac{1}{2}\sigma_0^{\mathrm{D}}\frac{q_\alpha q_\beta}{q^2}. \tag{8.47}$$

Collecting these results, we obtain the conductivity in q-space given by

$$\langle \sigma_{\alpha\beta}(q, \omega = 0) \rangle = \sigma_0^{\mathrm{D}} \left(\delta_{\alpha\beta} - \frac{q_\alpha q_\beta}{q^2} \right). \tag{8.48}$$

Expressing this in coordinate space, we have, according to Kane, Serota and Lee [1],

$$\langle \sigma_{\alpha\beta}(\boldsymbol{r}, \boldsymbol{r}'; \omega = 0) \rangle = \sigma_0^{\mathrm{D}} \left[\delta_{\alpha\beta} \overline{\delta(\boldsymbol{r} - \boldsymbol{r}')} - \nabla_\alpha \nabla'_\beta d(\boldsymbol{r}, \boldsymbol{r}') \right], \tag{8.49}$$

where $-\nabla^2 d(\boldsymbol{r}, \boldsymbol{r}') = \delta(\boldsymbol{r} - \boldsymbol{r}')$ and $\overline{\delta(\boldsymbol{r} - \boldsymbol{r}')}$ is a short-range function. The detailed structure of $\overline{\delta(\boldsymbol{r} - \boldsymbol{r}')}$ is not important to this formalism.

The above calculations can be carried out in two-dimensions when there is no SOI present. This directly leads to the result for the conductivity tensor determined by Goodman and Serota [4] with $C(q) = 1$, which is the result we obtained by setting $\alpha_{\mathrm{R}} = 0$ in our expressions in the presence of SOI.

References

1 Kane, C.L., Serota, R.A., and Lee, P.A. (1988) Long-range correlations in disordered metals. *Phys. Rev. B*, **37**, 6701.

2 Lee, P.A., Douglas Stone, A., and Fukuyama, H. (1987) Universal conductance fluctuations in metals: Effects of finite temperature, interactions, and magnetic field. *Phys. Rev. B*, **35**, 1039.

3 Serota, R.A., Feng, S., Kane, C., and Lee, P.A. (1987) Conductance fluctuations in small disordered conductors: Thin-lead and isolated geometries. *Phys. Rev. B*, **36**, 5031.

4 Goodman, B. and Serota, R.A. (2001) Polarizability and absorption of small conducting particles in a time-varying electromagnetic field. *Physica B*, **305**, 208.

5 Texier, C. and Montambaux, G. (2004) Weak localization in multiterminal networks of diffusive wires. *Phys. Rev. Lett.*, **92**, 186801.

6 Nikolic, B.K. (2001) Deconstructing Kubo formula usage: Exact conductance of a mesoscopic system from weak to strong disorder limit. *Phys. Rev. B*, **64**, 165303.

7 Gumbs, G. and Rhyner, J. (2003) Effect of an electric field on the conductance fluctuations of mesoscopic systems. *Superlattices Microstruct.*, **33**, 181.

8 Safi, I. and Schulz, H.J. (1999) Interacting electrons with spin in a one-dimensional dirty wire connected to leads. *Phys. Rev. B*, **59**, 3040.

9 van Rossum, M.C.W. and Nieuwenhuizen, T.M. (1999) Multiple scattering of classical waves: microscopy, mesoscopy, and diffusion. *Rev. Mod. Phys.*, **71**, 313.

10 Xiong, S., Read, N., and Stone, A.D. (1997) Mesoscopic conductance and its fluctuations at a nonzero Hall angle. *Phys. Rev. B*, **56**, 3982.

11 Zhang, X.G. and Butler, W.H. (1997) Landauer conductance in the diffusive regime. *Phys. Rev. B*, **55**, 10308.

12 Oreg, Y. and Finkelstein, A.M. (1996) dc transport in quantum wires. *Phys. Rev. B*, **54**, R14265.

13 Safi, I. and Schulz, H.J. (1995) Transport in an inhomogeneous interacting one-dimensional system. *Phys. Rev. B*, **52**, R17040.

14 van Rossum, M.C.W., Nieuwenhuizen, T.M., and Vlaming, R. (1995) Optical conductance fluctuations: Diagrammatic analysis in the Landauer approach and nonuniversal effects. *Phys. Rev. E*, **51**, 6158.

15 Pieper, J.B. and Price, J.C. (1994) Correlation functions for mesoscopic conduc-

tance at finite frequency. *Phys. Rev. B*, **49**, 17059.

16 Levinson, Y.B. and Shapiro, B. (1992) Mesoscopic transport at finite frequencies. *Phys. Rev. B*, **46**, 15520.

17 Ando, T. and Tamura, H. (1992) Conductance fluctuations in quantum wires with spin–orbit and boundary-roughness scattering. *Phys. Rev. B*, **46**, 2332.

18 Rammer, J. (1991) Quantum transport theory of electrons in solids: A single-particle approach. *Rev. Mod. Phys.*, **63**, 781.

19 Serota, R.A., Yu, J., and Kim, Y.H. (1990) Frequency-dependent conductivity of a finite disordered metal. *Phys. Rev. B*, **42**, 9724.

20 Baranger, H.U. and Stone, A.D. (1989) Electrical linear-response theory in an arbitrary magnetic field: A new Fermi-surface formation. *Phys. Rev. B*, **40**, 8169.

21 Shekhter, A., Khodas, M., and Finkelstein, A.M. (2005) Diffuse emission in the presence of an inhomogeneous spin–orbit interaction for the purpose of spin filtration. *Phys. Rev. B*, **71**, 125114.

22 Rashba, E.I. (1960) *Fiz. Tverd. Tela*, **2**, 1224.

23 Rashba, E.I. (1960) *Sov. Phys. Solid State*, **2**, 1109.

24 Bychkov, Y.A. and Rashba, E.I. (1984) Oscillatory effects and the magnetic susceptibility of carriers in inversion layers. *J. Phys. C*, **17**, 6039.

25 Dresselhaus, G. (1955) Spin–orbit coupling effects in zinc blende structures. *Phys. Rev.*, **100**, 580.

26 Zhang, C. and Ma, Z. (2005) Dynamic Hall resistivity of electronic systems in the presence of Rashba coupling at zero field. *Phys. Rev. B*, **71**, 121307(R).

27 Sinova, J., Culcer, D., Niu, Q., Sinitsyn, N.A., Jungwirth, T., and MacDonald, A.H. (2004) Universal intrinsic spin Hall effect. *Phys. Rev. Lett.*, **92**, 126603.

28 Kato, Y.K., Myers, R.C., Gossard, A.C., Awschalom, D.D. (2004) Observation of the spin Hall effect in semiconductors. *Science*, **306**, 1910.

29 Nitta, J., Akazaki, T., Takayanagi, H., and Enoki, T. (1997) Gate control of spin–orbit interaction in an inverted $In_{0.53}Ga_{0.47}As/In_{0.52}Al_{0.48}As$ heterostructure. *Phys. Rev. Lett.*, **78**, 1335.

30 Doniach, S. and Sondheimer, E.H. (1998) *Green's Functions for Solid State Physicists*, Benjamin, New York.

31 Mahan, G.D. (1981) *Many-Particle Physics*, Plenum, New York.

32 Gumbs, G. (2005) Polarization of interacting mesoscopic two-dimensional spintronic systems. *Phys. Rev. B*, **72** 165351.

33 Tang, H.S. and Fu, Y. (1992) Microwave response of mesoscopic rings. *Phys. Rev. B*, **46**, 3854.

34 Lee, P.A. and Ramakrishnan, T.V. (1985) Disordered electronic systems. *Rev. Mod. Phys.*, **57**, 287.

35 Bergmann, G. (1984) Weak localization in thin films: a time-of-flight experiment with conduction electrons. *Phys. Rep.*, **107**, 1.

36 Miller, J.B., Zumbühl, D.M., Marcus, C.M., Lyanda-Geller, Y.B., Goldhaber-Gordon, D., Campman, K., and Gossard, A.C. (2003) Gate-controlled spin–orbit quantum interference effects in lateral transport. *Phys. Rev. Lett.*, **90**, 076807.

37 Hikami, S., Larkin, A.I., and Nagaoka, Y. (1980) Spin–orbit interaction and magnetoresistance in the two dimensional random system. *Prog. Theor. Phys.*, **63**, 707.

38 Imry, J. (1997) *Introduction to Mesoscopic Physics*, Oxford University Press, New York.

39 Abrikosov, A.A., Gorkov, L.P., and Dzyaloshinski, I.E. (1963) *Methods of Quantum Field Theory in Statistical Physics*, Prentice-Hall, Englewood Cliffs, New Jersey.

40 Fetter, A.L. and Walecka, J.D. (1971) *Quantum Theory of Many-Particle Systems*, McGraw-Hill, New York.

9
Integer Quantum Hall Effect

9.1
Basic Principles of the Integer Quantum Hall Effect

The goal of this chapter is to give an overview of the integer quantum Hall effect which has received a considerable amount of attention since it was first reported in 1980 by von Klitzing [1]. von Klitzing was awarded the 1985 Nobel Prize in Physics for his discovery of the integer quantum Hall effect.

However, in 1975 Ando, Matsumoto and Uemura predicted the integer quantization of the Hall conductance [2] based on approximate calculations. Subsequently, several workers observed the effect in experiments carried out on the inversion layer of MOSFETs. The link between exact quantization and gauge invariance was subsequently elucidated by Robert Laughlin [3]. Currently, most integer quantum Hall experiments are performed on gallium arsenide heterostructures, though many other semiconductor materials can be used. The integer quantum Hall effect has also been found in graphene at room temperature, which is considered high [4]. More information on the quantization of the Hall conductance are in [5–11].

9.1.1
The Hall Effect

Figure 9.1 illustrates a conducting material with a magnetic field $\boldsymbol{B}$ applied in the z-direction and an electric field applied in the x-direction leading to current I flowing in the x-direction. A voltage drop V results across the sample in the y direction. In 1878, E.H. Hall discovered this phenomenon, called the Hall effect, as a graduate student at John's Hopkins University.

When an electron with charge $-e$ moves in a magnetic field, a Lorentz force $-e\boldsymbol{v} \times \boldsymbol{B}$, a result that is perpendicular to its velocity $\boldsymbol{v}$ and the magnetic field. The trajectories of the electrons are bent so that they move to a side boundary assumed in the y direction. Upon a sufficient accumulation of electrons on the side boundary, a static electric field builds up and balances the Lorentz force. Electrons then drift in their original intended direction. The resulting electric field gives rise to a potential difference along the y-direction.

Properties of Interacting Low-Dimensional Systems, First Edition. G. Gumbs and D. Huang.

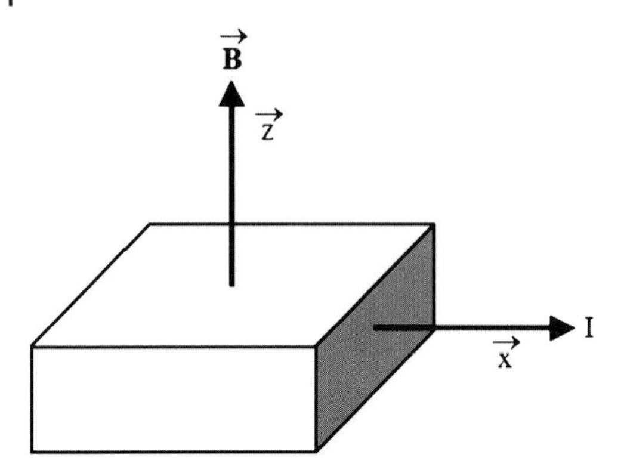

Figure 9.1 Standard geometry for the Hall effect.

For simplicity, we consider the gas of electrons moving with the same velocity v along the x-axis with unit vector $\hat{x}$ along this direction. The Lorentz force on each electron is given by $-evB\hat{y}$, where $\hat{y}$ is the unit vector along the y-axis. This force may be balanced by an electric field $E_y\hat{y}$ where $E_y = vB$. The current density is $j_x = en_{2D}v$, where n_{2D} is the electron density, thus

$$j_x = -n_{2D}e\frac{E_y}{B} . \tag{9.1}$$

This result is valid when the electric and magnetic fields are weak, that is, if we interpret $n_{2D}e$ as the charge density of the current carriers.

9.1.2
The Quantum Hall Effect

von Klitzing's discovery of the quantum Hall effect in 1980 led to a new and important area of research in condensed matter physics and his experiment was a pioneering contribution. The experiment sample used was a MOSFET. It is a semiconductor device that is a metal oxide silicon field effect transistor as shown in Figure 9.2. For a range of gate voltage V_g, a thin layer of electrons are attracted to the interface between the Si and SiO_2. This is the inversion layer. Because the inversion layer is so thin ($\sim$ 100 Å), electrons behave as a 2D gas. The 2D density can be changed by varying V_g.

In a quantum Hall effect experiment, the device is placed in a strong magnetic field whose direction is perpendicular to the 2D electron layer. A DC current is applied from the source to the drain and passes through the 2D electron layer. As shown in Figure 9.3, a voltage drop V_{ab} is measured between a and b, from which the Hall resistance is obtained as

$$R_H = \frac{V_{ab}}{I} . \tag{9.2}$$

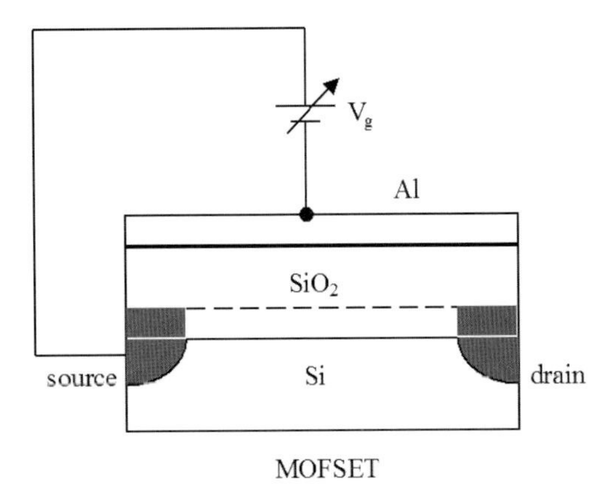

Figure 9.2 MOSFET used in the Quantum Hall Effect experiments.

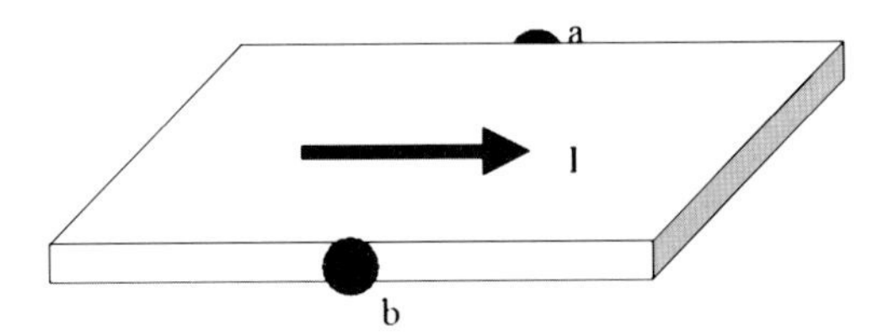

Figure 9.3 Diagram showing the direction of the current flow and the applied voltage in the Quantum Hall Effect experiments.

In general, R_{H} is dependent upon the material's temperature T, magnetic field B, electron density, and other physical properties. von Klitzing's remarkable discovery was performed at very low temperatures (a few degrees Kelvin) and a very high magnetic field (few Tesla). Under such extreme conditions, $\rho_{\mathrm{H}} = R_{\mathrm{H}}(\mathcal{A}/L_y)$ is a staircase function of V_g with extremely flat plateaus. Thus,

$$\rho_{\mathrm{H}} = \frac{h}{ne^2}, \tag{9.3}$$

where n is an integer, L_y is the sample width in the y-direction and $\mathcal{A}$ is the cross-sectional area for the current flow in the x-direction. Thus far, the accuracy of the plateaus have been approximately one part in 10^8. This phenomenon is the quantum Hall effect (QHE).

9.1.3
An Idealized Model

To explain the QHE, we use the following idealized model. Consider a non-interacting 2D electron gas in the presence of an impurity free environment that also has a uniform, perpendicular magnetic field $\boldsymbol{B}$. Compared to the free case with no external magnetic field, we will find a quite different energy spectrum. This problem was solved by Landau many years ago, but we will review the problem to establish a detailed physical illustration and mathematical language for later discussions. The magnetic field enters the Schrödinger equation through the vector potential $\boldsymbol{A}$, where $\boldsymbol{B} = \nabla \times \boldsymbol{A}$. A specific gauge must be chosen and we choose the so-called Landau gauge, $\boldsymbol{A} = (0, Bx, 0)$. Thus, the Schrödinger equation is

$$\frac{1}{2m^*}\left[\left(-i\hbar\frac{\partial}{\partial x}\right)^2 + \left(-i\hbar\frac{\partial}{\partial y} + eBx\right)^2\right]\psi = E\psi\,, \tag{9.4}$$

where m^* is the effective mass of electrons. Because the translational invariance is in the y-direction, we write $\psi = e^{iqy}\varphi(x)$. Thus,

$$\frac{1}{2m^*}\left[\left(-i\hbar\frac{\partial}{\partial x}\right)^2 + (\hbar q + eBx)^2\right]\varphi(x) = E\varphi(x)\,. \tag{9.5}$$

This equation is the same as that describing a harmonic oscillator and has solutions

$$E = E_n = \left(n + \frac{1}{2}\right)\hbar\omega_c\,, \tag{9.6}$$

for $n = 0, 1, 2, \cdots$, and

$$\varphi(x) = \varphi_n\left(x + \frac{\hbar q}{eB}\right), \tag{9.7}$$

where $\omega_c = eB/m^*$. The energy spectrum consists of a series of equally spaced levels and are called Landau levels. Each level is highly degenerate and has many states with the same energy. The number of states per level is proportional to the area of the 2D gas and is demonstrated in the following arguments. If there are periodic boundary conditions in the y-direction, with $\psi(y + L_y) = \psi(y)$, q is then quantized as $q = 2\pi j/L_y$ where j is an integer. Since the center of the wave function along the x-direction is at $-\hbar q/eB$, a restriction provided by the width L_x of the sample in the x-direction must be provided. Thus, the number of allowed q values are

$$\left(\frac{L_x eB}{\hbar}\right)\left(\frac{2\pi}{L_y}\right)^{-1} = \left(L_x L_y\right)\frac{eB}{h}\,. \tag{9.8}$$

This leads to the conclusion that each Landau level has a total number of eB/h states per unit area and also applies in the thermodynamic limit, that is, $L_x, L_y \to \infty$.

If an additional electric field E_y is applied in the y-direction, an extra term $-eE_y t$ must be added to eA_y. This leads to the conclusion that the center of each wave function moves with a velocity of E_y/B in the x-direction. When the lowest Landau level is full and the others are empty, the current density is

$$j_x = e\frac{eB}{h}\frac{E_y}{B} = \frac{e^2}{h}E_y \tag{9.9}$$

for which the Hall conductance is

$$\sigma_\mathrm{H} = \frac{j_x}{E_y} = \frac{e^2}{h} . \tag{9.10}$$

When the lowest n Landau levels are full and all others are empty, the current density increases by a factor of n and

$$\sigma_\mathrm{H} = \frac{ne^2}{h} . \tag{9.11}$$

The Hall resistance is

$$\rho_\mathrm{H} = \frac{1}{\sigma_\mathrm{H}} = \frac{h}{ne^2} . \tag{9.12}$$

9.1.4
Effect of Finite Temperature

In the aforementioned idealized model, the Hall conductance at finite temperature is

$$\sigma_\mathrm{H} = \frac{e^2}{h} \sum_{n=0}^{\infty} \frac{1}{e^{(E_n-\mu)/k_\mathrm{B}T} + 1} . \tag{9.13}$$

At $T = 0$, the Hall conductance is a staircase function of the magnetic field and chemical potential, and, as was observed experimentally, we have

$$\frac{\partial \sigma_\mathrm{H}}{\partial \mu} = \frac{e^2}{h} \sum_{n=0}^{\infty} \delta(E_n - \mu) .$$

At low temperatures ($k_\mathrm{B} T \ll \mu$),

$$\frac{\partial \sigma_\mathrm{H}}{\partial \mu} = \frac{e^2}{h} \frac{1}{k_\mathrm{B} T} \sum_{n=0}^{\infty} e^{-(E_n-\mu)/k_\mathrm{B}T} .$$

The slope at the center of a plateau with

$$\mu = \frac{E_n + E_{n-1}}{2} , \quad \text{i.e.,} \quad E_n - \mu = \frac{E_n - E_{n-1}}{2} = \frac{\hbar\omega_\mathrm{c}}{2}$$

is

$$\frac{\partial \sigma_\mathrm{H}}{\partial \mu} \approx \frac{e^2}{h} \frac{1}{k_\mathrm{B} T} e^{-\hbar\omega_\mathrm{c}/2k_\mathrm{B}T} . \tag{9.14}$$

For quantization, we have

$$\frac{\partial \sigma_\mathrm{H}}{\partial \mu} = \tan\theta = \frac{e^2}{h} \frac{f}{\hbar\omega_\mathrm{c}} \quad \text{with } f \ll 1 , \tag{9.15}$$

that is, θ is the slope angle of the plateau. Comparing Eqs. (9.14) and (9.15), we obtain

$$\frac{\hbar\omega_c}{k_B T} = -2\ln\left(\frac{f}{\frac{\hbar\omega_c}{k_B T}}\right) .$$

It has been found within experimental accuracy that $f = 10^{-7}$, for which we need

$$\frac{\hbar\omega_c}{k_B T} > 40 . \tag{9.16}$$

In the experiment, the electron effective mass is $m^* = 0.19 m_e$. For a magnetic field $B = 10\,\mathrm{T}$, it requires $T < 2\,\mathrm{K}$, as found experimentally.

Since $\omega_c \propto 1/m^*$, the quantizing condition for the Hall conductance is more easily achieved if the electron effective mass is reduced. This observation led Daniel Tsui, *et al.* [12] to use GaAs/AIGaAs heterojunctions to study the QHE, where $m^* \approx 0.068 m_e$.

9.1.5
Effect of Impurities

Let's now move on to the effect of impurities. Prior to the discovery of the QHE, some calculations were performed on the effect of impurities on the electron spectrum. The results revealed that the Landau levels become broadened. Subsequent calculations revealed that most of the states become localized except at the center of the band where the states are extended. von Klitzing demonstrated that in his experiment,

1. Changing the gate voltage V_g is equivalent to changing the chemical potential μ.
2. When μ lies in an energy gap or an energy range of localized states, the Hall conductance is quantized; σ_H jumps by a quantum e^2/h when μ goes across a level of extended states.

The question then is why does σ_H have to be quantized when μ is away from a region of extended states. This will be the central topic of later sections of this chapter.

9.1.6
Application of the Quantum Hall Effect

Quantization of the Hall conductance has been established to an accuracy of one part in 10^9 and has been used as a standard for electrical resistance since early 1990. Prior to this, the standard of resistance were wire-wound resistors subject to variations with temperature, aging, and other various problems. Another practical application was the standardization of the fine structure constant

$$\alpha = \frac{\mu_0 c}{2} \frac{e^2}{h} , \tag{9.17}$$

where $\mu_0 = 4\pi \times 10^{-7}\,\mathrm{V\,s\,A^{-1}\,m^{-1}}$ and $c = 299\,792\,458\,\mathrm{m/s}$.

9.2
Fundamental Theories of the IQHE

9.2.1
Energy Spectrum and Wave functions

In the presence of impurities, Landau levels are broadened into Landau bands. Ando and Uemura [13] calculated $n(E)$ for the white noise potential,

$$\begin{aligned} \langle V(\boldsymbol{r}) \rangle &= 0 \, , \\ \left\langle V(\boldsymbol{r}) V(\boldsymbol{r}') \right\rangle &= W^2 \delta(\boldsymbol{r} - \boldsymbol{r}'), \end{aligned} \tag{9.18}$$

with a self-consistent Born approximation. For each band, $n(E)$ is a semi-elliptic curve, as illustrated in Figure 9.4, with the width of the Landau levels given by

$$\Gamma \sim \frac{2W}{\ell_{\mathrm{H}}} , \tag{9.19}$$

where $\ell_{\mathrm{H}} = \sqrt{\hbar/eB}$.

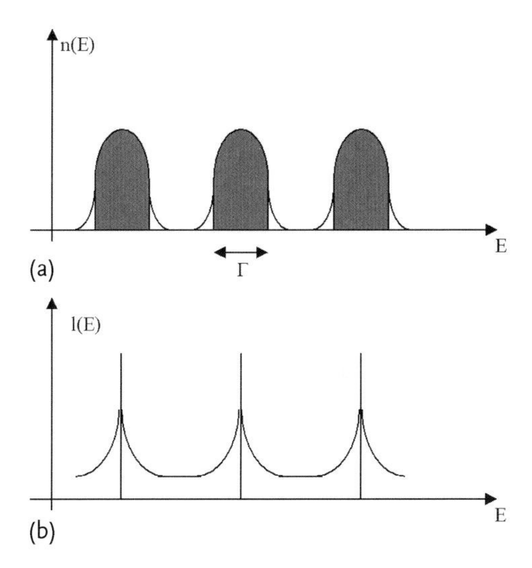

Figure 9.4 Broadened Landau levels. The density of states $n(E)$ for Landau bands is shown in (a) in the presence of impurities, The localization length $\ell(E)$ is also shown in (b) with its divergence indicated by vertical lines for extended states in Landau bands.

Subsequent calculations revealed that the band edges are smeared out into Gaussian tails. Wigner [14] also carried out an exact calculation for the lowest band in the large magnetic field limit. His exact result revealed that $n(E)$ is an analytic function of E, and has the same qualitative features as those for the above result. Similar results for $n(E)$ were obtained by Ando [15] and Trugman [16] for smoothly varying random potentials.

After the discovery of the QHE, the exact nature of the wave functions was understood. According to Anderson's theory of localization, all states in 2D are localized by any degree of disorder and this is also valid when a weak magnetic field is applied. However, if a strong magnetic field was applied, the Hall conductance would be zero, but this does not occur in the QHE experiments. Therefore, there must be some extended states in each Landau band.

Numerical calculations for short-ranged disordered potentials (so-called white noise) reveal that all states except those at the center E_c of a Landau band are localized, with a localization length

$$l(E) \sim |E - E_\mathrm{c}|^{-\nu} , \tag{9.20}$$

where $\nu \approx 2$ for the lowest Landau band [15]. In Ando's calculations, number analysis and finite size scaling methods of Thouless were employed. Approximate analytic results also show the existence of a singular energy, as illustrated in Figure 9.4 where the states are extended. However, the form of the divergence for the localization length is different with

$$l(E) \sim e^{\frac{\Gamma}{|E-E_\mathrm{c}|}} . \tag{9.21}$$

For smooth random potentials, an analogy can be made with a classical continuum percolation theory in 2D [16]. In the limit when the magnetic length ℓ_H is much smaller than the length scale of the potential variation, a wave function of energy E is concentrated along the contour of constant potential energy

$$V(\mathbf{r}) + \left(n + \frac{1}{2}\right)\hbar\omega_\mathrm{c} = E , \tag{9.22}$$

where n labels the Landau bands.

A qualitatively similar result is obtained for short-ranged potentials and the localization length diverges as $l(E) \sim |E - E_\mathrm{c}|^{-\nu}$ with $\nu = 4/3$ [16]. When the effect due to tunneling is included, we obtain $\nu = 7/3$ [17].

The electronic structure has also been analytically studied by Levine *et al.* [18]. They transformed the problem into a nonlinear sigma model by using the replica trick. The well-developed field theoretic results for this model led them to conclude that extended states must appear near or at the centers of the Landau bands. Based on this theoretical framework, two parameter scaling theory [3, 19] was developed and gave much more detailed information about the electronic structure: the states in a Landau band are localized everywhere except at a single energy at or near the center.

9.2.2
Perturbation and Scattering Theory

Ando *et al.* performed the first systematic treatment of impurities in a 2D electron gas in a strong quantizing magnetic field [2]. They used a self-consistent Born approximation and calculated the Hall conductivity σ_{H} for both short-ranged and long-ranged scatterers. One significant result was that if the Fermi level lies in an energy gap (not just a mobility gap), the deviation of the Hall conductivity from the classical formula

$$\Delta\sigma_{\mathrm{H}} = \sigma_{\mathrm{H}} + \frac{ne}{B} \tag{9.23}$$

was shown to vanish within their approximation. By the conservation law for the number of states below an energy gap, the Hall conductivity is quantized.

Apparently, this is the only theoretical prediction of the QHE in a non-ideal system before the experiment of von Klitzing et al. [1]. However, this theory cannot be considered satisfactory because it does not provide a mechanism for pinning the Fermi level in an energy gap for a finite range of the gate voltage or the magnetic field in order to account for the Hall plateaus. Also, this theory is too crude to explain the high accuracy of the experiment. The discovery of von Klitzing *et al.* motivated physicists to test and explain the quantization of the Hall resistance when the Fermi level lies in the energy range of localized states. If this can be done, the finite width of the Hall plateaus can be explained simultaneously since localized states can "pin" the Fermi level.

The first significant breakthrough was made by Aoki and Ando [20]. They used the Kubo formula in a form given by Ando *et al.* [2] prior to the discovery of the QHE. They demonstrated that the Hall conductivity is unchanged when the Fermi level is moved within the mobility gap. The quantization of σ_{H} was obtained in the strong magnetic field limit. The key result is that the correction from the finiteness of the magnetic field vanishes to order $(\Gamma/\hbar\omega_{\mathrm{c}})^2$, where Γ is the broadening of the Landau levels by the impurities. Later, Usov and Ulinich [21] proved that the third order correction vanishes as well.

Alternatively, Prange [6] considered the problem with a single impurity having a δ-potential. With an exact calculation, he proved that a localized state exists at the impurity. However, the remaining nonlocalized states carry a Hall current that exactly compensates for what was not carried by the localized state. An extensive treatment using various kinds of impurity potentials was given by Prange and Joynt [22]. Numerical calculations by Joynt [23] clearly demonstrate the compensating effect in which the electrons in the extended states move faster when they pass by electrons trapped in the localized states. Thouless [24] also studied the influence of impurities using a general Green's function analysis. He observed that as long as the Fermi energy lies in an energy gap or mobility gap (no extended states), the Green's function decays exponentially at large spatial separations. Using this property of the Green's function with the Kubo formula for the Hall conductance, he demonstrated that at zero temperature and with a Fermi energy not in the spectrum of extended states, the correction to the Hall conductance vanishes to all orders of the impurity potential strength. This finding is an extension of the results of Aoki and Ando, and Prange as mentioned above. Streda [25] also performed an interesting calculation that proved when the Fermi level lies in an energy gap, the Hall conductivity is given by

$$\sigma_H = e\frac{\partial \mathcal{D}(E_F)}{\partial B} , \tag{9.24}$$

where $\mathcal{D}(E_F)$ is the density of electron states whose energy lies below E_F. If the gap is between two Landau levels that can be broadened by the impurities, the quantization of σ_H is exactly derived from the conservation of the number of states below an energy gap. This exact result can be compared with the approximate result of Ando *et al.* mentioned at the beginning of this section. The Streda formula is also valid for other types of energy gap, but the quantization of σ_H is not obvious. However, Streda [26] demonstrated that for a periodic potential, Eq. (9.24) gives the same quantized Hall conductance as the theory of Thouless *et al.* [9] in which topological invariant arguments were employed.

9.2.3
Gauge Symmetry Approach

Using the fact that the QHE appears to be independent of the experimental details, Laughlin [27] suggested that this phenomenon might be explained by fundamental principles. He considered a 2D electron gas confined to the surface of a cylinder of finite length with a magnetic field perpendicular to the surface and a current I flowing in the azimuthal direction. The relationship between the current I and the potential difference ΔV between the two edges is established in the following way: suppose the magnetic flux through the center of the cylinder is increased and then the total energy of the electron system changes correspondingly. According to Faraday's law, the current I is given by the adiabatic derivative

$$I = \frac{\partial U}{\partial \phi} . \tag{9.25}$$

If the flux change is equal to a flux quantum h/e, then each localized state is unchanged, whereas the extended states go from one to another. If there are no extended states at the Fermi energy in the interior of the cylinder, the occupation of the electron states in the interior must be the same directly prior to and after the flux has been changed (at zero temperature). The net effect can only be a transfer of an integral number (assume, n) of electrons from one edge to the other, where there are extended states at the Fermi energy around the edges. The resulting energy change is

$$\Delta U = ne\Delta V . \tag{9.26}$$

Equations (9.25) and (9.26) jointly yield a quantized Hall conductance

$$\sigma_H = \frac{I}{\Delta V} = n\frac{e^2}{h} . \tag{9.27}$$

The integer n would be zero if there were no extended states below the Fermi level.

This derivation due to Laughlin is not restricted to any specific 2D electron system. The achievement of his theory is the utilization of the principle of gauge symmetry, that is, a quantum flux change can only map the electron states from one to another or back to themselves [28]. This was demonstrated in the numerical work of Aoki [29, 30], and Aers and MacDonald [11]. Laughlin's formulation does not need a real solenoid to produce the flux change. The flux change is a convenient construct. Also, the above equation does not indicate that a varying flux has to be added to generate the Hall current. However, the flux change may be viewed as being responsible for the Hall current when explaining the QHE. The role of edges and disorder in Laughlin's theory was analyzed by Halperin [31]. He pointed out that a difference in the chemical potential between the two edges can produce an imbalance in the diamagnetic currents localized along the two edges and the ratio of the resulting net current to the difference in chemical potentials (per electron charge) is still quantized in units of e^2/h. This is an important observation because the Hall potential measured in an experiment is actually the sum of the electric potential and the chemical potential.

9.2.4 The QHE in a Periodic Potential

The energy band structure of a 2D electron system in the presence of a uniform perpendicular magnetic field as well as a periodic electrostatic potential is well-known to have some very interesting properties. The case of a sinusoidal potential in the limit when its strength is weak or strong has been extensively studied in the semiclassical limit with the use of the Harper equation [32]. In both limits, the magnetic flux ϕ (measured in units of the flux quantum h/e) through a unit cell of the periodic potential plays an important role in the structure of the spectrum. Let us consider the case when $\phi = q/p$, where q and p are integers. In the limit of a strong periodic potential, each of the Bloch bands is split into p subbands. However, in the weak limit, each Landau level is split into q subbands.

The general formulation of the QHE in a periodic electrostatic potential was given by Thouless and coworkers [9]. Making use of the Kubo formula when the magnetic subbands are full, the Hall conductance is expressed as an integral over the magnetic Brillouin zone. The integral is then related to the phase change of the magnetic Bloch wave functions around the boundary of the zone. The single-value of the wave functions then results in the quantization of the Hall conductance that has an interesting dependence on the number of full subbands.

This result is important because the quantization of the Hall conductance is obtained in the presence of a strong modulation potential when the Fermi level lies between two subbands. Laughlin's theory is more general, but it is not convenient for performing calculations. Furthermore, this was the first demonstration of quantization based on a local theory (the Kubo formula). Via the integral expression for the quantum Hall conductance, we can study the topological properties of the magnetic Bloch wave functions in the magnetic Brillouin zone. Thouless [33] showed that a complete set of Wannier wave functions can be constructed if and only if the subband carries no Hall current.

9.2.5 Topological Equivalence of the Quantum Hall Conductance

Now, we will see that the topological structure of the wave functions characterizes the Hall conductance in a parameter space. Before we begin, let us look at two examples of topological invariance. The first is the winding number of a closed curve C

$$\frac{1}{2\pi}\oint_C \frac{x\,dy - y\,dx}{x^2+y^2} = n\,. \tag{9.28}$$

The second is the total solid angle of a closed surface S,

$$\frac{1}{4\pi}\oint_S d\mathcal{A}\kappa = n\,, \tag{9.29}$$

where κ is the Gaussian curvature of the surface. In Eq. (9.29), $n = 0$ for a torus and $n = 1$ for a sphere. In both Eqs. (9.28) and (9.29), the value of n does not get changed for any continuous deformation of the curve C or surface S.

As shown in a previous chapter, the transport coefficients, including the Hall conductance, can be calculated using the Kubo–Greenwood linear response formalism. In this theory, the external electric field $\boldsymbol{E}$ is treated as a small external perturbation and the current density is calculated to lowest order as

$$j = \overset{\leftrightarrow}{\sigma}\cdot\boldsymbol{E}\,, \quad \overset{\leftrightarrow}{\sigma} = \begin{pmatrix} \sigma & \sigma_{\mathrm{H}} \\ -\sigma_{\mathrm{H}} & \sigma \end{pmatrix}, \tag{9.30}$$

where σ_{H} is the Hall conductivity. Assuming $\sigma = 0$, the total current is

$$\begin{aligned} I_y &= \int dx\, j_y = -\int dx\,\sigma_{\mathrm{H}} E_x \\ &= -E_x L_x \int \frac{dx}{L_x}\sigma_{\mathrm{H}} \equiv -V_x\overline{\sigma}_{\mathrm{H}}\,, \end{aligned} \tag{9.31}$$

where $\overline{\sigma}_{\mathrm{H}}$ is the Hall conductance and in the Kubo formalism is given by

$$\overline{\sigma}_{\mathrm{H}} = \frac{ie^2\hbar}{L_x L_y}\sum_{n>0}\frac{(v_x)_{0n}(v_y)_{n0} - (v_y)_{0n}(v_x)_{n0}}{(E_n - E_0)^2}\,, \tag{9.32}$$

where $v_i = (1/e)\partial\mathcal{H}/\partial A_i$ for a vector potential with components A_i. Now, differentiating $\mathcal{H}|\psi_n\rangle = E_n|\psi_n\rangle$ with respect to A_i, we obtain

$$\langle\psi_0|\frac{\partial\mathcal{H}}{\partial A_i}|\psi_n\rangle + E_0\langle\psi_0|\frac{\partial}{\partial A_i}|\psi_n\rangle = E_n\langle\psi_0|\frac{\partial}{\partial A_i}|\psi_n\rangle\,, \tag{9.33}$$

which we can use to rewrite $\overline{\sigma}_{\mathrm{H}}$ as

$$\overline{\sigma}_{\mathrm{H}} = \frac{i\hbar}{L_x L_y}\left[\left\langle \frac{\partial \psi_0}{\partial A_x}\middle|\frac{\partial \psi_0}{\partial A_y}\right\rangle - \left\langle \frac{\partial \psi_0}{\partial A_y}\middle|\frac{\partial \psi_0}{\partial A_x}\right\rangle\right]. \tag{9.34}$$

If the 2D surface were a torus, A_i may be changed by varying the fluxes through the holes of the torus, that is, $L_x \delta A_x = \delta\varphi_1$ and $L_y \delta A_y = \delta\varphi_2$, so that

$$\overline{\sigma}_{\mathrm{H}} = i\hbar\left[\left\langle \frac{\partial \psi_0}{\partial \varphi_1}\middle|\frac{\partial \psi_0}{\partial \varphi_2}\right\rangle - \left\langle \frac{\partial \psi_0}{\partial \varphi_2}\middle|\frac{\partial \psi_0}{\partial \varphi_1}\right\rangle\right]. \tag{9.35}$$

It can be shown that $\overline{\sigma}_{\mathrm{H}}$ is independent of φ_1 and φ_2 in the thermodynamic limit $L_x, L_y \to \infty$ if there is an energy gap above E_0. Therefore, it follows that we can calculate $\overline{\sigma}_{\mathrm{H}}$ by averaging over a flux quantum in φ_1 and φ_2, that is,

$$\overline{\sigma}_{\mathrm{H}} = \frac{i\hbar}{\left(\frac{h}{e}\right)^2}\int_0^{h/e} d\varphi_1 \int_0^{h/e} d\varphi_2 \times \left[\left\langle \frac{\partial \psi_0}{\partial \varphi_1}\middle|\frac{\partial \psi_0}{\partial \varphi_2}\right\rangle - \left\langle \frac{\partial \psi_0}{\partial \varphi_2}\middle|\frac{\partial \psi_0}{\partial \varphi_1}\right\rangle\right] \equiv \frac{e^2}{h} C\,, \tag{9.36}$$

where

$$C = \frac{i}{2\pi}\int_0^{h/e} d\varphi_1 \int_0^{h/e} d\varphi_2 \left[\left\langle \frac{\partial \psi_0}{\partial \varphi_1}\middle|\frac{\partial \psi_0}{\partial \varphi_2}\right\rangle - \left\langle \frac{\partial \psi_0}{\partial \varphi_2}\middle|\frac{\partial \psi_0}{\partial \varphi_1}\right\rangle\right]. \tag{9.37}$$

This expression is a topological invariant describing the topological phase structure of the wave function $|\psi_0\rangle$ in the parameter space of (φ_1, φ_2). As a matter of fact, $2\pi C$ gives the phase change of $|\psi_0\rangle$ around the boundary of the rectangle $(0 < \varphi_1 < h/e, 0 < \varphi_2 < h/e)$ and C is therefore an integer. The following calculation demonstrates the aforementioned concept.

When φ_i changed by a flux quantum, the wave function must return to itself, except for a phase factor, that is,

$$\left|\psi_0\left(\varphi_1 + \frac{h}{e}, \varphi_2\right)\right\rangle = e^{i\alpha(\varphi_1,\varphi_2)}|\psi_0(\varphi_1, \varphi_2)\rangle\,, \tag{9.38}$$

$$\left|\psi_0\left(\varphi_1, \varphi_2 + \frac{h}{e}\right)\right\rangle = e^{i\beta(\varphi_1,\varphi_2)}|\psi_0(\varphi_1, \varphi_2)\rangle\,. \tag{9.39}$$

We then have

$$
\begin{aligned}
C &= \frac{i}{2\pi}\int_0^{h/e} d\varphi_1 \int_0^{h/e} d\varphi_2 \left[\frac{\partial}{\partial\varphi_1}\left\langle \psi_0 \middle| \frac{\partial\psi_0}{\partial\varphi_2}\right\rangle - \frac{\partial}{\partial\varphi_2}\left\langle \psi_0 \middle| \frac{\partial\psi_0}{\partial\varphi_1}\right\rangle\right] \\
&= \frac{i}{2\pi}\int_0^{h/e} d\varphi_2 \left[\left\langle \psi_0 \middle| \frac{\partial\psi_0}{\partial\varphi_2}\right\rangle\bigg|_{\frac{h}{e},\varphi_2} - \left\langle \psi_0 \middle| \frac{\partial\psi_0}{\partial\varphi_1}\right\rangle\bigg|_{0,\varphi_2}\right] \\
&\quad - \frac{i}{2\pi}\int_0^{h/e} d\varphi_1 \left[\left\langle \psi_0 \middle| \frac{\partial\psi_0}{\partial\varphi_1}\right\rangle\bigg|_{\varphi_1,\frac{h}{e}} - \left\langle \psi_0 \middle| \frac{\partial\psi_0}{\partial\varphi_1}\right\rangle\bigg|_{\varphi_1,0}\right] \\
&= \frac{i}{2\pi}\left[i\int_0^{h/e} d\varphi_2 \frac{\partial\alpha(0,\varphi_2)}{\partial\varphi_2} - i\int_0^{h/e} d\varphi_1 \frac{\partial\beta(\varphi_1,0)}{\partial\varphi_1}\right] \\
&= \frac{1}{2\pi}\left[\beta\left(\frac{h}{e},0\right) - \beta(0,0) - \alpha\left(0,\frac{h}{e}\right) + \alpha(0,0)\right] .
\end{aligned}
\tag{9.40}
$$

9.3 Corrections to the Quantization of the Hall Conductance

Now, let us look at determining possible corrections to the QHE. Basically, this is an effect occurring at zero temperature. The quantization can be destroyed at finite temperature by the inelastic scattering of the electrons off the phonons. However, this type of correction goes to zero exponentially with temperature and can be made to be extremely small. Our concern will be on the effects arising from the boundary as well as nonlinear effects beyond linear response theory. Also, we demonstrate that both of these effects are also exponentially small as a result of localization.

9.3.1 Properties of the Green's Function

For a Hamiltonian $\hat{\mathcal{H}}$, the Green's function in the coordinate representation is

$$
\begin{aligned}
G\left(\mathbf{r},\mathbf{r}';\lambda\right) &= \langle \mathbf{r}| \frac{1}{\lambda-\hat{\mathcal{H}}} |\mathbf{r}'\rangle \\
&= \sum_n \langle \mathbf{r}|\psi_n\rangle \langle\psi_n| \frac{1}{\lambda-\hat{H}} |\mathbf{r}'\rangle \\
&= \sum_n \langle \mathbf{r}|\psi_n\rangle \frac{1}{\lambda-E_n} \langle\psi_n|\mathbf{r}'\rangle \\
&= \sum_n \frac{\psi_n(\mathbf{r})\psi_n^*(\mathbf{r}')}{\lambda-E_n} .
\end{aligned}
\tag{9.41}
$$

If G is local, then

$$G\left(\boldsymbol{r}, \boldsymbol{r}'; \lambda\right) \approx e^{-\alpha|\boldsymbol{r}-\boldsymbol{r}'|} \,. \tag{9.42}$$

If λ is outside the spectrum of $\hat{\mathcal{H}}$, G is local and the length scale α^{-1} depends on how far away λ is from the spectrum. For example, for a one-dimensional free particle,

$$G(x, x'; \lambda) = -\sqrt{\frac{m^*}{2\hbar^2(-\lambda)}}\, e^{-\sqrt{\frac{2m^*(-\lambda)}{\hbar^2}}|x-x'|} \,, \tag{9.43}$$

and for a free particle moving in three dimensions,

$$G\left(\boldsymbol{r}, \boldsymbol{r}'; \lambda\right) = \frac{m^*}{2\pi\hbar^2}\frac{1}{|\boldsymbol{r}-\boldsymbol{r}'|}\, e^{-\sqrt{\frac{2m^*(-\lambda)}{\hbar^2}}|\boldsymbol{r}-\boldsymbol{r}'|} \,. \tag{9.44}$$

For a two-dimensional electron in a magnetic field, the Green's function is local in all directions and behaves asymptotically as

$$G\left(\boldsymbol{r}_{\|}, \boldsymbol{r}'_{\|}; \lambda\right) \approx e^{-\left|\frac{\boldsymbol{r}_{\|}-\boldsymbol{r}'_{\|}}{\ell_{\mathrm{H}}}\right|^2} \,, \tag{9.45}$$

where ℓ_{H} is the magnetic length and λ is not near the Landau level spectrum $(n + 1/2)\hbar\omega_{\mathrm{c}}$. When $\lambda = E + i\epsilon$ where ϵ is a positive infinitesimal and E is in the spectrum, the locality of G depends on the nature of the states at the energy E. As a matter of fact, if the states are localized, we have

$$G\left(\boldsymbol{r}_{\|}, \boldsymbol{r}'_{\|}; \lambda\right) \approx e^{-\frac{|\boldsymbol{r}_{\|}-\boldsymbol{r}'_{\|}|}{\xi}} \,, \tag{9.46}$$

where ξ is the localization length. When the states are extended, G may also be extended.

If an electron moving in the xy-plane is confined to the half space $x > 0$, then in the presence of a magnetic field B in the z-direction, the wave function in the Landau gauge can be written as $e^{iqy}\varphi(x)$ where $\varphi(x)$ is a solution of

$$\left[-\frac{\hbar^2}{2m^*}\frac{d^2}{dx^2} + \frac{1}{2m^*}\left(\hbar q - eBx\right)^2\right]\varphi(x) = E\varphi(x) \,,$$
$$\varphi(x = 0) = 0 \,. \tag{9.47}$$

When the guiding center $X_0 = \hbar q/eB$ is far from the boundary at $x = 0$, the energy has the values of bulk Landau levels $(n + 1/2)\hbar\omega_{\mathrm{c}}$. However, near the boundary, the degeneracy of the eigenstates is lifted, and the eigenenergy is increased as a result of the larger curvature of the wave function which causes the kinetic energy to become enhanced. If the Fermi energy E_{F} lies between two bulk Landau levels, then near the boundary, it intersects the spectrum. Since these eigenstates are localized in the x-direction, we expect that $G(E_{\mathrm{F}} \pm i\epsilon)$ is also local in this direction, where ϵ is real. Likewise, G is extended along the edge just like the energy eigenstates. However, a special property of these states is that their group velocity is solely unidirectional since

$$v_q = \frac{1}{\hbar}\frac{dE(q)}{dq} < 0\,. \tag{9.48}$$

Thus, from a classical point of view, in the presence of a magnetic field, the particle moves in a circular orbit far from the edge but as the guiding center approaches the edge, the particle bounces while traveling along the edge in one direction. This means that the propagator Green's function $G(\boldsymbol{r}, \boldsymbol{r}'; E_F + i\epsilon)$ decays exponentially for $y \gg y'$, but is plane wave-like for $y < y'$. This one-way local behavior of the Green's function G is referred to as semilocality.

References

1 von Klitzing, K., Dorda, G., and Pepper, M. (1980) New method for high-accuracy determination of the fine-structure constant based on quantized Hall resistance. *Phys. Rev. Lett.*, **45**, 494.

2 Ando, T., Matsumoto, Y., and Uemura, Y. (1975) Theory of Hall effect in a two-dimensional electron system. *J. Phys. Soc. Jpn.*, **39**, 279.

3 Laughlin, R.B. (1984) Anomalous quantum Hall effect: An incompressible quantum fluid with fractionally charged excitations. *Phys. Rev. Lett.*, **50**, 1395.

4 Novoselov, K.S. *et al.* (2007) Room-temperature quantum Hall effect in graphene. *Science*, **315**, 5817.

5 Prange, R.E., Girvin, S.M., and Joynt, R. (1988) The quantum Hall effect. *Am. J. Phys.*, **56**, 667.

6 Prange, R.E. (1981) Quantized Hall resistance and the measurement of the fine-structure constant. *Phys. Rev. B*, **23**, 4802.

7 Ando, T. (1974) Theory of quantum transport in a two-dimensional electron system under magnetic fields. III. Many-site approximation. *J. Phys. Soc. Jpn.*, **37**, 622.

8 Halperin, B.I. (1986) The quantized Hall effect. Scientific American, April.

9 Thouless, D.J., Kohmoto, M., Nightingale, M.P., and den Nijs, M. (1982) Quantized Hall conductance in a two-dimensional periodic potential. *Phys. Rev. Lett.*, **49**, 405.

10 MacDonald, A.H. and Streda, P. (1984) Quantized Hall effect and edge currents. *Phys. Rev. B*, **29**, 1616.

11 MacDonald, A.H. and Aers, G.C. (1984) Inversion-layer width, electron–electron interactions, and the fractional quantum Hall effect. *Phys. Rev. B*, **29**, 5976.

12 Tsui, D.C., Störmer, H.L., and Gossard, A.C. (1982) Two-dimensional magnetotransport in the extreme quantum limit. *Phys. Rev. Lett.*, **48**, 1559.

13 Ando, T. and Uemura, Y. (1974) *J. Phys. Soc. Jpn.* **36**, 959.

14 Wigner, E.P. (1953) Review of quantum mechanical measurement problem, in Quantum Optics, Experimental Gravity and Measurement Theory, (eds P. Meystre and M. O. Scully), Plenum Press, New York, p. 53.

15 Ando, T. (1984) *J. Phys. Soc. Jpn.*, **53**, 3101.

16 Trugman, S.A. (1983) *Phys. Rev. B*, **27**, 7539.

17 Milnikov, G.V. and Sokolov, I.M. (1988) *JETP Lett.*, **48**, 536.

18 Levine, H., Libby, S.B. and Pruisken, A.M.M. (1983) *Phys. Rev. Lett.*, **51**, 1915.

19 Khmelnitskii, D.E. (1983) *JETP Lett.*, **38**, 454.

20 Aoki, H. and Ando, T. (1981) *Solid State Commun.*, **38**, 1079.

21 Usov, A., Ulinich, F.R. and Grebenschikov, Yu.B. (1982) *Solid State Commun.*, **43**, 475.

22 Prange, R.E. and Joynt, R. (1982) *Phys. Rev. B*, **25**, 2943.

23 Joynt, R. and Prange, R.E. (1984) *Phys. Rev. B*, **29**, 3303.

24 Thouless, D.J. (1981) *J. Phys. C: Solid State Phys.*, **14**, 3475.

25 Streda, P.J. (1982) *Phys. C: Solid State Phys.*, **15**, L717.

26 Streda, P.J. (1982) *Phys. C: Solid State Phys.*, **15**, L1299.

27 Laughlin, B. (1981) *Phys. Rev. B*, **23**, 5632.

28 Byers, N. and Yang, C.N. (1961) *Phys. Rev. Lett.*, **7**, 46.

29 Aoki, H. (1982) *J. Phys. C*, **15**, L1227.

30 Aoki, H. (1983) *J. Phys. C: Solid State Phys.*, **16**, 1893.

31 Halperin, B.I. (1982) *Phys. Rev. B*, **25**, 2185.

32 Hofstadter, D. (1976) *Phys. Rev. B*, **14**, 2239.

33 Thouless, D.J. (1984) *J. Phys. C: Solid State Phys.*, **17**, L325.

10
Fractional Quantum Hall Effect

Shortly following the discovery of the IQHE by von Klitzing [1] in 1980 for a MOSFET, Tsui *et al.* [2] repeated the experiment on GaAs/AlGaAs heterostructures. The consequence of the small effective mass of the electrons in the 2DEG was that the QHE was observed at higher temperatures and lower magnetic fields. To their surprise, they found, as reported in their 1982 paper, that the Hall conductance exhibited a plateau at $1/3e^2/h$ to an accuracy of one part in 10^5. Further experiments revealed quantization of the Hall conductance at other simple fractions such as $2/3, 1/5, 2/5, 3/5, 4/5, 2/7, 3/7, 4/7, \ldots$ These results led physicists to believe that the 2DEG in a strong perpendicular magnetic field must behave extraordinarily at densities corresponding to fractions of the Landau level capacity.

10.1
The Laughlin Ground State

10.1.1
The Lowest Landau Level

In the Landau gauge [3] with vector potential $\boldsymbol{A} = (0, Bx, 0)$, the ground state is degenerate with each wave function

$$\psi_q(x, y) = e^{iqy} e^{-\frac{(x-q\ell^2)^2}{2\ell^2}} \tag{10.1}$$

and energy $1/2\hbar\omega_c$. Although each of these wave functions is localized in the x-direction, they can be recombined to form other types of wave functions such as

Properties of Interacting Low-Dimensional Systems, First Edition. G. Gumbs and D. Huang.

(for further reading, see [4–8])

$$\begin{aligned}
\tilde{\psi}_k(x, y) &= \int dq e^{-iqk\ell^2} e^{iqy} e^{-\frac{(x-q\ell^2)^2}{2\ell^2}} \\
&= e^{-\frac{x^2}{2\ell^2}} \int dq e^{iq\ell\left(\frac{y}{\ell}-k\ell\right)} e^{-\frac{(q\ell)^2}{2}} e^{qx} \\
&\propto e^{i\frac{x}{\ell}\left(\frac{y}{\ell}-k\ell\right)} e^{-\frac{\left(\frac{y}{\ell}-k\ell\right)^2}{2}} \\
&= e^{i\frac{xy}{\ell^2}} \left\{ e^{-ikx} e^{-\frac{(y-k\ell^2)^2}{2\ell^2}} \right\} .
\end{aligned} \tag{10.2}$$

The wave function in the curly brackets is the wave function in the gauge $\boldsymbol{A} = (-By, 0, 0)$. The k-independent prefactor is a gauge factor. It does not appear in making a transformation from the old gauge to the new gauge.

In order to obtain wave functions localized in the radial direction, we note that

$$\begin{aligned}
\psi_q &= e^{i\frac{xy}{2\ell^2}} e^{-\frac{x^2+y^2}{4\ell^2}} e^{-\frac{(x+iy)^2}{4\ell^2}} e^{iq(x+iy)} e^{-\frac{(q\ell)^2}{2}} \\
&= e^{i\frac{xy}{2\ell^2}} e^{-\frac{x^2+y^2}{4\ell^2}} \sum_{n=0}^{\infty} \frac{1}{n!} \left(\frac{(x+iy)}{2\ell} \right)^n i^n H_n(q\ell) e^{-\frac{(q\ell)^2}{2}} ,
\end{aligned} \tag{10.3}$$

from which we obtain

$$\begin{aligned}
\int dq H_n(ql) e^{-\frac{(q\ell)^2}{2}} \psi_q &\propto e^{i\frac{xy}{2\ell^2}} \left(\frac{(x+iy)}{2\ell} \right)^n e^{-\frac{x^2+y^2}{4\ell^2}} \\
&= e^{i\frac{xy}{2\ell^2}} (z)^n e^{-|z|^2} ,
\end{aligned} \tag{10.4}$$

where $z = (x + iy)/(2\ell)$. Except for the phase factor $e^{i\frac{xy}{2\ell^2}}$, the right-hand side of the wave function obtained is the same as one in the circular gauge $A_x = -By/2$, $A_y = Bx/2$ and $A_z = 0$.

10.1.2
Laughlin's Wave Function

The fractional quantum Hall effect was observed on samples of high mobility. Consequently, it is reasonable to exclude impurities as the cause of this effect. The solution in the absence of many-particle interactions is completely known, and there is no indication of any peculiarities at fractional fillings of the Landau levels. It was then suggested that the Coulomb interactions between electrons must play a crucial role.

The typical interaction energy per particle is $e^2/(4\pi\epsilon_0\epsilon_r d)$, where d is the average separation between neighboring electrons. At large magnetic fields, this is small compared to $\hbar\omega_c$, the spacing between two Landau levels. Therefore, the coupling between two Landau levels due to Coulomb interactions may be neglected, and all electrons should be in the lowest Landau level for filling factors less than one.

In the circular gauge, the lowest Landau level consists of states

$$(z)^n e^{-|z|^2} , \quad n = 0, 1, \ldots, \infty . \tag{10.5}$$

Therefore, any state in the lowest Landau level can be written as

$$f(z)e^{-|z|^2}\,, \tag{10.6}$$

where $f(z)$ is an analytic function of z. If there are N particles, the N-body wave function should be of the form

$$g(z_1, z_2, \ldots, z_N)\exp\left(-\sum_{i=1}^{N}|z_i|^2\right)\,, \tag{10.7}$$

where g is analytic in any of its variables. A Jastrow-type wave function has the following form

$$\psi\left(\boldsymbol{r}_1, \boldsymbol{r}_2, \ldots, \boldsymbol{r}_N\right) = \prod_{i=1}^{N} f_1\left(\boldsymbol{r}_i\right)\prod_{i<j}^{N} f_2\left(\boldsymbol{r}_i - \boldsymbol{r}_j\right)\,, \tag{10.8}$$

where the forms of f_1 and f_2 must be varied to minimize the energy of the system. The function $f_1(\boldsymbol{r})$ takes account of the response to an external perturbation and $f_2(\boldsymbol{r})$ takes care of correlations. The higher order effects arising from correlations among three particles can be included by another function, but this is not an important factor. Jastrow-like wave functions have been found to describe the ground states of liquid He^3 and He^4 very well. In the present problem, its form is

$$\psi(z_1, z_2, \ldots, z_N) = \prod_{i<j}^{N} f(z_i - z_j)\exp\left(-\sum_{i=1}^{N}|z_i|^2\right)\,. \tag{10.9}$$

Since all the single-particle states are in the lowest Landau level, the function $f(z)$ must be analytic and odd to satisfy antisymmetry with respect to particle exchange. Therefore,

$$f(z) = a_1 z + a_3 z^3 + a_5 z^5 + \ldots \tag{10.10}$$

The total angular momentum of the system

$$L = -i\hbar\sum_{i=1}^{N}\frac{\partial}{\partial\theta_i} \tag{10.11}$$

commutes with the Hamiltonian

$$H = \sum_{i=1}^{N}\frac{1}{2m^*}\left[\left(-i\hbar\frac{\partial}{\partial x_i} + \frac{1}{2}eBy_i\right)^2 + \left(-i\hbar\frac{\partial}{\partial y_i} - \frac{1}{2}eBx_i\right)^2\right] + \sum_{i<j}^{N}\frac{e^2}{4\pi\epsilon_0\epsilon_r|\boldsymbol{r}_i - \boldsymbol{r}_j|}\,. \tag{10.12}$$

Therefore, we choose the eigenstates of H as those of L. However, since each power of z_i in ψ contributes an angular momentum of $\hbar$ to L, the function $f(z)$ in ψ must be a single power in z, that is, $f(z) = z^m$. Thus, the wave function should have the form

$$\psi(z_1, z_2, \ldots, z_N) = \prod_{i<j}^{N} (z_i - z_j)^m \exp(-\sum_{i=1}^{N} |z_i|^2) , \tag{10.13}$$

where m is an odd integer.

10.1.3 Properties of the Laughlin Wave Function

The electron density at z is

$$\rho(z) = \frac{\int dz_1 \int dz_2 \ldots \int dz_N \, |\psi(z_1, \ldots, z_N)|^2 \sum_{i=1}^{N} \delta(z - z_i)}{\int dz_1 \int dz_2 \ldots \int dz_N \, |\psi(z_1, \ldots, z_N)|^2} , \tag{10.14}$$

where, in this notation, we have $\int dz_i \ldots \equiv \int dx_i \int dy_i \ldots$ and $\delta(z - z_i) \equiv \delta(x - x_i)\delta(y - y_i)$. We may write

$$|\psi(z_1, \ldots, z_N)|^2 = e^{-2m\phi(z_1, \ldots, z_N)} , \tag{10.15}$$

where

$$\phi(z_1, \ldots, z_N) = -\sum_{i<j}^{N} \ln |z_i - z_j| + \frac{1}{m} \sum_{i=1}^{N} |z_i|^2 . \tag{10.16}$$

We can consider ϕ as the potential energy of 2D particles each with "charge" unity, repelling each other through the logarithmic term and attracted to a circular disk with uniform charge density $\rho_0 = 2/(\pi m)$. It can be shown that the particles will have a charge density $+2/(\pi m)$ near the center of the disk. In terms of the original length scale, the particle density will be $(1/m)1/(2\pi\ell^2)$, which is just $(1/m)$th the Landau level capacity. The total occupied area is $Nm(\pi\ell^2)$, with a radius $\sqrt{2Nm}\ell$. The electron density $\rho(r)$ decays like a Gaussian function when r is larger than this radius.

The pair density distribution function

$$\rho(z, z') = \int dz_1 \int dz_2 \ldots \int dz_N \, |\psi(z_1, \ldots, z_N)|^2 \sum_{i<j}^{N} \delta(z - z_i)\delta(z' - z_j) \times \left[\int dz_1 \int dz_2 \ldots \int dz_N \, |\psi(z_1, \ldots, z_N)|^2 \right]^{-1} , \tag{10.17}$$

is another important characteristic of the wave function. It has been shown that when z and z' are not near the boundary,

$$\rho(z, z') = \left(\frac{2}{\pi m} \right)^2 g\left(|z - z'|\right) , \tag{10.18}$$

where (i) $g(|z|) \to 0$ for $|z| \gg \ell$ and (ii) $g(|z|) \sim |z|^{2m}$ as $|z| \to 0$. The first result implies that two particles become uncorrelated when their separation exceeds the magnetic length. The second result means that two particles tend to avoid each other. However, we note that the Pauli exclusion principle in itself gives

$$g(|z|) \sim |z|^2 , \quad \text{as } |z| \to 0 . \tag{10.19}$$

The additional correlations make the short-ranged Coulomb repulsion ineffective and therefore lowers the interaction energy considerably. As a matter of fact, the total interaction energy is

$$NU = \frac{e^2}{8\pi\epsilon_0\epsilon_r\ell}\left\{\frac{\int dz_1 \int dz_2 \ldots \int dz_N \, |\psi(z_1,\ldots,z_N)|^2 \sum_{i<j}^{N} \frac{1}{|z_i - z_j|}}{\int dz_1 \int dz_2 \ldots \int dz_N \, |\psi(z_1,\ldots,z_N)|^2} - \int dz \int dz' \rho(z) \frac{1}{|z-z'|}\rho_0\right\} . \tag{10.20}$$

Since, away from the boundary, we have

$$\begin{aligned} \rho(z,z') &= \rho_0^2 g(|z-z'|) \\ \rho(z) &= \rho_0 , \end{aligned} \tag{10.21}$$

we may rewrite Eq. (10.20) as

$$\begin{aligned} NU &= \frac{e^2}{8\pi\epsilon_0\epsilon_r\ell}\rho_0^2 \int dz \int dz' \frac{1}{|z-z'|}[g(|z-z'|)-1] \\ &= \frac{e^2}{8\pi\epsilon_0\epsilon_r\ell}\rho_0^2 \int dz' \int dz \frac{1}{|z|}[g(|z|)-1] \\ &= N\frac{e^2}{8\pi\epsilon_0\epsilon_r\ell}\rho_0 \int dz \frac{1}{|z|}[g(|z|)-1] . \end{aligned} \tag{10.22}$$

The interaction energy per particle is thus

$$U = \frac{e^2}{8\pi\epsilon_0\epsilon_r\ell}\rho_0 \int dz \frac{1}{|z|}[g(|z|)-1] . \tag{10.23}$$

The long-range contribution from $1/|z|$ is included because of the rapid decay of $[g(|z|)-1]$.

10.1.4
Justification of the Laughlin State

Numerical calculations have shown that U is low at the filling factor $\nu = 1/3$. Standard theories give energies larger than the numerical results and show no peculiarity for $\nu = 1/3$. The Laughlin wave function for $m = 3$ agrees with the numerical results for U at $\nu = 1/3$ with an accuracy of about 99.95 %. Also, the overlap of Laughlin's wave function with the ground state wave function is almost equal to one.

The Laughlin wave function only applies for densities equal to $1/m$ of that of a Landau level. This means that the states at other densities must contain inhomogeneities and that correlations may exist among more than two particles. Investigations have shown that the ground state at other densities consist of the Laughlin state together with a number of quasiparticles or quasiholes. Under certain conditions, these quasiparticles and quasiholes form a Laughlin state and gives rise to a stable state at filling fractions such as $\nu = 2/5$.

The Laughlin wave function describes a circular liquid drop. We can modify the shape of the liquid drop by multiplying the Laughlin wave function by $\exp[\sum_{i=1}^{N} p(z_i)]$, where $p(z)$ is an analytic function of z. This can be understood as follows. In the plasma analogy, the potential term becomes

$$V(z) = \frac{1}{m}\left(|z|^2 - Re\left[p(z)\right]\right). \tag{10.24}$$

The corresponding charge density for this potential is

$$\rho(z) = \frac{1}{2\pi}\nabla^2 V(z) = \left(\frac{1}{2\pi}\right)\left(\frac{4}{m}\right) = \rho_0\,, \tag{10.25}$$

where we assumed $\nabla^2 p(z) = 0$ for any analytic function $p(z)$. Thus, the background density is unchanged. The potential is modified because of the location of charges at infinity. Consequently, the particles will still form a liquid drop of the same density, though with its boundary lying on a contour of constant $V(z)$ whose shape may be arbitrarily modified by the form of $p(z)$. It may also be shown that the pair distribution function $\rho(z, z')$ is unchanged for z and z' within the boundary.

We note that the Laughlin wave function can be written in other gauges. For example, in the Landau gauge, we have

$$\psi(z_1, z_2, \ldots, z_N) = e^{-\frac{1}{2\ell^2}\sum_i x_i^2} \prod_{i<j}^{N} \left(e^{\frac{2\pi}{L_y} z_i} - e^{\frac{2\pi}{L_y} z_j}\right)^m, \tag{10.26}$$

where L_y is the length of the system in the y direction. It has been shown that this state describes a liquid of density $1/(2\pi\ell^2 m)$ on a cylinder.

10.2
Elementary Excitations

10.2.1
Fractional Charge

A hole state may be created at the origin by multiplying the Laughlin wave function in Eq. (10.13) with $\prod_{i=1}^{N} z_i$, that is,

$$\psi_{\text{hole}}(z_1, z_2, \ldots, z_N) = \prod_{i=1}^{N} z_i \prod_{i<j}^{N} (z_i - z_j)^m \exp\left(-\sum_{i=1}^{N} |z_i|^2\right), \tag{10.27}$$

that has a momentum $N\hbar$ larger than the ground state and is thus orthogonal to it. A hole exists at the origin because the wave function vanishes when any particle is at the origin. To determine the charge of the hole, we proceed as follows. Let us regard the hole state as adiabatically evolved from the ground state and, during the process, the momentum of each single-particle state increases by $\hbar$. In particular, by expanding the ground state as

$$\psi(z_1, z_2, \ldots, z_N) = \sum_{k_1, k_2, \ldots, k_N} a_{k_1 k_2 \ldots k_N} z_1^{k_1} z_2^{k_2} \ldots z_N^{k_N} \times \exp\left(-\sum_{i=1}^{N} |z_i|^2\right), \tag{10.28}$$

the hole state may be written as

$$\psi_{hole}(z_1, z_2, \ldots, z_N) = \sum_{k_1, k_2, \ldots, k_N} a_{k_1 k_2 \ldots k_N} z_1^{k_1+\lambda} z_2^{k_2+\lambda} \ldots z_N^{k_N+\lambda} \times \exp\left(-\sum_{i=1}^{N} |z_i|^2\right), \tag{10.29}$$

where $\lambda = 1$, that is, the ground state becomes a hole state when λ increases from zero to one.

The number density of the single-particle state $z^{k+\lambda} e^{-|z|^2}$ is maximum at $|z| = \sqrt{(k+\lambda)/2}$ with a width $1/\sqrt{12}$. The area encircled by the peak is $\pi/2(k+\lambda)$. As λ increases by one, this area increases by $\pi/2$. In the many-body state, every single-particle state moves outward by an area $\pi/2$. However, since the total charge density is $2/(\pi m)$ far from the origin, there is a charge of $1/m$ flowing out a circle far from the origin. Charge conservation then ensures there is a hole at the origin with a charge $-1/m$. This is a fractional charge.

If λ is decreased by one, the new state should have an excess charge at the origin of $1/m$. This is a quasiparticle state with wave function

$$\exp\left(-\sum_{i=1}^{N} |z_i|^2\right) \prod_{i=1}^{N} \left(\frac{\partial}{\partial z_i}\right) \prod_{i<j}^{N} (z_i - z_j)^m . \tag{10.30}$$

The quasihole has an area of circle $\pi/2$. There is a total charge density

$$\delta\rho(z) = \delta\rho_{\text{h}}(z) + \delta\rho_{\text{b}}(z) \tag{10.31}$$

near the origin and around the boundary with

$$\int dz \delta\rho_{\text{h}}(z) = -\int dz \delta\rho_{\text{b}}(z) = -\frac{1}{m} . \tag{10.32}$$

The energy required to create a hole is essentially equal to the electrostatic energy of these charges

$$\frac{1}{2} \frac{e^2}{8\pi\epsilon_0\epsilon_r\ell} \int dz \int dz' \frac{\delta\rho(z)\delta\rho(z')}{|z-z'|} . \tag{10.33}$$

Since $\delta\rho_{\mathrm{b}}$ is smeared out on the boundary, its contribution vanishes when the thermodynamic limit is taken. The creation energy is thus

$$\frac{e^2}{16\pi\epsilon_0\epsilon_r\ell}\int dz\int dz'\frac{\delta\rho_{\mathrm{h}}(z)\delta\rho_{\mathrm{h}}(z')}{|z-z'|}\,, \tag{10.34}$$

which can be estimated by assuming that $\delta\rho_{\mathrm{h}}(z)$ is uniformly distributed on a disk of area $\pi/2$, yielding

$$\frac{\left(\frac{e}{m}\right)^2}{4\pi\epsilon_0\epsilon_r\ell}\frac{4\sqrt{2}}{3\pi^2}=0.201\frac{1}{m^2}\frac{e^2}{4\pi\epsilon_0\epsilon_r\ell}\,. \tag{10.35}$$

10.2.2
The Complete Set of Quasi-Hole States

We can also create a quasi-hole state at z with wave function

$$\psi_z=\exp\left(-\frac{1}{m}|z|^2\right)\prod_{i=1}^{N}(z_i-z)\prod_{i<j}^{N}(z_i-z_j)^m\exp\left(-\sum_{i=1}^{N}|z_i|^2\right). \tag{10.36}$$

We choose the factor of $\exp(-(1/m)|z|^2)$ in ψ_z to make the normalization of ψ_z independent of z. Treating z and z^* as independent variables, we have

$$\begin{aligned}\frac{\partial}{\partial z}\ln\langle\psi_z|\psi_z\rangle &= -\frac{2}{m}z^*+\left\langle\sum_i\frac{1}{z-z_i}\right\rangle\\ &= -\frac{2}{m}z^*+\int dz'\rho(z')\frac{1}{z-z'}\,.\end{aligned} \tag{10.37}$$

Let $\rho(z')=\rho_0(z')+\rho_{\mathrm{h}}(z'-z)$ denote the charge density in the Laughlin state with $\rho_{\mathrm{h}}(z'-z)$ the charge density associated with the hole. There is no contribution from $\rho_{\mathrm{h}}(z'-z)$ because of the rotational symmetry, whereas the contribution from ρ_0 is

$$\begin{aligned}\rho_0\int dr r\int d\theta\frac{1}{z-re^{i\theta}} &= \rho_0\int dr r\oint_{|z'|=r}\frac{dz'}{iz'}\frac{1}{z-z'}\\ &= \rho_0\int_0^{|z|}dr r2\pi\frac{1}{z}=\pi\rho_0 z^*=\frac{2}{m}z^*\,,\end{aligned} \tag{10.38}$$

which means that $\langle\psi_z|\psi_z\rangle$ is independent of z. We may similarly show that $\langle\psi_z|\psi_z\rangle$ is independent of z^*.

In general, we can write $\langle\psi_{z'}|\psi_z\rangle$ as $\exp[-(1/m)(|z|^2+|z'|^2)]F(z'^*,z)$, where F is an analytic function of its two variables. At $z'=z$, we have $F(z^*,z)=C\exp[-(2/m)z^*z]$, where C is a constant. By analytic continuation, we generally have $F(z'^*,z)=C\exp[-(2/m)z'^*z]$. Therefore, $\langle\psi_{z'}|\psi_z\rangle=C\exp[-(1/m)|z-z'|^2]\exp[(1/m)(z'^*z-z'z^*)]$, where the second factor is a phase factor.

The nature of the overlap $\langle\psi_{z'}|\psi_z\rangle$ implies that two holes are independent if their separation exceeds $\sqrt{m}$. However, the non-zero value of the overlap does not allow us to label the hole states by z. To obtain the orthogonal set of hole states, we proceed as follows. Let us expand ψ_z as

$$\psi_z = \exp\left(-\frac{1}{m}|z|^2\right)\sum_{n=0}^{N}(-z)^n f_{N-n}(z_1,\ldots,z_N)\prod_{i<j}^{N}(z_i-z_j)^m \times \exp\left(-\sum_{i=1}^{N}|z_i|^2\right), \quad (10.39)$$

where $f_n(z_1,\ldots,z_N)$ is a symmetric product of n factor of different z_i's like $\sum z_{i_1}z_{i_2}\ldots z_{i_n}$. Thus, the components must be orthogonal to each other because they involve different powers of the z_i's and therefore have different total angular momentum, that is, we have

$$\langle\psi_{j'}|\psi_j\rangle = 0 \quad \text{for } j \neq j', \quad (10.40)$$

where

$$\psi_{N-n} = (-1)^n f_{N-n}(z_1,\ldots,z_N)\prod_{i<j}^{N}(z_i-z_j)^m \times \exp\left(-\sum_{i=1}^{N}|z_i|^2\right). \quad (10.41)$$

To obtain the normalization constant of ψ_{N-n}, we note that

$$\psi_{N-n} = \frac{1}{\pi m^{n+1}n!}\int dz(z^*)^n\psi_z . \quad (10.42)$$

Therefore,

$$\begin{aligned}\langle\psi_{N-n}|\psi_{N-n}\rangle &= \frac{1}{(\pi m^{n+1}n!)^2}\int dz\int dz'(z'z^*)^n\langle\psi_{z'}|\psi_z\rangle\\ &= \frac{C}{(\pi m^{n+1}n!)^2}\int dz\int dz'(z'z^*)^n\\ &\quad\times\exp\left[-\frac{1}{m}(|z|^2+|z'|^2)\right]\exp\left(\frac{2}{m}z'^*z\right)\\ &= C\left(\frac{2}{m}\right)^n . \end{aligned} \quad (10.43)$$

However, since $C = \langle\psi_z|\psi_z\rangle_{z=0}$, we have

$$\langle\psi_{j'}|\psi_j\rangle = \delta_{j'j}\left(\frac{2}{m}\right)^{N-j}\langle\psi_z|\psi_z\rangle_{z=0} . \quad (10.44)$$

10.3
The Ground State: Degeneracy and Ginzburg–Landau Theory

10.3.1
Ground State Degeneracy

We have made several assumptions about the topological invariant theory of the integer quantum Hall effect. First, the electrons are on a torus. Second, the ground state of the system of electrons is nondegenerate. An additional assumption is that the ground state is separated from the excited states by an energy gap that is finite when the thermodynamic limit is taken. In this limit, the Hall conductance is given by

$$\sigma_{\mathrm{H}} = \frac{e^2}{h}\frac{i}{2\pi}\int_0^{h/e} d\varphi_1 \int_0^{h/e} d\varphi_2 \left[\left\langle \frac{\partial \psi_0}{\partial \varphi_1}\middle| \frac{\partial \psi_0}{\partial \varphi_2}\right\rangle - \left\langle \frac{\partial \psi_0}{\partial \varphi_2}\middle| \frac{\partial \psi_0}{\partial \varphi_1}\right\rangle\right], \tag{10.45}$$

which is quantized as integer multiples of e^2/h. To obtain fractional conductance, one of the above assumptions has to be broken down. Laughlin's ground state is separated from the excited states by a finite gap because the quasiparticles and quasi-holes have finite creation energies. Furthermore, there is no difficulty in adapting Laughlin's theory to the torus geometry. We have already demonstrated how to write Laughlin's wave function on a cylinder. Therefore, if Laughlin's theory is correct, the ground state must be degenerate for us to obtain a fractionally quantized Hall conductance.

Tao and Wu [4] first pointed out the degeneracy of the ground state and attempted to generalize the gauge symmetry argument of Laughlin to the case of the fractional QHE. The ground state degeneracy was also proven by Haldane and Rezayi [9] using group theory and was demonstrated numerically by Su [6]. If the electron density is p/q of that of a Landau level, then there are q independent ground states. As an example, suppose that $p/q = 1/3$, then there are three ground states ψ_1, ψ_2 and ψ_3, say. The Hall conductance should be averaged over the contributions from each of these states,

$$\sigma_{\mathrm{H}} = \frac{e^2}{3h}\frac{i}{2\pi}\int_0^{h/e} d\varphi_1 \int_0^{h/e} d\varphi_2 \times \sum_{j=1}^{3} \left[\left\langle \frac{\partial \psi_j}{\partial \varphi_1}\middle| \frac{\partial \psi_j}{\partial \varphi_2}\right\rangle - \left\langle \frac{\partial \psi_j}{\partial \varphi_2}\middle| \frac{\partial \psi_j}{\partial \varphi_1}\right\rangle\right]. \tag{10.46}$$

Because of the degeneracy, ψ_j may not return to itself as φ_1, or φ_2 is increased by a flux quantum $\frac{h}{e}$. As a matter of fact, we have

$$\psi_j\left(\varphi_1 + \frac{h}{e}, \varphi_2\right) = \sum_{k=1}^{3} U_{jk}(\varphi_1, \varphi_2)\psi_k(\varphi_1, \varphi_2) \,.$$

$$\psi_j\left(\varphi_1, \varphi_2 + \frac{h}{e}\right) = \sum_{k=1}^{3} V_{jk}(\varphi_1, \varphi_2)\psi_k(\varphi_1, \varphi_2) \,, \tag{10.47}$$

where U_{jk} and V_{jk} are 3×3 unitary matrices. The sum of the contributions from the states is quantized as integers although the contribution from each state is quantized as integers. This means that $\sigma_{\text{H}} = ne^2/(3h)$, where n is an integer. For a basis in which U_{jk} is diagonal, the structure of V_{jk} is such that ψ_j becomes ψ_{j+1}, except for a possible phase factor when φ_2 is increased by h/e. We then have

$$\sigma_{\text{H}} = \frac{e^2}{3h}\frac{i}{2\pi}\int_0^{h/e} d\varphi_1 \int_0^{h/e} d\varphi_2 \times \left[\left\langle \frac{\partial \psi_1}{\partial \varphi_1}\middle| \frac{\partial \psi_1}{\partial \varphi_2}\right\rangle - \left\langle \frac{\partial \psi_1}{\partial \varphi_2}\middle| \frac{\partial \psi_1}{\partial \varphi_1}\right\rangle\right]. \tag{10.48}$$

This can be shown as an integer multiple of $e^2/(3h)$ by using arguments similar to those for the integer QHE.

Tao and Haldane [7], and Wen and Niu [8] demonstrated that the ground state degeneracy is not lifted by weak impurity scattering in the thermodynamic limit. Therefore, the above argument is still valid in the presence of impurities. Furthermore, when the electron density is slightly different from a rational fraction p/q of that of a Landau level, the extra density of particles is believed to exist as quasi-particles or quasi-holes that are bound to the impurities. The many-particle system still has a q-fold ground state degeneracy with an energy separated from the excited states by a finite gap. This explains the fractional plateaus in the Hall conductance as a function of the magnetic field or gate voltage. However, numerical calculations on a spherical geometry by Haldane [5] and Haldane and Rezayi [9] have not shown a ground state degeneracy. This myster2y was clarified by Wen and Niu using a Ginzburg–Landau formalism of the quantum Hall effect.

10.3.2 Ginzburg–Landau Theory of the Quantum Hall Effect

Girvin and MacDonald noticed many similarities between the theory for the FQHE and those for superfluids and superconductors. In particular, they found a long-range off-diagonal order in the bosonized Laughlin state

$$\psi_L^B(z_1, z_2, \ldots, z_N) = C\prod_{i<j}^{N}|z_i - z_j|^m \exp\left(-\sum_{i=1}^{N}|z_i|^2\right), \tag{10.49}$$

which is the absolute value of the Laughlin wave function. They found that the density matrix

$$\rho(z,z') = N \int dz_2 \ldots \int dz_N \times \psi_L^B(z, z_2, \ldots, z_N)\psi_L^B(z', z_2, \ldots, z_N) \tag{10.50}$$

has a long-range behavior

$$\rho(z,z') \propto \left|z - z'\right|^{-\frac{m}{2}} \quad \text{as } \left|z - z'\right| \to \infty \,. \tag{10.51}$$

The density matrix of the Laughlin wave function decays as a Gaussian at large distances. Consequently, it was concluded that a bosonized Ginzburg–Landau theory was in order.

The Hamiltonian for the interacting electron system is

$$H = \int d^2r\psi^\dagger(\mathbf{r})\left[\frac{1}{2m^*}(-i\hbar\nabla - e\mathbf{A})^2 + eA^0\right]\psi(\mathbf{r}) + \frac{1}{2}\int d^2\mathbf{r}\int d^2\mathbf{r}'\psi^\dagger(\mathbf{r})\,\psi^\dagger(\mathbf{r}')V(\mathbf{r}-\mathbf{r}')\psi(\mathbf{r}')\psi(\mathbf{r})\,, \tag{10.52}$$

where $\mathbf{A}$ is the vector potential, A^0 is the scalar potential, and $V(\mathbf{r} - \mathbf{r}')$ is the interacting potential energy. The fermion field operators satisfy

$$\psi(\mathbf{r})\,\psi^\dagger(\mathbf{r}') + \psi^\dagger(\mathbf{r}')\psi(\mathbf{r}) = \delta(\mathbf{r} - \mathbf{r}') \,. \tag{10.53}$$

The field $\psi(\mathbf{r})$ may be bosonized by a singular gauge transformation:

$$\phi(\mathbf{r}) = \exp\left[-iq\int d^2\mathbf{r}'\theta(\mathbf{r}-\mathbf{r}')n(\mathbf{r}')\right]\psi(\mathbf{r})\,, \tag{10.54}$$

where q is an odd integer, $\theta(\mathbf{r})$ is the angle that $\mathbf{r}$ makes with the real axis, and $n(\mathbf{r}) = \psi^\dagger(\mathbf{r})\psi(\mathbf{r}) = \phi^\dagger(\mathbf{r})\phi(\mathbf{r})$ is the number density operator. It can be explicitly shown that

$$\phi(\mathbf{r}')\,\phi^\dagger(\mathbf{r}) - \phi^\dagger(\mathbf{r})\,\phi(\mathbf{r}') = \delta(\mathbf{r} - \mathbf{r}')\,. \tag{10.55}$$

The Hamiltonian then becomes

$$H[\phi, \mathbf{a}] = \int d^2\mathbf{r}\phi^\dagger(\mathbf{r})\left[\frac{1}{2m^*}(-i\hbar\nabla - e\mathbf{A} - e\mathbf{a})^2 + eA^0\right]\phi(\mathbf{r}) + \frac{1}{2}\int d^2\mathbf{r}\int d^2\mathbf{r}'\phi^\dagger(\mathbf{r})\,\phi^\dagger(\mathbf{r}')\,V(\mathbf{r}-\mathbf{r}')\,\phi(\mathbf{r}')\,\phi(\mathbf{r})\,, \tag{10.56}$$

where

$$\mathbf{a}(\mathbf{r}) = -\frac{\hbar q}{e}\int d^2\mathbf{r}'\nabla\theta(\mathbf{r}-\mathbf{r}')n(\mathbf{r}') = -\frac{\hbar q}{e}\int d^2\mathbf{r}'\frac{\hat{z}\times(\mathbf{r}-\mathbf{r}')}{|\mathbf{r}-\mathbf{r}'|^2}n(\mathbf{r}')\,. \tag{10.57}$$

The gauge transformation is singular because $\theta(\boldsymbol{r}-\boldsymbol{r}')$ is a multi-valued function. However, the exponential in the gauge transformation is well-defined because $\int d^2\boldsymbol{r}'\ n(\boldsymbol{r}')$ is an integer. Closely related to the multi-value of $\theta(\boldsymbol{r}-\boldsymbol{r}')$ is the fact that $\boldsymbol{a}(\boldsymbol{r})$ has a non-zero curl, that is,

$$\nabla\times\boldsymbol{a}(\boldsymbol{r}) = -\frac{hq}{e}n(\boldsymbol{r})\,\hat{z}\,. \tag{10.58}$$

If the density is $1/q$ of the Landau level capacity, that is,

$$n(\boldsymbol{r}) = \frac{1}{q}\frac{eB}{h}\,, \tag{10.59}$$

it follows that the total gauge field $\nabla\times(\boldsymbol{a}+\boldsymbol{A})$ is zero. Under this condition, Bose condensation can occur and $\phi(\boldsymbol{r})$ acquires a non-zero expectation value. It is a reasonable approximation to replace $\phi(\boldsymbol{r})$ by a classical field $\overline{\phi}(\boldsymbol{r})$ in the long wavelength and low-frequency limit.

A natural way of deriving an effective classical theory is with the use of the path integral method of quantum field theory in which the central object is the amplitude

$$Z[A^\mu] = \int[d\phi]e^{\frac{i}{\hbar}\int dt\mathcal{L}[\phi]}\,, \tag{10.60}$$

where

$$\mathcal{L}[\phi] = \int d^2\boldsymbol{r}\left[\phi^* i\hbar\frac{\partial}{\partial t}\phi + \mu\phi^*\phi\right] - H[\phi,\boldsymbol{a}]\,. \tag{10.61}$$

In this approximation, ϕ is treated as a classical field, and $[d\phi]$ means that the integration is carried out over all possible values of the field $\phi(\boldsymbol{r},t)$. A term which depends on the chemical potential (denoted by μ) has been added in the Lagrangian as a constraint on the number of particles. The quantity $\boldsymbol{a}$ in $H[\phi,\boldsymbol{a}]$ is an explicit function of ϕ. However, we may treat $\boldsymbol{a}$ as an independent field if we introduce an auxiliary field $a_0(\boldsymbol{r})$ as follows:

$$Z[A^\mu] = \int[d\phi][d\boldsymbol{a}][da_0]e^{\frac{i}{\hbar}\int dt\mathcal{L}[\phi,\boldsymbol{a},a_0]}\,, \tag{10.62}$$

where

$$\begin{aligned}\mathcal{L}[\phi,\boldsymbol{a},a_0] = &\int d^2\boldsymbol{r}\left[\phi^* i\hbar\frac{\partial}{\partial t}\phi + \mu\phi^*\phi\right] - H[\phi,\boldsymbol{a}]\\ &-\int d^2\boldsymbol{r}a_0(\boldsymbol{r})\left[e\phi^*\phi + \frac{e^2}{hq}\nabla\times\boldsymbol{a}(\boldsymbol{r})\right].\end{aligned} \tag{10.63}$$

We assumed that $\nabla\cdot\boldsymbol{a}=0$ so that the constraint $e\phi^*\phi + \frac{e^2}{hq}\nabla\times\boldsymbol{a}(\boldsymbol{r}) = 0$, which is a consequence of the integration over $a_0(\boldsymbol{r})$, leading us to the original explicit expression for $\boldsymbol{a}$ in terms of ϕ.

The above path integral is formally equivalent to that for the matter field ϕ interacting with the gauge fields $(\boldsymbol{a}, a_0)$. The equation $\nabla \cdot \boldsymbol{a} = 0$ is no more than a gauge-fixing condition that is needed when that path integration quantizes the fields. If we choose a different gauge, such as $a_0 = 0$, and replace ϕ with $\tilde{\phi} \exp\left(-\frac{ie}{\hbar} \int dt \tilde{a}_0\right)$, the Lagrangian becomes

$$\mathcal{L}[\tilde{\phi}, \tilde{\boldsymbol{a}}, \tilde{a}_0] = \int d^2\boldsymbol{r} \left[\tilde{\phi}^* \left(i\hbar \frac{\partial}{\partial t} - eA^0 \right) \tilde{\phi} + \mu \tilde{\phi}^* \tilde{\phi} \right] - H\left[\tilde{\phi}, \tilde{\boldsymbol{a}}\right]$$
$$- \frac{1}{2} \frac{e^2}{hq} \int d^2\boldsymbol{r} \tilde{a}_0(\boldsymbol{r})\, \hat{z} \cdot [\nabla \times \tilde{\boldsymbol{a}}(\boldsymbol{r})] , \qquad (10.64)$$

where $\tilde{\boldsymbol{a}}(\boldsymbol{r}) = \boldsymbol{a}(\boldsymbol{r}) - \int dt \nabla \tilde{a}_0$, and $a_0(\boldsymbol{r}) = 0$.

The quantum amplitude may be written in terms of the new fields as

$$Z[A^\mu] = \int [d\tilde{\phi}][d\tilde{\boldsymbol{a}}][d\tilde{a}_0]\, e^{\frac{i}{\hbar} \int dt \mathcal{L}[\tilde{\phi}, \tilde{\boldsymbol{a}}, \tilde{a}_0]} . \qquad (10.65)$$

There is no constraint in these expressions. The purpose of the effective classical field theory is to evaluate the path integral by taking the extremal of the action. The Ginzburg–Landau equations then become the equations for $(\tilde{\phi}, \tilde{\boldsymbol{a}})$ that minimize the action. Specifically, we have

$$\left(i\hbar \frac{\partial}{\partial t} - eA^0 \right) \tilde{\phi}(\boldsymbol{r}) + \mu \tilde{\phi}(\boldsymbol{r}) - \frac{1}{2m^*} [-i\hbar\nabla - e\boldsymbol{A} - e\tilde{\boldsymbol{a}}]^2 \tilde{\phi}(\boldsymbol{r})$$
$$- \left[\int d^2\boldsymbol{r}' \tilde{\phi}^*(\boldsymbol{r}') V(\boldsymbol{r} - \boldsymbol{r}') \tilde{\phi}(\boldsymbol{r}') \right] \tilde{\phi}(\boldsymbol{r}) = 0 . \qquad (10.66)$$

There may be some important quantum fluctuations around these classical solutions. The effect of these fluctuations is to renormalize parameters, such as the interaction parameter, in these equations.

10.4 Problems

1. Show that when $m = 1$ in Eq. (10.13), the wave function ψ is simply a determinant

$$\text{Det}\left\{ z_i^{j-1} \right\}_{N \times N} \exp\left(- \sum_{i=1}^{N} |z_i|^2 \right) .$$

2. Let the electron density be $\rho(r) = -\rho_0$ for $r < r_c$ and $\rho(r) = 0$ for $r > r_c$. Show that the potential

$$V(\boldsymbol{r}) = - \int d^2\boldsymbol{r}' \ln|\boldsymbol{r} - \boldsymbol{r}'| \rho(\boldsymbol{r}')$$

is given by

$$V(\boldsymbol{r}) = \frac{\pi}{2} \begin{cases} \rho_0 r^2 & \text{for r} < \text{r}_c \\ 2\rho_0 r_c^2 \left[\frac{1}{2} + \ln\left(\frac{r}{r_c}\right) \right] & \text{for r} > \text{r}_c \end{cases} . \qquad (10.67)$$

3. Show that $\psi(z_1, z_2, \ldots, z_N)$ in Eq. (10.26) is the gauge transformation of Laughlin's state in Eq. (10.13).

4. Show that ψ_z in Eq. (10.36) has a hole of charge $-1/m$ at z.

References

1 von Klitzing, K., Dorda, G., and Pepper, M. (1980) New method for high-accuracy determination of the fine-structure constant based on quantized Hall resistance. *Phys. Rev. Lett.*, **45**, 494.

2 Tsui, D.C., Störmer, H.L., and Gossard, A.C. (1982) Two-dimensional magnetotransport in the extreme quantum limit. *Phys. Rev. Lett.*, **48**, 1559.

3 Laughlin, R.B. (1984) Anomalous quantum Hall effect: An incompressible quantum fluid with fractionally charged excitations. *Phys. Rev. Lett.*, **50**, 1395.

4 Tao, R. and Wu, Y.S. (1984) Gauge invariance and fractional quantum Hall effect. *Phys. Rev. B*, **30**, 1097.

5 Haldane, F.D.M. (1983) Fractional quantization of the Hall effect: A hierarchy of incompressible quantum fluid states. *Phys. Rev. Lett.*, **51**, 605.

6 Su, P. (1984) Ground-state degeneracy and fractionally charged excitations in the anomalous quantum Hall effect. *Phys. Rev. B*, **30**, 1069.

7 Tao, R. and Haldane, F.D.M. (1986) Impurity effect, degeneracy, and topological invariant in the quantum Hall effect. *Phys. Rev. B*, **33**, 3844.

8 Wen, X.G. and Niu, Q. (1990) Ground-state degeneracy of the fractional quantum Hall states in the presence of a random potential and on high-genus Riemann surfaces. *Phys. Rev. B*, **41**, 9377.

9 Haldane, D.M. and Rezayi, E.H. (1985) *Phys. Rev. Lett.*, **54**, 237.

11
Quantized Adiabatic Charge Transport in 2D Electron Systems and Nanotubes

11.1
Introduction

A considerable amount of effort has been devoted to the study and successful operation of devices that employ the principles of quantized adiabatic charge transport [1–11]. The goal is to produce a device capable of delivering N electrons (or holes) in each cycle of a moving quantum dot in a controlling signal, thus giving rise to a current $I = Nef$, where f is the frequency of the signal [3–8]. However, thus far, only the surface acoustic wave (SAW) on a piezoelectric substrate [3, 4] and the recently developed GHz single-electron pumps [8, 9] operate at high enough frequencies for the measured current to be suitable as a current standard. As of today, the accuracy of the quantized current on the plateaus is one part in 10^4 for the SAW and GHz single-electron pumps [3, 4, 8]. Robinson and Talyanskii [12] measured the noise produced by an approximate 3 GHz SAW single-electron pump. When the current is close to the quantized plateau corresponding to one electron being transported per SAW cycle, the noise in the current is dominated by shot noise. Away from the plateau, that is, either above or below a specified plateau under consideration, the noise is due to electron traps in the material. Various suggestions exist on how to reduce the noise in single-electron pumps, as this would have important applications in setting an alternative and universal means of establishing a charge standard. The flatness of the quantized current plateaus is one part in 10^4 and, as such, makes it a rather crude standard of measurement. One needs to further refine its operation in order for it to be accepted as a standard that seeks an accuracy of one part in 10^8. In spite of this, there are proposals to use the SAW and high frequency GHz single-electron pumps in single photon sources, single photon detectors, quantum cryptography and quantum computation [12–15].

The operation of the single-electron SAW pump is as follows. A negative voltage is applied to a split gate on the surface of a piezoelectric heterostructure such as GaAs/AlGaAs. This forms a narrow depleted channel between two regions of 2DEG and the SAW is launched toward this channel from an interdigital transducer. Because the substrate is piezoelectric, a wave of electrostatic potential accompanies the SAW. The screening of the SAW potential by the metal gate is negligible and may be neglected in a model calculation [6, 7]. Electrons can be captured by the

Properties of Interacting Low-Dimensional Systems, First Edition. G. Gumbs and D. Huang.

SAW from the source 2DEG and transported to the drain 2DEG. We obtain a perfectly flat plateau if the number of electrons transported in each cycle of the signal is held constant. This number of electrons is determined by the electron–electron interaction and the smallest size achieved by the moving quantum dot whose energy levels are calculated from the instantaneous Hamiltonian. There are ranges of gate voltage where this minimum sized dot is not changed appreciably, leading to quantized plateaus for the current at $I = Nef$. In the acoustoelectric current measurements reported to this point, the plateaus have a slope of one part in 10^4. The many-particle interactions, the variations of the size of the quantum dot minimum on the plateau and the shot noise, that is, the statistical fluctuations in the occupation of the dot, all make this problem very challenging to work with, theoretically. In this chapter, we address the many-body physics in a simple model and make various suggestions to reduce the shot noise in the measured current.

Additional to the turnstile and single-electron pumps, Thouless [10] suggested quantized adiabatic charge transport that involves the utilization of a one-dimensional (1D) electron system under the influence of a periodic potential. It is assumed that the periodic potential varies slowly in time and with the aid of a gate, the Fermi level can be varied so that when it lies within a minigap of the instantaneous Bloch Hamiltonian, there will be an integral multiplicity of charge Ne, transported across the system during a single time period [11]. If realized experimentally, such a device could also provide an important application as a current standard. We will study the mechanisms of quantized adiabatic charge transport in carbon nanotubes using SAWs, as suggested by Talyanskii, *et al.* [11]. Also, we will consider this principle when the electron density in a two-dimensional (2D) graphene sheet is modulated by a SAW in the presence of a parallel magnetic field through the cross-section of the nanotube. The occurrence and reduction of noise in the quantized currents will also be addressed

11.2
Theory for Current Quantization

Let us assume that a narrow channel is formed in a 2D electron gas (EG) lying in the xy-plane. Neglect the finite thickness of the quantum well at the heterointerface in the z-direction and consider the electron motion as strictly 2D. We will employ one of the simplest models for the gate-induced or etched [16] confining electrostatic potential which forms the channel in the 2DEG and take it as

$$V_g(x, y) = \frac{V_0}{\cosh^2 \frac{x}{a}} + \frac{1}{2} m^* \Omega^2 y^2 , \qquad (11.1)$$

where the x-axis is chosen along the channel and the y-axis is across the channel. The parameter V_0 in Eq. (11.1) determines the effective height of the 1D potential barrier which arises in the quasi-1D channel because of the applied gate voltage. The more negative the gate voltage becomes, the value of V_0 will be larger. Beyond the pinch-off voltage, the energy V_0 is greater than the electron Fermi energy in the wide 2D region. The parameter a determines the effective length of the channel, $\ell_{\text{eff}} = 2a$, in the x-direction. Experimentally, the SAW wavelength, λ, is comparable to the length of the channel. Thus, in our numerical calculations we set $a = \lambda/2$ and the electrostatic potential assumes a parabolic profile in the y-direction. This simple approximation is reasonable near the bottom of the well for a narrow channel. The parameter $\ell_0 = \sqrt{\hbar/m^* \Omega}$ may be interpreted as an effective width of the channel. In our numerical calculations, we chose $\ell_0/a = 4 \times 10^{-2}$, $\lambda = 1.0$ μm and $m^* = 0.067\, m_e$, appropriate for GaAs.

The SAW launched on the surface of the GaAs heterostructure along the x-axis ([011] crystallographic direction) is accompanied by the piezoelectric potential sliding along the surface with the speed of sound v_s. The piezoelectric potential accompanying the SAW was previously calculated [6, 7]. If the 2D layer is located at a distance d below the GaAs surface, then the additional SAW induced potential affecting electron motion has the form

$$V_{\text{SAW}}(x,t) = \frac{8\pi e_{14} C}{\epsilon_b} \left[A_1 e^{-\kappa q d} + A_1^* e^{-\kappa q^* d} + A_2 e^{-\kappa d} \right] e^{i\kappa(x - v_s t)} , \quad (11.2)$$

where C is the amplitude of the SAW and κ is its wave vector, e_{14} and ϵ_b are the piezoelectric modulus and dielectric constant of AlGaAs, respectively. The dimensionless coefficients $A_{1,2}$ and q are determined by the elastic constants of the host material [6, 7]. Our numerical calculations were done for $Al_xGa_{1-x}As$ ($x = 0.3$) for which non-zero elastic constants $c_{11}(x) = 11.88 + 0.14x$, $c_{12}(x) = 5.38 + 0.32x$ and $c_{44}(x) = 5.94 - 0.05x$ in units of 10^{10} N/m^2. For this density, $m(x) = (5.36 - 1.6x) \times 10^3$ kg/m^3, which gives $v_s = 2981$ m/s, $A_1 = 1.446 + 0.512i$, $A_2 = -1.524 - 2.654i$ and $q = 0.498 + 0.482i$. In Eq. (11.2), we neglected the screening effect of the gate electrodes on the piezoelectric potential. As shown in [6], the screening reduces the amplitude of the piezoelectric potential but has little effect on its shape. The SAW-induced potential in Eq. (11.2), superimposed on the gate potential in Eq. (11.1), leads to the formation of local quantum wells in the x-direction. Taking into account the confinement in the y-direction, one can conclude that local quantum dots are formed in the channel. Statistical fluctuations in the occupancy of successive quantum dots during the course of one cycle would lead to shot noise. We now present a formalism for calculating the tunneling probability through a moving quantum dot.

Let us consider two electrons in the quantum dot formed by the SAW-induced potential in a narrow electron channel. The Hamiltonian for two interacting electrons takes the form

$$\begin{aligned} H &= -\frac{\hbar^2}{2m^*} \left(\nabla_1^2 + \nabla_2^2 \right) \\ &\quad + V_1(x_1) + V_1(x_2) + V_2(y_1) + V_2(y_2) + V_3(z_1) + V_3(z_2) \\ &\quad + \frac{e^2}{4\pi\epsilon_0\epsilon_b \sqrt{(x_1 - x_2)^2 + (y_1 - y_2)^2 + (z_1 - z_2)^2}} \\ &\equiv H_0 + H_1 , \end{aligned} \quad (11.3)$$

where H_0 describes two electrons, each with effective mass m^*, in the quantum dot without interaction and H_1 is the interaction term. Here, V_i $(i = 1, 2, 3)$ are potentials in the x-, y- and z-directions, respectively, described above. The Hamiltonian in Eq. (11.3) does not include any spin-dependent effects and is therefore diagonal in the spin variables of both electrons. In the absence of electron–electron interactions, a single-electron problem for the Hamiltonian H_0 can be easily solved. The single-electron wave functions diagonalizing the Hamiltonian H_0 take the form

$$\langle \boldsymbol{r}|i, \alpha\rangle = \Phi_{i,\alpha}(x, y, z, \sigma) = \varphi_i(x)\psi(y)Z(z)\chi_\alpha(\sigma) . \tag{11.4}$$

In this notation, $\varphi_i(x)$ is the localized wave function in the x-direction and the subscript $i = 1, 2, 3$ labels the quantized energy levels in the SAW-induced quantum well. Also, $\psi(y)$ is a localized wave function in the y-direction and we assume that the electrons will be in the lowest subband in the transverse direction, that is, $\psi(y)$ is the ground state wave function of a 1D harmonic oscillator and $|Z(z)|^2 = \delta(z)$. Here, $\chi(\sigma)$ is the spin component of the wave function with σ representing the spin, and $\alpha =\uparrow$ or $\downarrow$ is the spin eigenvalue.

The wave function in Eq. (11.4) can be used to construct two-electron basis functions in the form of a Slater determinant, namely,

$$\begin{aligned}&\langle \boldsymbol{r}_1\boldsymbol{r}_2|i_1, \alpha_1; i_2, \alpha_2\rangle \\ &= \frac{1}{\sqrt{2}}\left[\Phi_{i_1,\alpha_1}(\boldsymbol{r}_1, \sigma_1)\Phi_{i_2,\alpha_2}(\boldsymbol{r}_2, \sigma_2) - \Phi_{i_1,\alpha_1}(\boldsymbol{r}_2, \sigma_2)\Phi_{i_2,\alpha_2}(\boldsymbol{r}_1, \sigma_1)\right] .\end{aligned} \tag{11.5}$$

The wave functions in Eq. (11.5) are normalized and orthogonal to each other. These functions form a complete set if the indices i_1, i_2 take all possible values, both discrete and continuous. From Eq. (11.4), quantum number i labels the discrete eigenstates in the 1D quantum well along the x-axis. This index may also change continuously if the energy level lies above the top of the confining potential. In general, when the Hamiltonian H in Eq. (11.3) is written in terms of the representation of functions in Eq. (11.5), we have an infinite-order matrix. However, the problem can be simplified if we exclude the continuous spectrum from our consideration and assume that the number of discrete energy levels within the quantum well is finite. In general, if $i = 1, \cdots, n$, then the number of independent basis functions in Eq. (11.5) is $n(2n - 1)$. The Hamiltonian in Eq. (11.3), written in the basis representation of the functions in Eq. (11.5), is an $n \times n$ matrix that then must be diagonalized to get the energy eigenvalues of the two-electron problem. Calculation of the matrix elements of H with wave functions in Eq. (11.5) is straightforward and can be easily accomplished if we take into consideration that the Hamiltonian is diagonal in the spin variables and the single-electron functions in Eq. (11.4) are eigenfunctions of the unperturbed Hamiltonian H_0.

If there are just two discrete single electron levels in the well, that is, $i = 1, 2$, then the following six functions in Eq. (11.5) form the basis set

$$|1\uparrow; 2\uparrow\rangle , \quad |1\downarrow; 2\downarrow\rangle , \quad |1\uparrow; 1\downarrow\rangle , \quad |1\uparrow; 2\downarrow\rangle , \quad |2\uparrow; 1\downarrow\rangle , \quad |2\uparrow; 2\downarrow\rangle . \tag{11.6}$$

In this case, the Hamiltonian is a 6×6 matrix having the form [17]

$$\hat{H} = \begin{pmatrix} E_1 & 0 & 0 & 0 & 0 & 0 \\ 0 & E_1 & 0 & 0 & 0 & 0 \\ 0 & 0 & & & & \\ 0 & 0 & & \mathbf{A}_{4\times 4} & & \\ 0 & 0 & & & & \\ 0 & 0 & & & & \end{pmatrix}, \tag{11.7}$$

where $E_1 \equiv \varepsilon_1 + \varepsilon_2 + H^{(1)}_{12;12} - H^{(1)}_{12;21}$ with $H^{(1)}_{mn;kl} = \langle m|\langle n|H_1|k\rangle|l\rangle$ and the matrix $\mathbf{A}_{4\times 4}$ is given by

$$\mathbf{A}_{4\times 4} = \begin{pmatrix} 2\varepsilon_1 + H^{(1)}_{11;11} & H^{(1)}_{11;12} & H^{(1)}_{11;21} & H^{(1)}_{11;22} \\ H^{(1)}_{12;11} & \varepsilon_1 + \varepsilon_2 + H^{(1)}_{12;12} & H^{(1)}_{12;21} & H^{(1)}_{12;22} \\ H^{(1)}_{21;11} & H^{(1)}_{21;12} & \varepsilon_1 + \varepsilon_2 + H^{(1)}_{21;21} & H^{(1)}_{21;22} \\ H^{(1)}_{22;11} & H^{(1)}_{22;12} & H^{(1)}_{22;21} & 2\varepsilon_2 + H^{(1)}_{22;22} \end{pmatrix}. \tag{11.8}$$

In Eqs. (11.7) and (11.8), ε_1 and ε_2 are the energy eigenvalues of the ground and first excited states for a single electron in the SAW-induced quantum well and $|i\rangle$ is defined in Eq. (11.2). Following from Eqs. (11.7) and (11.8), the Hamiltonian $\hat{H}$ is a block matrix. The two elements along the diagonal in the first and second rows correspond to the quantum states where both electrons have spin up or spin down, thus the total spin $S = 1$ and $S_z = \pm 1$. The energies of these are equal because the total Hamiltonian does not depend on the total spin orientation. The matrix $\mathbf{A}_{4\times 4}$ describes the states where two electrons have opposite orientations, that is, $S = 0$, $S_z = 0$ or $S = 1$, $S_z = 0$. The energy eigenvalues will correspond to spin singlet or spin triplet states.

11.3 Tunneling Probability and Current Quantization for Interacting Two-Electron System

Suppose there are two interacting electrons localized in the moving quantum dot. We Now calculate the probability for one of these electrons escaping while the other one remains in the well. This problem can be solved if we can determine the effective single electron potential for each electron and then we can use the procedure for single electron tunneling. We can accomplished this by using a simplified version of the density functional method if the two interacting electrons in the ground state have opposite spin.

11.3.1
Spin Unpolarized Case

Let us denote the wave function describing two interacting electrons in the well with energy E and coordinates x_1 and x_2 in the ground state by $\Psi(x_1, x_2)$. The electron density in the well is given by

$$\rho(x) = 2 \int_{-\infty}^{\infty} dx_1 \, |\Psi(x, x_1)|^2 . \tag{11.9}$$

Making use of the single electron description for non-interacting electrons, each electron has the same energy $\varepsilon = E/2$ and wave function $\varphi(x)$ since they have opposite spin. Because the wave function $\varphi(x)$ gives the same electron density distribution as the two-electron wave function $\Psi(x_1, x_2)$, we have $|\varphi(x)|^2 = N(x)/2$ where $\varphi(x)$ is a solution of the one electron Schrödinger equation with effective electric potential

$$V_{\text{eff}}(x) = \varepsilon + \frac{\hbar^2}{2m^*} \frac{1}{\varphi(x)} \frac{d^2\varphi(x)}{dx^2} . \tag{11.10}$$

Now, we can calculate the probability of tunneling as we have previously done [6, 7]. When two electrons are captured in the local SAW induced quantum well, they can escape from the well in two possible ways. First, one of electrons can tunnel back to the source and the probability of the tunneling $P_2(\tau, \beta)$ can be calculated for arbitrary τ and β using

$$P_2(\tau, \beta) = \exp\left(-\frac{1}{\hbar} \int_{X_1}^{X_2} dx \sqrt{2m^*[V_{\text{eff}}(x) - \varepsilon]} \right) , \tag{11.11}$$

where $\varepsilon \equiv E_0/2$ with E_0 equal to the ground state energy of the two-electron system, $0 \le \tau \le 2\pi$, and $\beta = 8\pi e_{14} C/\varepsilon_b V_0$. When this electron (electron 2, say) escapes, the potential profile returns to its original shape $V_c(x)$ and the remaining electron (electron 1, say) is in its ground state with energy ε_0. The probability for this electron to tunnel out of the well to the source is $P_1(\tau, \beta) P_2(\tau, \beta)$ where

$$P_1(\tau, \beta) = \exp\left(-\frac{1}{\hbar} \int_{X_1}^{X_2} dx \sqrt{2m^*[V_c(x) - \varepsilon_0]} \right) . \tag{11.12}$$

This calculation is based on the assumption that two electrons never tunnel out of the well simultaneously, but always one after another. The total probability for electron 2 to escape from the well in one SAW cycle is

$$P_{2\text{T}}(\beta) = \int_0^{2\pi} d\tau \, P_2(\tau, \beta) , \tag{11.13}$$

whereas the corresponding probability for electron 1 is

$$P_{1T}(\beta) = \int_0^{2\pi} d\tau\, P_1(\tau,\beta)\, P_2(\tau,\beta)\,. \tag{11.14}$$

As a function of β, the acoustoelectric current is given by

$$I(\beta) = \left[1 - P_{1T}(\beta)\right] e f + \left[1 - P_{2T}(\beta)\right] e f\,. \tag{11.15}$$

We solved the single-particle problem numerically and obtained the wave functions $\varphi_i(x)$ for each value of τ in the interval $0 \le \tau \le 2\pi$ and different values of the dimensionless parameter β, which is the ratio of the SAW potential amplitude to the height of the gate induced potential barrier in the channel. The probability distributions for the two lowest eigenstates, along with the ground state energy ε_0 for one electron captured in the dot, are shown in Figure 11.1a. Using the procedure described above, the ground state energies E_0 and the x components of the ground state eigenfunctions $\Psi(x_1, x_2)$ were found numerically for each value of

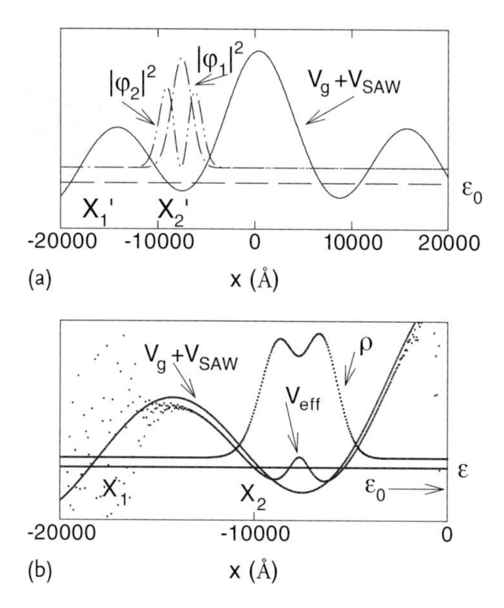

Figure 11.1 (a) Schematic plots for fixed β and τ of the probability distributions for the ground (φ_1) and first excited (φ_2) states for one electron captured in the local quantum dot and the potential $V_c = V_g + V_{SAW}$ in the channel. The ground state energy ε_0 is shown as a horizontal line. The boundaries of the classically forbidden region under the barrier at energy ε_0 are denoted by X_1' and X_2'. (b) The effective potential $V_{eff}(x)$ and the density distribution ρ for a pair of interacting electrons are shown. For comparison, $V_g + V_{SAW}$ is plotted. The horizontal line is $\varepsilon = E_0/2$ where E_0 is the ground state energy for the two-electron system. The boundaries of the classically forbidden region under the barrier at energy ε are denoted by X_1 and X_2; the position of ε_0 is also indicated.

τ with $0 \leq \tau \leq 2\pi$ and for different values of β in the range $0 < \beta < 2.0$. An example of the electron density distribution $\rho(x) = 2\int_{-\infty}^{\infty} dx_1 |\Psi(x, x_1)|^2$ within the local well in the x-direction for fixed β and τ is shown in Figure 11.1b.

11.4 Adiabatic Charge Transport in Carbon Nanotubes

An intrinsic carbon nanotube has a linear band structure for the valence and conduction bands that intersect at the Fermi level ($E_F = 0$). When a magnetic field is applied, minigaps open close to the Fermi level and is simulated through a parameter Δ. The chemical potential can be aligned with one of the minigaps with the use of a gate or via doping. Talyanskii, *et al.* [11] employed a magnetic field-dependent Hamiltonian to study the effects of electrostatic modulation due to a SAW on the band structure of a carbon nanotube. The SAW velocity $v_s \approx 3 \times 10^3$ m/s is much less than $v_F \approx 10^5$ m/s. Thus, the 1D energy spectrum can be treated in a stationary/adiabatic approximation. In this approximation, the low-lying eigenstates are solutions of a Dirac equation for massless particles [11]

$$\hat{H} = -i\hbar v_F \sigma_3 \frac{d}{dx} + \sigma_1 \Delta + A\sin(kx) , \tag{11.16}$$

where σ_1 and σ_3 are Pauli matrices. The amplitude of the SAW is denoted by A, and $2\pi/k$ is its wavelength. The Fermi velocity is denoted by v_F. Since the wavelength of the SAW $2\pi/k$ (~ 1 μm) is much larger than the diameter (~ 1 nm) of the nanotube, we are justified in ignoring the variation of the SAW potential over the cross-section of the nanotube. The role played by a parallel magnetic field will produces a gap near the Fermi level in the energy bands of the nanotube, denoted by the parameter Δ, couples the two components of the spinor wave function. Writing the two-component wave function as

$$\begin{bmatrix} \psi_1(x) \\ \psi_2(x) \end{bmatrix} = \exp\left[\tfrac{i}{2}\sigma_3 \lambda_k \cos(kx)\right] \begin{bmatrix} \Psi_1(x) \\ \Psi_2(x) \end{bmatrix} \tag{11.17}$$

with $\lambda_k = (2A)/(\hbar v_F k)$, we obtain the following pair of coupled ordinary differential equations when the Hamiltonian in Eq. (11.16) operates on the column vector in Eq. (11.17)

$$\begin{aligned} &-i\hbar v_F \frac{d}{dx}\Psi_1(x) + \Delta e^{-i\lambda_k \cos(kx)}\Psi_2(x) = \varepsilon \Psi_1(x) \\ &i\hbar v_F \frac{d}{dx}\Psi_2(x) + \Delta e^{i\lambda_k \cos(kx)}\Psi_1(x) = \varepsilon \Psi_2(x) . \end{aligned} \tag{11.18}$$

This Eq. (11.18) was solved numerically and the energy bands have minigaps between which the Fermi energy can lie to produce current quantization.

In Figure 11.2, we present our results obtained by solving Eq. (11.18) for ε/E_α as a function of A/E_α for fixed $\Delta/E_\alpha = 0.6$. Here, energies are measured in units

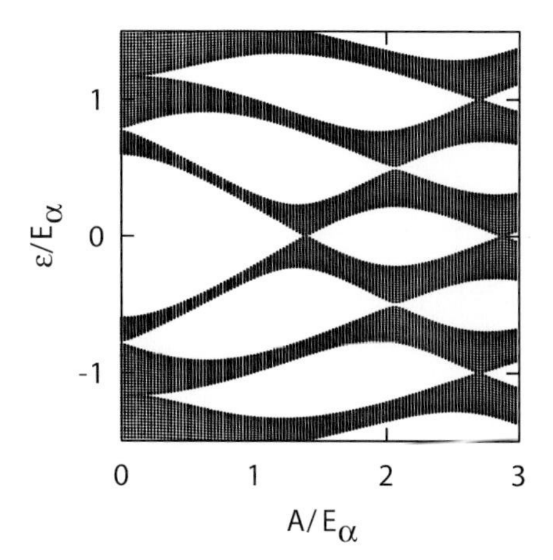

Figure 11.2 Electron energy bands obtained by solving Eq. (11.18) as a function of the SAW field A/E_α in units of $E_\alpha = \hbar v_F k$ with $2\pi/k$ being the SAW wavelength and v_F the Fermi velocity. We chose $\Delta/E_\alpha = 0.6$ for the parameter representing the effect of a parallel magnetic field on the electronic band structure of the nanotube.

of $E_\alpha = \hbar v_F k$. In Figure 11.3, these calculations were repeated for fixed $\Delta/E_\alpha = 1.2$. Figure 11.4 is a plot of the energy spectrum as a function of Δ/E_α for fixed $A/E_\alpha = 0.8$. Shown in Figures 11.2 through 11.4 is that minigaps are formed in the energy spectrum. These minigaps oscillate as a function of the SAW amplitude A and vanish at specific values that are determined by the choice of the magnetic field parameter Δ. The widths of the minibands are reduced as Δ is increased. Additionally, Figure 11.4 demonstrates that there are no minigaps when $\Delta = 0$. Since we have quantized charge transport when the chemical potential falls within one of the minigaps, the widths of the plateaus depend on the SAW power and the magnetic field. The value of the quantized current will remain the same within

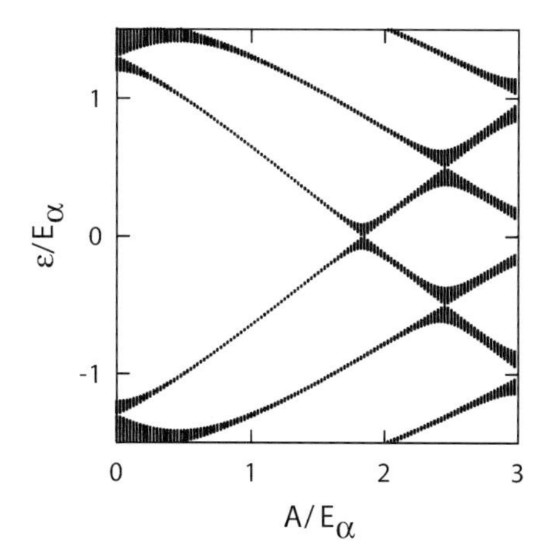

Figure 11.3 This plot was identically produced as that in Figure 11.2, except that we chose $\Delta/E_\alpha = 1.2$.

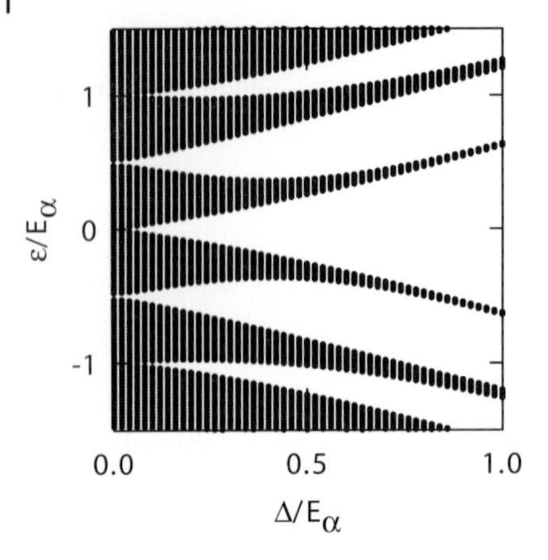

Figure 11.4 Calculated electron energy spectrum for solutions of Eq. (11.18) as a function of the magnetic field parameter $\varDelta$ in units of the energy parameter $E_\alpha = \hbar v_F k$. The SAW field is chosen as $A/E_\alpha = 0.8$.

a whole range of values of $\varDelta$ and A that stay within a gap. The energy bands in Figures 11.2 through 11.4 have $\varepsilon \to -\varepsilon$ electron–hole symmetry. This is due to the symmetry of the Hamiltonian for left and right-traveling carriers when $\varDelta = 0$.

11.5 Summary and Remarks

As stated above, the accuracy of the SAW pumps has to be substantially improved in order to have usable applications, for example, as in electrical metrology To achieve such an improvement will require a reduction of the *shot* noise in the current produced by the SAW or high-frequency single-electron pumps. This means improving the probability of transporting an electron in a cycle.

In a recent paper, Huang, *et al.* [14] considered the overall efficiency for using SAWs in a photon detector. The process would involve photon absorption, the degeneration of a photon into an electron–hole pair and the capture and transport of the electron and hole by the SAW. In our dual-charged-fluid model for the steady-state transport of SAWs, dragged photocurrents of 1D confined-state carriers were considered. This model incorporated the effects of the quantum confinement, the tunneling escape of SAW-dragged 1D carriers, the effects of the inelastic capture of 2D continuous-state carriers and the self-consistent space-charge field. Our results reveal that as a function of SAW power and frequency, as well as temperature, the heavier hole was more likely to be transported during a SAW cycle. Based on this model, the shot noise from the SAW-dragged hole currents should diminish from their experimentally measured levels involving electrons.

As we demonstrated by the minigaps in a 1D conductor, quantized charge transport can be observed in a non-interacting model of electrons. However, for a strongly interacting Lüttinger liquid in which the second-order perturbations are treated as bosons, the current quantization may not be so sharp. Unlike the Fermi liquid's quasiparticles that carry spin and charge, the elementary excitations of the Lüttinger liquid are charge and spin waves for which impurities and other types of mechanisms producing *backscattering* are important. Even at low temperature, the distribution function for the particle momentum does not have a sharp jump. This is in contrast with the Fermi liquid where this jump indicates the Fermi surface. In the momentum-dependent spectral function, there is no "quasiparticle peak" whose width becomes much narrower than the excitation energy above the Fermi level such as what occurs for the Fermi liquid. Instead, there is a power-law singularity with a "non-universal" exponent that depends on the strength of the interaction.

References

1. Kouwenhoven, L.P., Johnson, A.T., van der Vaart, N.C., Harmans, C.J.P.M., and Foxon, C.T. (1991) Quantized current in a quantum-dot turnstile using oscillating tunnel barriers. *Phys. Rev. Lett.*, **67**, 1626.
2. Keller, M.W., Eichenberger, A.L., Martinis, J.M., and Zimmerman, N.M. (1999) A Capacitance standard based on counting electrons. *Science*, **285**, 1706.
3. Shilton, J.M., Talyanskii, V.I., Pepper, M., Ritchie, D.A., Frost, J.E.F., Ford, C.J.B., Smith, C.G., and Jones, G.A.C. (1996) High-frequency single-electron transport in a quasi-one-dimensional GaAs channel induced by surface acoustic waves. *J. Phys.: Condens. Matter*, **8**, L531.
4. Talyanskii, V.I., Shilton, J.M., Pepper, M., Smith, C.G., Ford, C.J.B., Linfield, E.H., Ritchie, D.A., and Jones, G.A.C. (1997) Single-electron transport in a one-dimensional channel by high-frequency surface acoustic waves. *Phys. Rev. B*, **56**, 15180.
5. Azin, G.R., Gumbs, G., and Pepper, M. (1998) Screening of the surface-acoustic-wave potential by a metal gate and the quantization of the acoustoelectric current in a narrow channel. *Phys. Rev. B*, **58**, 10589.
6. Gumbs, G., Azin, G.R., and Pepper, M. (1999) Coulomb interaction of two electrons in the quantum dot formed by the surface acoustic wave in a narrow channel. *Phys. Rev. B*, **60**, R13954.
7. Gumbs, G., Azin, G.R., and Pepper, M. (1998) Interaction of surface acoustic waves with a narrow electron channel in a piezoelectric material. *Phys. Rev. B*, **57**, 1654.
8. Blumenthal, M.D., Kaestner, B., Li, L., Giblin, S., Janssen, T.J.B.M., Pepper, M., Anderson, D., Jones, G., and Ritchie, D.A. (2007) Gigahertz quantized charge pumping. *Nat. Phys.*, **3**, 343.
9. Kaestner, B., Kashcheyevs, V., Amakawa, S., Blumenthal, M.D., Li, L., Janssen, T.J.B.M., Hein, G., Pierz, K., Weimann, T., Siegner, U., and Schumacher, H.W. (2008) Single-parameter nonadiabatic quantized charge pumping. *Phys. Rev. B*, **77**, 153301.
10. Thouless, D.J. (1983) Quantization of particle transport. *Phys. Rev. B*, **27**, 6083.
11. Talyanskii, V.I., Novikov, D.S., Simons, B.D., and Levitov, L.S. (2001) Quantized adiabatic charge transport in a carbon nanotube. *Phys. Rev. Lett.*, **87**, 276802.
12. Robinson, A.M. and Talyanskii, V.I. (2005) Shot Noise in the current of a surface acoustic-wave-driven single-electron pump. *Phys. Rev. Lett.*, **95**, 247202.
13. Gumbs, G. (2008) Semiconductor photodetectors, bio-material sensors and quantum computers using high frequency sound waves. *AIP Conf. Proc.*, **991**, 57.

14 Huang, D.H., Gumbs, G., and Pepper, M. (2008) Effects of inelastic capture, tunneling escape, and quantum confinement on surface acoustic wave-dragged photocurrents in quantum wells. *J. Appl. Phys.*, **103**, 083714.

15 Barnes, C.H.W., Shilton, J.M., and Robinson, A.M. (2000) Quantum computation using electrons trapped by surface acoustic waves. *Phys. Rev. B*, **62**, 8410.

16 Cunningham, J., Talyanskii, V.I., Shilton, J.M., and Pepper, M. (2000) Single-electron acoustic charge transport on shallow-etched channels in a perpendicular magnetic field. *Phys. Rev. B*, **62**, 1564.

17 Gumbs, G. (2003) Tunneling of two Interacting electrons in a Narrow Channel. *Solid State Commun.*, **128**, 443.

12
Graphene

12.1
Introduction

Carbon-based nanoelectronics have become a fast growing field. Chemical bonding between carbon atoms is flexible but strong enough to provide a variety of different structures with strikingly different physical and electronic properties. These largely depend on the dimensionality of the lattice structures. For instance, the reader may be familiar with expensive diamonds, rather than with cheap graphite and mysterious artificial structures such as carbon nanotubes, nanoribbons and fullerenes. Judging from the lattice structure, diamonds truly are 3D objects. Graphite is a stack of carbon planes weakly bounded by van der Waals forces. In carbon nanotubes, the atoms are arranged on a cylindrical surface and once chemically unzipped and flattened, they are transformed into 2D nanoribbons. Fullerenes are molecules where carbon atoms are arranged spherically, and thus, are 0D objects. Apart from diamonds, the rest of the structures are made of carbon allotrope known as graphene and its electronic and transport properties are the subject of this chapter.

Graphene is a 2D object, whereas carbon atoms are self-organized into a honeycomb structure made of benzene rings stripped of their hydrogen atoms. Released chemical bonds are used to hold the lattice together. This oversimplified graphene model will be refined in Section 12.2. Graphite, a 3D object made of stacks of graphene layers, has been known since the invention of the pencil in 1564. However, as late as 2004, Novoselov isolated and observed a single graphene layer via some subtle optical effects it creates on top of SiO_2 substrate.

Neutral graphene was found to be a semi-metal with unusual linearly dispersive electronic excitations called Dirac electrons. These are massless, chiral fermions moving at a fraction 1/300 of the speed of light. The chemical potential crosses the dispersion curves exactly at Dirac points as shown in Figure 12.1.

As a result of their massless nature, Dirac fermions behave differently compared with massive electrons in a variety of experiments. The massless nature of Dirac fermions or charge carriers leads to anomalies such as the Klein paradox and Zitterbewegung effect. The effects are not due to the massless nature but due to the Dirac equation since regular electrons or positrons or any half integer spin Dirac

Properties of Interacting Low-Dimensional Systems, First Edition. G. Gumbs and D. Huang.

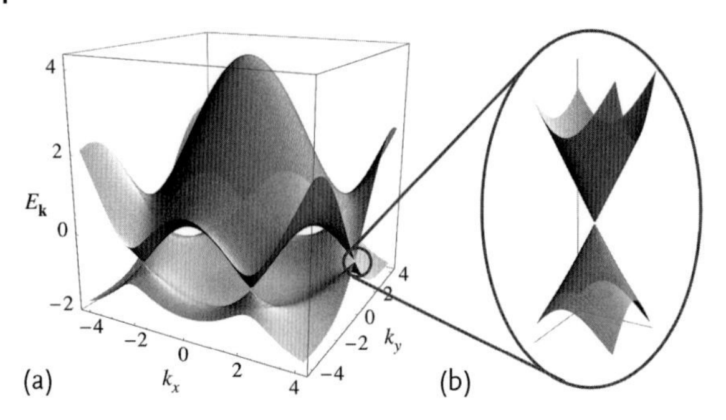

Figure 12.1 (a) Energy spectrum (in units of t) for finite values of t and t', with $t = 2.7$ eV and $t' = 0.2t$. (b) zoom-in of the energy bands close to one of the Dirac points.

particle exhibits it regardless of mass. The Hall effect shows itself when graphene is placed in a perpendicular magnetic field and becomes observable even at room temperature due to the large cyclotron energies of the Dirac particles. The Klein paradox signifies insensitivity of the Dirac fermions to an external electrostatic potential. That is, they can be transmitted unimpeded through classically forbidden regions. The Zitterbewegung phenomenon, or jittery motion, is due to the unusual behavior of the way Dirac fermions behave in the presence of a confining potential, such as, for example, those generated by disorder.

The last two effects result in high electron mobility under applied gate voltage, making ballistic transport a reality of contemporary nanoelectronics. Prior to the existence of graphene, ballistic transport was not easy to achieve because of graphene's novel properties. Prior to the discovery of graphene ballistic transport was not easy to achieve. Because of graphenes novel properties ballistic transport is feasible. Many-body effects in graphene by themselves are a wide topic beyond the scope of this chapter. Readers interested in those phenomena can find detailed discussion in the literature.

Mesoscopic graphene samples, such as, graphene nanoribbons (GNR) and nanotubes, can act as the connectors between nanodevices. Electrons and holes in the nanoribbons are quasi-one-dimensional (1D) since their motion are confined by the edges of the nanoribbons. The shape of the edge where a graphene sheet is terminated plays a very important role in the electronic and transport properties. In fact, it even challenges the notion of graphene being semi-metallic. Some GNR configurations may exhibit semiconducting properties, making them suitable for field effect transistors and switches. The most studied edges, zigzag (ZNR) and armchair (ANR), have drastically different electronic properties. Even more unusual, these strongly depend on the number of carbon atoms across the GNR.

Detailed discussion of the GNR with various edges using the Dirac formalism as well as sharp boundary conditions is presented in Section 12.3. For the ZNR edge, we will prove that the confinement leads to the formation of electron-like and hole-like evanescent wave functions confined to the edges of a ribbon. For the ANR edge, the band structure may be either metallic or insulating, depending on the ribbon width. Unlike for conventional semiconducting quasi-1D structures, the low-energy bands of the ANR have a linear momentum. For the ZNR, the density of states shows a peak at the Fermi energy. Besides sharp edges as alternative models of GNR, there are other well-known models that utilize position dependent electron mass, which will also be briefly discussed.

Our focus is on the Klein paradox in zigzag (ZNR) and anti-zigzag (AZNR) graphene nanoribbons. Specifics of the Klein effect in GNR's will be discussed in Section 12.4. When treated by the Dirac equation, ZNR (N, the number of lattice sites across the nanoribbon, is even) and AZNR (N is odd) configurations are indistinguishable. Thus, the conventional Dirac model must be supplemented with a pseudo-parity operator whose eigenvalues are sensitive to the number of carbon atoms across the ribbon, in agreement with the tight-binding model. We will show that the Klein tunneling in zigzag nanoribbons is related to conservation of the pseudo-parity rather than pseudo-spin in infinite graphene, or for that matter, in ANR. The perfect transmission in the case of head-on incidence is replaced by perfect transmission at the center of the ribbon and the chirality is interpreted as the projection of pseudo-parity on momentum at different corners of the Brillouin zone.

Section 12.5 is designated for the discussion of the dramatic effects due to external fields on the dispersion spectra, local density of states and ballistic transport. A strong electric field across the ribbon induces multiple chiral Dirac points, closing the semiconducting gap in armchair GNR's. A perpendicular magnetic field induces partially formed Landau levels as well as dispersive surface-bound states. Each of the applied fields on its own preserves the even symmetry $E_k = E_{-k}$ of the subband dispersion. When applied together, they reverse the dispersion parity to be odd and gives $E_{e,k} = -E_{h,-k}$ and mix the electron and hole subbands within the energy range corresponding to the change in potential across the ribbon. This leads to oscillations of the ballistic conductance within this energy range. These amazing transport properties of GNR allow for their use in a variety of applications that range from chemical single molecule detectors to single electron transmission and spin injection.

The limited scope of this book does not allow a plethora of GNR properties to be covered. However, recent scientific and technological progress in their fabrication and study has opened up seemingly unlimited possibilities.

12.2
Electronic Properties of Graphene

Graphene is composed of carbon atoms arranged in a hexagonal structure as shown in Figure 12.2. The sp^2 hybridization between one s-orbital and two p-orbitals leads to a trigonal planar structure (gray and black triangles in Figure 12.2) with a formation of a σ-bond between carbon atoms that are separated by $a = 1.42$ Å. The σ-band is responsible for the robustness of the lattice structure in all allotropes. Due to the Pauli principle, these bonds have a filled shell and thus form a deep valence bond. The unaffected p-orbital that is perpendicular to the pla-

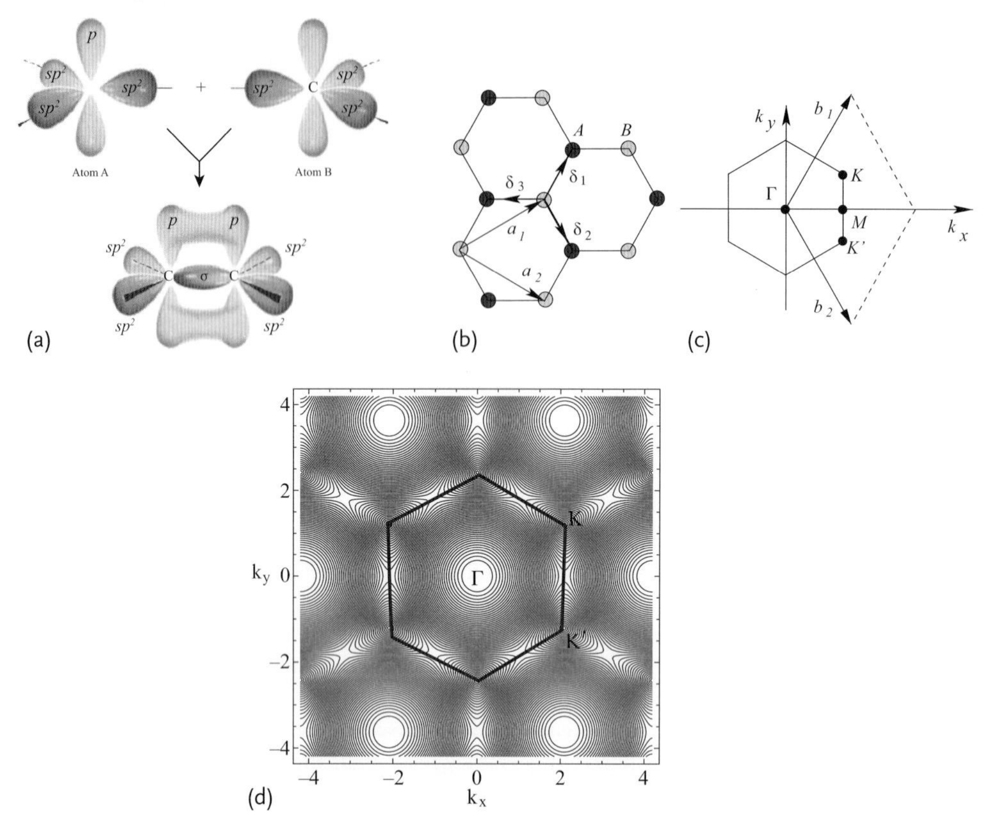

Figure 12.2 (a) The graphene lattice is held together by σ bonds that are in turn formed by sp^2 electron orbitals hybridization. p electrons provide the tight-binding model in Eq. (12.5). (b) Lattice structure of graphene, composed of two inter-penetrating triangular lattices ($\boldsymbol{a}_1$ and $\boldsymbol{a}_2$ are the lattice unit vectors, and $\boldsymbol{d}_i$ with $i = 1, 2, 3$ are the nearest neighbor vectors); (c) Corresponding Brillouin zone. The Dirac cones are located at the $\boldsymbol{K}$ and $\boldsymbol{K}'$ points. (d) Formation of the honeycomb lattice in momentum space obtained from the tight-binding model solution.

nar structure and can covalently bond with neighboring carbon atoms leading to the formation of a π-bond. Since each p-orbital has one extra electron, the π-band is half-filled. The resultant structure is not a Bravais lattice but can be seen as two intermingled triangular lattices with a basis of two atoms per unit cell.

The lattice vectors can be written as

$$\boldsymbol{a}_1 = \frac{a}{2}\left(3, \sqrt{3}\right), \quad \boldsymbol{a}_2 = \frac{a}{2}\left(3, -\sqrt{3}\right), \tag{12.1}$$

where $a \approx 1.42$ Å is the carbon–carbon distance. The reciprocal lattice vectors are given by

$$\boldsymbol{b}_1 = \frac{2\pi}{3a}\left(1, \sqrt{3}\right), \quad \boldsymbol{b}_2 = \frac{2\pi}{3a}\left(1, -\sqrt{3}\right). \tag{12.2}$$

Of particular importance for the physics of graphene are the two points K and K' at the corners of the graphene Brillouin zone (BZ). These are named Dirac points for reasons that will become clear later. Their positions in momentum space are given by

$$K = \left(\frac{2\pi}{3a}, \frac{2\pi}{3\sqrt{3}a}\right), \quad K' = \left(\frac{2\pi}{3a}, -\frac{2\pi}{3\sqrt{3}a}\right). \tag{12.3}$$

The three nearest-neighbor vectors in real 2D space are given by

$$\boldsymbol{d}_1 = \frac{a}{2}(1, \sqrt{3}), \quad \boldsymbol{d}_2 = \frac{a}{2}(1, -\sqrt{3}), \quad \boldsymbol{d}_3 = -a(1, 0). \tag{12.4}$$

The tight-binding Hamiltonian for electrons in graphene, considering that electrons can hop both to nearest and next nearest neighbor atoms, has the form (we use units such that $\hbar = 1$ here)

$$\hat{H} = -t \sum_{\langle\langle i,j\rangle\rangle,\sigma} a^{\dagger}_{i,\sigma} a_{j,\sigma} + b^{\dagger}_{i,\sigma} b_{j,\sigma} + \text{h.c.}, \tag{12.5}$$

where $a_{i,\sigma}$ ($a^{\dagger}_{i,\sigma}$) annihilates (creates) an electron with spin σ ($\sigma = \uparrow, \downarrow$) on lattice site $\boldsymbol{R}_i$ on sublattice A (an analogous definition of b_j is used for sublattice B), ($t \approx 2.8$ eV) is the nearest neighbor hopping energy (hopping between different sublattices).

It is convenient to write down the eigenfunctions of the above Hamiltonian in the form of a spinor (two component wave function). Those components correspond to the amplitudes of the p electrons on the A and B atoms within the unit cell labeled by the reference point $\boldsymbol{R}_j$

$$\begin{bmatrix} \psi_A(\boldsymbol{k}, \boldsymbol{r}) \\ \psi_B(\boldsymbol{k}, \boldsymbol{r}) \end{bmatrix} = \sum_j \exp(i\boldsymbol{k} \cdot \boldsymbol{R}_j) \begin{bmatrix} a^{\dagger}_j e^{-i\boldsymbol{k}\cdot(\boldsymbol{d}_2/2 - \boldsymbol{r})} \\ b^{\dagger}_j e^{-i\boldsymbol{k}\cdot(-\boldsymbol{d}_2/2 - \boldsymbol{r})} \end{bmatrix} |0\rangle . \tag{12.6}$$

The Hamiltonian equation (12.5) in Eq. (12.6) representation is purely off-diagonal with

$$\hat{H}_{\boldsymbol{k}} = \begin{bmatrix} 0 & \Delta^*_{\boldsymbol{k}} \\ \Delta_{\boldsymbol{k}} & 0 \end{bmatrix}, \tag{12.7}$$

where $\Delta_{\boldsymbol{k}} = -t \sum_{j=1}^{3} \exp(i\boldsymbol{k} \cdot \boldsymbol{d}_j)$. The off-diagonal nature of the above Hamiltonian illustrates the fact that the electron jumps occur only between the A and B sublattices. The energy bands derived from this Hamiltonian along with the explicit expressions for the nearest neighbors coordinates in (12.4) have the form:

$$E_{\pm}(\boldsymbol{k}) = \pm\sqrt{|\Delta_{\boldsymbol{k}}|^2} = \pm t\sqrt{3 + f(\boldsymbol{k})},$$
$$f(\boldsymbol{k}) = 2\cos\left(\sqrt{3}k_y a\right) + 4\cos\left(\frac{\sqrt{3}}{2}k_y a\right)\cos\left(\frac{3}{2}k_x a\right), \tag{12.8}$$

where the plus sign applies to the upper (π-electron like) and the minus sign to the lower (π^* hole-like) band. Eq. (12.8) clearly illustrates that the spectrum is symmetric around zero energy only if the nearest neighbors are taken into account. Otherwise, the electron–hole symmetry is broken and the π and π^* bands become asymmetric. In Figure 12.1 and Figure 12.2d, the band structure of infinite graphene is shown. In Figure 12.1, we also show a zoom-in of the band structure close to one of the Dirac points (at the $\boldsymbol{K}$ or $\boldsymbol{K}'$ point at the edges of the BZ) where $E(\boldsymbol{K}) = E(\boldsymbol{K}') = 0$. This dispersion can be obtained by expanding the full band structure in Eq. (12.8) close to the $\boldsymbol{K}$ (or $\boldsymbol{K}'$) vector in Eq. (12.3) as: $\boldsymbol{k} \to \boldsymbol{K} + \boldsymbol{k}$, with $|\boldsymbol{k}| \ll |\boldsymbol{K}|$. Also, $\Delta_{\boldsymbol{K}+\boldsymbol{k}} \approx -(3ta/2\hbar)(ik_x - k_y)\exp(-iK_x a)$. One may neglect an overall constant factor $-i\exp(-iK_x a)$ that does not affect Eq. (12.8) and thus recasts it in the following form

$$\Delta_K(\boldsymbol{k}) = \hbar v_F \left[k_x + ik_y + \mathcal{O}\frac{k^2}{K^2} \right], \tag{12.9}$$

where $v_F = 3ta/2\hbar \approx 10^6$ m/s is the velocity of the Dirac particles. Using the analogous expansion around $\boldsymbol{K}' = (K_x, -K_y)$, we obtain:

$$\Delta_{K'}(\boldsymbol{k}) = \hbar v_F \left[k_x - ik_y + \mathcal{O}\frac{k^2}{K^2} \right] = \Delta_K^*(\boldsymbol{k}) . \tag{12.10}$$

Substituting Eq. (12.9) into (12.7), the Dirac Hamiltonian around the Dirac point $\boldsymbol{K}$ reads in momentum space as

$$\hat{H}_K = \hbar v_F \begin{bmatrix} 0 & k_x - ik_y \\ k_x + ik_y & 0 \end{bmatrix} = \hbar v_F \boldsymbol{\sigma} \cdot \boldsymbol{k} , \tag{12.11}$$

and around the $\boldsymbol{K}'$, it is

$$\hat{H}_{K'} = \hbar v_F \begin{bmatrix} 0 & k_x + ik_y \\ k_x - ik_y & 0 \end{bmatrix} = \hbar v_F \boldsymbol{\sigma}^* \cdot \boldsymbol{k} , \tag{12.12}$$

where $\boldsymbol{\sigma} = (\overleftrightarrow{\sigma}_x, \overleftrightarrow{\sigma}_y)$, $\boldsymbol{\sigma}^* = (\overleftrightarrow{\sigma}_x, -\overleftrightarrow{\sigma}_y)$ are the Pauli matrices. Equations (12.12) and (12.12) can be combined into a single Hamiltonian, that is,

$$\hat{H} = \hbar v_F \begin{bmatrix} 0 & -k_x + ik_y & 0 & 0 \\ -k_x - ik_y & 0 & 0 & 0 \\ 0 & 0 & 0 & k_x + ik_y \\ 0 & 0 & k_x - ik_y & 0 \end{bmatrix} . \tag{12.13}$$

The wave function in real space for the sublattice is given by

$$\begin{aligned} \Psi_A(\boldsymbol{r}, \boldsymbol{k}) &= e^{i\boldsymbol{K}\cdot\boldsymbol{r}} \psi_A(\boldsymbol{r}, \boldsymbol{k}) + e^{i\boldsymbol{K}'\cdot\boldsymbol{r}} \psi'_A(\boldsymbol{r}, \boldsymbol{k}) \\ \Psi_B(\boldsymbol{r}, \boldsymbol{k}) &= e^{i\boldsymbol{K}\cdot\boldsymbol{r}} \psi_B(\boldsymbol{r}, \boldsymbol{k}) + e^{i\boldsymbol{K}'\cdot\boldsymbol{r}} \psi'_B(\boldsymbol{r}, \boldsymbol{k}) , \end{aligned} \tag{12.14}$$

where ψ_A and ψ_B are the components of the spinor wave function of Hamiltonian in Eq. (12.11) and ψ'_A and ψ'_B have the identical meaning but relatively to

Eq. (12.12). Alternatively, they can be combined into a four component wave function:

$$\Psi_{\pm} = \begin{bmatrix} \psi_A \\ \psi_B \\ \psi'_A \\ \psi'_B \end{bmatrix} = \frac{1}{\sqrt{2}} \begin{bmatrix} e^{-i\theta_k/2} \\ \pm e^{i\theta_k/2} \\ e^{i\theta_k/2} \\ \pm e^{-i\theta_k/2} \end{bmatrix} \exp(i\boldsymbol{k} \cdot \boldsymbol{r}) \,. \tag{12.15}$$

Here, θ_k is the angle between k_x and k_y.

A relevant quantity used to characterize the eigenfunctions is their chirality (helicity) defined as the projection of the momentum operator along the pseudo-spin direction. The quantum mechanical operator for the chirality has the form

$$\hat{h} = \frac{1}{2}\boldsymbol{\sigma} \cdot \frac{\boldsymbol{k}}{|\boldsymbol{k}|} \,. \tag{12.16}$$

From the definition of $\hat{h}$, it is clear that the states $\psi_{A,B}(\boldsymbol{k}, \boldsymbol{r})$ and $\psi'_{A,B}(\boldsymbol{k}, \boldsymbol{r})$ are also eigenstates of $\hat{h}$

$$\hat{h}\psi_{A,B}(\boldsymbol{k}, \boldsymbol{r}) = \pm\frac{1}{2}\psi_{A,B}(\boldsymbol{k}, \boldsymbol{r}) \,, \tag{12.17}$$

and an analogous equation holds for $\psi'_{A,B}(\boldsymbol{k}, \boldsymbol{r})$ with opposite sign. Therefore, electrons (holes) have a positive (negative) chirality that are opposite at two different K and K' points. Equation (12.17) implies that $\boldsymbol{\sigma}$ has its two eigenvectors either in the direction of or opposite to the momentum $\boldsymbol{k}$ and thus the name pseudo-spin. Notice that chirality is not defined with regard to the intrinsic spin of the electron (that has been neglected in this problem), but rather with respect to a pseudo-spin variable associated with the two components of the wave function for two sublattices. The chirality values are good quantum numbers as long as the Hamiltonian in Eq. (12.13) does not constrain the boundary conditions. Otherwise, they are no longer a good quantum number. The existence of such quantum numbers hold only as an asymptotic property that is well defined close to the Dirac points $\boldsymbol{K}$ and $\boldsymbol{K}'$. Either for larger energies or due to the presence of higher order hopping beyond the nearest neighbor one, the concept of chirality breaks down.

12.3 Graphene Nanoribbons and Their Spectrum

The energy spectrum of graphene nanoribbons very much depends on the nature of their edges – zigzag or armchair. In Figure 12.3, we show a honeycomb lattice having zigzag edges along the x-direction and armchair edges along the y-direction. If we choose the ribbon to be infinite in the x-direction, we produce a graphene nanoribbon with zigzag edges. On the other hand, choosing the ribbon to be macroscopically large along the y but finite in the x-direction, we produce a graphene nanoribbon with armchair edges.

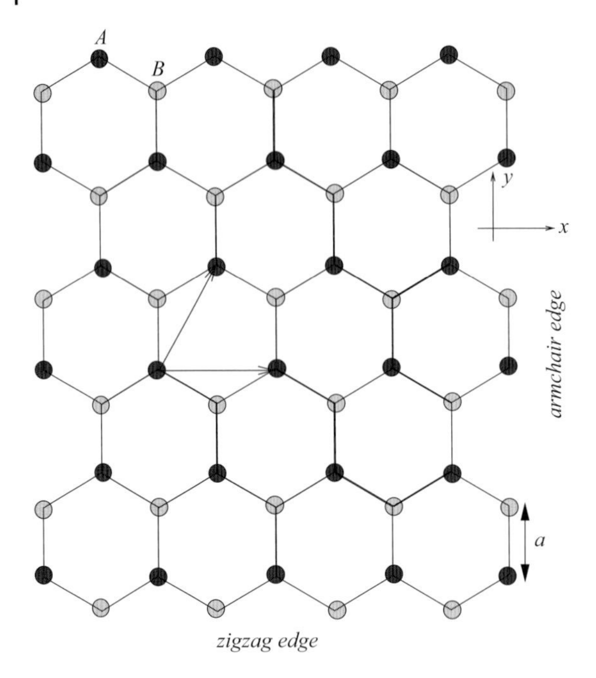

Figure 12.3 A piece of a honeycomb lattice displaying both zigzag and armchair edges. This is just a schematic. In our calculations, we assume that the ribbon edge runs along the x-direction always.

The edges are hydrogen passivated and can be modeled in the tight-binding approach by modifications of the hopping energies or via additional phases in the boundary conditions. Theoretical modeling of edge passivation indicates that this has a strong effect on the electronic properties at the edge of graphene nanoribbons. Currently, graphene nanoribbons have a high degree of roughness at the edges. Such edge disorder can significantly change the properties of edge states and lead to Anderson localization, anomalies in the quantum Hall effect and Coulomb blockade effects.

In Figure 12.4, we show energy levels for ideal edges, calculated in the tight-binding approximation, closest to zero energy for a nanoribbon with zigzag and armchair edges and with $N = 51$ carbon atoms across the ribbon. We can see that:

1. ANR can be either metallic or semiconducting depending on N.
2. ZNR presents a band of zero energy modes that is absent in the armchair case.

In the next two sections, we will derive the spectrum for both types of edges by using the Dirac formalism. Let us assume that the edges of the nanoribbons are parallel to the x-axis. In this case, the translational symmetry guarantees that the

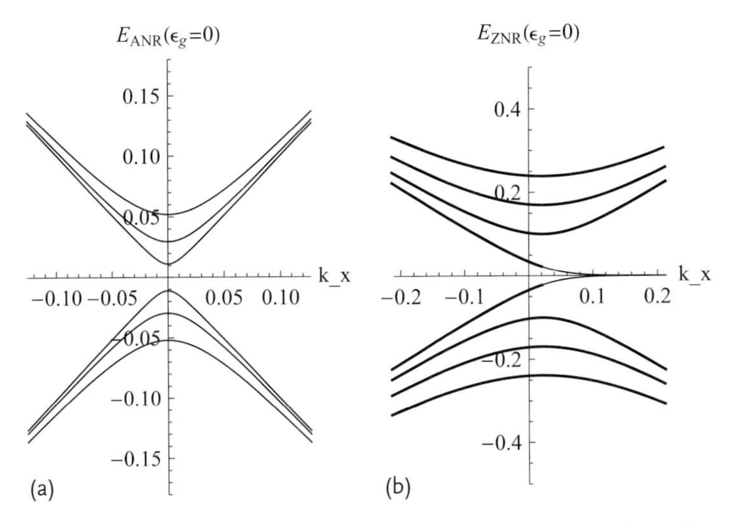

Figure 12.4 The energy dispersions for (a) armchair and (b) zigzag edges with $N = 51$ carbon atoms across the nanoribbon.

spinor wave function can be written as

$$\begin{bmatrix} \psi_A(\boldsymbol{r},\boldsymbol{k}) \\ \psi_B(\boldsymbol{r},\boldsymbol{k}) \\ \psi'_A(\boldsymbol{r},\boldsymbol{k}) \\ \psi'_B(\boldsymbol{r},\boldsymbol{k}) \end{bmatrix} = e^{ik_x x} \begin{bmatrix} \phi_A(y) \\ \phi_B(y) \\ \phi'_A(y) \\ \phi'_B(y) \end{bmatrix} . \tag{12.18}$$

To find the components of the above wave function, boundary conditions at the given edges has to be specified as follows.

12.3.1 Zigzag Edge

For zigzag edges, the boundary conditions at the edge of the ribbon (located at $y = 0$ and $y = W$, where W is the ribbon width) are given by

$$\Psi_A(y = W) = \Psi_B(y = 0) = 0 , \tag{12.19}$$

leading to

$$e^{iKx}e^{ik_x x}\phi_A(W) + e^{-iKx}e^{ik_x x}\phi'_A(W) = 0 , \tag{12.20}$$

$$e^{iKx}e^{ik_x x}\phi_B(0) + e^{-iKx}e^{ik_x x}\phi'_B(0) = 0 . \tag{12.21}$$

The boundary conditions in Eqs. (12.20) and (12.21) are satisfied for any x when we choose

$$\phi_A(W) = \phi'_A(W) = \phi_B(0) = \phi'_B(0) = 0 . \tag{12.22}$$

The next step is to find out the explicit form of the envelope functions. The eigenfunction around the point $\boldsymbol{K}$ has the form:

$$\begin{bmatrix} 0 & -k_x + \frac{d}{dy} \\ -k_x - \frac{d}{dy} & 0 \end{bmatrix} \begin{bmatrix} \phi_A(y) \\ \phi_B(y) \end{bmatrix} = \tilde{\varepsilon} \begin{bmatrix} \phi_A(y) \\ \phi_B(y) \end{bmatrix} \tag{12.23}$$

with $\tilde{\varepsilon} = \varepsilon/\hbar v_{\mathrm{F}}$ and ε is the energy eigenvalue. The eigen-problem can be written as two linear differential equations of the form

$$\begin{cases} \left(-k_x + \frac{d}{dy}\right) \phi_B(y) = \tilde{\varepsilon}\phi_A(y) \,, \\ \left(-k_x - \frac{d}{dy}\right) \phi_A(y) = \tilde{\varepsilon}\phi_B(y) \,. \end{cases} \tag{12.24}$$

Applying the operator $(k_x + \frac{d}{dy})$ to the first equation of Eqs. (12.24) leads to

$$\left(-\frac{d^2}{dy^2} + k_x^2\right) \phi_B(y) = \tilde{\varepsilon}^2 \phi_B(y) \tag{12.25}$$

with ϕ_A given by

$$\phi_A(y) = \frac{1}{\tilde{\varepsilon}} \left(-k_x + \frac{d}{dy}\right) \phi_B(y) \,. \tag{12.26}$$

The solution of Eq. (12.25) has the form

$$\phi_B(y) = A e^{\kappa_y y} + B e^{-\kappa_y y} \,, \tag{12.27}$$

leading to an energy:

$$\tilde{\varepsilon}^2 = k_x^2 - \kappa_y^2 \,. \tag{12.28}$$

The boundary conditions for a zigzag edge require that $\phi_A(y = W) = 0$ and $\phi_B(y = 0) = 0$, leading to

$$\begin{cases} A + B = 0 \,, \\ (k_x + \kappa_y) A e^{\kappa_y W} + (k_x - k_y) B e^{-\kappa_y W} = 0 \,, \end{cases} \tag{12.29}$$

which reduce to the following eigenvalue equation

$$e^{-2\kappa_y W} = \frac{k_x - \kappa_y}{k_x + \kappa_y} \,. \tag{12.30}$$

I get the following conditions $kx > |y|$, $kx < |y|$. Equation (12.30) has real solutions for κ_y whenever k_x is positive. These solutions correspond to surface waves (edge states) existing near the edges of the zigzag graphene ribbon. These states may also be obtained from the tight-binding model. In addition to real solutions for κ_y, Eq. (12.30) also supports complex roots of the form $\kappa_y = i k_y$, leading to

$$k_x = \frac{k_y}{\tan(k_y W)} \,. \tag{12.31}$$

The solutions of Eq. (12.31) correspond to confined modes in the graphene ribbon.

In both cases, the eigenstates in Eq. (12.18) can be written as two component wave functions given by

$$\psi_{\lambda,K}(\boldsymbol{r},\boldsymbol{k}) = \frac{2\kappa_y C}{\sinh(2\kappa_y W) - 2\kappa_y W}\begin{bmatrix} -i\lambda\sinh(\kappa_y y) \\ \sinh[\kappa_y(W-y)] \end{bmatrix} e^{ik_x x} . \qquad (12.32)$$

Here, the appropriate quantum numbers are $\lambda = \pm 1$ for the electron (hole) solutions of Eq. (12.28), K is the wave number at the $\boldsymbol{K}$ point around which the solution holds, k_x is the wave vector along the edge of the ZNR, κ_y is not technically an independent quantum number since it is connected to k_x by Eq. (12.30). The quantity C is a normalization constant.

If we follow the same procedure to the Dirac equation around the Dirac point $\boldsymbol{K}'$, we obtain a different eigenvalue equation given by

$$e^{-2\kappa_y W} = \frac{k_x + \kappa_y}{k_x - \kappa_y} . \qquad (12.33)$$

I get the following conditions $kx > |y|$, $kx < |y|$. This equation supports real solutions for κ_y if k_x is negative. Therefore, we have edge states for negative values of k_x with momentum around $\boldsymbol{K}'$. As in the case of $\boldsymbol{K}$ point, the system also supports confined modes given by

$$k_x = -\frac{k_y}{\tan(k_y W)} . \qquad (12.34)$$

One should note that the eigenvalue equations for $\boldsymbol{K}'$ are obtained from those in Eq. (12.32) by inversion, that is, $k_x \to -k_x$.

12.3.2 Armchair Nanoribbon

Let us now consider an armchair nanoribbon with armchair edges along the y-direction. In Eq. (12.18), we have to change $y \longleftrightarrow x$. The boundary conditions at the edges of the ribbon (located at $x = 0$ and $x = W$) are

$$\Psi_A(x=0) = \Psi_B(x=0) = \Psi_A(x=W) = \Psi_B(x=W) = 0 . \qquad (12.35)$$

In a similar way, these give rise to:

$$e^{ik_y y}\phi_A(0) + e^{ik_y y}\phi'_A(0) = 0 , \qquad (12.36)$$

$$e^{ik_y y}\phi_B(0) + e^{ik_y y}\phi'_B(0) = 0 , \qquad (12.37)$$

$$e^{iKW}e^{ik_y y}\phi_A(W) + e^{-iKW}e^{ik_y y}\phi'_A(W) = 0 , \qquad (12.38)$$

$$e^{iKW}e^{ik_y y}\phi_B(W) + e^{-iKW}e^{ik_y y}\phi'_B(W) = 0 , \qquad (12.39)$$

and are satisfied for any y if

$$\phi_\alpha(0) + \phi'_\alpha(0) = 0\,, \tag{12.40}$$

and

$$e^{iKW}\phi_\alpha(W) + e^{-iKW}\phi'_\alpha(W) = 0 \tag{12.41}$$

with $\alpha = A, B$. Clearly, these boundary conditions mix states from the two Dirac points. Now, we must find the form of the envelope functions obeying the boundary conditions in Eqs. (12.40) and (12.41). As it was done before, the functions ϕ_B and ϕ'_B obey a second order differential equation in Eq. (12.25) (with y replaced by x) and the function ϕ_A and ϕ'_A are determined from Eq. (12.26). The solutions of Eq. (12.25) have the following form

$$\phi_B = Ae^{ik_n x} + Be^{-ik_n x}\,, \tag{12.42}$$

$$\phi'_B = Ce^{ik_n x} + De^{-ik_n x}\,. \tag{12.43}$$

Applying the boundary conditions in Eqs. (12.40) and (12.41), we obtain

$$A + B + C + D = 0\,, \tag{12.44}$$

$$Ae^{i(k_n+K)W} + De^{-i(k_n+K)W} + Be^{-i(k_n-K)W} + Ce^{i(k_n-K)W} = 0\,. \tag{12.45}$$

From this, we see that the boundary conditions are satisfied with the choice

$$A = -D\,, \quad B = C = 0\,, \tag{12.46}$$

that leads to $\sin[(k_n + K)W] = 0$. Therefore, the allowed values of k_n are given by

$$k_n = \frac{n\pi}{W} - \frac{4\pi}{3a_0}\,, \tag{12.47}$$

and the energies are given by

$$\tilde{\varepsilon}^2 = k_y^2 + k_n^2\,. \tag{12.48}$$

No surface states exist in this case. If $KW \equiv 4\pi W/3a_0 = n\pi$, we have a metallic ANR. Otherwise, it has an insulating gap.

Each energy level $\tilde{\varepsilon}_{n,\lambda}(k_y) = \lambda\sqrt{k_n^2 + k_y^2}$ corresponds to a single wave function given by

$$\psi_{n,\lambda}(\mathbf{r},\mathbf{k}) = \frac{\exp(ik_y y)}{\sqrt{W}}\begin{bmatrix} \frac{k_y+ik_n}{\sqrt{k_n^2+k_y^2}}\exp(-ik_n x) \\ \lambda\exp(-ik_n x) \\ -\frac{k_y+ik_n}{\sqrt{k_n^2+k_y^2}}\exp(ik_n x) \\ \lambda\exp(ik_n x) \end{bmatrix}\,. \tag{12.49}$$

In the following section, we discuss the effect of electron scattering in GNR by a square barrier using the results obtained above and compare this with the scattering of chiral fermions in infinite graphene.

12.4 Valley-Valve Effect and Perfect Transmission in GNR's

The field of nanoelectronics based on graphene has been growing rapidly [1–3] mostly because of the high mobility two-dimensional electron or hole gas can be created by applying a gate voltage. Such high mobility is a striking consequence of the massless nature of the carriers and gives rise to the so-called Klein paradox that is the unimpeded penetration through high and wide potential barriers [2, 4]. In quantum electrodynamics, the phenomenon is related to the high probability of a relativistic particle tunneling through a potential barrier by means of conversion into the conjugated antiparticle. Its observation requires large potential drop ($> 10^{16}$ V cm^{-1}) and makes the effect purely theoretical. Whereas graphene provides a convenient medium for experimental verification of this effect.

One of the simplest experimental setups is shown schematically in Figure 12.5 where the rectangular barrier can be provided by an external electric field using an underlying insulating strip, carbon nanotube or by local chemical doping creating a *p–n–p* junction [5, 6]. The barrier is defined via the potential drop of width D (in units of carbon–carbon distance a_0) as $V(x) = V_0[\theta(x) - \theta(x - D)]$. This alternatively doped configuration provides the scattering (*p*-doped) region sandwiched between left (right) leads (*n*-doped). The electric current through the potential barrier is given by the incident electron of energy E in the range of $[0, \Delta]$, where 2Δ is the energy separation between the top of the next highest valence and the bottom of the next lowest conduction band (parabolically shaped $\mathcal{V}(\mathcal{V}^*)$ bands in Figure 12.5). To stay in the single channel regime where the scattering occurs within the lowest conduction π and highest valence π^* bands, it requires the barrier height V_0 in the range of $|E - V_0| \leq \Delta$.

Owing to the massless nature of the carriers, even in disordered graphene samples, a barrier field of $< 10^5$ V cm^{-1} should be sufficient. The underlying physics of Klein tunneling is based on the notion of chirality as a symmetry between electrons and holes in graphene. Formally, each of the branches of the dispersion conical sections is characterized by a specific sign of the pseudo-spin projection on to the momentum. The Klein effect is related to conservation of this quantity across the junction.

We shall concentrate on scattering around the $\boldsymbol{K}$ point assuming that the scattering does not mix the momenta from $\boldsymbol{K}'$ point. Through a gauge transformation, the wave function in Eq. (12.15) can be written as

$$\psi(\boldsymbol{r}, \boldsymbol{k}) = \frac{1}{\sqrt{2}} \begin{bmatrix} 1 \\ \pm e^{i\theta_k} \end{bmatrix} e^{ik_x x + ik_y y} . \tag{12.50}$$

We further assume that the scattering does not mix the momenta around $\boldsymbol{K}$ and $\boldsymbol{K}'$ points. The wave functions in the various regions can be written in terms of

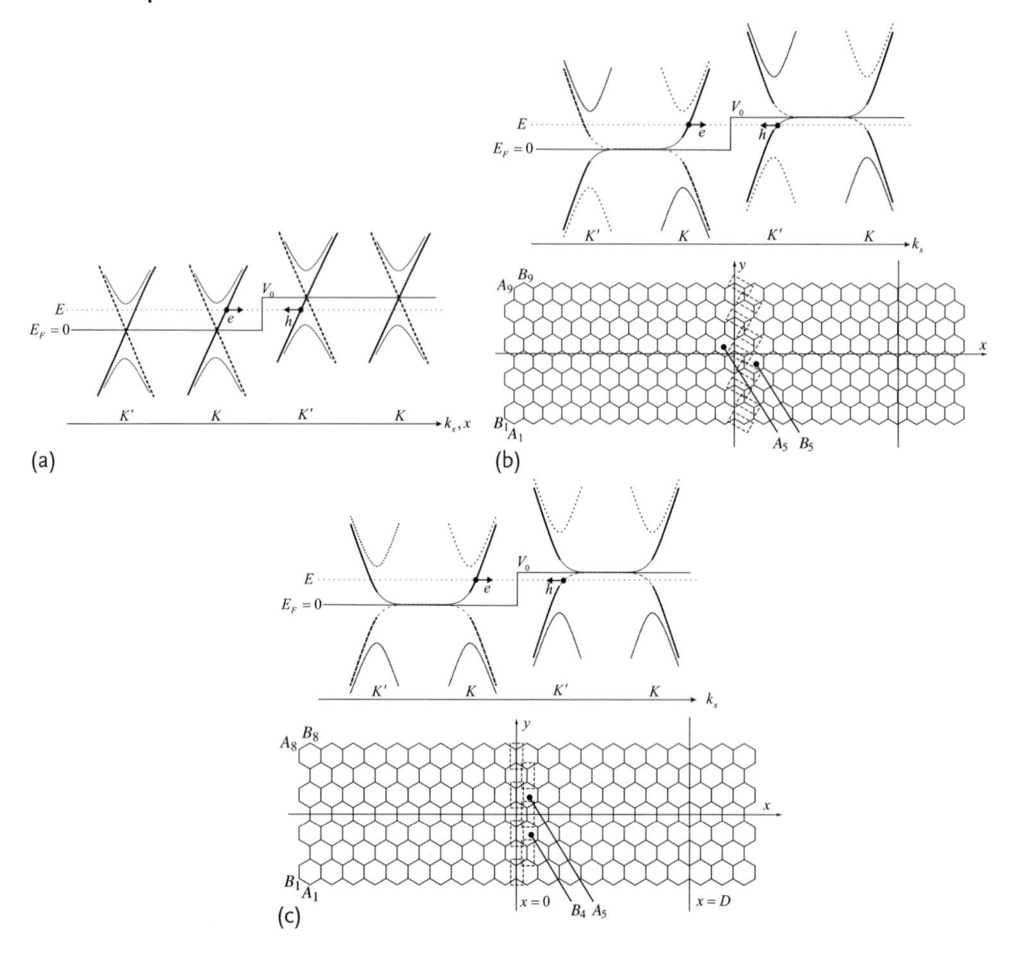

Figure 12.5 (a) Infinite graphene *n–p–n* junction and dispersion curves formed by two sublattices. The lowest conduction and highest valence bands are denoted as π and π^*, respectively. The parabolic shaped curves denote $\mathcal{V}, \mathcal{V}^*$ bands. The different values of the pseudo-spin projected on the momentum are shown in solid and dashed curves. (b) 9-AZNR *n–p–n* junction. The direction of the unit cells (dashed rectangles) satisfies the invariance of the ribbon under integer (by the hexagon plaquette) longitudinal translations. The two different values of pseudo-parity projection on the pseudo-spin $1 - \mathbf{s}^* \cdot \boldsymbol{\eta}$ are presented in solid (dashed) curves. Perfect transmission occurs at the center of the ribbon since there one has pure wave functions in the "electron–hole" basis given by Eq. (12.62). (c) 8-ZNR configuration. The perfect reflection is another manifestation of the Klein effect due to opposite pseudo-parity in the *p* and *n* regions.

incident and reflected waves. In region I, we have

$$\psi_s^{\mathrm{I}}(\mathbf{r}, \mathbf{k}) = \frac{1}{\sqrt{2}} \begin{bmatrix} 1 \\ s e^{i\phi} \end{bmatrix} e^{i k_x x + i k_y y} + \frac{r}{\sqrt{2}} \begin{bmatrix} 1 \\ s e^{i(\pi - \phi)} \end{bmatrix} e^{-i k_x x + i k_y y} \tag{12.51}$$

with $\phi = \arctan(k_y/k_x)$, $k_x = k_F \cos\phi$, $k_y = k_F \sin\phi$ and k_F the Fermi momentum. In region II, we have

$$\psi_{s'}^{II}(\mathbf{r}) = \frac{a}{\sqrt{2}}\begin{bmatrix} 1 \\ s' e^{i\theta} \end{bmatrix} e^{iq_x x + ik_y y} + \frac{b}{\sqrt{2}}\begin{bmatrix} 1 \\ s' e^{i(\pi-\theta)} \end{bmatrix} e^{-iq_x x + ik_y y} \tag{12.52}$$

with $\theta = \arctan(k_y/q_x)$ and

$$q_x = \sqrt{\frac{(V_0 - E)^2}{\hbar^2} v_F^2 - k_y^2}\,, \tag{12.53}$$

and finally in region III, we only have a transmitted wave, that is,

$$\psi_s^{III}(\mathbf{r}, \mathbf{k}) = \frac{t}{\sqrt{2}}\begin{bmatrix} 1 \\ s e^{i\phi} \end{bmatrix} e^{ik_x x + ik_y y} \tag{12.54}$$

with $s = \text{sgn}(E)$ and $s' = \text{sgn}(E - V_0)$. The coefficients r, a, b and t are determined from the continuity of the wave function this implies that the wave function has to obey the conditions given by $\psi^{I}(x = 0, y) = \psi^{II}(x = 0, y)$ and $\psi^{II}(x = D, y) = \psi^{III}(x = D, y)$. Unlike the Schrödinger equation, we only need to match the wave function and not its derivative. The transmission through the barrier is obtained from $\mathcal{T}(\phi) = tt^*$ and has the form

$$\mathcal{T}(\phi) = \frac{\cos^2\theta \cos^2\phi}{\cos^2(q_x D)\cos^2\phi\cos^2\theta + \sin^2(q_x D)(1 - ss'\sin\phi\sin\theta)^2}\,. \tag{12.55}$$

Note that $\mathcal{T}(\phi) = \mathcal{T}(-\phi)$ and for values of $q_x D$ satisfying the relation $q_x D = n\pi$ with n integer, the barrier becomes completely transparent since $\mathcal{T}(\phi) = 1$, independent of the value of ϕ. Also, for normal incidence ($\phi \to 0$ and $\theta \to 0$) and for any value of $q_x D$, we obtain $\mathcal{T}(0) = 1$, and the barrier is again totally transparent. This result is a manifestation of the Klein paradox and does not occur for non-relativistic electrons. In this latter case and for normal incidence, the transmission is always smaller than one. In the limit $|V_0| \gg |E|$ ($\theta \approx 0$), Eq. (12.55) has the following asymptotic form

$$\mathcal{T}(\phi) \simeq \frac{\cos^2\phi}{1 - \cos^2(q_x D)\sin^2\phi}\,. \tag{12.56}$$

In Figure 12.6, we present the angular dependence of $\mathcal{T}(\phi)$ for two values of the potential V_0. Clearly, there are several directions for which the transmission is unity.

In Figure 12.7, we present the transmission as a function of the transverse and longitudinal components of the wave vector in n- and p-doped regions. The conversion is simply given by $\lambda k_{x,1} = 2\pi \sin\phi_1 \cot\phi_2$, and $\lambda k_{x,2} = 2\pi\cos\phi_1$.

Now let us move on to the Klein scattering in graphene nanoribbons. As shown above, their energy band structure depends a lot on whether the shape of their

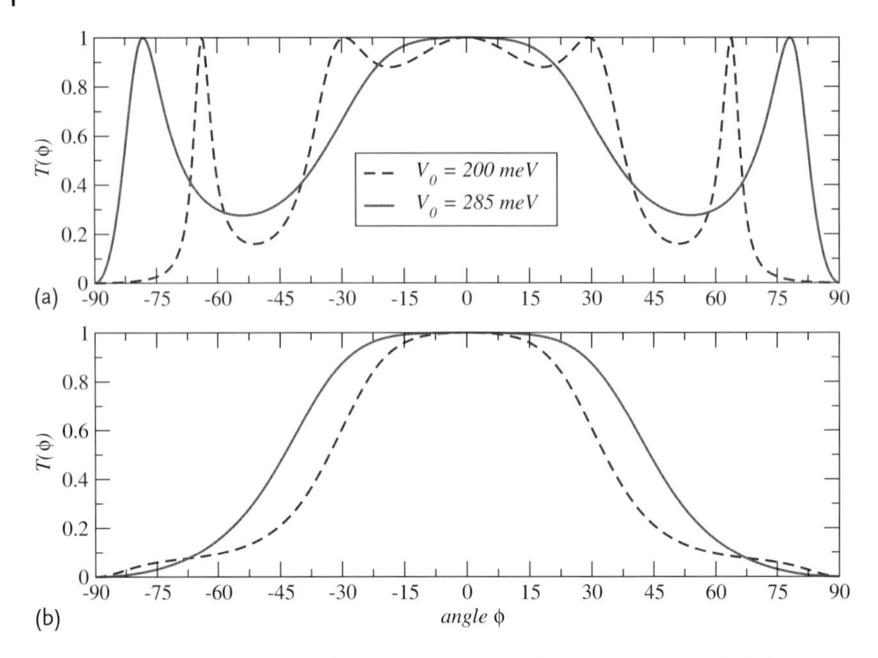

Figure 12.6 Angular dependence of $\mathcal{T}(\phi)$ for two values of V_0: $V_0 = 200$ meV (dashed curve) and $V_0 = 285$ meV (solid curve). The parameters used in the calculation are $D = 110$ nm (a), $D = 50$ nm (b), $E = 80$ meV, $k_F = 2\pi/\lambda$, and $\lambda = 50$ nm.

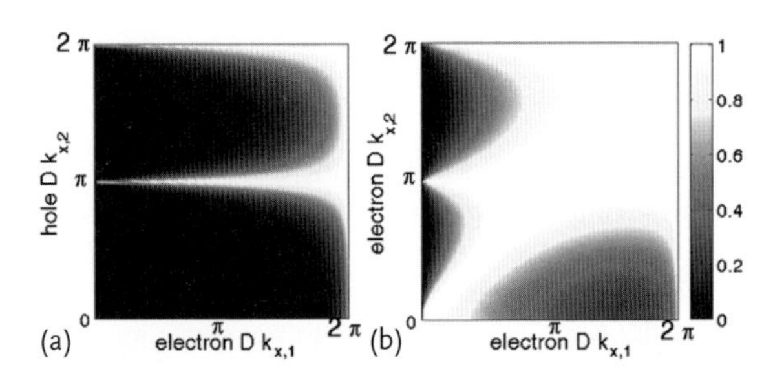

Figure 12.7 The rectangular potential barrier transmission $|t(k_{x,1}, k_{x,2})|^2$ for infinite graphene. The incoming electron wavelength is taken equal to the width of the *p*-doped region. (a) Through the barrier for $E \leq V_0$, transmission of the chiral electron transformed into a hole state. (b) Over the barrier transmission for $E \geq V_0$ for the chiral electron that is being projected. The transmission is ideal on the diagonal $k_{x,1} = k_{x,2}$ and corresponds to the absence of the gating potential. In both panels, the Klein effect is related to the perfect transmission at $Dk_{x,1} = 2\pi$ (head-on incidence).

edges is either zigzag, armchair or bearded. Here, our focus is on the ZNR configuration since this geometry does not possess chiral fermions. In other words, at both the $\boldsymbol{K}$ ($\boldsymbol{K}'$) points, there is an excess right (left) moving channel. We will confine our attention to these channels (single channel regime) [7]. We model the conducting channels by the $\boldsymbol{k} \cdot \boldsymbol{p}$ approximation for the tight-binding model usually used for the band structure descriptions [2]. It yields the following electron (with "+" sign) and hole (with "−" sign) dispersion valleys: $V(x) - E = \pm\hbar v_F \sqrt{k_x^2 - \kappa_y^2}$, where the hard wall boundary conditions mix the transverse and the longitudinal electron momenta through

$$k_x = \kappa_y \coth(\kappa_y W) . \tag{12.57}$$

Here, the ribbon width is of the form $W/a_0 = 2N - 1$ for even number N of carbon atoms across the ribbon and $W/a_0 = 3/2(N - 1)$. The corresponding eigenfunctions for the two sublattices (A and B) can be factorized into the Hadamard product of longitudinal, transverse, pseudo-parity and electron–hole parity two component wave functions as

$$\Psi_{\pm K,N}(x, y) = \begin{bmatrix} \Psi_A \\ \Psi_B \end{bmatrix} = \phi_{\pm K,N}(x)\chi(y)\eta_{\pm K} s . \tag{12.58}$$

Here, the longitudinal wave function depends on the parity X_N of the number of carbon atoms across the ribbon:

$$\phi_{\pm K,N}(x) = \frac{1}{\sqrt{2\pi}} \begin{bmatrix} e^{\pm i(K+k_x)x} \\ e^{\pm i(K+k_x)(x+X_N)} \end{bmatrix} , \tag{12.59}$$

where $X_{\text{even}} = 0$ and $X_{\text{odd}} = \sqrt{3}a_0/2$. Its presence stems from the suitable shape of the unit cells as depicted in Figure 12.5. The choice of $\pm$ defines the right (transmitted) and left (reflected) movers (direction of the electron group velocity for $E - V(x) > 0$), respectively. The shape of the unit cells and, consequently, the form of the longitudinal wave function ascertains translational symmetry of the wave function in Eq. (12.58) along the ribbon.

The transverse wave function component is given by

$$\chi(y) = \sqrt{\frac{2k_y}{\sinh(2\kappa_y W) - 2\kappa_y W}} \begin{bmatrix} \sinh\left(\kappa_y \frac{W}{2+y}\right) \\ \sinh\left(\kappa_y \frac{W}{2-y}\right) \end{bmatrix} . \tag{12.60}$$

We note that Eq. (12.60) satisfies the Dirac equation at $\pm K = \pm 2\pi/3\sqrt{3}a$ and satisfies the hard-wall boundary conditions across the ZNR. In contrast to infinite graphene or the ANR, the lowest conduction and highest valence (π and π^*) bands describe localized bound edge states for $\text{Im}(\kappa_y W) = 0$ and delocalized bulk states for $\text{Re}(\kappa_y W) = 0$ and $\text{Im}(\kappa_y W) \leq \pi$. The $\mathcal{V}$ ($\mathcal{V}^*$) bands correspond to $\text{Re}(\kappa_y W) = 0$ and $\text{Im}(\kappa_y W) > \pi$. An interesting note is that in contrast to infinite graphene, head-on incidence ($\text{Im}(\kappa_y) \ll k_x$ occurring at $\text{Im}(\kappa_y W) \to n\pi$) is prohibited due to mixing of the longitudinal and transverse wave vectors in Eq. (12.57).

There are two types of electron (hole) states in the nanoribbons given by the bonding and anti-bonding mixing of the sublattice wave functions. These strongly depend on the mirror symmetry (or lack of it) of the GNR that follows from the tight-bing model [8]. To incorporate this feature into our model, we introduce the concept of pseudo-parity similar in nature to that in [9], that is,

$$\eta_{\pm K} = \left[\frac{1}{\left(\pm 3\sqrt{3}\frac{K}{2\pi}\right)^N e^{i\,\mathrm{Im}(\kappa_y W)}} \right] . \tag{12.61}$$

The pseudo-parity component of the wave function in Eq. (12.58) determines whether the electron (hole) wave functions are even or odd, forming bonding or anti-bonding states. Formally, these can be defined by the following transformation [10]:

$$\begin{bmatrix} \Psi_b \\ \Psi_a \end{bmatrix} = \frac{1}{\sqrt{2}} \begin{bmatrix} 1 & 1 \\ -i & i \end{bmatrix} \begin{bmatrix} \Psi_A \\ \Psi_B \end{bmatrix} . \tag{12.62}$$

The pseudo-spin Eq. (12.61) ensures alternation of the wave function sign for different bands as shown in Figure 12.5. The parity of N determines if the bonding and anti-bonding wave functions in Eq. (12.62) switch between conduction and valence bands at $\boldsymbol{K}$ ($\boldsymbol{K}'$) points.

The electron–hole parity is comprised of the eigenvalues of the time-reversal operator

$$S = \left[\frac{1}{\mathrm{sgn}[E - V(x)]} \right] , \tag{12.63}$$

where the positive sign corresponds to the electrons in the lower conduction π band and the negative sign applies to the holes in the highest valence π^* band.

The problem of tunneling through the n–p–n barrier is described as follows: an incident particle belonging to the right moving channel ($\boldsymbol{K}$ point) for $x < 0$ strikes the potential barrier at $x = 0$. The nature of the scattering determines the transmission probability density by employing continuity of the wave functions in Eq. (12.58) at the junctions and conservation of the pseudo-parity in Eq. (12.61). The reflected wave contributes to the left-moving channel (at $\boldsymbol{K}'$). Clearly, the necessary condition for non-zero transmission is the change of the wave function parity under reflection. Therefore, for N even (ZNR configuration in Figure 12.5), the through-the-barrier transmission vanishes identically, that is, $t \equiv 0$. That is, the sublattice wave functions in Eq. (12.58) for π band in the n-doped region and for π^* band in the p-doped region have opposite sign for the pseudo-parity (opposite bonding for the particle–hole representation in Eq. (12.62)) [11].

The situation drastically changes for AZNR (odd N) configuration. The reflected wave function changes its parity while the transmitted one keeps the parity of the incident particle (π band at $\boldsymbol{K}$, and π^* at $\boldsymbol{K}'$ have the same pseudo-parity). The transmission amplitude is given by

$$\begin{aligned} t(\gamma) = {} & \left(1 - e^{i2\pi D/3a_0}\right)^2 e^{i(k_{x,2} - k_{x,1})D}\, S \\ & \times \left[e^{i2\pi D/3a_0} \left(e^{i2k_{x,2}D} - 1\right) \left(\chi_{A,1}^2 \chi_{B,2}^2 + \chi_{A,2}^2 \chi_{B,1}^2\right) \right. \\ & \left. - \left(2e^{i2(3k_{x,2} + \pi/a_0)D/3} + e^{i4\pi D/3a_0} + 1\right) S \right]^{-1} . \end{aligned} \tag{12.64}$$

Here, $S = s_1 s_2 \chi_{A,1} \chi_{A,2} \chi_{B,1} \chi_{B,2}$ and the sub-indices A and B on the wave function components denote the corresponding sublattices. The gated p-doped and n-doped regions are specified by the sub-indices 2 and 1, respectively. For example, $k_{x,2}$ and $k_{x,1}$ represent the wave numbers along the ribbon in the p and n doped regions. For brevity, $s_{1,2}$ denote two s_B components of the electron–hole wave function in Eq. (12.63).

The transmission probability (and hence, in the single-channel regime, the conductance [12] in units of $2e^2/h$) is given by

$$|t(\kappa_{y,1}, \kappa_{y,2})|^2 = \frac{1}{W} \int_0^W dy\, t(y) t^*(y) \tag{12.65}$$

and is shown in Figure 12.8. The transmission probability jumps up to one at $\mathrm{Re}(\kappa_{y,2} D) = n\pi$ for any integer n for the case of surface bound holes, and at $\mathrm{Im}(\kappa_{y,2} D) = n\pi/2$ for bulk-like holes. In the latter, there are additional resonances when $\mathrm{Im}(\kappa_{y,2} D) \to n\pi$. This situation corresponds to the quantum-dot-like states in the gated region [13]. These resonances are similar to those in infinite graphene as shown in Figure 12.7.

In summary, we analytically demonstrated that when the Dirac fermion picture is supplemented with the concept of pseudo-spin for ZNR (or AZNR), we remove the ambiguities arising for transmission through the barrier and correctly describe the valley-valve effect as well as current blocking. Now, let us turn our attention to the role played by pseudo-parity in the Klein undamped transmission in ZNR. The Klein paradox in infinite graphene or metallic ANR is related to the conservation of

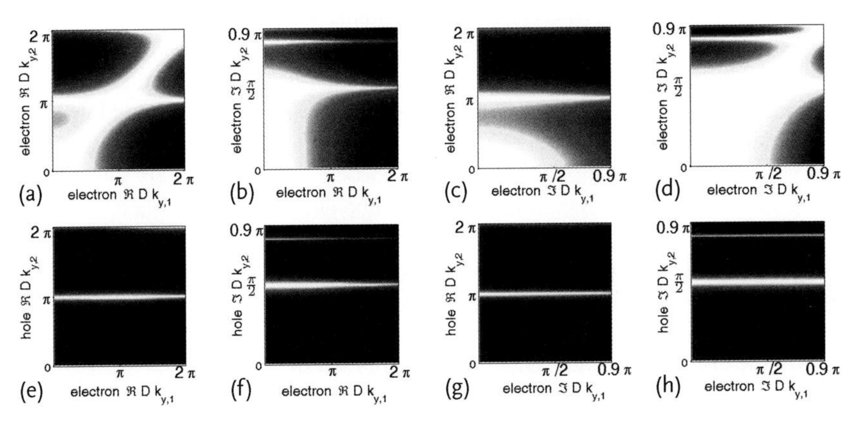

Figure 12.8 The rectangular potential barrier transmission $|t(\kappa_{y,1}, \kappa_{y,2})|^2$ for 9-AZNR given by Eq. (12.65). Panels (a–d) correspond to over the barrier transmission $E > V_0$ of π band electrons ($s_2 = 1$). Panels (e–h) represent through the barrier $E < V_0$ transmission of the electrons transformed into π^* hole states ($s_2 = -1$). In both cases, the π and π^* bands are comprised of real ($|E - V(x)| < \hbar k_F v_F$) and imaginary part ($|E - V(x)| \geq \hbar k_F v_F$) of the transverse wave vector κ_y, thus making four distinctive transmission cases (the four columns). In particular, $\mathrm{Im}(\kappa_{y,1(2)})$ implies that the electron (hole) is in the bulk state, while $\mathrm{Re}(\kappa_{y,1(2)})$ signifies the surface-bound states. For simplicity, the incoming particle of wave vector $k_{x,1}$ is in the π band.

pseudo-spin projection onto the momentum (chirality). That is, the through barrier unimpeded transmission occurs for head-on incidence as indicated in Figure 12.7a for $k_{x,1}D = 2\pi$ for all values of the gate potential V_0 (any value of $k_{x,2}D$). Transformation from the sublattice basis to the bonding-antibonding representation in Eq. (12.62) yields that the head-on incident electron is in the purely bonding state ($\Psi_a = 0$) [2].

However, the fermions in the π and π^* bands for ZNR and AZNR do not satisfy the conventional (pseudo-spin based) definition of chirality and are thus referred to as non-chiral. Furthermore, the net transmission in Eq. (12.65) does not demonstrate such an effect. We now turn our attention to the transmission *amplitude* given by Eq. (12.64). This shows that at the center of the nanoribbon, the transmission amplitude is always perfect $t(y = W/2) = 1$, regardless of the gate potential V_0. This effect is intimately related to the Klein paradox in infinite graphene. Since at the nanoribbon center, the incoming wave functions from the sublattices A and B form the bonding electron state, $[\Psi_a(y = W/2) = 0]$ in accordance with Eq. (12.62). This holds true for both edge-bound and bulk states of the π and π^* bands. Now, we can relate to the unimpeded transition amplitude to the conservation of the pseudo-parity rather than conservation of the pseudo-spin in infinite graphene. The pseudo-parity itself defines the chirality of the fermions in ZNR. The valley-valve effect of the blockage current is the other manifestation of the Klein paradox and pseudo-parity defined chirality in AZNR.

12.5
GNR's Electronic and Transport Properties in External Fields

This section explains the individual and combined effects of an electric and magnetic field on the band structure and conductance of GNRs. Since we neglect spin, the action of the time reversal operator $\hat{T}$ amounts to reversing the direction of the wave vector propagation. The even/odd particle energy symmetry may be defined as $E_{n,k} = \pm\hat{T}(E_{n,k}) = \pm E_{n,-k}$. If only one of the fields is applied, it is well known that time reversal symmetry of the energy bands for electrons and holes is preserved for all types of GNRs as listed above. However, the combined effect of an electric and a magnetic field on the energy dispersion is to break the time reversal symmetry for both electrons and holes and mix the energy bands. The effect of mixing on the differential conductance and local density of states (LDOS) is presented below and our results are compared to those obtained when only one of the two external fields is applied to an ANR with quantum point contacts as schematically illustrated in Figure 12.9. The ribbon is attached to left (L) and right (R) leads, and serves as infinite electron reservoirs. The R-lead is assumed to be the drain held at chemical potential μ. The L-lead is held at DC biased chemical potential $\mu + eV$ (e is the electron charge and V is the bias potential) and serves as the source. We choose coordinate axes so that the nanoribbon is along the x-axis in the xy-plane. Mutually perpendicular static electric field $\mathcal{E}_y$ along the y-axis and magnetic field $\mathcal{B}_z$ along the z-axis are applied, as shown in Figure 12.9.

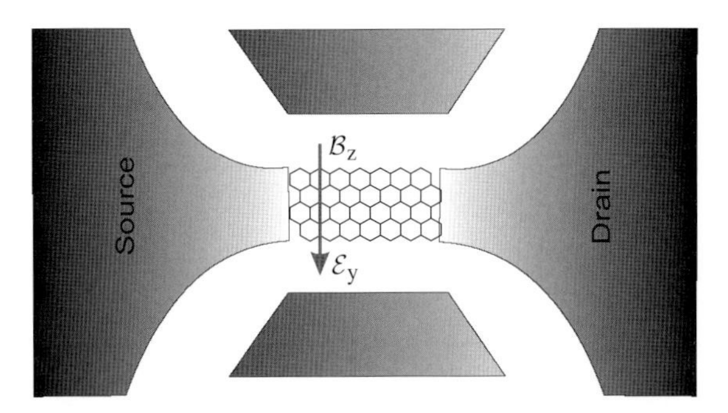

Figure 12.9 Schematic of an ANR in the presence of an in-plane electric field $\mathcal{E}_y$ along the y-axis and a perpendicular magnetic field $\mathcal{B}_z$ along the z-axis.

Following [2, 14], we calculated the energy bands for graphene with sublattices A and B in the tight-binding model. These are then separated into hole ($\{h\} = \{1 \le n < N\}$) and electron ($\{e\} = \{N \le n < 2N\}$) energy bands. The two component wave function is a normalized $2N$ vector $\langle \Psi(x)|_{n,k} = \{\langle \Psi_A(x)|_{n,k}, \langle \Psi_B(x)|_{n,k}\}$. The electric field induces a potential across the ribbon $U(y) = e\mathcal{E}_y(y - W/2) = U_0[(y/W) - 1/2]$, where W is the ribbon width and $U_0 = e\mathcal{E}_y W$. The magnetic field modifies the wave vector as $k \to k - e\mathcal{B}_z y/\hbar$, amounting to the Peierls phase [15] in the hopping integrals in Hamiltonian equation (12.5):

$$t_{i,j} \to \exp\left(i\frac{2\pi}{\phi_0}\int_i^j \mathbf{A}\cdot d\boldsymbol{\ell}\right) t_{i,j} \,. \tag{12.66}$$

Here, $\phi_0 = h/e$ and $\mathbf{A}$ is the vector potential. The magnetic field strength is assumed weak so we could take the energy levels as spin degenerate. The dispersion curves can be experimentally observed via a scanning tunneling microscopy [16]. The tunneling current flowing through the microscope tip is proportional to the LDOS given by

$$\text{LDOS}(E, x) = \sum_{n,k} |\Psi_{n,k}(x)|^2 \delta(E - E_{n,k}) \,. \tag{12.67}$$

The energy dispersion determines the ballistic charge current I through the ribbon, at temperature T, by

$$I(V, \mu, T; \mathcal{E}_y, \mathcal{B}_z) = -2e \sum_n \int \frac{dk}{2\pi} v_{n,k} \times \left[\theta(-v_{n,k}) f_{n,k}^{>}(1 - f_{n,k}^{<}) + \theta(v_{n,k}) f_{n,k}^{<}(1 - f_{n,k}^{>})\right], \tag{12.68}$$

where $v_{n,k} = \hbar^{-1} dE_{n,k}/dk$ is the carrier group velocity. At $T = 0$, the Fermi function at the source contact is $f_{n,k}^{<} = 1 - \theta(E_{n,k} - \mu - eV)$ and for the drain, we have $f_{n,k}^{>} = 1 - \theta(E_{n,k} - \mu)$. We note that Eq. (12.68) does not assume any symmetry for the energy dispersion relation. If the energy satisfies $E_{n,k} = E_{n,-k}$, we obtain the well-known Landauer–Büttiker formula [17]. The differential conductance $G(\mu; \mathcal{E}_y, \mathcal{B}_z) = (\partial I/\partial V)_{V=0}$ is determined by the number of right-moving carriers through $v_{n,k}/|v_{n,k}| > 0$ at the chemical potential $E_{n,k} = \mu$. Alternatively, one can take the difference between the local minima and maxima below the chemical potential $E_{n,k} < \mu$ [18].

Our numerical results for the energy bands, LDOS and conductance for semiconducting ANR ($N = 51$) in the presence of an electric and/or a magnetic field are presented in Figure 12.10. When either only an electric or a magnetic field is applied, $\mathcal{E}_y \mathcal{B}_z = 0$, the electron–hole energy bands are symmetric with $E_{\text{h},k} = -E_{\text{e},k}$ and time reversal symmetry is satisfied with $E_{n,k} = E_{n,-k}$ around the $k = 0$ Dirac point in Figure 12.7b. This means that if the time for the particle is reversed, the particle retraces its path along the same electron–hole branch. The LDOS also demonstrates the wave function symmetry with respect to the ribbon center $\text{LDOS}(E, x) = \text{LDOS}(-E, x) = \text{LDOS}(E, -x)$. In accordance with the Landauer–Büttiker formalism, the conductivity illustrates the well-known pattern. The magnetic field by itself distorts the weak dispersion (n close to N) so that the partially formed Landau levels $E_{n,0} \sim \sqrt{\mathcal{B}_z n}$ shows itself up as the flat portions in the dispersion curves. The lowest Landau level provides the single conducting channel (along the ribbon edges), while the other states are doubly degenerate. When the wave vector evolves from the Dirac point, the degeneracy is lifted and the lowest subband acquires a local minimum. Of these two effects, the first one can be observed in the LDOS, while the second reveals itself as sharp spikes in the conductance as depicted in the third panel of Figure 12.10b. For the high energy subbands when the radii of the Landau orbits (spread of the wave function in the second panel of Figure 12.10b) become comparable to the ribbon width, the confinement effects dominate and the spectra become linear in magnetic field with $E_{n,0} \sim \mathcal{B}_z/n$. These subbands are not degenerate.

The main effect that the electric field has on the energy dispersion is to fracture the Fermi surface into small pockets for $k \neq 0$, and thereby closing the semiconducting energy gap. These zero energy points where the group velocity abruptly changes sign represent new Dirac points, which follows from the chirality of the wave function in their vicinity [19]. The rapid changes in the group velocity cause the appearance of spikes in the conductance near $|\mu| \leq U_0/2$ and its step-like pattern is affected. Due to the Dirac symmetry of the problem, the electron–hole band structure remains symmetric. The energy dispersion is not affected by the magnetic field at the original Dirac point $k = 0$. Time reversal symmetry also persists. The LDOS shows that at high energies the electric field confines the electrons and holes near opposite boundaries. However, at low energies, the LDOS does not change across the ribbon, which is a manifestation of the Zitterbewegung effect (attempt to confine Dirac fermions causes wave function delocalization [2]). With respect to the three cases considered above, we point out that the hallmark of Dirac fermions is the even symmetry of the dispersion with respect to the wave vector, and stems from time reversal symmetry. Even though an attempt to confine them may lead to broken electron–hole symmetry [20], the wave vector symmetry is preserved.

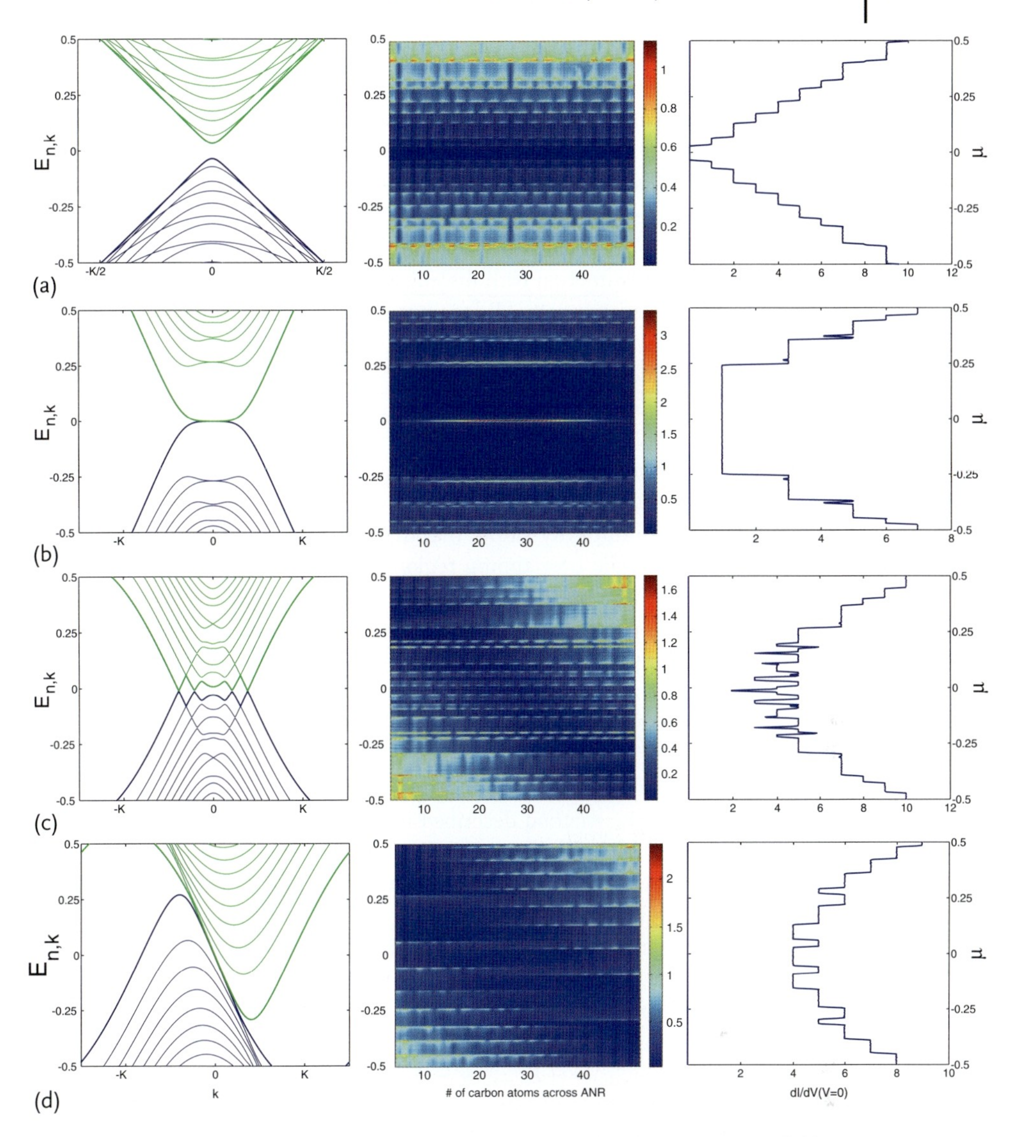

Figure 12.10 Column 1 represents the dispersion curves for the electrons (green curves) in the conduction band and holes (blue curves) in the valence band. The lowest conduction and highest valence subbands are given by the thick curves. Column (2) shows local density of states. Column 3 gives the corresponding ballistic conductance in units of $2e^2/h$. (a) gives results in the absence of the electromagnetic field; (b) is for the sole magnetic field with flux through a single hexagon placket $\phi/\phi_0 = 1/150$; (c) demonstrates the effect of applying only an electric field of strength $U_0/t = 1/2$; (d) is a result of the combined application of an electric and magnetic field with the same strength as employed in (b) and (c).

We now turn our attention to a noteworthy case when "both" electric and magnetic fields are applied together. Concurrent action of the electric field dragging force, the Lorentz force and confinement by the ribbon edges *destroys* the Dirac symmetry of the problem so that $E_{n,k} \neq E_{n,-k}$ as shown in the first panel of Figure 12.10d. The dispersion distortion is different for the electrons and holes, so the symmetry between the conduction and valence bands is also broken. On one hand, the partially formed Landau levels are distorted by the confinement due to the electric field in conjunction with the edges. Their degeneracy is also lifted. On the other hand, the magnetic field does not allow formation of additional Dirac points and wave function delocalization. At high energies, where the group velocity is decreased and the drag due to the electric field prevails, the electrons and holes are gathered at the opposite ribbon edges (Figure 12.10d). At lower energies, in the region $|E_{n,k}| \leq U_0/2$, the electron–hole dispersions overlap. The electron bands only have local minima, whereas the hole bands only have local maxima. Regardless of the broken Dirac k symmetry of the dispersion, our numerical simulation of the differential conductivity shows that the Landauer–Büttiker expression still applies. Therefore, in the overlapping region $|\mu| \leq U_0/2$, the conductivity oscillates since the minimum of the electron band is followed by the maximum on the hole band when the chemical potential is increased. As for possible applications of the broken Dirac symmetry, the ribbon subjected to mutually transverse electric and magnetic fields, may serve as a field-effect transistor with a tunable operating point. An interesting feature of our results is that there is not only a breakdown in the even-k symmetry of the energy dispersion relation, but also the energy bands are reversed with odd symmetry, satisfying $E_{e,k} = -E_{h,-k}$. This effect can be explained by adopting the method described in [19, 21]. We concentrate on the energy region close to the original Dirac point $k = 0$, where the unperturbed wave functions are governed by the conventional Dirac equation. Both applied fields are treated perturbatively. The effect of magnetic field is included through the wave vector replacement $k \to k - \mathcal{B}_z e y/\hbar$. The electric field is treated by a chiral gauge transformation. This transformation shows that the spectrum at $k = 0$ is *affected* by the electric field in the presence of the magnetic field. Regardless of the metallic or semiconducting ANR, the electron and hole dispersions become degenerate around $k = 0$ with $E_{e,k} = -E_{h,-k} \sim -(\mathcal{E}_y/\mathcal{B}_z)k$. A similar result was reported for ZNR configuration around Dirac points using Lorentz transformation for the Dirac fermions, but it was not attributed to the broken chirality since the fermions are not chiral in ZNR.

In conclusion, we demonstrated that when GNRs are placed in mutually perpendicular electric and magnetic fields, there are dramatic changes in their band structure and transport properties. The electric field across the ribbon induces multiple chiral Dirac points, whereas a perpendicular magnetic field induces partially formed Landau levels accompanied by dispersive surface-bound states. Each field preserves the original even parity of the subband dispersion, that is, $E_{n,k} = E_{n,-k}$, maintaining the Dirac fermion symmetry. When applied together, their combined effect is to reverse the dispersion parity to being odd with $E_{e,k} = -E_{h,-k}$ and to mix electron and hole subbands within an energy range equal to the potential drop across the ribbon. Broken Dirac symmetry suppresses the wave function delocalization and the Zitterbewegung effect. The Büttiker formula for the conductance holds true for the odd k symmetry. This, in turn, causes the ballistic conductance to oscillate within this region that can be used to design tunable field-effect transistors.

12.6 Problems

1. Derive the following formula for the dielectric function:

$$\epsilon_{i,j;m,n} = \delta_{i,m}\delta_{j,n} - V_{i,j,m,n}(k_x, k'_x; q) \times \frac{f_0(E_m(k_x + q)) - f_0(E_n(k_x))}{E_m(k_x + q) - E_n(k_x) - \hbar(\omega + i\delta)}, \tag{12.69}$$

where $f_0(E)$ is the Fermi function, $E_n(k_x)$ is the subband energy of GNR, and

$$V_{i,j,m,n} = \int_0^W dy_1 \int_0^W dy_2 \psi^*_{i,k_x}(y_1)\psi_{j,k_x+q}(y_1) \times v_q(y_1 - y_2)\psi^*_{m,k'_x}(y_2)\psi_{n,k'_x+q}(y_2) \tag{12.70}$$

is the Coulomb matrix element. Prove that $v_q(y_2 - y_1) = K_0(q|y_2 - y_1|)$. Consider

$$\Pi^0_{i,j;m,n} = \frac{f_0(E_m(k_x + q)) - f_0(E_n(k_x))}{E_m(k_x + q) - E_n(k_x) - \hbar(\omega + i\delta)} \tag{12.71}$$

for $T = 0$ and for a very little non-zero temperature $T > 0$. Show that the Fermi–Dirac distribution function becomes Heaviside step function when $T \to 0$. Show and explain to what changes in the Coulomb matrix element this temperature leads. Find the matrix element for infinite graphene wavefunction in Eq. (12.15).

2. Find the Brillouin zone for graphene. Write down the tight-binding Hamiltonian for infinite graphene and the corresponding band structure. Show that for $|\boldsymbol{K} - \boldsymbol{k}| \ll K$ and $|\boldsymbol{K}' - \boldsymbol{k}| \ll K'$, it leads to a Dirac-like linear spectrum.

3. Consider the wave function in Eq. (12.32) for zigzag nanoribbons. Show that it is properly normalized, satisfies the Dirac equation and the boundary conditions. What happens to the normalization when $k_y \to 0$? Show that this wavefunction may describe either surface states or bulk states depending on the κ_y.

4. For both zigzag and armchair nanoribbons, find how its width in terms of carbon–carbon distance a depends on the number of carbon atoms across.

5. For ANR (Armchair Nanoribbon), find the subband structure (dispersion $\varepsilon(k_x)$) using the both tight-binding model and Dirac equation. Study the gap

between conduction and valence bands and how it depends on the number of carbon atoms across.
Hint: You are expected to be able to observe a very interesting effect called Modulo 3: the gap is opened for $N_{\text{across}} = 3n$ and $N_{\text{across}} = 3n + 1$, and is closed for $N_{\text{across}} = 3n + 2$. For the latter case, you will also obtain a linear dispersion. Here, n is an arbitrary integer. Prove that the dispersion obtained using $\boldsymbol{k} \cdot \boldsymbol{p}$ approximation is a limiting case of the dispersion obtained from the next-neighbor tight-binding model.

6. Prove directly that the armchair nanoribbon wavefunction in Eq. (12.49) is the eigenfunction of the helicity operator. What about the zigzag wave function in Eq. (12.32)? Is the part in Eq. (12.58) an eigenfunction of any operator? If yes, provide it explicitly. What are the eigenvalues?

7. In the presence of circularly polarized light, the Dirac Hamiltonian [22] becomes

$$\mathcal{H} = \hbar\omega\,\hat{a}^{\dagger}\,\hat{a} + \mathcal{V}_{\mathcal{F}}\;\hat{\sigma}\cdot\hat{p} - e\sqrt{\frac{4\pi\mathcal{V}_{\mathcal{F}}^2}{\omega_0 V}}\,(\hat{\sigma}_{+}\hat{a}_{+} + \hat{\sigma}_{-}\hat{a}_{-})\,. \tag{12.72}$$

Here, the first term describes the number of photons, the second – Dirac electrons (standard Dirac Hamiltonian) and the last term is the interaction between electrons and photons. Show that the energy eigenvalues are given by

$$\epsilon_{N_0}^{\pm}(k) = \hbar\omega_0 \pm \sqrt{\frac{\epsilon_{\text{g}}}{2}^2 + (\hbar\mathcal{V}_{\mathcal{F}}k)^2} \tag{12.73}$$

where

$$\epsilon_{\text{g}} = \sqrt{W_0^2 + (\hbar\omega_0)^2} - \hbar\omega_0 \tag{12.74}$$

and

$$\alpha = \frac{W_0}{\hbar\omega_0} = \frac{2\mathcal{V}_{\mathcal{F}}eE}{\hbar\omega_0^2} \tag{12.75}$$

satisfies the condition $\alpha \ll 1$.

8. For graphene, the polarization function to lowest order in the electron interaction is given by

$$P(q_{||},\omega) \equiv \chi^{(0)}(q_{||},\omega) = \frac{2g_V}{A}\sum_{\boldsymbol{k}_{||}}\sum_{s,s'=\pm1}\frac{f_0(\varepsilon_{k_{||},s}) - f_0(\varepsilon_{|\boldsymbol{k}_{||}+\boldsymbol{q}_{||}|,s'})}{\hbar(\omega + i0^{+}) - (\varepsilon_{|\boldsymbol{k}_{||}+\boldsymbol{q}_{||}|,s'} - \varepsilon_{k_{||}},s)}$$
$$\times\, F_{s,s'}\left(\boldsymbol{k}_{||},\boldsymbol{k}_{||} + \boldsymbol{q}_{||}\right), \tag{12.76}$$

where $\varepsilon_{k_{||},s} = s\hbar v_{\text{F}} k_{||}$ and the sums over s and s' are carried out over the valence and conduction bands, and the form factor comes from the wavefunction and is defined as

$$F_{s,s'}\left(\boldsymbol{k}_{||},\boldsymbol{k}_{||} + \boldsymbol{q}_{||}\right) = \frac{1}{2}\left\{1 + \frac{k_{||} + q_{||}\cos\varphi}{\left|\boldsymbol{k}_{||} + \boldsymbol{q}_{||}\right|}\right\} \tag{12.77}$$

where φ is the angle between $\boldsymbol{k}_{||}$ and $\boldsymbol{q}_{||}$. For non-zero doping, that is, when the chemical potential $\mu > 0$, show that at $T = 0$ with $q_{||} \to q$, we have [23]

$$P(q,\omega) = P_0(q,\omega) + \varDelta P(q,\omega) \,, \tag{12.78}$$

where

$$P_0(q,\omega) = -i\pi \frac{F(q,\omega)}{\hbar^2 v_F^2} \tag{12.79}$$

$$\begin{aligned}\varDelta P(q,\omega) = &-\frac{g_v \mu}{\pi \hbar^2 v_F^2} + \frac{F(q,\omega)}{\hbar^2 v_F^2} \left\{ G\left(\frac{\hbar\omega + 2\mu}{\hbar v_F q}\right) - \theta\left(\frac{2\mu - \hbar\omega}{\hbar v_F q} - 1\right) \right. \\ &\left. \times \left[G\left(\frac{2\mu - \hbar\omega}{\hbar v_F q}\right) - i\pi \right] - \theta\left(\frac{\hbar\omega - 2\mu}{\hbar v_F q} + 1\right) G\left(\frac{\hbar\omega - 2\mu}{\hbar v_F q}\right) \right\}\end{aligned} \tag{12.80}$$

where

$$P(q,\omega) = \frac{g_v}{8\pi} \frac{\hbar v_F^2 q^2}{\sqrt{\omega^2 - v_F^2 q^2}} \,, \quad G(x) = x\sqrt{x^2 - 1} - \ln\left(x + \sqrt{x^2 - 1}\right) . \tag{12.81}$$

In the RPA, the renormalized polarization function is

$$P_{\text{RPA}}(q,\omega) = \frac{P(q,\omega)}{1 - v(q) P(q,\omega)} \,, \tag{12.82}$$

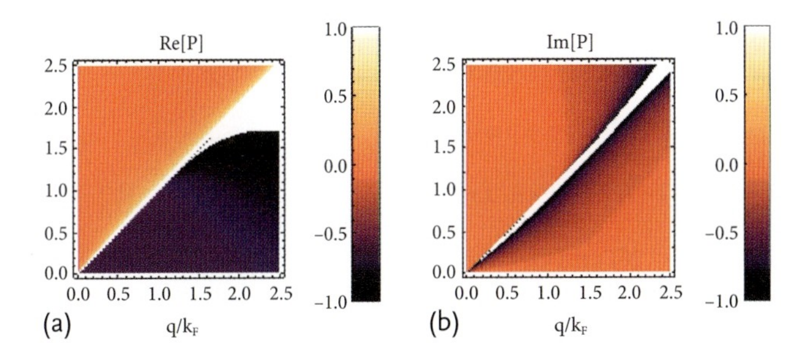

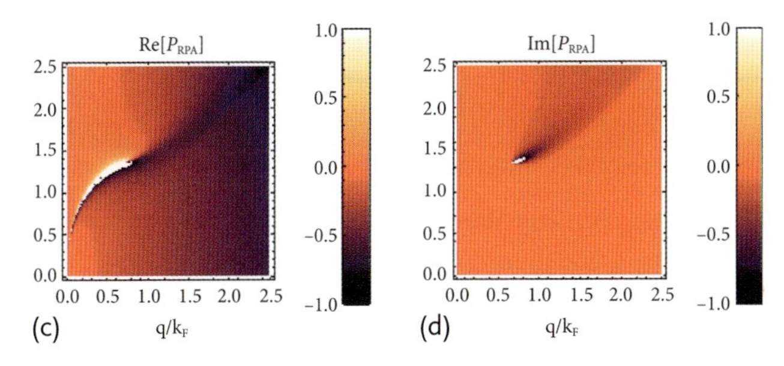

Figure 12.11 Density plots for the (a) real part of $P(q,\omega)$, (b) imaginary part of $P(q,\omega)$, and the (c) real part of $P_{\text{RPA}}(q,\omega)$ and (d) imaginary part of $P_{\text{RPA}}(q,\omega)$.

where $v(q) = 2\pi e^2/(\epsilon_s q)$, where $\epsilon_s = 4\pi\epsilon_0\epsilon_b$, with the background dielectric constant for graphene $\epsilon_b = 12$. Verify the plots in Figure 12.11 for the real and imaginary parts of the polarization function.

Appendix 12.A
Energy Eigen States

In this appendix, we deal with the problem of finding eigenvalues and eigenfunctions of both types of nanoribbons when an electromagnetic field is applied as shown in Figure 12.9. We employ the tight-binding model, where the magnetic field is included via the Peierls phase added to the hopping coefficients between two sublattices. Graphene nanoribbons are made of two equivalent sublattices A and B of carbon atoms with sublattice constant a. The π and π^* hybridization of p valence electrons at each carbon atom $|A(B); m, n\rangle$ provides the electron wave function in the Bloch form along the ribbon:

$$|\Psi(k)\rangle = 2 \sum_{0<n<N} \Psi_A(n,k)|A; n, k\rangle + \Psi_B(n,k)|B; n, k\rangle \,, \tag{12.A.1}$$

where $k = k_y a$ is the particle wave vector along the ribbon.

It is convenient to group coefficients in Eq. (12.A.1) into component wave function:

$$\Psi(n,k) = \begin{bmatrix} \Psi_A(n,k) \\ \Psi_B(n,k) \end{bmatrix} . \tag{12.A.2}$$

This satisfies the following Schrödinger equation:

$$E_{n,k}\Psi(n,k) = \left[\mathcal{H}(\phi) + \mathcal{U}(U)\right]\Psi(n,k) \,. \tag{12.A.3}$$

Here, $E_{n,k}$ is the hole $(1 < n < N)$ and electron $(N + 1 < n < 2N)$ energies given in units of hopping energies between the sublattices, that is, $t \approx 2.8$ eV (in the absence of magnetic field). Also, ϕ is the magnetic flux through a plaquette in units of the quantum magnetic flux ϕ_0 and U is the modulation field across the ribbon. The first term in Eq. (12.A.3) is sensitive to the nanoribbon edge configuration and we have

$$\mathcal{H}_{\text{ZNR}} = \begin{bmatrix} 0 & Q_{\text{ZNR}} \\ Q^T_{\text{ZNR}} & 0 \end{bmatrix} , \quad \mathcal{H}_{\text{ANR}} = \begin{bmatrix} 0 & Q_{\text{ANR}} \\ Q^\dagger_{\text{ANR}} & 0 \end{bmatrix} . \tag{12.A.4}$$

The $N \times N$ Q matrix for the zigzag nanoribbon is dual-diagonal, with matrix elements given by

$$Q^{\text{ZNR}}_{n,n'} = 2\delta_{n,n'}\cos(\alpha_{n,k}) + \delta_{n,n'-1} \,, \tag{12.A.5}$$

where the magnetic field dependent Peierls phase assumes the form $\alpha_{n,k} = \sqrt{3}k/2 + (n + 1/6)\pi\phi$. The $N \times N$ Q matrix corresponding to the armchair configuration is tri-diagonal with

$$Q^{\text{ANR}}_{n,n'} = \delta_{n,n'} e^{-i(2\alpha_{n,k} - \pi\phi/3)} + \delta_{n,n'-1} e^{i\alpha_{n,k}} + \delta_{n,n'+1} e^{i\alpha_{n-1,k}} \tag{12.A.6}$$

and the Peierls phase has the form of $\alpha_{n,k} = \sqrt{3}k/2 + (n/3 + 1/6)\pi\phi$. The forms of the Q matrices in Eqs. (12.A.5) and (12.A.6) ensure that the hard wall boundary conditions $\Psi(0,k) = \Psi(N+1,k) = 0$ are satisfied. For our convenience, we also change the coordinate origin to the center of the ribbon with $n \to (N-1)/2 - n + 1$.

The modulation part of the Hamiltonian in Eq. (12.A.3) is of block-diagonal form

$$\mathcal{U} = \begin{bmatrix} U & 0 \\ 0 & U \end{bmatrix}, \tag{12.A.7}$$

where the $N \times N$ sub-matrices have the diagonal form

$$U_{n,n'} = \delta_{n,n'} \begin{cases} U_0 \sin\left[\left(k\left(\frac{N}{2-n}\right)\right)\right] & \text{for modulation} \\ U_0\left(\frac{1}{2} - \frac{n}{N}\right) & \text{for DC bias} \end{cases} . \tag{12.A.8}$$

Appendix 12.B
The Conductance

For symmetric dispersion, Eq. (12.68) reduces to the famous Landauer–Büttiker formula. That is, the conductivity is given by the amount of right moving channels $v_{n,k}/|v_{n,k}| > 0$ at the chemical potential $E_{n,k} = \mu$. The symmetric bands are characterized in terms of the following self-explanatory conditions:

$$E_{n,k} = E_{n,-k}, \quad f_{n,k} = f_{n,-k}, \quad v_{n,k} = -v_{n,-k}. \tag{12.B.1}$$

Therefore, we can rewrite Eq. (12.68) in the form which retains only positive values of the wave vector:

$$\begin{aligned} I = -\frac{2e}{\hbar} \sum_{n,k\geq 0} v_{n,k} \Big[& \theta(-v_{n,k}) f^{>}_{n,k}\left(1 - f^{<}_{n,k}\right) + \theta(v_{n,k}) f^{<}_{n,k}\left(1 - f^{>}_{n,k}\right) \\ & - \theta(v_{n,k}) f^{>}_{n,k}\left(1 - f^{<}_{n,k}\right) - \theta(-v_{n,k}) f^{<}_{n,k}\left(1 - f^{>}_{n,k}\right) \Big]. \end{aligned} \tag{12.B.2}$$

All terms of the sort $\sim f^{<}_{n,k} f^{>}_{n,k}$ cancel each other and we are left with

$$\begin{aligned} I = -\frac{2e}{\hbar} \sum_{n,k\geq 0} v_{n,k} \big[& \theta(-v_{n,k}) f^{>}_{n,k} + \theta(v_{n,k}) f^{<}_{n,k} \\ & - \theta(v_{n,k}) f^{>}_{n,k} - \theta(-v_{n,k}) f^{<}_{n,k} \big]. \end{aligned} \tag{12.B.3}$$

Now, we employ the following identity

$$|v_{n,k}| = v_{n,k}\theta(v_{n,k}) - v_{n,k}\theta(-v_{n,k}). \tag{12.B.4}$$

Combining Eqs. (12.B.3) and (12.B.4), we obtain

$$I = -\frac{2e}{\hbar} \sum_{n,k \geq 0} |v_{n,k}| \left(f^{<}_{n,k} - f^{>}_{n,k} \right) . \tag{12.B.5}$$

In the linear approximation with respect to the DC bias, we have

$$f^{<}_{n,k} - f^{>}_{n,k} = \left(\frac{d f^{>}_{n,k}}{d E_{n,k}} \right) eV . \tag{12.B.6}$$

Summation over the wave vector can be changed to integration over the energy regions where the group velocity keeps its sign:

$$\sum_k = \frac{1}{\pi} \int dk = \frac{1}{\pi} \left[\int_{E_{n,k0}}^{E_{n,k1}} \frac{d E_{n,k}}{|v_{n,k}|} + \int_{E_{n,k1}}^{E_{n,k2}} \frac{d E_{n,k}}{|v_{n,k}|} + \cdots \right] . \tag{12.B.7}$$

Applying Eqs. (12.B.6) and (12.B.7) to Eq. (12.B.5), we obtain the conductance in the form

$$G = \frac{I}{V} = \frac{2e^2}{\hbar} \sum_{n,\gamma} \left[f^{>}_{n,\gamma}(\text{min}) - f^{>}_{n,\gamma}(\text{max}) \right] . \tag{12.B.8}$$

This expression suggests the following interpretation of the conductance. For each of the energy subbands (n), we have to count the number of minima below the chemical potential (first term in Eq. (12.B.8)) and subtract the number of local maxima (second term in Eq. (12.B.8)). For example, in [9, Figure 1], we have

$$\begin{aligned}
G(\mu_3) &= \frac{2e^2}{\hbar} \left[2 f^{>}_{1,k1} \right] = 2 \frac{2e^2}{\hbar} , \\
G(\mu_2) &= \frac{2e^2}{\hbar} \left[2 f^{>}_{1,k1} - f^{>}_{1,k0} \right] = \frac{2e^2}{\hbar} , \\
G(\mu_1) &= \frac{2e^2}{\hbar} \left[2 f^{>}_{1,k1} - f^{>}_{1,k0} + f^{>}_{2,k0} \right] = 2 \frac{2e^2}{\hbar} .
\end{aligned} \tag{12.B.9}$$

Now, let us connect Eq. (12.B.8) to the Büttiker formula which says that you have to count the net number of right-moving channels or, alternatively, the net number of Fermi points for $k \geq 0$. To do so, we again open up summation in Eq. (12.B.3) to the whole range of the wave vector space, that is,

$$\begin{aligned}
I &= -\frac{2e}{\hbar} \left[\sum_{n,k<0} -v_{n,k} \theta(v_{n,k}) \left(f^{>}_{n,k} - f^{<}_{n,k} \right) \right. \\
&\quad \left. + \sum_{n,k \geq 0} v_{n,k} \theta(v_{n,k}) \left(f^{<}_{n,k} - f^{>}_{n,k} \right) \right] \\
&= -\frac{2e}{\hbar} \sum_{n,k} v_{n,k} \theta(v_{n,k}) \left(f^{<}_{n,k} - f^{>}_{n,k} \right) ,
\end{aligned} \tag{12.B.10}$$

where we take Eq. (12.B.1) into account. Note that

$$f^{<}_{n,k} - f^{>}_{n,k} \approx -eV\delta(E_{n,k} - \mu)\,, \tag{12.B.11}$$

and therefore the conductance given by Eq. (12.B.10) is

$$\begin{aligned} G &= \frac{2e^2}{\hbar}\sum_{n,k}\int dE_{n,k}\theta(v_{n,k})\delta(E_{n,k}-\mu) \\ &= \frac{2e^2}{\hbar}\sum_{n,k}\theta(v_{n,k})\mid_{E_{n,k}=\mu}\,. \end{aligned} \tag{12.B.12}$$

This is an exact Büttiker formulation of the conductance in the form of the number of right movers as reported in [11].

References

1 Geim, A.K. (2009) Graphene: Status and prospects. *Science*, **324**, 1530.

2 Neto, A.H.C., Guinea, F., Peres, N.M.R., Novoselov, K.S., and Geim, A.K. (2009) The electronic properties of graphene. *Rev. Mod. Phys.*, **81**, 109.

3 Novoselov, K.S., Geim, A.K., Morozov, S.V., Jiang, D., Zhang, Y., Dubonos, S.V., Grigorieva, I.V., and Firsov, A.A. (2004) Electric field effect in atomically thin carbon films. *Science*, **306**, 666.

4 Katsnelson, M.I., Novoselov, K.S., and Geim, A.K. (2006) Chiral tunnelling and the Klein paradox in graphene. *Nat. Phys.*, **2**, 620.

5 Cheianov, V.V. and Fal'ko, V.I. (2006) Selective transmission of Dirac electrons and ballistic magnetoresistance of n-p junctions in graphene. *Phys. Rev. B*, **74**, 041403.

6 Cayssol, J., Huard, B., and Goldhaber-Gordon, D. (2009) Contact resistance and shot noise in graphene transistors. *Phys. Rev. B*, **79**, 075428.

7 Nakabayashi, J., Yamamoto, D., and Kurihara, S. (2009) Band-selective filter in a zigzag graphene nanoribbon. *Phys. Rev. Lett.*, **102**, 066803.

8 Cresti, A., Grosso, G., and Parravicini, G.P. (2008) Valley-valve effect and even-odd chain parity in p-n graphene junctions. *Phys. Rev. B*, **77**, 233402.

9 Rainis, D., Taddei, F., Dolcini, F., Polini, M. and Fazio, R. (2009) Andreev reflection in graphene nanoribbons. *Phys. Rev. B*, **79**, 115131.

10 Wakabayashi, K., Takane, Y., Yamamoto, M., and Sigrist, M. (2009) Edge effect on electronic transport properties of graphene nanoribbons and presence of perfectly conducting channel. *Carbon*, **47**, 124.

11 Li, Z., Qian, H., Wu, J., Gu, B.-L., and Duan, W. (2008) Role of symmetry in the transport properties of graphene nanoribbons under bias. *Phys. Rev. Lett.*, **100**, 206802.

12 Akhmerov, A.R., Bardarson, J.H., Rycerz, A., and Beenakker, C.W.J. (2008) Theory of the valley-valve effect in graphene nanoribbons. *Phys. Rev. B*, **77**, 205416.

13 Trauzettel, B., Bulaev, D.V., Loss, D., and Burkard, G. (2007) Spin qubits in graphene quantum dots. *Nat. Phys.*, **3**, 192.

14 Lin, H.H., Hikihara, T., Jeng, H.T., Huang, B.L., Mou, C.Y., and Hu, X. (2009) Ferromagnetism in armchair graphene nanoribbons. *Phys. Rev. B*, **79**, 35405.

15 Liu, J., Ma, Z., Wright, A.R., and Zhang, C. (2008) Orbital magnetization of graphene and graphene nanoribbons. *J. Appl. Phys.*, **103**, 103711.

16 Simon, L., Bena, C., Vonau, F., Aubel, D., Nasrallah, H., Habar, M., and Peruchetti, J.C. (2009) Symmetry of standing waves generated by a point defect in epitaxial graphene. *Europ. Phys. B*, **69**, 351.

17 Büttiker, M. (1988) Absence of backscattering in the quantum Hall effect in multiprobe conductors. *Phys. Rev. B*, **38**, 9375.

18 Lyo, S.K. and Huang, D.H. (2004) Quantized magneto-thermopower in tunnel-coupled ballistic channels: sign reversal and oscillations. *J. Phys.: Condens. Matt.*, **16**, 3379.

19 Brey, L. and Fertig, H.A. (2009) Emerging zero modes for graphene in a periodic potential. *Phys. Rev. Lett.*, **103**, 46809.

20 Peres, N.M.R., Neto, A.H.C., and Guinea, F. (2006) Dirac fermion confinement in graphene. *Phys. Rev. B*, **73**, 241403.

21 Novikov, D.S. and Levitov, L.S. (2006) Energy anomaly and polarizability of carbon nanotubes. *Phys. Rev. Lett.*, **96**, 36402.

22 Kibis, O.V. (2010) Metal-insulator transition in graphene induced by circularly polarized photons. *Phys. Rev. B*, **81**, 165433.

23 Wunsch, B., Stauber, T., Sols, F., and Guinea, F. (2006) Dynamical polarization of graphene at finite doping. *New J. Phys.*, **8**, 318.

13
Semiclassical Theory for Linear Transport of Electrons

Recently, much attention has been focused on low temperature ballistic quantum transport through a single narrow constricted channel (or wire), the so-called quantum point contact [1–3] and the tunnel-coupled double wires [4–7]. A single quantum well wire (SQWR) is schematically shown in Figure 13.1a. This channel consists of an electron gas, for example, in a thin, highly conducting GaAs layer (~100 Å) confined between $Al_xGa_{1-x}As$ layers in the growth (z) direction. The current flows in the y-direction through a narrow quasi-one-dimensional (1D) wire region that is formed by further constricting the current in the perpendicular (x) direction by applying a negative bias to the split metallic gate on top of the $Al_xGa_{1-x}As$ layer, as shown in Figure 13.1a. In this structure, only the ground sublevel is occupied in the z-direction. However, the confinement in the x-direction is much less severe and produces many closely separated sublevels (to be defined as channel sublevels). For a channel width of the order of which can be controlled by an applied bias, the energy separation for the low-lying sublevels is a small fraction of an meV. The energy dispersion curves of these sublevels are illustrated in Figure 13.1b.

A tunnel-coupled double well wire (DQWR) is illustrated in Figure 13.1c [4–7]. In this DQWR structure, the two GaAs conducting channels are separated by a thin $Al_xGa_{1-x}As$ barrier that allows the electrons to tunnel between the two GaAs channels. The channel constriction in the x-direction is achieved in both channels independently through the top and bottom split gates. Electron tunneling deforms the electronic structure in the channel direction dramatically in the presence of an in-plane magnetic field B due to the anti-crossing effect as illustrated in Figure 13.1d [8–10]. Here two thick solid curves represent the lower and upper branches of the tunnel-split ground-state doublet separated by the anti-crossing gap for the ground channel sublevel. These branches are made of two ground-state parabolas from each well which are displaced in k_y space relative to each other, with the degeneracy at the intersecting point split and the curves near this point deformed by the anti-crossing gap as shown [11]. The humps in Figure 13.1d develop at a sufficiently high B and the gap passes through the chemical potential as B increases. The thin curves are replicas of these curves: each pair represents a higher channel sublevel $n = 1, 2, \dots$

Properties of Interacting Low-Dimensional Systems, First Edition. G. Gumbs and D. Huang.

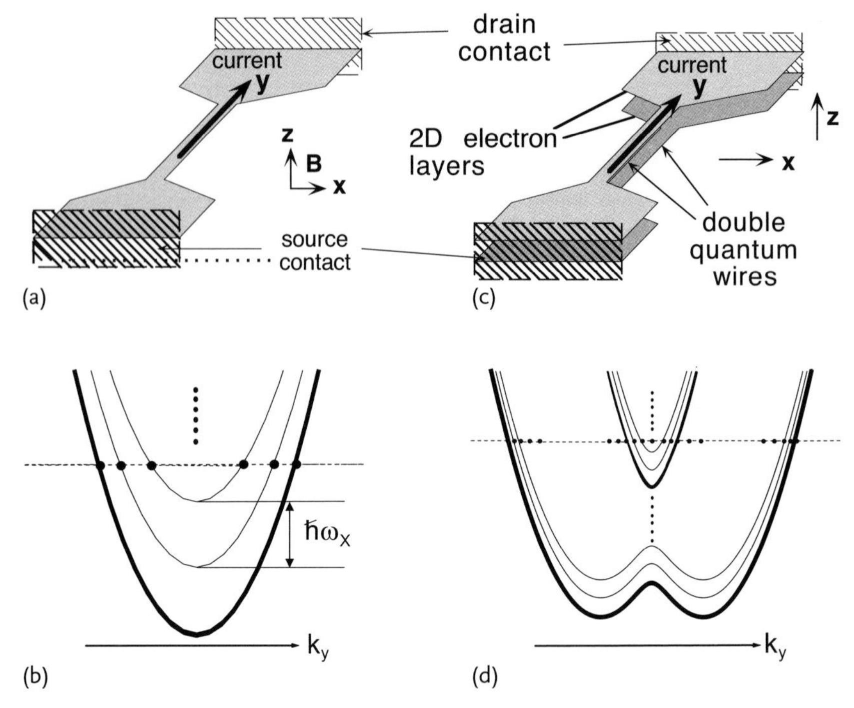

Figure 13.1 (a) A schematic diagram of a SQWR. (b) Parallel energy-dispersion curves of the channel sublevels of a SQWR. (c) A schematic diagram of DQWRs. (d) The energy-dispersion curves of tunnel-coupled symmetric DQWRs. The tunnel-split ground doublet for the ground ($n = 0$) channel sublevel is shown in thick curves for upper and lower branches. The thin curves (including the higher-energy levels represented by the vertical dots) are replicas of these curves shifted uniformly by $\hbar\omega_x$ in the harmonic channel confinement model and belong to the ground doublet. The horizontal black dots represent the Fermi points. The current flows in the y-direction. A magnetic field $\boldsymbol{B}$ is in the x-direction for the DQWRs and is either in the x- or in the z-direction for the SQWR.

When the bias is low, electrons are only weakly driven away from their equilibrium state. In this case, the non-equilibrium part of electron distribution function is proportional to the bias. Therefore, the semiclassical Boltzmann transport equation can be linearized with respect to the non-equilibrium part of electron distribution function. As a result, the mobility of electrons turns out to be independent of the bias that becomes a characteristic quantity of the system studied. However, it can still be affected by an external magnetic field.

13.1
Roughness Scattering

In this section, we consider two systems consisting of either a SQWR or a DQWRs and are schematically illustrated in Figure 13.1a,c in the presence of a magnetic field. While the treatment is specifically given for interface-roughness scattering, the same basic formalism is applicable to scattering by ionized impurities.

13.1.1
Model for Elastic Scattering

For both systems, the linearized Boltzmann transport equation for a steady state is given by [12, 13]

$$v_j + \frac{2\pi}{\hbar}\sum_{j'} |I_{j',j}|^2 (g_{j'} - g_j)\delta\left(\varepsilon_j - \varepsilon_{j'}\right) = 0 , \tag{13.1}$$

where $j = \{n, m, k_y\}$ represents a set of quantum numbers, k_y is the wave number of an electron along the wire, $n = 0, 1, \ldots$ is the channel-sublevel quantum number associated with the quantization in the x-direction, ε_j is the kinetic energy of an electron, and $v_j = \hbar^{-1} d\varepsilon_j / dk_y$ is the electron group velocity along the wire. The quantity m is the sublevel index quantum number associated with the quantization in the z-direction and is restricted to the ground-state level $m = 0$ for a SQWR and $m = 0, 1$ for the ground-state doublet of DQWRs at low T. The quantity g_j characterizes the non-equilibrium component of the distribution function $f_j = f_0(\varepsilon_j) + g_j[-f_0'(\varepsilon_j)]e\mathcal{E}_0$, where the second term represents the linear deviation from the equilibrium Fermi function $f_0(\varepsilon_j)$, $\mathcal{E}_0$ is a DC electric field and $f_0'(\varepsilon_j)$ is the first derivative of the Fermi function with respect to electron energy.

The energy of an electron is given by $\varepsilon_j = (n + 1/2)\hbar\omega_x + E^z_{mk_y}$ and is shown in Figure 13.1b. The energy $E^z_{mk_y}$ and the wave function $\xi_{mk_y}(z)$ in the z-direction can be calculated numerically through the Schrödinger equation with the following Hamiltonian

$$\mathcal{H}_z = -\frac{\hbar^2}{2}\frac{d}{dz}\left[\frac{1}{m^*(z)}\frac{d}{dz}\right] + V(z) + \frac{\hbar^2}{2m^*(z)}\left(k_y - \frac{z}{\ell_c^2}\right)^2 \quad \text{for } B \| x . \tag{13.2}$$

Here, $\ell_c = \sqrt{\hbar/eB}$, $m^*(z) = m_W = 0.067 m_0$ in GaAs, $m^*(z) = m_B = 0.073 m_0$ in $Al_xGa_{1-x}As$ with m_0 being the free-electron mass, and $V(z) = V_{SQW}(z)$ for SQWRs and $V(z) = V_{DQW}(z)$ for DQWRs are the quantum-well potential that is V_c within the barrier layers and zero inside the well. The total wave function related to the electron energy ε_j is given by

$$\Psi_j(x, y, z) = \phi_n\left(\frac{x}{\ell_x}\right)\frac{\exp(i k_y y)}{\sqrt{L_y}}\xi_{mk_y}(z) , \tag{13.3}$$

where $\ell_x = \sqrt{\hbar/m^*\omega_x}$, $\phi_n(x/\ell_x)$ is the harmonic oscillator wave function in the x-direction and L_y is the wire length.

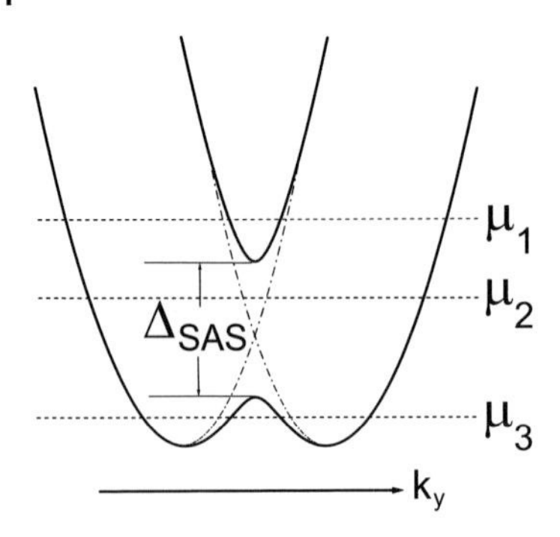

Figure 13.2 Energy dispersion (thick solid curves) for the ground-state doublet with $n = 0$ and $m = 0, 1$ in symmetric double quantum well wires. The dashed horizontal lines show the position of the chemical potential $\mu = \mu_1$ above the gap at low B, $\mu = \mu_2$ inside the gap at medium B, and $\mu = \mu_3$ below the gap at high B. The dashed-dotted curves are the energy parabolas in the absence of inter-layer tunneling. An anti-crossing gap on the order of Δ_{SAS} at $B = 0$ is opened at the intersecting point of two parabolas by electron tunneling.

In the following, we obtain g_j's and the conductance. Equation (13.1) can be rewritten after carrying out the k'_y integration as

$$s_\nu + \sum_{\nu'=1}^{N_F} u_{\nu',\nu}(g_{\nu'} - g_\nu) = 0 , \tag{13.4}$$

where N_F is a large even number representing the total number of Fermi points, $s_\nu = v_\nu/|v_\nu|$, and ν represents each intersecting point of the energy parameter ε with the dispersion curve labeled by the quantum numbers $\{n, m\}$. These points become the Fermi points at zero temperature. The set of the quantum numbers $\{n, m, k_y\}$ at the energy ε will still be called the "Fermi points" in this chapter at finite T for convenience. In addition, the $N_F \times N_F$ scattering matrix $\overleftrightarrow{\mathbf{u}}$ introduced in Eq. (13.4) is related to the quantity $|I_{\nu',\nu}|^2$ in Eq. (13.1) and is given by [13]

$$u_{\nu',\nu} \equiv \frac{L_y}{\hbar^2}\frac{|I_{\nu',\nu}|^2}{|v_{\nu'}v_\nu|} = \frac{\Lambda_0^2 V_c^2 \delta b^2}{\hbar^2|v_{\nu'}v_\nu|\ell_x}\exp\left[-\frac{1}{4}\left(k_y - k'_y\right)^2\Lambda_0^2\right]|\xi_{m'k'_y}(z_1)\xi_{mk_y}(z_1)|^2$$
$$\times \int_0^\infty d\eta \exp(-\eta^2/2) L_n(\eta^2/2) L_{n'}(\eta^2/2) \quad \text{for } \nu \neq \nu' , \tag{13.5}$$

where $L_n(x)$ is the nth-order Laguerre polynomial, δb is the amplitude of the layer fluctuation $b(\mathbf{r}_\parallel)$, Λ_0 is the correlation length defined by $\langle b(\mathbf{r}_\parallel)b(\mathbf{r}'_\parallel)\rangle = \delta b^2 \exp\{-[(x-x')^2 + (y-y')^2]/\Lambda_0^2\}$. Here, the angular brackets signify the spatial average. The quantity z_1 is the position of the rough $Al_xGa_{1-x}As/GaAs$ interface. For DQWRs, the contribution from the other rough $Al_xGa_{1-x}As/GaAs$ interface at $z = z_2$ is added to the expression in Eq. (13.5). The $GaAs/Al_xGa_{1-x}As$ interfaces are known to be smooth.

By defining the diagonal elements for $\overleftrightarrow{\mathbf{u}}$ as

$$u_{\nu,\nu} = -\sum_{\nu' \neq \nu} u_{\nu,\nu'} \quad \text{for } \nu = 1, 2, \ldots, N_{\mathrm{F}} \,, \tag{13.6}$$

and introducing

$$\mathbf{g} = \begin{bmatrix} g_1 \\ g_2 \\ \vdots \\ g_{N_{\mathrm{F}}} \end{bmatrix} \quad \text{and} \quad s = \begin{bmatrix} s_1 \\ s_2 \\ \vdots \\ s_{N_{\mathrm{F}}} \end{bmatrix} , \tag{13.7}$$

we can cast Eq. (13.4) into a linear matrix equation

$$\overleftrightarrow{\mathbf{u}} \otimes \mathbf{g} = -s \,, \tag{13.8}$$

where the notation $\otimes$ corresponds to the product of two matrices. Unfortunately, $\overleftrightarrow{\mathbf{u}}$ does not have an inverse [i.e., $\det(\overleftrightarrow{\mathbf{u}}) = 0$] since the sum of all the rows of $\overleftrightarrow{\mathbf{u}}$ vanishes for each column [13]. To avoid this problem, we discard the last row in Eq. (13.8) and obtain the coupled equations:

$$\overleftrightarrow{\mathrm{U}} \otimes \boldsymbol{G} + g_{N_{\mathrm{F}}} \boldsymbol{U}_{N_{\mathrm{F}}} = -\boldsymbol{S} \,, \tag{13.9}$$

where $\overleftrightarrow{\mathrm{U}}$ is a $(N_{\mathrm{F}} - 1) \times (N_{\mathrm{F}} - 1)$ sub-matrix obtained by discarding the last row and the last column of $\overleftrightarrow{\mathbf{u}}$, $\boldsymbol{U}_{N_{\mathrm{F}}}$ is the last column vector of $\overleftrightarrow{\mathbf{u}}$ without the last element and $\boldsymbol{S}$, $\boldsymbol{G}$ are obtained from s, $\mathbf{g}$ by truncating the last elements $S_{N_{\mathrm{F}}}$ and $g_{N_{\mathrm{F}}}$, respectively. Further introducing a new column vector

$$\tilde{\mathbf{g}} = \begin{bmatrix} \tilde{g}_1 \\ \tilde{g}_2 \\ \vdots \\ \tilde{g}_{N_{\mathrm{F}}-1} \end{bmatrix} = \boldsymbol{G} - g_{N_{\mathrm{F}}} \begin{bmatrix} 1 \\ 1 \\ \vdots \\ 1 \end{bmatrix} , \tag{13.10}$$

where $\tilde{g}_\ell$ is related to the original g_ℓ by $\tilde{g}_\ell = g_\ell - g_{N_{\mathrm{F}}}$, we obtain from Eq. (13.9)

$$\overleftrightarrow{\mathrm{U}} \otimes \tilde{\mathbf{g}} = -\boldsymbol{S} \,, \tag{13.11}$$

yielding the solution $\tilde{\mathbf{g}} = -(\overleftrightarrow{\mathrm{U}})^{-1} \otimes \boldsymbol{S}$.

Finally, the conductances for both SQWR and DQWR systems can be analytically calculated by [13]

$$\begin{aligned} G_{yy}(B) &= \frac{2e^2}{hL_y} \int_0^\infty d\varepsilon \left[-f_0'(\varepsilon) \right] \boldsymbol{S}^\dagger \otimes \tilde{\mathbf{g}} \\ &= \frac{2e^2}{hL_y} \int_0^\infty d\varepsilon \left[-f_0'(\varepsilon) \right] \sum_{\nu=1}^{N_{\mathrm{F}}-1} s_\nu(\varepsilon) \tilde{g}_\nu(\varepsilon) \,, \end{aligned} \tag{13.12}$$

where $S^\dagger$ is the complex-transpose of S.

For the ballistic motion of electrons along quantum wires, the quantized conductance $G_0(B)$ is found to be [14]

$$G_0(B) = \frac{2e^2}{h} \sum_s \sum_\gamma C_{s,\gamma} \, f_0[\varepsilon_s(k_\gamma)] \,, \tag{13.13}$$

where s- and γ-summation indicate summing over all the energy dispersion curves and over all the energy extremum points $k_y = k_\gamma$ on each curve, respectively. The quantity $\varepsilon_s(k_\gamma)$ is the extremum energy. For a given curve s, $C_{s,\gamma} = 1$ for a local energy minimum point and $C_{s,\gamma} = -1$ for a local energy maximum point. At $T = 0$ K, the quantity that follows $2e^2/h$ in Eq. (13.13) represents the total number of the Fermi points.

13.1.2
Numerical Results for Roughness Scattering Effect

The parameters employed for the sample in our calculation are listed in Table 13.1. For this sample, we use $V_0 = 280$ meV, $m_W = 0.067$, and $m_B = 0.073$. The level-broadening parameters are chosen to be $\gamma_\nu = 0.1\Delta_{SAS}$ for symmetric double-quantum-wire samples where Δ_{SAS} is the splitting between the symmetric and antisymmetric states at $B = 0$. The other roughness-related parameters are $\Lambda_0 = 30$ Å and $\delta b = 5$ Å. The double QW's are assumed to be rectangular wells in the z-direction. In the following application, only the ground tunnel-split doublet is populated for double wires. Other parameters will be given in the corresponding figure captions.

We show in Figure 13.3 the conductance ratio $G_{\gamma\gamma}(B)/G_{\gamma\gamma}(0)$ (lower thick curves) and the quantized conductance $G_0(B)$ (upper thin curves) at $T = 0$ K as a function of B for the DQWR sample, listed in Table 13.1, for several electron densities $n_{1D} = 1 \times 10^7$ cm^{-1} (red curves), 2×10^7 cm^{-1} (blue curves), and 3×10^7 cm^{-1} (green curves). $G_0(B)$ exhibits a V-shape as a function of B. This B-dependence was explained earlier in detail [15] and can be understood with the following simple argument and is also useful for understanding the B-dependence of the diffusive conductance to be presented below. At $B = 0$, each $m = 0, 1$ pair of the doublet consists of two parallel parabolas and generates four Fermi points except for a few large-n top sublevels near the chemical potential $\mu = \mu_1$ in Figure 13.2, assuming a high density of electrons. As B increases, the upper and

Table 13.1 Double-quantum-well wires with well depth of $V_0 = 280$ meV, widths L_{z1}, L_{z2}, center-barrier width L_B, ground-doublet tunnel splitting Δ_{SAS} at $B = 0$, and the uniform channel sublevel separation $\hbar\omega_x$.

L_{z1}/L_{z2} (Å)	L_B (Å)	Δ_{SAS} (meV)	$\hbar\omega_x$ (meV)
80/80	50	1.6	0.02

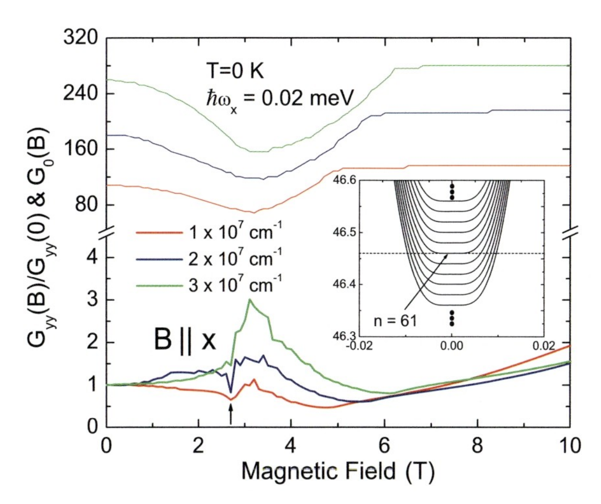

Figure 13.3 $G_{yy}(B)/G_{yy}(0)$ (thick curves) and $G_0(B)$ (thin curves) in unit of $2e^2/h$ for the sample listed in Table 13.1 with $n_{1D} = 1 \times 10^7\ \text{cm}^{-1}$ (red curves), $2 \times 10^7\ \text{cm}^{-1}$ (blue curves) and $3 \times 10^7\ \text{cm}^{-1}$ (green curves) at $T = 0\,\text{K}$ as a function of B in the x-direction. Here, $G_{yy}(0) = 17.7e^2/h$ (thick red curve), $28.4e^2/h$ (thick blue curve) and $32.5e^2/h$ (thick green curve) for $L_y = 1\,\mu\text{m}$. Both branches are occupied for blue and green curves, while only the lower branch is occupied for the red curve at $B = 0$. The arrow indicates the dips near $B = 2.7\,\text{T}$ where the bottom of the lower branch becomes flat just before the hump develops as shown in the inset. The latter presents $\varepsilon_{n,1,k_y}$ in units of meV as a function of k_y (in $0.1\pi\,\text{Å}^{-1}$) at $B = 2.7\,\text{T}$. The horizontal dashed line indicates the Fermi level nested at the sublevel $n = 61$.

lower branches of each sublevel n deform from a pair of parallel parabolas into the anticrossing structure with a gap shown in Figure 13.1d by thick curves, for example, for $n = 0$. At high fields, the gaps sweep through the chemical potential successively starting from large n. For each pair, the number of the Fermi points decreases from four to two when the chemical potential $\mu = \mu_2$ is in the gap and increases back to four when the gap moves above the chemical potential $\mu = \mu_3$, as shown in Figure 13.2. Therefore, the minimum $G_0(B)$ is obtained when the chemical potential lies in the middle of the anticrossing gaps of the majority of the channel sublevels. The minimum of $G_0(B)$ shifts to a higher B for a higher-density sample.

An interesting note is that the maximum of $G_{yy}(B)/G_{yy}(0)$ is aligned with the minimum of $G_0(B)$ in Figure 13.3 for each density. This behavior is readily understood if we first consider an extremely narrow channel where $\hbar\omega_x$ is very large and assume that only the ground ($n = 0$) doublet is occupied [16]. In this case, $G_0(B)$ is minimum when the chemical potential lies inside the gap with two Fermi points as explained above. Also, the conductance becomes very large due to the fact that back scattering occurs between the initial (k_i) and final ($k_f = -k_i$, say, in a symmetric structure) Fermi points in the lower branch ($m = 0$). For these two points, the wave functions $\xi_{mk_i}(z)$ and $\xi_{mk_f}(z)$ are localized in the opposite wells, yielding very small scattering matrix $u_{j',j}$ and a large conductance [16]. For a wide channel with many sublevels ($N_F \gg 1$) populated at high density, however, there are some sublevels for which the Fermi level is outside their gaps, although the majority of the sublevels have the Fermi level inside their gaps at the $G_0(B)$ minimum. The wave functions of the Fermi points outside the gap have significant amplitudes in both wells, yielding large scattering matrices and reducing the enhancement [16]. Thus, only a moderate enhancement is obtained for the diffusive conductance as shown in Figure 13.3. This figure indicates that the effective back scattering is weakest when the number of the Fermi points is minimum and yields a maximum $G_{yy}(B)/G_{yy}(0)$. The above B-induced separation of the initial and final scattering states and the concomitant weakening of the scattering rate is still significant for the Fermi points above the gaps of the sublevels at low B and is responsible for the initial rise of the diffusive conductance at high densities (thick blue and green curves). Note that the diffusive conductance shown by the thick red curve for the lowest density initially decreases in contrast to the other two curves. This behavior occurs when only the lower branch is occupied at $B = 0$. Note that the peak enhancement is larger for the thick green curve (with a larger electron density) than the thick blue curve because the chemical potential enters the gaps at higher B where the separation of the initial and final scattering states is more complete in the former case. The minimum of $G_{yy}(B)/G_{yy}(0)$ in the range $4.5 < B < 6.5\,\mathrm{T}$ arises when the chemical potential passes through the last few humps in the lower branches with a large DOS that increases the scattering rate. At high B where all the gaps are above the chemical potential, the two wells behave as independent single wells. Therefore, $G_{yy}(B)/G_{yy}(0)$ increases gradually as a function of B [13].

Also notice that $G_{yy}(B)/G_{yy}(0)$ has a dip at $B = 2.7\,\mathrm{T}$ in Figure 13.3 indicated by an arrow. The position of the dip is insensitive to the electron density of the samples. This dip is associated with the flat bottoms of the lower branch of the dispersion curves of the sublevels (see the inset) that pin the Fermi level to the divergence in the DOS. The latter yields rapid scattering of the electrons and thus a small conductance. These flat bottoms are the consequence of the balanced competition between the B-induced rise of the crossing point arising from the increasing displacement δk_y ($= d_{\mathrm{eff}}/\ell_c^2$, and d_{eff} is the center-to-center separation of two QWs) between the two parabolas and the downward repulsion from the upper level. These flat bottoms eventually develop into humps at higher fields [17]. Other rugged structures arise from the sublevel depopulation effect.

13.2
Phonon Scattering

In Section 13.1, we studied the effects of interface-roughness scattering on the conductance. In this section, we further study the effects of electron–phonon scattering on conductances for the systems shown in Figure 13.1a,c. We study two noteworthy cases with a magnetic field in the z-direction for high mobility SQWR and in the x-direction for DQWRs.

13.2.1
Model for Inelastic Scattering

For SQWRs with a magnetic field $B\|z$, the energy of an electron is $\varepsilon_{nk_y} = (n + 1/2)\hbar\Omega_x + \hbar^2 k_y^2/2m^{**} + E_0^z$, $\Omega_x = \sqrt{\omega_x^2 + \omega_c^2}$, $\omega_c = eB/m^*$, and $m^{**} = m^*/[1-(\omega_c/\Omega_x)^2]$ is the modified effective mass of electrons in the wire direction. The ground-state electron energy E_0^z and wave function $\xi_0(z)$ in the z-direction can be calculated through the Schrödinger equation with the following Hamiltonian

$$\mathcal{H}_z = -\frac{\hbar^2}{2}\frac{d}{dz}\left[\frac{1}{m^*(z)}\frac{d}{dz}\right] + V_{\text{SQW}}(z) \quad \text{for } B\|z\,. \tag{13.14}$$

The total wave function is

$$\Psi_j(x, y, z) = \phi_n\left(\frac{x + \Delta x_{k_y}}{\ell_{cx}}\right)\frac{\exp(ik_y y)}{\sqrt{L_y}}\xi_0(z)\,, \tag{13.15}$$

where $\Delta x_{k_y} = k_y\ell_c^2(\omega_c/\Omega_x)^2$ and $\ell_{cx} = \sqrt{\hbar/m^*\Omega_x}$.

As a generalization of Eq. (13.1), the Boltzmann transport equation with the inclusion of electron–phonon interaction for electrons diffusing along the wire (y-direction) is given by [18]

$$v_j + \frac{2\pi}{\hbar}\sum_{j'}|I_{j',j}|^2(g_{j'} - g_j)\delta(\varepsilon_j - \varepsilon_{j'}) + P_j = 0\,, \tag{13.16}$$

where $|I_{j',j}|^2$ has already been defined in Eq. (13.5). In addition to the quantities introduced in Eq. (13.1), P_j in Eq. (13.16) represents the contribution from the electron–phonon interaction given by [18]

$$P_j = \frac{2\pi}{\hbar}\sum_{j',s,q,\pm}\left|V_{j'j}^{sq}\right|^2(f_{j'}^{(\mp)} + n_{sq})(g_{j'} - g_j) \times \delta(\varepsilon_j - \varepsilon_{j'} \pm \hbar\omega_{sq})\delta_{k_y',k_y\pm q_y}\,, \tag{13.17}$$

where the upper (lower) sign in Eq. (13.17) corresponds to the one-phonon emission (absorption) process, $f_j^{(-)} = f_0(\varepsilon_j)$, $f_j^{(+)} = 1 - f_0(\varepsilon_j)$, $\hbar\omega_{sq}$ is the energy of the phonon of mode s and wave vector $\boldsymbol{q}$, and $n_{sq} = n_0(\omega_{sq})$ is the Boson function for the distribution of phonons. The electron–phonon interaction matrix $|V_{j'j}^{sq}|^2$ in Eq. (13.17) will be given explicitly in Eqs. (13.22) and (13.25) below for SQWR and DQWRs, respectively.

In order to solve Eq. (13.16), we draw dense lines parallel to the k_y-axis around the Fermi surface on top of dispersion curves. These horizontal lines are uniformly spaced by a sufficiently small energy interval $\delta\varepsilon$. Each horizontal line corresponds to a specific value of $\varepsilon_{j'}$. The intersecting points of dispersion curves with horizontal lines are labeled by composite index $t = \{l, \beta\}$, where $l = \{n, k_y\}$ counts the k_y-points of dispersion curves from the left to the right "horizontally" for a given energy, and β is the index of the energy branch in the "vertical" $\varepsilon_{j'}$ subdivision. Equation (13.17) yields the relationship $W_{t,t'}$ in Eq. (13.21) between t and t'.

In comparison with Eq. (13.11), we find

$$\tilde{g} = -(\overleftrightarrow{\mathbf{U}} + \overleftrightarrow{\tilde{\mathbf{W}}})^{-1} \otimes \mathbf{S} \,, \tag{13.18}$$

where the matrix $\overleftrightarrow{\mathbf{U}}$ has been defined after Eq. (13.9), and the matrix $\overleftrightarrow{\tilde{\mathbf{W}}}$ related to the phonon scattering is calculated as

$$\tilde{W}_{t,t'} = \left(W_{t,t'} - \delta_{t,t'} W_t^{(0)} \right) \left(1 - \frac{1}{2} \delta_{l',1} \right) . \tag{13.19}$$

In Eq. (13.19), we have introduced the notations

$$W_t^{(0)} = \sum_{t'} W_{t,t'} \quad \text{for } t = \{l, \beta\} \text{ and } t' = \{l', \beta'\} \,, \tag{13.20}$$

$$W_{t,t'} = \frac{\mathcal{V} \delta \varepsilon}{(2\pi)^2 \hbar^2} \sum_{s,\pm} \frac{q \theta \left(\pm \varepsilon' \mp \varepsilon \right)}{\hbar c_s} \times \int_0^{2\pi} d\phi \frac{\left| V_{t',t}^{sq} \right|^2}{|v_t v_{t'}|} \left[f_{t'}^{(\mp)} + n_0(\omega_{sq}) \right] , \tag{13.21}$$

where $\varepsilon = \varepsilon_j$ and $\varepsilon' = \varepsilon_{j'}$. Equation (13.18) is obtained by replacing the k_y' integration in Eq. (13.17) with a summation over the intersecting points. The quantity $\mathcal{V}$ in Eq. (13.21) is the sample volume, $\theta(x)$ is the unit step function, the upper (lower) sign corresponds to the one-phonon emission (absorption) process, $q_y = k_y' - k_y$, $\hbar \omega_{sq} = |\varepsilon_t - \varepsilon_{t'}|$, $q_x = q_\perp \cos \phi$, $q_z = q_\perp \sin \phi$, and $q_\perp = \sqrt{q^2 - q_y^2}$.

The electron–phonon interaction matrix $|V_{t',t}^{sq}|^2$ in Eq. (13.21) for the SQWR with $B \| x$ is given by [19]

$$\left| V_{t',t}^{sq} \right|^2 = \frac{V_{sq}^2}{\epsilon_n \epsilon_{n'}} \left(\frac{n_<!}{n_>!} \right) Q^{(n_> - n_<)} \exp(-Q) \times \left[L_{n_<}^{\{n_> - n_<\}}(Q) \right]^2 \left| \int_{-\infty}^{\infty} dz e^{i q_z z} |\xi_0(z)|^2 \right|^2 , \tag{13.22}$$

where $Q = [(\Delta x_{q_y})^2 + q_x^2 \ell_{cx}^4]/2\ell_{cx}^2$, $n_>$ $(n_<)$ are the larger (lesser) of n and n' and $L_n^{\alpha}(x)$ is the associated Laguerre polynomial. In Eq. (13.22), the inverted dielectric function ϵ_n^{-1} is given by the diagonal element $[\epsilon_{\text{TF}}^{-1}(q_y)]_{n,n}$ in the static Thomas–Fermi approximation [19]. For the longitudinal ($s = \ell$) and transverse ($s = t$) modes of acoustic phonons, the quantity V_{sq}^2 in Eq. (13.22) is

$$V_{\ell q}^2 = \frac{\hbar \omega_{\ell q}}{2 \mathcal{V} \rho_i c_\ell^2} \left[D^2 + (e h_{14})^2 \frac{A_\ell(\mathbf{q})}{q^2} \right] \quad \text{for } s = \ell \,, \tag{13.23}$$

$$V_{tq}^2 = \frac{\hbar\omega_{tq}}{2\mathcal{V}\rho_i c_t^2}(eh_{14})^2 \frac{A_t(\mathbf{q})}{q^2} \quad \text{for } s = t \,, \tag{13.24}$$

where ρ_i, c_s, D and h_{14} are the ion density, sound velocity, deformation-potential coefficient, and the piezoelectric constant, respectively. In addition, the form factors $A_s(\boldsymbol{q})$ in Eqs. (13.23) and (13.24) are given by [19] $A_\ell(\boldsymbol{q}) = 36q_x^2q_y^2q_z^2/q^6$ for $s = \ell$ and $A_t(\boldsymbol{q}) = 2[q^2(q_x^2q_y^2 + q_y^2q_z^2 + q_z^2q_x^2) - 9q_x^2q_y^2q_z^2]/q^6$ for $s = t$.

Whereas for DQWRs, the electron–phonon interaction matrix in Eq. (13.21) with $B \| x$ is given by [19]

$$\left|V_{t',t}^{sq}\right|^2 = \frac{V_{sq}^2}{\epsilon_n\epsilon_{n'}}\left(\frac{n_<!}{n_>!}\right) Q_1^{(n_>-n_<)}\left[L_{n_<}^{\{n_>-n_<\}}(Q_1)\right]^2$$
$$\times \exp(-Q_1)\left|\int_{-\infty}^{\infty} dz e^{iq_z z}\xi_{m'k_y'}^*(z)\xi_{mk_y}(z)\right|^2 , \tag{13.25}$$

where $Q_1 = q_x^2\ell_x^2/2$, $\ell_x = \sqrt{\hbar/m^*\omega_x}$, $l = \{n, m, k_y\}$ and $l' = \{n', m', k_y'\}$ label the Fermi points for a given energy. The wave function $\xi_{mk_y}(z)$ is defined in Eq. (13.3) for DQWRs and $\varepsilon_{n,m,k_y} = (n + 1/2)\hbar\omega_x + E_{mk_y}^z$ is the corresponding electron energy. Finally, the conductance of the system at finite T is obtained from

$$G_{yy}(B) = \left(\frac{2e^2}{hL_y}\right)\delta\varepsilon\sum_t\left[-f_0'(\varepsilon_t)\right]S_t\tilde{g}_t \,, \tag{13.26}$$

where both $\boldsymbol{S}$ and $\tilde{\mathbf{g}}$ are $(N_F - 1)$-dimensional column vectors, identical to those in Eq. (13.12).

13.2.2 Numerical Results for Phonon Scattering Effect

For the numerical calculation, we must have the knowledge of a GaAs wire sandwiched by $Al_{0.3}Ga_{0.7}As$ barriers and assume the following parameters: electron effective mass $m_B = 0.073m_0$ in the barrier with free-electron mass m_0, $m^* = 0.067m_0$ in the well, bulk dielectric constant $\epsilon_b = 12$, longitudinal sound velocity $c_\ell = 5.14 \times 10^5$ cm/s, transverse sound velocity $c_t = 3.04 \times 10^5$ cm/s, GaAs mass density $\rho = 5.3\,\mathrm{g/cm^3}$, piezoelectric constant $h_{14} = 1.2 \times 10^7$ V/cm, deformation-potential coefficient $D = -9.3$ eV, and level-broadening $\gamma_j = 0.16$ meV [17]. The broadening is employed at the bottom of the sublevels where the one-dimensional density of states diverges. The Debye approximation is employed for the acoustic phonons at low temperatures.

Let us also assume that elastic scattering is dominated by the interface-roughness of a mono-layer fluctuation $\delta b = 5$ Å with a Gaussian correlation length given by $\Lambda_0 = 30$ Å [15]. The interface roughness is assumed to exist only at one (two) interface(s) between GaAs QW and the $Al_{0.3}Ga_{0.7}As$ barrier in the growth sequence in the z-direction for single (double) QWs, a well known fact. We assume a conduction-band offset of 270 meV in the z-direction. For DQWR sample, the sublevels are due to the tunnel-split doublet in the z-direction in a rectangular potential, while the electrons are in the ground sublevel in the x-direction in a parabolic potential. The parameters for the DQWR sample are given in Table 13.2. Here, $\hbar\omega_x = 1$ meV corresponds to the size of the wave function $\sim 2\ell_x = 2\sqrt{\hbar/m^*\omega_x} = 477$ Å in the x-direction for $m^* = 0.067 m_0$. Other parameters will be given in the corresponding figure captions.

Figure 13.4 shows the conductance of the sample, listed in Table 13.2, in the presence of roughness and phonon scattering at several temperatures. By comparing the results with only elastic scattering, we find that roughness scattering dominates over phonon scattering for $T \leq 4$ K. At high temperatures $T \geq 8$ K, however, the large conductance enhancement due to suppression of the elastic back-scattering in the gap region is suppressed by the phonon scattering effect, making the conductance peak smaller for $T = 16$ K than for 8 K in Figure 13.4 in contrast to only the result with elastic scattering.

Table 13.2 DQWRs with the wire width 42 Å in the x-direction, QW widths L_{z1}, L_{z2}, center-barrier width L_B, total one-dimensional electron density n_{1D}, ground-doublet tunnel splitting Δ_{SAS} (at $B = 0$), the wire length L_y, and the chemical potential μ (at $T = 0$ K and $B = 0$) relative to the bottom of the ground level.

L_{z1}/L_{z2} (Å)	L_B (Å)	n_{1D} (10^5 cm^{-1})	Δ_{SAS} (meV)	L_y (μm)	μ (meV)
80/80	40	6.5	3.6	1	3.79

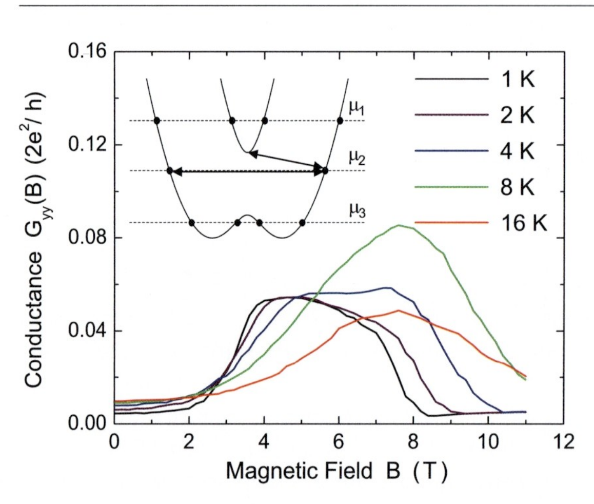

Figure 13.4 Conductance $G(B)$ vs. B for several temperatures for the DQWR sample, listed in Table 13.2, including both roughness and phonon scattering. The inset displays the energy dispersion of two tunnel-split branches. The dashed horizontal lines in the inset show the position of the chemical potential $\mu = \mu_1$ above the gap, $\mu = \mu_2$ inside the gap, and $\mu = \mu_3$ below the gap. The thick horizontal (tilted) double-headed arrow illustrates roughness (phonon) scattering.

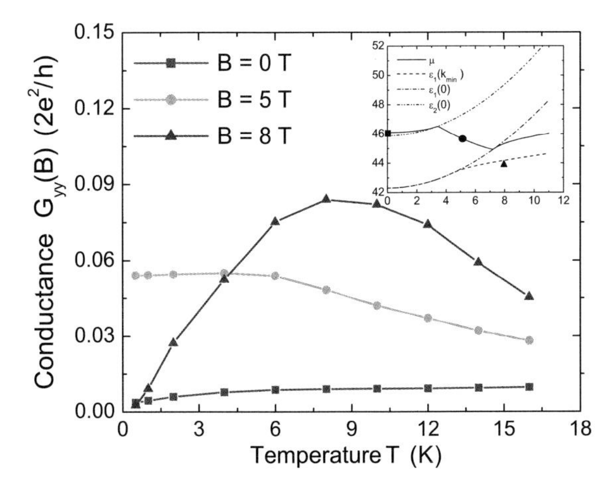

Figure 13.5 Conductance $G(B)$ vs. T for several fields for the DQWR sample, listed in Table 13.2, including roughness and phonon scattering. The inset of the figure shows the positions of the chemical potential (solid curve), the bottom of the upper branch (dash-double-dotted curve), the lower gap edge (dash-dotted curve), and the bottom of the lower branch (dashed curve).

In order to further elucidate phonon scattering's role, we present in Figure 13.5 the temperature dependence of the conductance of the sample for several fields. The conductance at $B = 5\,\mathrm{T}$ is relatively large at $T = 1\,\mathrm{K}$ because the chemical potential is near the mid-gap as seen from the inset by the solid circle. As the temperature increases, the conductance drops at this field because the electrons can be scattered and thermally activated to the bottom of the upper branch, where the conductance is much smaller as discussed earlier. The inset shows the position of the chemical potential relative to the gap edges shown by dash-dotted and dash-double-dotted curves. The dashed curve indicates the bottom of the lower branch. At $B = 8\,\mathrm{T}$, the chemical potential is just below the gap as shown by the solid triangle in the inset. The electrons are activated to the highly-conducting mid-gap region with temperature increase, thus explaining the rapid rise of the conductance with the temperature. However, the conductance drops after a peak above $T = 8\,\mathrm{K}$ due to phonon scattering.

13.3 Thermoelectric Power

This section examines electron-diffusion and phonon-drag thermoelectric powers resulting from non-equilibrium phonon distribution. We only consider a SQWR shown in Figure 13.6 in the presence of magnetic field in the z-direction.

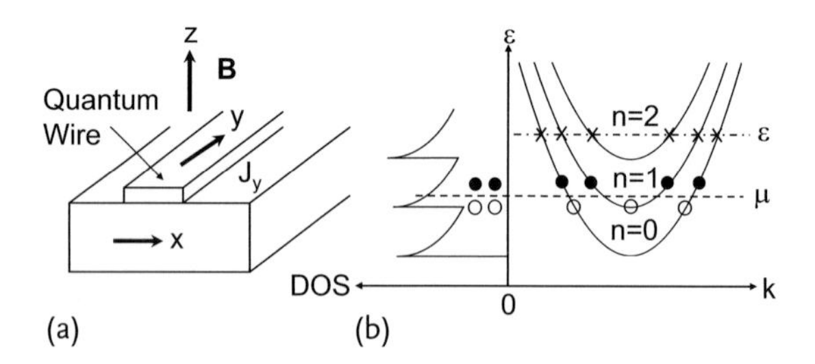

Figure 13.6 Illustration of (a) a quasi-1D wire structure with an external perpendicular magnetic field B and (b) energy dispersion curves and the DOS of for the first three sublevels. Solid and empty dots indicate quasi-particles and quasi-holes near the chemical potential μ. The crosses signify the points in k space at energy ε for the sublevel dispersion curves.

13.3.1
Model for Non-equilibrium Phonons

For non-equilibrium phonons, their distribution function n_{sq} is written as

$$n_{sq} = n_0(\omega_{sq}) - n_0'(\omega_{sq})\delta \mathcal{W}_{sq} \,, \tag{13.27}$$

where $n_0(\omega_{sq})$ is the Boson function, and $n_0'(\omega_{sq})$ is the first derivative with respect to $\hbar\omega_{sq}$. Here, $\hbar\omega_{sq}$ is the energy of the phonon of mode s and wave number q. The deviation function $\delta\mathcal{W}_{sq}$ is induced through electron–phonon interaction and is linear in the electric field $\mathcal{E}_0$. This quantity is related to the electron deviation function through the steady-state condition of the phonon distribution: $\dot{n}_{sq}\,|_{\text{scatt}} + \dot{n}_{sq}\,|_{\text{pe}} = 0$. Here, $\dot{n}_{sq}\,|_{\text{pe}}$ and $\dot{n}_{sq}\,|_{\text{scatt}}$ are the rates of the change of the phonon population through scattering from the electron–phonon interaction and other sources, respectively. For $\dot{n}_{sq}\,|_{\text{scatt}}$, we assume a relaxation-time approximation: [20, 21] $\dot{n}_{sq}\,|_{\text{scatt}} = [n_0(\omega_{sq}) - n_{sq}]/\tau_{sq} = n_0'(\omega_{sq})\delta\mathcal{W}_{sq}/\tau_{sq}$, where the relaxation time τ_{sq} of the phonons arises dominantly from the boundary scattering as will be discussed below. On the other hand, $\dot{n}_{sq}\,|_{\text{pe}}$ is given by [19]

$$\dot{n}_{sq}\,|_{\text{pe}} = \frac{2\pi}{\hbar}\sum_{j,j',\pm}(\pm)\left|V_{j'j}^{sq}\right|^2 f_j\left(1-f_{j'}\right)\left(n_{sq}+\frac{1}{2}\pm\frac{1}{2}\right)$$
$$\times\,\delta(\varepsilon_j' - \varepsilon_j \pm \hbar\omega_{sq})\delta_{k',k\mp q_y}\,, \tag{13.28}$$

where the upper (lower) sign corresponds to the one-phonon emission (absorption) process except for the sign. Linearizing Eq. (13.28) according to $f_j = f_0(\varepsilon_j) - f_0'(\varepsilon_j)\delta\mathcal{D}_j$ and Eq. (13.27), where $\delta\mathcal{D}_j = g_j e\mathcal{E}_0$ is the deviation function and is also linear in $\mathcal{E}_0$, using the steady-state condition, we find, after additional algebra [22, 23],

$$\delta\mathcal{W}_{sq} = \left(\frac{2}{\tau_{sq}^{-1} + a_{sq}}\right)\frac{2\pi}{\hbar}\sum_{j,j'}\left|V_{j',j}^{sq}\right|^2\left(\delta\mathcal{D}_{j'} - \delta\mathcal{D}_j\right)$$
$$\times\left[f_0(\varepsilon_j) - f_0(\varepsilon_{j'})\right]\delta_{k',k+q_y}\delta\left(\varepsilon_j + \hbar\omega_{sq} - \varepsilon_{j'}\right), \tag{13.29}$$

where the spin degeneracy is included. The quantity

$$a_{sq} = \frac{4\pi}{\hbar}\sum_{j,j'}\left|V^{sq}_{j',j}\right|^2\left[f_0(\varepsilon_j) - f_0(\varepsilon_{j'})\right] \times \delta_{k',k+q_y}\delta\left(\varepsilon_j + \hbar\omega_{sq} - \varepsilon_{j'}\right) \tag{13.30}$$

in Eq. (13.29) originates from the linear term $\propto \mathcal{W}_{sq}$ contained in n_{sq} in Eq. (13.28) and indicates the contribution to the phonon relaxation rate via scattering by electrons. The electron scattering is negligibly small compared with the contribution from boundary scattering in our structure due to the scarcity of the electrons and will be neglected [23]. The mean-free-path of a phonon due to the boundary scattering depends on the nature of the sample surface and has the size of the sample dimension for absorbing (i.e., rough) boundaries. The mean-free path can be increased by polishing the sample surface [24]. Assume absorbing boundaries and take the mean-free-path as a typical sample size, independent of the mode for a numerical evaluation, thereby estimating the lower limit of the phonon-drag thermoelectric power.

The heat current density due to the electron-diffusion is given, accounting for the spin degeneracy, by

$$Q_{\rm d} = \frac{2e\mathcal{E}_0}{\mathcal{V}}\sum_j v_j g_j \int_0^\infty d\varepsilon\,(\varepsilon - \mu)\left[-f_0'(\varepsilon)\right]\delta(\varepsilon_j - \varepsilon)\,. \tag{13.31}$$

The j summation includes summation on the sublevels and the energy integration over ε_{nk} with the DOS $\propto 1/|v_{nk}|$. This leads to

$$\begin{aligned} Q_{\rm d} &= \frac{2e\mathcal{E}_0}{\mathcal{S}h}\int_0^\infty d\varepsilon\left[-f_0'(\varepsilon)\right]\sum_{\nu=1}^{N_{\rm F}} s_\nu g_\nu(\varepsilon_\nu - \mu) \\ &= \frac{2e\mathcal{E}_0}{\mathcal{S}h}\int_0^\infty d\varepsilon\left[-f_0'(\varepsilon)\right]\sum_{\nu=1}^{N_{\rm F}-1} s_\nu g_\nu'(\varepsilon_\nu - \mu)\,, \end{aligned} \tag{13.32}$$

where $\mathcal{S}$ is the sample cross-sectional area, ε_ν is the energy at the νth Fermi point. The second equality follows from $g_\nu' = g_\nu - g_{N_{\rm F}}$, $\sum_{\nu=1}^{N_{\rm F}} s_\nu = 0$, and $\sum_{\nu=1}^{N_{\rm F}} s_\nu\varepsilon_\nu = 0$. The electron-diffusion thermoelectric power is then given by

$$S_{\rm d} = \frac{Q_{\rm d}}{J_y T} = \frac{1}{eT\mathcal{G}}\int_0^\infty d\varepsilon\left[-f_0'(\varepsilon)\right]\sum_{\nu=1}^{N_{\rm F}-1} s_\nu g_\nu'(\varepsilon_\nu - \mu)\,, \tag{13.33}$$

where $\mathcal{G} = G_{yy}/(2e^2/hL_y)$. The result in Eq. (13.33) reduces to Mott's formula $S_{\rm d} = (\pi^2 k_{\rm B}^2 T/3e)\partial\ln\mathcal{G}/\partial\mu$ to the first order in T at extremely low temperatures.

The heat current density due to phonon transport is

$$Q_{\rm ph} = \frac{1}{\mathcal{V}}\sum_{s,q}\hbar\omega_{sq}n_{sq}\frac{\partial\omega_{sq}}{\partial q_y}\,. \tag{13.34}$$

Inserting Eqs. (13.27) and (13.29) into Eq. (13.34) and using the symmetry relations $\varepsilon_{n,-k} = \varepsilon_{n,k}$, $g_{n,-k} = -g_{n,k}$, we rewrite the heat current density in Eq. (13.34) as

$$Q_{\text{ph}} = \sum_{j=1}^{N_{\text{F}}} g_j Z_j \,, \tag{13.35}$$

$$Z_j = \frac{4\pi}{\hbar} \sum_{s,q,j';\pm} \frac{\partial \omega_{sq}}{\partial q_y} F_{sq} \left| V^{sq}_{j',j} \right|^2 \left[f_0(\varepsilon_j) - f_0(\varepsilon_{j'}) \right] \times \delta_{k',k+q_y}(\pm) \delta\left(\varepsilon_{j'} \pm \hbar\omega_{s,q} - \varepsilon_j \right), \tag{13.36}$$

and $F_{sq} = e\mathcal{E}_0 \tau_{sq} \hbar\omega_{sq}[-n_0'(\omega_{sq})]/\mathcal{V}$. Using the anti-symmetry relation $Z_{n,-k} = -Z_{n,k}$, we further reduce Eq. (13.35) to

$$Q_{\text{ph}} = \sum_{j=1}^{N_{\text{F}}-1} g_j' Z_j \,. \tag{13.37}$$

After employing the following identity

$$\mp \left[f_0(\varepsilon_{n,k}) - f_0(\varepsilon_{n',k+q_y}) \right] = f_0(\varepsilon_{n,k}) \left[1 - f_0(\varepsilon_{n',k+q_y}) \right] \times \left[n_0(\omega_{sq}) + \frac{1}{2} \mp \frac{1}{2} \right]^{-1} \tag{13.38}$$

for $\varepsilon_{n',k+q_y} \pm \hbar\omega_{sq} - \varepsilon_{n,k} = 0$ in the expression in Eq. (13.36), we find the phonon-drag thermoelectric power from Eq. (13.37)

$$S_{\text{ph}} = \frac{Q_{\text{ph}}}{J_y T} = -\left[\frac{k_{\text{B}} \mathcal{S} h}{e\mathcal{V}(k_{\text{B}}T)^2 \mathcal{G}} \right] \frac{2\pi}{\hbar} \sum_{s,q;\pm} \hbar\omega_{sq} \frac{\partial \omega_{sq}}{\partial q_y} \tau_{sq} \left[n_0(\omega_{sq}) + \frac{1}{2} \pm \frac{1}{2} \right] \times \sum_{n,n';k} g'_{n,k} f_0(\varepsilon_{n,k}) \left[1 - f_0(\varepsilon_{n',k+q_y}) \right] \left| V^{sq}_{n',n} \right|^2 \times \delta\left(\varepsilon_{n',k+q_y} \pm \hbar\omega_{sq} - \varepsilon_{n,k} \right). \tag{13.39}$$

It is important to note that G_{yy}, S_{d}, and S_{ph} only depend on $(N_{\text{F}} - 1)$ dimensional $\mathbf{g}'$ instead of N_{F} dimensional $\mathbf{g}$.

Now, we perform the integrations over k and $\mathbf{q}$ in Eq. (13.39). For this purpose, it is convenient to use the following identity between the Fermi–Dirac distribution $f_0(x)$ and the Bose–Einstein distribution $n_0(z)$:

$$f_0(x) \left[1 - f_0(y) \right] \delta(x + z - y) = z \left[\frac{f_0(x) - f_0(y)}{z} \right] \left[1 + n_0(z) \right] \delta(x + z - y) \,. \tag{13.40}$$

At low temperatures, the right-hand side of this identity can be approximated for small z as [25]

$$f_0(x) \left[1 - f_0(y) \right] \delta(x + z - y) \approx z \left[1 + n_0(z) \right] \delta(x - \mu) \delta(x - y) \,. \tag{13.41}$$

Inserting $x = \varepsilon_{n,k}$, $y = \varepsilon_{n',k'}$ into Eq. (13.41), using $z = -\hbar\omega_{sq}$ for the upper sign and $z = \hbar\omega_{sq}$ for the lower sign in Eq. (13.39) and carrying out the integration over k and $\boldsymbol{q}$, we find

$$S_{\text{ph}} = \int_0^{+\infty} d\varepsilon \left[-f_0'(\varepsilon)\right] \mathcal{P}(\varepsilon) \,. \tag{13.42}$$

where

$$\begin{aligned}\mathcal{P}(\varepsilon) = &-\left[\frac{\frac{8k_{\text{B}}}{e}}{(2\pi k_{\text{B}}T)^2 \mathcal{G}\hbar^2}\right] \int_0^{+\infty} dq_x \int_0^{+\infty} dq_z \sum_s (\hbar\omega_{sq})^2 \frac{q_y}{q} \Lambda_{sq} \\ &\times \sum_{j=1}^{N_{\text{F}}-1} \sum_{j'=1}^{N_{\text{F}}} n_0(\omega_{sq}) \left[n_0(\omega_{sq}) + 1\right] g_j' \mathcal{V} \left|V_{j',j}^{sq}\right|^2 \frac{1}{|v_j v_{j'}|} \delta_{q_y, k'-k} \,.\end{aligned} \tag{13.43}$$

Here, $j = (n, k)$ represents the states at the intersection of the energy ε and the dispersion curves. In Eq. (13.43), the Debye approximation is employed for the long-wave length phonons $\omega_{sq} = c_s q$ and $\Lambda_{sq} = c_s \tau_{sq}$ is the phonon mean-free-path. The wave numbers k, k' in Eq. (13.43) at points j, j' are given by

$$\begin{aligned} k &= -\frac{q_y}{2} + \frac{\mp\hbar\omega_{sq} + (n - n')\hbar\Omega_x}{\frac{\hbar^2 q_y}{m^{**}}} \,, \\ k' &= \frac{q_y}{2} + \frac{\mp\hbar\omega_{sq} + (n - n')\hbar\Omega_x}{\frac{\hbar^2 q_y}{m^{**}}} \,. \end{aligned} \tag{13.44}$$

13.3.2
Numerical Results for Thermoelectric Power

For the numerical calculation, we must have the knowledge of a GaAs wire embedded in AlGaAs and assume: $L_z = 100$ Å (quantum-well width), $V = 280$ meV (barrier height), $m_{\text{B}} = 0.073 m_0$ (mass in the barrier with free-electron mass m_0), $\epsilon_{\text{b}} = 12$, $m^* = 0.067 m_0$, $c_\ell = 5.14 \times 10^5$ cm/s, $c_{\text{t}} = 3.04 \times 10^5$ cm/s, $\rho_i = 5.3\,\text{g/cm}^3$, $h_{14} = 1.2 \times 10^7$ V/cm, $D = -9.3$ eV, $\Lambda_{\text{sq}} = 0.3$ mm, and $\gamma_j = 0.16$ meV. We also assume that elastic scattering for the conductance is dominated by the interface-roughness of a mono-layer fluctuation $\delta b = 5$ Å with a correlation length given by $\Lambda_0 = 30$ Å as studied earlier. Other parameters will be given in the corresponding figure captions.

In this subsection, we learn of the situation where $\hbar\omega_x$ is 10 meV with $N_{\text{1D}} = 10^6\,\text{cm}^{-1}$. In this case, the second sublevel $n = 1$ is populated near the bottom with a small Fermi energy $\varepsilon_{\text{1F}} = 0.29$ meV. The T dependence of S_{ph} (solid, dashed curves, left scale) and S_{d} (dotted curve, right scale) is shown at $B = 0$ in Figure 13.7. The exponential T-dependence $\exp(-\hbar\omega_{\text{F}}/(k_{\text{B}}T))$ for S_{ph} is seen only at

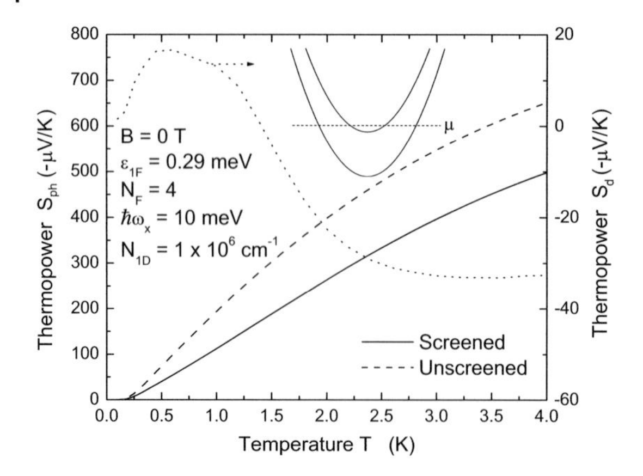

Figure 13.7 *T* dependence of S_{ph} (solid, dashed curves, left axis) and S_{d} (dotted curve, right axis) at $B = 0$ for two-sublevel occupation $N_{\mathrm{F}} = 4$. Other parameters are given in the text.

extremely low $T < 0.25$ K because $q_y = 2k_{1\mathrm{F}}$ is small. Here, $\hbar\omega_{\mathrm{F}}$ is approximately the minimum phonon energy necessary to scatter an electron across the Fermi surface, namely, the transverse phonon energy corresponding to the wave number $q_y = 2k_{1\mathrm{F}}$. This activation behavior is unique to a one-dimensional structure where the Fermi surface consists of two discrete points. We have also considered "vertical" phonon scattering processes around each Fermi point corresponding to $j = j'$ in Eq. (13.43). This contribution arises from the small-q_y solution in Eq. (13.44) with $n = n'$, namely, $q_y \simeq \mp k\hbar\omega_{s,q}/2\varepsilon_{0\mathrm{F}} - k(\hbar\omega_{s,q}/\varepsilon_{0\mathrm{F}})^2/8$ for the absorption and emission processes. Clearly the first term cancels out for one-phonon emission and absorption processes, and leaves only the second term $q_y \sim -k(k_{\mathrm{B}}T/\varepsilon_{0\mathrm{F}})^2/8$, which is very small (i.e., $k_{\mathrm{B}}T/\varepsilon_{0\mathrm{F}} \ll 1$). The quantity $-S_{\mathrm{d}}$ rises linearly below 0.15 K and steeper between $0.15\mathrm{K} < T < 0.5$ K with a negative sign for S_{d} and then decreases rapidly. Amazingly, S_{d} changes sign and yields a positive value. Its magnitude is larger than the screened $|S_{\mathrm{ph}}|$ below 0.27 K. A positive sign for S_{d} from the electron heat transport is unusual because of the following situation: Due to the proximity of μ to the bottom of the sublevel $n = 1$, the DOS near the thermal layer is very large above μ for the quasi-particles with a small G_{yy} and small for the quasi-holes in the sublevel $n = 0$ below μ with a large G_{yy}. Therefore, the heat is carried mainly by the quasi-holes and yields a S_{d}.

We display the $N_{1\mathrm{D}}$ dependence of S_{ph} (solid, dashed curves, left scale), S_{d} (dotted curve, right scale) in Figure 13.8 at $T = 2$ K and $B = 0$ for $\hbar\omega_x = 10$ meV. The vertical arrow indicates the density $N_{1\mathrm{D}} = 0.92 \times 10^6\,\mathrm{cm}^{-1}$ where the second sublevel begins to be populated. The quantity $q_y = 2k_{0\mathrm{F}}$ reaches its maximum value just prior to the thermal occupation of the $n = 1$ sublevel, yielding the minimum in $-S_{\mathrm{ph}}$ around $N_{1\mathrm{D}} = 0.8 \times 10^6\,\mathrm{cm}^{-1}$, as discussed earlier. The quantity $q_y = 2k_{1\mathrm{F}}$ becomes small just above the arrow, increases with $N_{1\mathrm{D}}$ and yields a sharp peak in $-S_{\mathrm{ph}}$ around $N_{1\mathrm{D}} = 1.02 \times 10^6\,\mathrm{cm}^{-1}$, as shown in Figure 13.8. This sharp minimum-maximum structure is rounded at higher temperatures. S_{d}

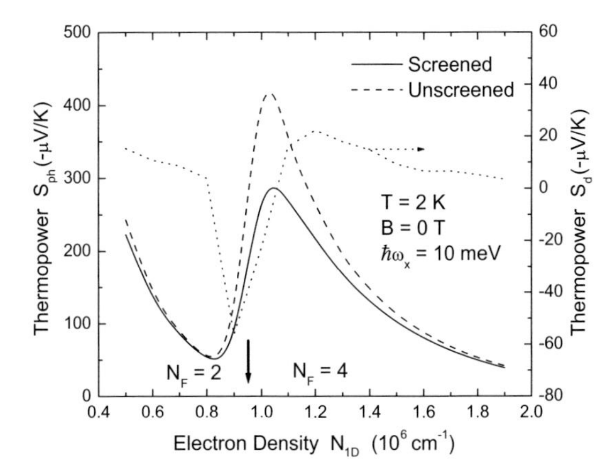

Figure 13.8 N_{1D} dependence of S_{ph} (solid, dashed curves, left axis) and S_d (dotted curve, right axis) at $T = 2$ K and $B = 0$ for $\hbar\omega_x = 10$ meV. Other parameters are given in the text. The vertical arrow in the figure indicates N_{1D} where the second sublevel begins to be populated.

changes to a positive value and becomes large between the dip and peak of $-S_{ph}$. Interestingly, S_d nearly cancels out S_{ph} near $N_{1D} = 0.9 \times 10^6\ \text{cm}^{-1}$. A similar level depopulation can be created by a magnetic field. A much shallower dip is expected in two dimensions due to the absence of the van Hove singularity, as observed in [26].

13.4 Electron–Electron Scattering

In this section, we learn about the effect of electron–electron scattering. Consider only a SQWR as shown in Figure 13.1a in the absence of magnetic field.

13.4.1 Model for Pair Scattering

As a further generalization of Eq. (13.16), the Boltzmann transport equation with the inclusion of electron–electron scattering for electrons diffusing along the wire (y-direction) is given by [27]

$$v_j + \frac{2\pi}{\hbar} \sum_{j'} \left| I_{j',j} \right|^2 \left(g_{j'} - g_j \right) \delta\left(\varepsilon_j - \varepsilon_{j'} \right) + P_j + Q_j = 0\,, \qquad (13.45)$$

where $|I_{j',j}|^2$ and P_j have already been defined in Eqs. (13.5) and (13.17), respectively. Additionally, the quantity Q_j represents the contribution from the electron–

electron scattering, given to the leading order by [27]

$$Q_j = \frac{4\pi}{\hbar k_B T}\left[-\frac{\partial f_j^{(-)}}{\partial \varepsilon_j}\right]^{-1} \sum_{j',j_1,j_1'} \left|\bar{K}_{j,j_1}^{j',j_1'}\right|^2 f_j^{(-)} f_{j_1}^{(-)} f_{j'}^{(+)} f_{j_1'}^{(+)}$$
$$\times \left(g_{j'} + g_{j_1'} - g_j - g_{j_1}\right) \delta\left(\varepsilon_j + \varepsilon_{j_1} - \varepsilon_{j'} - \varepsilon_{j_1'}\right). \tag{13.46}$$

Here, the electron–electron interaction matrix $|\bar{K}_{j,j_1}^{j',j_1'}|^2$ that includes direct and exchange Coulomb interactions describes scattering from an initial two-particle state (j, j_1) to a final state (j', j_1').

Equation (13.45) can be solved in a similar way to Eq. (13.16). In comparison with Eq. (13.18) and in the presence of additional electron–electron scattering, we find

$$\tilde{g} = -(\overleftrightarrow{\mathbf{U}} + \overleftrightarrow{\tilde{\mathbf{W}}} + \overleftrightarrow{\tilde{\mathbf{Z}}})^{-1} \otimes \boldsymbol{S}\,, \tag{13.47}$$

where the matrices $\overleftrightarrow{\mathbf{U}}$ and $\overleftrightarrow{\tilde{\mathbf{W}}}$ have been defined after Eq. (13.9) and in Eq. (13.19), respectively, and the matrix $\overleftrightarrow{\tilde{\mathbf{Z}}}$ related to the electron–electron scattering is calculated to be

$$\tilde{Z}_{j,j_1} = \left(Z_{j,j_1} - \delta_{j,j_1} Z_j^{(0)}\right)\left(1 - \frac{1}{2}\delta_{l_1,1}\right). \tag{13.48}$$

In Eq. (13.48), we have introduced the notations

$$Z_j^{(0)} = \sum_{j_1} Z_{j,j_1}^{(1)} \quad \text{for } j = \{l, \beta\} \text{ and } j_1 = \{l_1, \beta_1\}\,, \tag{13.49}$$

$$Z_{j,j_1} = Z_{j,j_1}^{(2)} - Z_{j,j_1}^{(1)}\,, \tag{13.50}$$

where j and j_1 correspond to t and t' in Eq. (13.20), and the quantum numbers $\{l, \beta\}$ and $\{l_1, \beta_1\}$ have the same meaning as those in Section 13.2. The explicit expressions for scattering-in and scattering-out terms in Eqs. (13.49) and (13.50) are given by [27]

$$Z_{j,j_1}^{(1)} = -\frac{m^* L_y^2 \delta\varepsilon}{\pi\hbar^4 f_j^{(+)}} \sum_{n',n_1',\alpha=\pm} \frac{f_{j_1}^{(-)}}{|v_j v_{j_1}|} \left|\bar{K}_{j,j_1}^{j',j_1'}\right|^2 f_{j'}^{(+)} f_{j_1'}^{(+)} \frac{\theta\left[D(k_{1y} - k_y, \nu)\right]}{\sqrt{D(k_{1y} - k_y, \nu)}}$$
$$\times \delta_{k_{1y}', k_\alpha(k_{1y}, k_y, \nu)} \delta_{k_y', k_{-\alpha}(k_{1y}, k_y, \nu)} \quad \text{for } \nu \neq 0\,, \tag{13.51}$$

$$Z^{(2)}_{j,j_1} = \frac{m^* L_y^2 \delta\varepsilon}{\pi\hbar^4 f_j^{(+)}} \sum_{n',n'_1,\alpha=\pm} \frac{f_{j_1}^{(+)}}{|v_j v_{j_1}|} \left\{ \left|\bar{K}^{j_1,j'_1}_{j,j'}\right|^2 f_{j'}^{(-)} f_{j'_1}^{(+)} \right.$$

$$\times \frac{\theta\left[D\left(k'_{1y} - k_{1y}, -\nu_1\right)\right]}{\sqrt{D\left(k'_{1y} - k_{1y}, -\nu_1\right)}} \delta_{k'_y,k_\alpha(k_{1y},k'_{1y},-\nu_1)} \delta_{k'_{1y},k_y-\nu_1 k_\Delta^2/[2(k_{1y}-k_y)]}$$

$$+ \left|\bar{K}^{j',j_1}_{j,j'_1}\right|^2 f_{j'}^{(+)} f_{j'_1}^{(-)} \frac{\theta\left[D\left(k_{1y} - k'_y, -\nu_2\right)\right]}{\sqrt{D(k_{1y} - k'_y, -\nu_2)}} \delta_{k'_{1y},k_\alpha(k'_y,k_{1y},-\nu_2)}$$

$$\left. \times \delta_{k'_y,k_y-\nu_2 k_\Delta^2/[2(k_{1y}-k_y)]} \right\} \quad \text{for } \nu_1 \neq 0,\ \nu_2 \neq 0\,, \tag{13.52}$$

where $l = \{n, k_y\}$, $\nu = n+n_1-n'-n'_1$, $\nu_1 = n-n_1+n'-n'_1$, $\nu_2 = n-n_1-n'+n'_1$, $k_\pm(k_{1y}, k_y, \nu) = [k_{1y} + k_y \pm \sqrt{D(k_{1y} - k_y, \nu)}]/2$, $D(k_{1y} - k_y, \nu) = \left(k_{1y} - k_y\right)^2 + 2\nu k_\Delta^2$, and $k_\Delta^2 = 2m^*\hbar\omega_x/\hbar^2$. $f_j^{(\pm)}$ and $\delta\varepsilon$ still have the same meaning as those in Section 13.2. Moreover, the Coulomb interaction matrix, including the direct (first term) and exchange (second term) interactions, in Eqs. (13.51) and (13.52) is $\bar{K}^{j',j'_1}_{j,j_1} = K^{j',j'_1}_{j,j_1} - K^{j'_1,j'}_{j,j_1}/2$, and

$$K^{j',j'_1}_{j,j_1}(q_y) = \delta_{k_y+k_{1y},k'_y+k'_{1y}} \left(\frac{2e^2}{\epsilon_0\epsilon_r L_y}\right) \sqrt{\frac{n^<!n_1^<!}{n^>!n_1^>!}} \int_0^\infty d\eta\, \frac{\mathcal{F}(\eta, q_y)}{\sqrt{\eta^2 + q_y^2}}$$

$$\times (-P_2)^{(n^>-n^<+n_1^>-n_1^<)/2} e^{-P_2} L_{n^<}^{(n^>-n^<)}(P_2) L_{n_1^<}^{(n_1^>-n_1^<)}(P_2)$$

$$\times \left(\delta_{n^>-n^<,n^{\rm ev}} \delta_{n_1^>-n_1^<,n_1^{\rm ev}} - 2\delta_{n^>-n^<,n^{\rm od}} \delta_{n_1^>-n_1^<,n_1^{\rm od}}\right), \tag{13.53}$$

where $n^{\rm od}$ and $n_1^{\rm od}$ ($n^{\rm ev}$ and $n_1^{\rm ev}$) are arbitrary odd (even) integers, $q_y = k'_y - k_y$, $n^< = \min(n, n')$, $n^> = \max(n, n')$, $n_1^< = \min(n_1, n'_1)$, $n_1^> = \max(n_1, n'_1)$, and $P_2 = \eta^2\ell_x^2/2$. The form factor of the Coulomb interaction in the z-direction in Eq. (13.53) is

$$\mathcal{F}(\eta, q_y) = \int_{-\infty}^{\infty} dz \int_{-\infty}^{\infty} dz_1 \xi_0^2(z) e^{-\sqrt{\eta^2+q_y^2}|z-z_1|} \xi_0^2(z_1)\,, \tag{13.54}$$

where $\xi_0(z)$ is the ground-state wave function in the z-direction at zero magnetic field.

13.4.2
Numerical Results for Coulomb Scattering Effect

In our numerical calculations, we have chosen the following parameters. For the quantum well in the z-direction, we employ $m^* = 0.067 m_0$ in the well (m_0 being the free electron mass), $m_B = 0.073m_0$ in the barrier, $V_0 = 280$ meV for the barrier height, $L_W = 50$ Å for the well width, and $\epsilon_b = 12$ for the average dielectric constant. These parameters yield $E_1^z = 84$ meV for the ground-state energy and $E_2^z - E_1^z = 188$ meV for the energy separation between the first excited level and the ground state. For the parabolic confinement in the x-direction, we use $\hbar\Omega_0 = 2.7$ meV for the energy-level separation. For the interface roughness, we assume $\delta b = 5$ Å for the average layer fluctuation in the z-direction and $\Lambda_0 = 10$ Å for the Gaussian correlation length in the x, y-directions. For the acoustic phonons in the bulk, the following parameters are employed: $c_l = 5.14 \times 10^5$ cm/s for the longitudinal mode, $c_t = 3.04 \times 10^5$ cm/s for the transverse mode, $D = -9.3$ eV for the deformation potential, $h_{14} = 1.2 \times 10^7$ V/cm for the piezoelectric field, and $\rho_i = 5.3$ g/cm^3 for the ion mass density. The wire length in the y-direction is assumed to be $L_y = 1$ µm except for the indicated cases.

Figure 13.9 shows the effect of electron–electron scattering on the resistivity in the presence of strong roughness scattering. The contribution from the phonon or electron–electron scattering can be separated as clearly seen in the figure that presents the scaled resistance differences $\Delta R/R_0$ as a function of the temperature. The contribution from phonon scattering is shown as $R_2 - R_1$ (dashed curve), while the contribution from electron–electron scattering is shown as $R_3 - R_2$ (solid curve). The phonon scattering part $R_2 - R_1$ displays a rapid rise above $T > T_0 = 2c_s\sqrt{2m^*(\mu - \hbar\Omega_0)}/k_B \sim 2.7$ K. The energy $k_B T_0$ equals the energy of the phonon with wave number $q_y = 2k_{1F}$ corresponding to the momentum transfer between the Fermi points in the sublevel $n = 1$. For $T \gg T_0$, $R_2 - R_1$ is found to exhibit

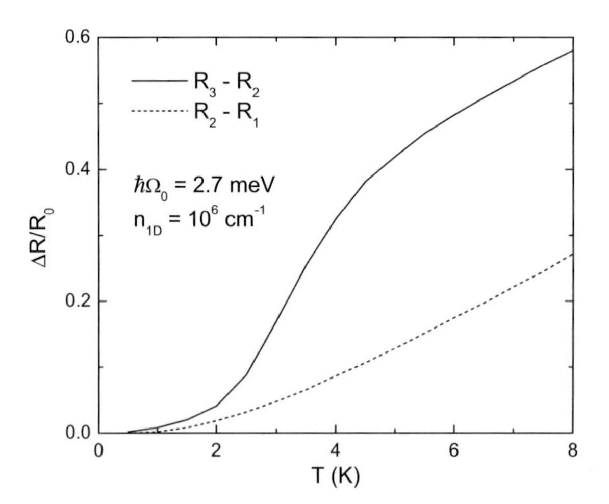

Figure 13.9 Scaled resistance difference $\Delta R/R_0$ as a function of the temperature T. Here, we introduce the notations: Roughness scattering only (R_1); Roughness plus phonon scattering (R_2); Roughness plus phonon plus electron–electron scattering (R_3). Two different cases are compared to each other with $R_0 = 0.106(h/2e^2)$ and $L_y = 0.1$ µm: (1) $R_2 - R_1$ (dashed curve, effect of phonon scattering); (2) $R_3 - R_2$ (solid curve, effect of electron–electron scattering).

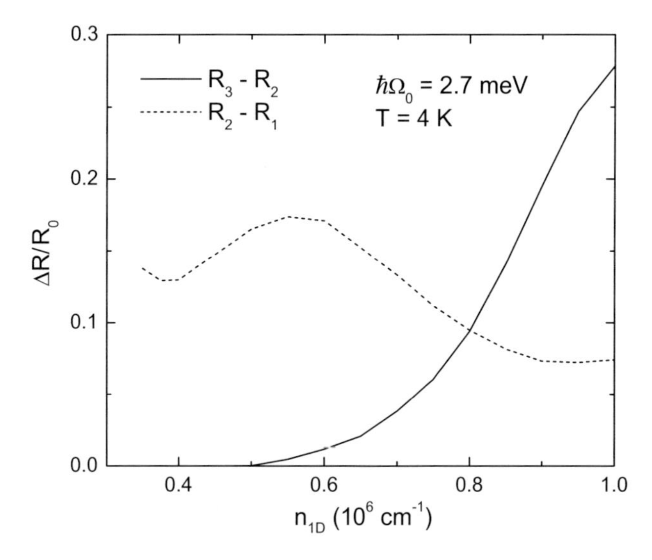

Figure 13.10 Scaled resistance difference $\Delta R/R_0$ as a function of the density n_{1D}. Roughness scattering is included in the figure. This corresponds to the situation studied in Figure 13.9 with the same definitions for R_1, R_2, R_3, but not for R_0. Two different cases are compared to each other with $R_0 = 0.124(h/2e^2)$ and $L_y = 0.1\,\mu\text{m}$: (1) $R_2 - R_1$ (dashed curve, effect of phonon scattering); (2) $R_3 - R_2$ (solid curve, effect of electron–electron scattering).

a linear behavior in T/T_0. The electron–electron scattering contribution $R_3 - R_2$ vanishes at $T = 0$ K due to the restricted phase space available for scattering. However, $R_3 - R_2$ increases rapidly with T, roughly $\propto T^2$ below 2 K, a well-known behavior from the Umklapp scattering process [20, 21]. In the high-temperature nondegenerate regime, we can analytically show that $(R_3 - R_2)/R_1$ approaches a constant $(4 - \pi)/\pi$ for the roughness-scattering dominated system in the presence of strong electron–electron scattering following the method given earlier [28]. In this high temperature regime, the increase of $R_3 - R_2$ slows down to a quasilinear dependence of R_1 as seen from Figure 13.9.

Figure 13.10 highlights the effects of electron–electron and phonon scattering by including the roughness scattering from our system. The contribution from the phonon or electron–electron scattering can be separated as clearly seen in the figure which presents the scaled resistance differences $\Delta R/R_0$ shown as a function of the density. In this figure, the contribution from phonon scattering is shown as $R_2 - R_1$ (dashed curve), while the contribution from electron–electron scattering is shown as $R_3 - R_2$ (solid curve). For phonon scattering, $R_2 - R_1$ displays a broad peak around $n_{1D} = 0.5 \times 10^6\,\text{cm}^{-1}$ where the $n = 1$ sublevel starts to be populated. Just above the threshold density for the occupation of the $n = 1$ sublevel, the Fermi wave vector k_{1F} of this sublevel is small, allowing for efficient emission and absorption of small-momentum phonons. With further increasing of n_{1D}, k_{1F} becomes larger and leads to a suppression of phonon scattering or a decrease of

$R_2 - R_1$ with n_{1D}. The electron–electron scattering contributions $R_3 - R_2$ vanishes at low density due to the complete cancellation of the intrasubband electron–electron scattering until the $n = 1$ sublevel is populated at $n_{1D} = 0.5 \times 10^6\ \text{cm}^{-1}$. When n_{1D} increases above this value, $R_3 - R_2$ increases dramatically with n_{1D} because more and more electrons are available for Coulomb scattering, in sharp contrast to the decrease of $R_2 - R_1$ with n_{1D}. As a result, $R_3 - R_2$ eventually dominates $R_2 - R_1$ above $n_{1D} > 0.8 \times 10^6\ \text{cm}^{-1}$.

References

1 Beenaker, C.V.J. and van Houton, H. (1991) in *Solid State Physics: Semiconductor Heterostructures and Nanostructures* (eds H. Ehrenreich and D. Turnbull), Academic, New York, Vol. **44**, pp.1–228. And references therein.

2 van Wees, B.J., Kouwenhoven, L.P., van Houten, H., Beenakker, C.W.J., Mooij, J.E., Foxon, C.T., and Harris, J.J. (1988) *Quantized conductance of magnetoelectric subbands in ballistic point contacts, Phys. Rev. B*, **38**, 3625.

3 Wharam, D.A., Thornton, T.J., Newbury, R., Pepper, M., Ahmed, H. Frost, J.E.F., Hasko, D.G., Peacock, D.C., Ritchie, D.A., and Jones, G.A.C. (1988) One-dimensional transport and the quantization of the ballistic resistance. *J. Phys. C*, **21**, L209.

4 Moon, J.S., Blount, M.A., Simmons, J.A., Wendt, J.R., Lyo, S.K., and Reno, J.L. (1999) Magnetoresistance of one-dimensional subbands in tunnel-coupled double quantum wires. *Phys. Rev. B*, **60**, 11530.

5 Lyo, S.K. (1999) Magnetic-field-induced V-shaped quantized conductance staircase in a double-layer quantum point contact. *Phys. Rev. B*, **60**, 7732.

6 Stoddart, S.T., Main, P.C., Gompertz, M.J., Nogaret, A., Eaves, L., Henini, M., and Beaumont, S.P. (1998) Phase coherence in double quantum well mesoscopic wires. *Physica B*, **258**, 413.

7 Thomas, K.J., Nicholls, J.T., Simmons, M.Y., Tribe, W.R., Davies, A.G., and Pepper, M. (1999) Controlled wave-function mixing in strongly coupled one-dimensional wires. *Phys. Rev. B*, **59**, 12252.

8 Simmons, J.A., Lyo, S.K., Harff, N.E., and Klem, J.K. (1994) Conductance modulation in double quantum wells due to magnetic field-induced anticrossing. *Phys. Rev. Lett.*, **73**, 2256.

9 Lyo, S.K. (1994) Transport and level anticrossing in strongly coupled double quantum wells with in-plane magnetic fields. *Phys. Rev. B*, **50**, 4965.

10 Kurobe, A., Castleton, I.M., Linfield, E.H., Grimshaw, M.P., Brown, K.M., Ritchie, D.A., Pepper, M., and Jones, G.A.C. (1994) Wave functions and Fermi surfaces of strongly coupled two-dimensional electron gases investigated by in-plane magnetoresistance. *Phys. Rev. B*, **50**, 4889.

11 Lyo, S.K. (1994) Transport and level anticrossing in strongly coupled double quantum wells with in-plane magnetic fields. *Phys. Rev. B*, **50**, 4965.

12 Huang, D.H. and Lyo, S.K. (2000) Suppression of impurity and interface-roughness back-scattering in double quantum wires: theory beyond the Born approximation. *J. Phys.: Condens. Matter*, **12**, 3383.

13 Lyo, S.K. and Huang, D.H. (2001) Multisublevel magnetoquantum conductance in single and coupled double quantum wires. *Phys. Rev. B*, **64**, 115320.

14 Huang, D.H., Lyo, S.K., Thomas, K.J., and Pepper, M. (2008) Field-induced modulation of the conductance, thermoelectric power, and magnetization in ballistic coupled double quantum wires under a tilted magnetic field. *Phys. Rev. B*, **77**, 085320.

15 Lyo, S.K. (1999) Magnetic-field-induced V-shaped quantized conductance stair-

case in a double-layer quantum point contact. *Phys. Rev. B*, **60**, 7732.

16 Lyo, S.K. (1996) Magnetic quenching of back scattering in coupled double quantum wires: giant mobility enhancement. *J. Phys.: Condens. Matter*, **8**, L703.

17 Lyo, S.K. (1994) Transport and level anticrossing in strongly coupled double quantum wells with in-plane magnetic fields. *Phys. Rev. B*, **50**, 4965.

18 Lyo, S.K. and Huang, D.H. (2003) Temperature-dependent magnetoconductance in quantum wires: Effect of phonon scattering. *Phys. Rev. B*, **68**, 115317.

19 Lyo, S.K. and Huang, D.H. (2002) Magnetoquantum oscillations of thermoelectric power in multisublevel quantum wires. *Phys. Rev. B*, **66**, 155307.

20 Ziman, J.M. (1967) *Electrons and Phonons*, Oxford University Press, London, p. 298.

21 Ziman, J.M. (1972) *Principles of the Theory of Solids*, 2nd edn, Cambridge Press, pp. 212–215.

22 Lyo, S.K. (1989) Magnetoquantum oscillations of the phonon-drag thermoelectric power in heterojunctions. *Phys. Rev. B*, **40**, 6458.

23 Lyo, S.K. (1991) Low-temperature lattice-scattering mobility in multiple heterojunctions: Phonon-drag enhancement. *Phys. Rev. B*, **43**, 2412.

24 Wybourne, M.N., Eddison, C.G., and Kelly, M.J. (1984) Phonon boundary scattering at a silicon-sapphire interface. *J. Phys. C*, **17**, L607.

25 Cantrell, D.G. and Butcher, P.N. (1986) A calculation of the phonon drag contribution to thermopower in two dimensional systems. *J. Phys. C*, **19**, L429.

26 Fletcher, R., Harris, J.J., and Foxon, C.T. (1991) The effect of second subband occupation on the thermopower of a high mobility GaAs-$Al_{0.33}Ga_{0.67}As$ heterojunction. *Semiconduct. Sci. Technol.*, **6**, 54.

27 Lyo, S.K. and Huang, D.H. (2006) Effect of electron–electron scattering on the conductance of a quantum wire studied with the Boltzman transport equation. *Phys. Rev. B*, **73**, 205336.

28 Lyo, S.K. (1986) Electron–electron scattering and mobilities in semiconductors and quantum wells. *Phys. Rev B*, **34**, 7129.

Part Two Nonlinear Response of Low Dimensional Quantum Systems

The optical response of confined excitons in semiconductor quantum dots (QDs) has been extensively studied. The ability to control their electronic properties makes them ideal candidates for studying fundamental many-body effects. Additionally, they are promising candidates for a broad range of applications, including fluorescence labels of bio-molecules, lasers, solar cells and quantum computing. An arrays of QDs were proposed as building blocks in quantum information applications, e.g., as a quantum register for noiseless information encoding. Biexcitons have been suggested as a source of entangled photons. Much effort has been devoted to the investigation of the coupling between the dots, either due to exciton or carrier migration. However, both coupling mechanisms may coexist and exploring their relative role and interplay will contribute to the understanding of inter-dot interactions.

The main experimental challenges in observing such biexcitons, and in general multi-excitonic states, comes from two factors. First, the multi-exciton binding energies are few orders of magnitude smaller than that of a single exciton. Therefore, one has a strong ground state bleaching and any reliable multi-exciton classification becomes impossible. Second, the multi-exciton complexes are highly unstable, due to rapid non-radiative relaxation to the single exciton state and hot electron–hole plasma (Auger recombination). Various nonlinear spectroscopic techniques have been designed to study such ultra-fast processes and resolve the thin structure of the multi-excitons.

The first chapter of this part is devoted to the study of nonlinear transport for a non-thermal distribution of electrons under a high bias in quantum-dot superlattices and in quantum wires, including both elastic and inelastic scattering of electrons. The second chapter of this part addresses the QDs as an artificial atom with given multi-exciton energies and corresponding dipole transitions. Assorted nonlinear techniques are unified by the fully quantum mechanical approach and superoperator formalism for the interacting optical field and excitons. By employing QD distribution functions, we classify the spontaneous signals according to the coherence range. The signal dependence on wave vector, number of QD and their density is discussed for sparse and polymer assembled QD's solutes. Several two-photon induced signals: second harmonic generation, hyper-Rayleigh scattering, two photon fluorescence and hyper-Raman are described within the same framework, keeping the focus on the frequency resolved spectra. The third chapter of this part will allow for realistic QD parameters. We calculate two-dimensional photon-echo and double-quantum-coherence spectra of two coupled InAs/InGaAs quantum dots at various distances, taking into account electron, hole and exciton hopping. We will look for signatures of direct and indirect excitons in two-exciton resonances, electron delocalization and its contribution to the creation of new biexcitonic peaks as well as dipole-dipole interactions to the shift the two-exciton energies. The last chapter attempts to establish a quantum kinetic theory for resonant optical excitation of electrons by an ultra-fast laser pulse as well as ultra-fast carrier scattering including photon-assisted absorption by non-thermal electrons.

14
Theory for Nonlinear Electron Transport

A quantum-dot superlattice (QDS) is schematically shown in Figure 14.1. Each quantum-dot (QD) disk (GaAs) with the radius R consists of an electron gas. The disks that are separated by a barrier layer ($Al_xGa_{1-x}As$) are stacked in the growth (x) direction with a period d. The tunneling current flows in the x-direction. In this structure, only the ground miniband is assumed to be occupied in the x-direction. Moreover, the confinement in the transverse yz-plane is also severe and produces well separated sublevels, leading to a single occupied sublevel.

14.1
Semiclassical Theory

In this section, we assume that electrons are subject to a strong AC electric field

$$F(t) = F_0 + F_1 \cos(2\pi \nu_{\rm ac} t) , \tag{14.1}$$

where F_0 is the DC field, and F_1 and ν_1 represent the AC-field amplitude and frequency.

14.1.1
Transient Boltzmann Equation

Let us begin by considering a simple system consisting of a degenerate electron gas moving in a quantum-dot superlattice with a period d in the superlattice (x) direction. For this system with a strong transverse confining potential, only the lowest subband is occupied by electrons in the low-density and low-temperature regimes, thus, a single subband model is adequate. Moreover, the so-called Umklapp scattering process can be neglected due to the existence of a large miniband gap and small phonon energies allowed at low temperatures. The original form of the Boltzmann transport equation for a time-dependent distribution function $f(k, t)$ of electrons

Properties of Interacting Low-Dimensional Systems, First Edition. G. Gumbs and D. Huang.

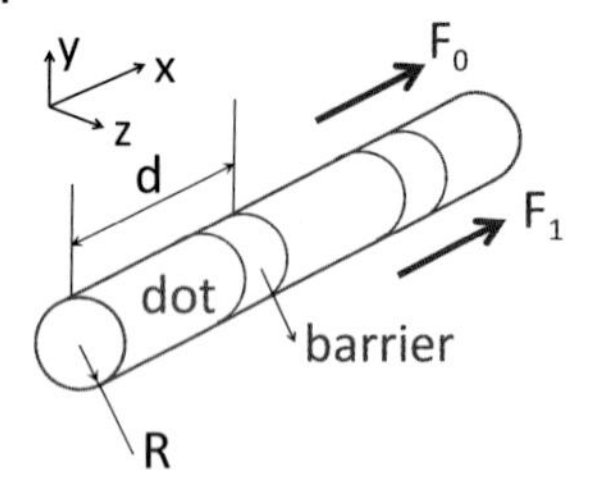

Figure 14.1 Schematic portrayal of a quantum dot superlattice whose axis is in the x-direction. The period is d and R is the radius of quantum dots separated by a barrier layer. Both a DC electric field $\mathbf{F}_0$ and an optical probe field $\mathbf{F}_1$ are applied along the superlattice direction.

is given by [1]

$$\begin{aligned}\frac{\partial f(k,t)}{\partial t} &= \frac{eF(t)}{\hbar}\frac{\partial f(k,t)}{\partial k} + \sum_{k'} P_e(k,k')\left[f(k',t) - f(k,t)\right] \\ &+ \sum_{k',q,\lambda}\left(P_+(k,k')\left\{(n_{q\lambda}+1)\, f(k',t)\left[1 - f(k,t)\right]\right.\right. \\ &- \left. n_{q\lambda} f(k,t)\left[1 - f(k',t)\right]\right\} + P_-(k,k')\left\{n_{q\lambda} f(k',t)\left[1 - f(k,t)\right]\right. \\ &- \left.\left.(n_{q\lambda}+1)\, f(k,t)\left[1 - f(k',t)\right]\right\}\right), \end{aligned} \tag{14.2}$$

where k is the electron wave vector along the x-direction, $\boldsymbol{q}$ is the three-dimensional wave vector of phonons, $\lambda = \ell$ ($\lambda = t$) corresponds to the longitudinal (transverse) acoustic phonons and $n_{q\lambda} = N_0(\omega_{q\lambda})$ is the Bose function for the equilibrium phonons with the frequency $\omega_{q\lambda}$ in the mode of $\boldsymbol{q}$ and λ. In Eq. (14.2), we have defined the electron–phonon inelastic scattering rate as [2]

$$P_{\pm}(k,k') = \frac{2\pi}{\hbar}|V_{k,k'}|^2 \delta(\varepsilon_k - \varepsilon_{k'} \pm \hbar\omega_{q\lambda})\delta_{k',k\pm q_x}\,, \tag{14.3}$$

and the electron–impurity elastic scattering rate as [3]

$$P_e(k,k') = \frac{2\pi}{\hbar}|U_{k,k'}|^2 \delta(\varepsilon_k - \varepsilon_{k'})\,, \tag{14.4}$$

where $\varepsilon_k = W(1 - \cos kd)/2$ is the electron kinetic energy with a bandwidth W in the single-band tight-binding model, $V_{k,k'}$ represents the interaction between electrons and phonons and $U_{k,k'}$ is the interaction between electrons and impurities.

We define $f(k,t) = f_k^{(0)} + g_k(t)$ with a dynamical non-equilibrium part $g_k(t)$ for the total distribution function, where $f_k^{(0)} = f_0(\varepsilon_k)$ is the Fermi function for the equilibrium electrons. We then note that the terms of zeroth order in g_k in the scattering parts of Eq. (14.2) cancel out due to detailed balance. After incorporating these considerations, we can algebraically simplify Eq. (14.2) into

$$\begin{aligned}\frac{\partial g_k(t)}{\partial t} &= -eF(t)v_k\left[-\frac{\partial f_k^{(0)}}{\partial \varepsilon_k}\right] + \frac{eF(t)}{\hbar}\frac{\partial g_k(t)}{\partial k} + \left[g_{-k}(t) - g_k(t)\right] \\ &\times \sum_{k' \in -k} P_e(k,k') - g_k(t)\sum_{k',q,\lambda}\Big\{P_+(k,k')\left(n_{q\lambda} + f_{k'}^{(0)}\right) + P_-(k,k') \\ &\times \left(n_{q\lambda} + 1 - f_{k'}^{(0)}\right) + g_{k'}(t)\left[P_+(k,k') - P_-(k,k')\right]\Big\} + \sum_{k',q,\lambda} g_{k'}(t) \\ &\times \left[P_+(k,k')\left(n_{q\lambda} + 1 - f_k^{(0)}\right) + P_-(k,k')\left(n_{q\lambda} + f_k^{(0)}\right)\right], \end{aligned} \tag{14.5}$$

where $v_k = (1/\hbar)d\varepsilon_k/dk$ is the group velocity of electrons with the wave vector k and the k'-sum for $k' \in -k$ is carried out near the resonance at $k' = -k$.

The transport relaxation rate from electron–impurity scattering equals [3]

$$\frac{1}{\tau_k} \equiv 2\sum_{k' \in -k} P_e(k,k') = \frac{\pi}{\hbar}|U_{k,-k}|^2\mathcal{D}(\varepsilon_k). \tag{14.6}$$

Here, the factor 2 arises from the $2k_F$ (k_F is the Fermi wave vector at zero temperature) scattering across the Fermi sea and $\mathcal{D}(\varepsilon)$ is the total density-of-states given by

$$\mathcal{D}(\varepsilon) = \frac{2L_x}{\pi d}\frac{1}{\sqrt{(W-\varepsilon)\varepsilon}}, \tag{14.7}$$

where L_x is the length of the superlattice.

For electron–phonon scattering, we define scattering rates:

$$W_\pm(k,k') \equiv \sum_{q,\lambda} P_\pm(k,k'). \tag{14.8}$$

Using Eq. (14.3) and writing $\hbar\omega_{q\lambda} = \alpha_\lambda\sqrt{(k-k')^2 + q_\perp^2}$ from the Debye model, we find [2, 4]

$$\begin{aligned} W_\pm(k,k') &= \frac{\mathcal{S}}{\hbar}\sum_\lambda \theta\left[(\varepsilon_k - \varepsilon_{k'})^2 - \alpha_\lambda^2(k-k')^2\right] \\ &\times \theta(\pm\varepsilon_{k'} \mp \varepsilon_k)\frac{\hbar\omega_{q\lambda}|V_\lambda(\mathbf{q})|^2}{\alpha_\lambda^2}\Phi(q_\perp), \end{aligned} \tag{14.9}$$

where $\theta(x)$ is the unit step function, $\mathcal{S}$ is the sample cross-sectional area, $\alpha_\lambda = \hbar c_\lambda$, c_λ is the sound velocity for the phonon of λ mode, $q_x = |k' - k|$,

$$q_\perp = \sqrt{q_y^2 + q_z^2} = \frac{1}{\alpha_\lambda}\sqrt{(\varepsilon_k - \varepsilon_{k'})^2 - \alpha_\lambda^2(k-k')^2}, \tag{14.10}$$

$$q = \sqrt{q_x^2 + q_\perp^2} = \frac{|\varepsilon_k - \varepsilon_{k'}|}{\alpha_\lambda}, \tag{14.11}$$

and the form factor in Eq. (14.9) takes the form:

$$\Phi(q_\perp) = e^{-q_\perp^2 R^2/2} = \exp\left\{-\left[(\varepsilon_k - \varepsilon_{k'})^2 - \alpha_\lambda^2 (k-k')^2\right]\frac{R^2}{2\alpha_\lambda^2}\right\} \tag{14.12}$$

for a transverse parabolic confinement with the localization radius R. The quantity $\Phi(q_\perp)$ arises from the consequence of momentum conservation for $q_\perp$. A smaller value of R in Eq. (14.12) implies a stronger electron–phonon scattering in Eq. (14.9).

Assuming an axial symmetry, the strength of the electron–phonon interaction $|V_\lambda(\mathbf{q})|^2$ in Eq. (14.9) is given by [4]

$$\hbar\omega_{q\lambda}|V_\lambda(\mathbf{q})|^2 = \frac{(\hbar\omega_{q\lambda})^2}{2\epsilon_s(q_x)^2\rho_i c_\lambda^2 \mathcal{V}}\left[D^2\delta_{\lambda,\ell} + \left(\frac{eh_{14}}{q}\right)^2 A_{q\lambda}\right], \tag{14.13}$$

where $\mathcal{V} = \mathcal{S}L_x$ is the sample volume, ρ_i is the mass density, D is the deformation-potential coefficient, h_{14} is the piezoelectric constant, $\epsilon_s(q_x)$ is the static dielectric function [4], and $A_{q\lambda}$ is the structure factor for acoustic phonons [4] due to anisotropic electron–phonon coupling. We further note from Eq. (14.9) that only one of $W_\pm(k,k')$ is non-zero, that is, either $W_+(k,k') > 0$ for $\varepsilon_{k'} > \varepsilon_k$ or $W_-(k,k') > 0$ for $\varepsilon_{k'} < \varepsilon_k$. This directly yields the result

$$W_\pm(k,k') = \theta(\pm\varepsilon_{k'} \mp \varepsilon_k)\sum_\lambda W_\lambda(k,k'), \tag{14.14}$$

where $W_\lambda(k,k') = W_\lambda(k',k)$ is given by

$$W_\lambda(k,k') = \frac{\mathcal{S}}{\hbar}\theta\left[(\varepsilon_k - \varepsilon_{k'})^2 - \alpha_\lambda^2(k-k')^2\right]|V_\lambda(\mathbf{q})|^2\frac{q}{\alpha_\lambda}\Phi(q_\perp). \tag{14.15}$$

14.1.2 Numerical Procedure

Using the scattering rates introduced in Eqs. (14.6) and (14.9), we rewrite Eq. (14.5) into a concise form

$$\begin{aligned}
\frac{\partial g_k(t)}{\partial t} &= eF(t)v_k\frac{\partial f_k^{(0)}}{\partial \varepsilon_k} + \frac{eF(t)}{\hbar}\frac{\partial g_k(t)}{\partial k} + \frac{1}{2\tau_k}\left[g_{-k}(t) - g_k(t)\right] \\
&\quad - g_k(t)\sum_{k'}\left\{W_+(k,k')\left(n_{k,k'} + f_{k'}^{(0)}\right) + W_-(k,k')\left(n_{k,k'} + 1 - f_{k'}^{(0)}\right)\right. \\
&\quad \left. + g_{k'}(t)\left[W_+(k,k') - W_-(k,k')\right]\right\} + \sum_{k'} g_{k'}\left[W_+(k,k')\right. \\
&\quad \times \left.\left(n_{k,k'} + 1 - f_k^{(0)}\right) W_-(k,k')\left(n_{k,k'} + f_k^{(0)}\right)\right],
\end{aligned} \tag{14.16}$$

where $n_{k,k'} = N_0(|\varepsilon_k - \varepsilon_{k'}|/\hbar)$ is independent of $\lambda = \ell, t$. By introducing the notations $f_k^- = f_k^{(0)}$ and $f_k^+ = 1 - f_k^{(0)}$, we obtain the discrete form for the total inelastic scattering rate

$$W_j = \frac{L_x}{2\pi}\delta k \sum_{j',\pm} W_\pm(k,k')\left(n_{k,k'} + f_{k'}^{\mp}\right), \tag{14.17}$$

where $k \equiv j\,\delta k$ and $k' \equiv j'\delta k$ (with $\delta k \to 0$), and

$$W_{j,j'} = \sum_{\pm} W_\pm(k,k')\left(n_{k,k'} + f_{k'}^{\pm}\right) . \tag{14.18}$$

Alternatively,

$$W_j^g(t) = \frac{L_x}{2\pi}\delta k \sum_{j'} g_{k'}(t)\left[W_+(k,k') - W_-(k,k')\right], \tag{14.19}$$

and obtain a discrete form of Eq. (14.16)

$$\begin{aligned}\frac{dg_k(t)}{dt} &= eF(t)v_k \frac{\partial f_k^{(0)}}{\partial \varepsilon_k} + \frac{eF(t)}{\hbar}\frac{\partial g_k(t)}{\partial k} - g_k(t)\left[W_j + W_j^g(t)\right] \\ &+ \frac{1}{2\tau_k}\left[g_{-k}(t) - g_k(t)\right] + \frac{L_x}{2\pi}\delta k \sum_{j'} g_{k'}(t)\,W_{j,j'} .\end{aligned} \tag{14.20}$$

In Eq. (14.20), the quantity $g_k W_j^g$ is the only nonlinear term with respect to g_k.
Equation (14.20) is equivalent to the following matrix equation for $1 \le j \le N$

$$\frac{dg_j(t)}{dt} = b_j(t) - \sum_{j'=1}^{N} a_{j,j'}(t) g_{j'}(t) , \tag{14.21}$$

where $k \equiv j\delta k$, $k' \equiv j'\delta k$, and $\delta k = 2\pi/(N-1)d$ with $N \gg 1$ being an odd integer. The elements for the vector $\boldsymbol{b}(t)$ in Eq. (14.21) are

$$b_j(t) = eF(t)v_j \frac{\partial f_j^{(0)}}{\partial \varepsilon_j} , \tag{14.22}$$

and the elements for the matrix $\overleftrightarrow{\mathbf{a}}(t)$ in Eq. (14.21) are

$$\begin{aligned}a_{j,j'}(t) &= \delta_{j,j'}\left[W_j + W_j^g(t) + \frac{1-\delta_{j,(N+1)/2}}{2\tau_j}\right] \\ &- \delta_{j+j',N+1}\left[\frac{1-\delta_{j,(N+1)/2}}{2\tau_j}\right] - \frac{eF(t)}{\hbar\delta k}c_{j,j'} - \frac{L_x}{2\pi}\delta k\, W_{j,j'} .\end{aligned} \tag{14.23}$$

By assuming that the field $F(t)$ is turned on at the moment of $t = 0$, the differential equation (14.21) can be solved in combination with the initial condition $g_j(0) = 0$

for $j = 1, 2, \cdots, N$. Additionally, using the simplest so-called 3-point differential formula, we can write $c_{j,j'}$ in Eq. (14.23) in the form of

$$c_{j,j'} = \frac{1}{2}\left(\delta_{j,j'-1} - \delta_{j,j'+1}\right). \tag{14.24}$$

Unfortunately, the quantities $\{g_1, g_2, \cdots, g_N\}$ are not linearly independent of each other [2, 3, 5], implying that the matrix $\overleftrightarrow{\mathbf{a}}(t)$ in Eq. (14.21) is a singular. In fact, the condition for the particle-number conservation requires that

$$\sum_{j=1}^{N} g_j(t) = 0\,. \tag{14.25}$$

Combined with the fact that $\varepsilon_k = \varepsilon_{-k}$ and $v_k = -v_{-k}$, for $1 \le j \le N$, this yields

$$g_1(t) = g_N(t) = -\frac{1}{2}\sum_{j=2}^{N-1} g_j(t)\,, \tag{14.26}$$

where $j = 1$ and N are the two k-space points at the first Brillouin-zone boundary. As a result of Eq. (14.26), we can renormalize the singular $(N \times N)$-matrix $\overleftrightarrow{\mathbf{a}}(t)$ in Eq. (14.21) into a regular $[(N-1) \times (N-1)]$-matrix $\overleftrightarrow{\mathbf{a}}'(t)$ through the following relation:

$$a'_{j,j'}(t) = a_{j,j'}(t) - \left\langle a_j(t)\right\rangle\,, \tag{14.27}$$

where $\langle a_j(t)\rangle = [a_{j,1}(t) + a_{j,N}(t)]/2$ and $j, j' = 2, 3, \cdots, N-1$. Consequently, Eq. (14.21) is renormalized into

$$\frac{dg'_j(t)}{dt} = b'_j(t) - \sum_{j'=2}^{N-1} a'_{j,j'}(t) g'_{j'}(t)\,, \tag{14.28}$$

where the vectors $\mathbf{g}'(t)$ and $\mathbf{b}'(t)$ are the same as the vectors $\mathbf{g}(t)$ and $\mathbf{b}(t)$, respectively, without the first and last elements with $j = 1$ and N. Similarly, we correspondingly rewrite Eq. (14.19) by excluding the elements with $j = 1$ and N as

$$W_j^g(t) = \frac{L_x}{2\pi}\delta k \sum_{j'=2}^{N-1} g'_{j'}(t) \times \left[W_+(j,j') - W_-(j,j') - \left(\langle W_+(j)\rangle - \langle W_-(j)\rangle\right)\right], \tag{14.29}$$

where $\langle W_\pm(j)\rangle = [W_\pm(j,1) + W_\pm(j,N)]/2$.

The time-dependent tunneling current of the system can be found from

$$I_{1D}(t) = -\frac{eWd\delta k}{2\pi\hbar}\sum_{j=2}^{N-1} g'_j(t) \sin\left[\left(j - \frac{N+1}{2}\right)\delta k d\right], \tag{14.30}$$

and the steady-state current I_0 for a DC electric field (with $F_1 = 0$) is given by $I_{1D}(t)$ at the $t \to \infty$ limit. In the presence of an AC electric field (with $F_1 \neq 0$), the AC current $I_{1D}(t)$ quickly establishes regular periodic oscillations after the scattering time. Therefore, we can define a long-time average current $\langle I_{1D} \rangle$ from the transient current $I_{1D}(t)$ through its limiting behavior:

$$\langle I_{1D} \rangle = \lim_{T_0 \to \infty} \frac{1}{T_0} \int_0^{T_0} I_{1D}(t) dt \,, \tag{14.31}$$

where T_0 is much larger than the scattering time. The long-time-averaged quantity $\langle I_{1D} \rangle$ includes contributions from both the DC and AC electric fields. Mathematically, $\langle I_{1D} \rangle$ represents the time-independent term after the Fourier transform with respect to t is done to $I_{1D}(t)$.

14.1.3 Numerical Results for Bloch Oscillations and Dynamical Localization

For the simplicity of the notation, we define $\tau_j = \tau_k$ at the energy $\varepsilon_j = \varepsilon_k$ in Eq. (14.6):

$$\frac{1}{\tau_j} = \gamma_0 \sqrt{\frac{(W - \mu)\mu}{(W - \varepsilon_j)\varepsilon_j}} \,, \tag{14.32}$$

where γ_0 is the value of $1/\tau_j$ at $\varepsilon_j = \mu$. For a fixed electron temperature T and linear density n_{1D}, the electron chemical potential μ is determined from the following relation:

$$n_{1D} = \frac{\delta k}{\pi} \sum_{j=1}^{N} \frac{1}{\exp\left(\frac{\varepsilon_j - \mu}{k_B T}\right) + 1} \,. \tag{14.33}$$

The Fermi wave vector equals $k_F = \pi n_{1D}/2$ at $T = 0$ K. We choose GaAs as the host material for our numerical calculations and employ the following parameters: the sound velocities $c_\ell = 5.14 \times 10^5$ cm/s, $c_t = 3.04 \times 10^5$ cm/s, the mass density $\rho_i = 5.3$ g/cm^3, the piezoelectric constant $h_{14} = 1.2 \times 10^7$ V/cm, and the deformation-potential coefficient $D = -9.3$ eV [6, 7]. For the Bloch oscillations, we consider a DC field. Whereas for dynamical localization, a combination of DC and AC fields is considered. The other parameters will be directly indicated in the figure captions. Figure 14.2 exhibits I_0 as a function of F_0 for two values of R, where the field scales as $1/d$. For a stronger phonon scattering with $R = 10$ Å (solid squares), we first find a linear-transport behavior $I_0 \propto F_0$ in the weak-field regime $F_0 \leq 25$ V/cm corresponding to $\omega_B \ll \gamma_0, \gamma_{ph}$, where $\omega_B = eF_0 d/\hbar$ is the frequency of Bloch oscillations and γ_{ph} represents the effective phonon scattering rate introduced earlier by Lyo [8, 9]. Whereas, when $F_0 > 25$ V/cm, I_0 decreases with F_0 and evolves into a $1/\omega_B$-dependence for $F_0 \gg 25$ V/cm, where $\omega_B \gg \gamma_{ph}, \gamma_0$. A sharp current peak is observed at $F_0 = 25$ V/cm. When the phonon scattering is

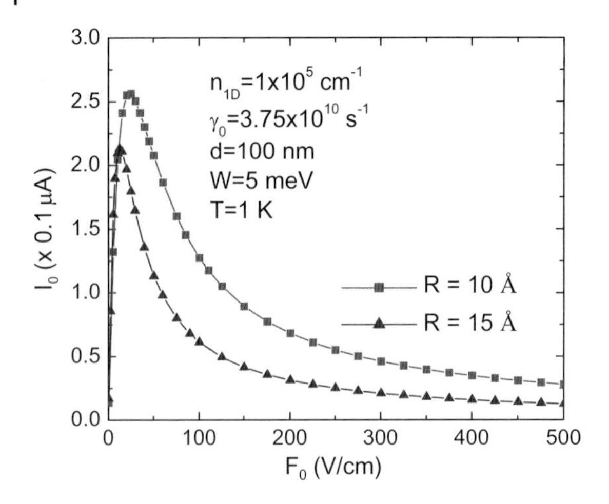

Figure 14.2 Calculated steady-state currents I_0 as a function of the DC electric field F_0 for two values of R. We choose $R = 10$ Å (solid squares) and $R = 15$ Å (solid triangles). Here, we take $F_1 = 0$, $n_{1D} = 10^5\,\text{cm}^{-1}$, $d = 100$ nm, $W = 5$ meV, $T = 1$ K, and $\gamma_0 = 3.75 \times 10^{10}\,\text{s}^{-1}$.

reduced with $R = 15$ Å (solid triangles), the peak current I_p is shifted to a lower field because the field F_p at the peak value of I_0 is proportional to γ_{ph} for small γ_0 [8, 9]. Moreover, the magnitude of I_p is also reduced since it is approximately proportional to γ_{ph}/γ_0 [8, 9]. Furthermore, the high-field current at $F_0 = 500$ V/cm decreases when γ_{ph} is reduced. This can be attributed to the fact that the steady-state current is proportional to γ_{ph}/ω_B in the high-field limit $\omega_B \gg \gamma_0, \gamma_{ph}$ [8, 9].

In Figure 14.3, we present I_0 as a function of F_0 for two values of γ_0. For a large $\gamma_0 = 3.75 \times 10^{12}\,\text{s}^{-1}$ (solid triangles), the low-field I_0 is drastically reduced in comparison with that for a small $\gamma_0 = 3.75 \times 10^{10}\,\text{s}^{-1}$ (solid squares) because $I_0 \propto 1/\sqrt{\gamma_0}$ for $\gamma_0 \gg \gamma_{ph}$ [8, 9]. However, the high-field I_0 at $F_0 = 500$ V/cm remains the same, independent of γ_0. Additionally, F_p shifts to a high field because $F_p \propto \sqrt{\gamma_0}$ for $\gamma_0 \gg \gamma_{ph}$ [8, 9].

A comparison of the results for the calculated $\langle I_{1D} \rangle$ is given in Figure 14.4 as a function of the amplitude of F_0 for $F_1 = 100$ V/cm (solid squares), $F_1 = 300$ V/cm (gray triangles on gray curve) and $F_1 = 500$ V/cm (inverted solid triangles), where $\nu_{ac} = 2$ THz is assumed. When F_1 is approached in the limit of $F_1 = 0$, the I–V curve is obtained as depicted in Figure 14.2. A DC-field-induced Esaki–Tsu peak [10] is gradually built up at a DC field F_0 that decreases with decreasing F_1: the dynamical-localization intersection point moves to the left at the same time. Moreover, the negative minimum also shifts to the left, disappearing at $F_0 = 0$. There is only one Esaki–Tsu-type peak in this figure, which qualitatively agrees with the early experimental observation by Winnerl, *et al.* [11]. Eventually, the three $\langle I_{1D} \rangle$ curves for different values of F_1 merge into one for large F_0 and the dynamical localization in the system is completely suppressed by the strong Bloch oscillations.

The results for $\langle I_{1D} \rangle$ are presented as a function of F_{dc} in Figure 14.5 for $\nu_{ac} = 0.1$ THz (solid squares), $\nu_{ac} = 1$ THz (solid triangles on dark gray curve), $\nu_{ac} =$

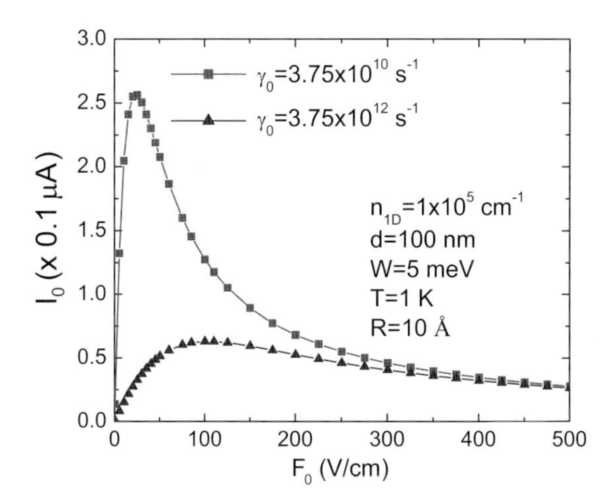

Figure 14.3 Calculated steady-state currents I_0 as a function of the DC electric field F_0 for two values of γ_0. We assume $\gamma_0 = 3.75 \times 10^{10}\,\text{s}^{-1}$ (solid squares) and $\gamma_0 = 3.75 \times 10^{12}\,\text{s}^{-1}$ (solid triangles). Here, we take $F_1 = 0$, $n_{1D} = 10^5\,\text{cm}^{-1}$, $d = 100$ nm, $W = 5$ meV, $T = 1$ K and $R = 10$ Å.

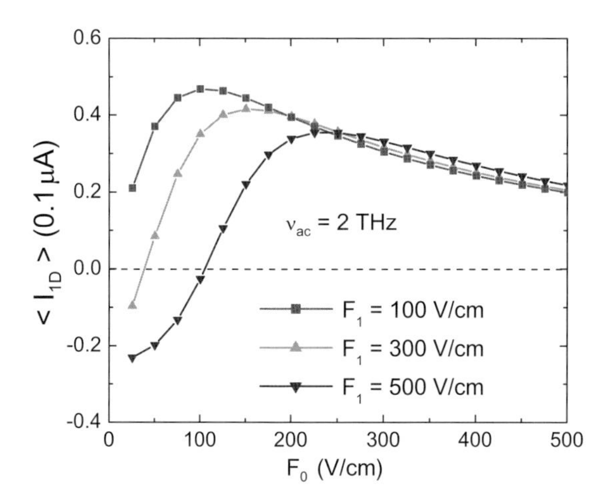

Figure 14.4 Calculated long-time average currents $\langle I_{1D}\rangle$ as a function of the amplitude of a DC-field component F_0 for three values of F_1. We choose $F_1 = 100\,\text{V/cm}$ (solid squares), $F_1 = 300\,\text{V/cm}$ (gray triangles on gray curve) and $F_1 = 500\,\text{V/cm}$ (inverted solid triangles). Here, we set $\nu_{ac} = 0.1\,\text{THz}$, $n_{1D} = 10^5\,\text{cm}^{-1}$, $\gamma_0 = 3.75 \times 10^{12}\,\text{s}^{-1}$, $d = 80\,\text{nm}$, $W = 5\,\text{meV}$, $T = 1\,\text{K}$, and $R = 10\,$Å. The horizontal black dashed lines are used as a guideline for $\langle I_{1D}\rangle = 0$.

2 THz (inverted gray triangles on gray curve) and for $F_1 = 300\,\text{V/cm}$. Here, the negative minimum completely suppresses when ν_{ac} reduces to 0.1 THz. At the same time, the Esaki–Tsu-type peak pushes to a high DC field for $\nu_{ac} = 0.1\,\text{THz}$. Again, $\langle I_{1D}\rangle$ curves for three values of ν_{ac} merge into one as F_0 becomes very large.

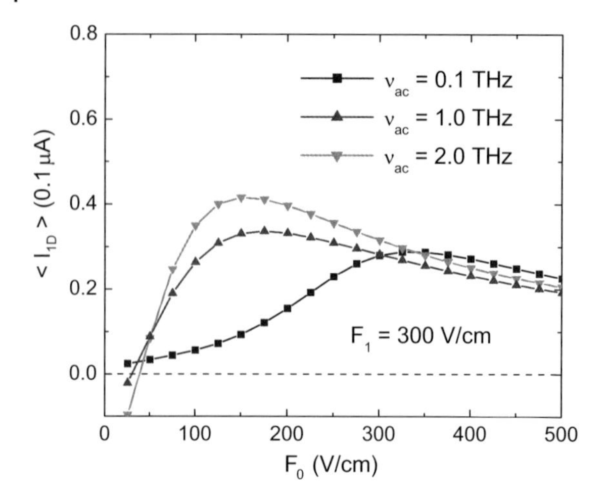

Figure 14.5 Calculated long-time average currents $\langle I_{1D} \rangle$ as a function of the amplitude of a DC-field component F_0 for three values of ν_{ac}. We assume $\nu_{ac} = 0.1\,\text{THz}$ (solid squares), $\nu_{ac} = 1\,\text{THz}$ (solid triangles on dark gray curve) and $\nu_{ac} = 2\,\text{THz}$ (inverted solid triangles). Here, we set $F_1 = 300\,\text{V/cm}$, $n_{1D} = 10^5\,\text{cm}^{-1}$, $\gamma_0 = 3.75 \times 10^{12}\,\text{s}^{-1}$, $d = 80\,\text{nm}$, $W = 5\,\text{meV}$, $T = 1\,\text{K}$, and $R = 10\,\text{Å}$. The horizontal black dashed lines are used as a guideline for $\langle I_{1D} \rangle = 0$.

14.2
Quantum Theory

For a quantum theory in this section, as opposed to a semiclassical theory in the previous section, we still consider, as a sample, the same QDS shown in Figure 14.1.

14.2.1
Force Balance Equation

The x-component of the many-body Hamiltonian, $\mathcal{H}_x$, for N interacting electrons in a QDS under a spatially-uniform electric field F_0 along the QDS growth (x) direction can be written as

$$\mathcal{H}_x = -\frac{\hbar^2}{2m^*}\sum_{j=1}^{N}\frac{\partial^2}{\partial x_j^2} + \sum_{j=1}^{N} V_{\text{SL}}(x_j) - eF_0\sum_{j=1}^{N} x_j + \mathcal{H}_{\text{ph}} + \mathcal{H}_{\text{e-i}} + \mathcal{H}_{\text{e-p}} + \mathcal{H}_{\text{e-e}}\,, \tag{14.34}$$

where m^* is the effective mass of electrons in host materials, x_j is the jth electron's position in the growth direction, and $V_{\text{SL}}(x) = V_{\text{SL}}(x+d)$ is a one-dimensional periodic potential in the x-direction with a period d, as shown in Figure 14.1. In Eq. (14.34), $\mathcal{H}_{\text{ph}}$, $\mathcal{H}_{\text{e-i}}$, $\mathcal{H}_{\text{e-p}}$ and $\mathcal{H}_{\text{e-e}}$ represent different parts of the Hamiltonian from phonons (ph), electron–impurity interaction (e–i), electron–phonon interaction (e–p), and electron–electron interaction (e–e), respectively. For all Bloch electrons in the presence of $V_{SL}(x)$ QDS potential, We introduce an ensemble-averaged inverse of effective mass $1/m_x^* V_{SL}(x)$. The explicit expression for $1/m_x^*$ will be given by Eq. (14.39). In this model, we can write down the center-of-mass position and momentum operators along the x-direction

$$X = \frac{1}{N} \sum_{j=1}^{N} x_j \, , \quad \hat{P} = \sum_{j=1}^{N} \hat{p}_j \, , \tag{14.35}$$

where $\hat{p}_j = -i\hbar\partial/\partial x_j$ is the jth electron's momentum operator. For simplicity, we assume that the quantum confinements in both the y and z-directions are so strong that only the ground states in these two directions will be occupied. Therefore, we only consider the relevant single (lowest) miniband for the QSD at low electron linear densities n_{1D} and lattice temperatures T with a thermal contact to an external heat bath.

For resonant electron tunnelings in QSD without a bias, the electron momentum p in the x-direction is a good quantum number and the electron group velocity becomes finite. Therefore, we can introduce in an effective Hamiltonian based on Eq. (14.34) as below:

$$\mathcal{H}_{\text{eff}} = \sum_{j=1}^{N} E_{SL}(\hat{p}_j) - eF_0 \sum_{j=1}^{N} x_j + \mathcal{H}_{\text{ph}} + \mathcal{H}_{\text{e-i}} + \mathcal{H}_{\text{e-p}} + \mathcal{H}_{\text{e-e}} \, , \tag{14.36}$$

where

$$E_{SL}(p) = \frac{W}{2} \left[1 - \cos\left(\frac{pd}{\hbar} \right) \right] \tag{14.37}$$

is the energy dispersion for the QDS's lowest miniband in the tight-binding approximation with $p = \hbar k$, $|k| \le \pi/d$ within the first Brillouin zone, and W being the width of the miniband. This imposes an upper limit to F_0 by requiring $eF_0 d \le W$ for Bloch electrons in QDS. Additionally, we write the ensemble-averaged inverse of effective mass of Bloch electrons along the x-direction

$$\frac{1}{m_x^*} = \frac{1}{N} \left\langle \sum_{j=1}^{N} \frac{d^2 E_{SL}(\hat{p}_j)}{dp_j^2} \right\rangle_{\text{av}} \, , \tag{14.38}$$

where $\langle \hat{A} \rangle_{\text{av}}$ represents the quantum-statistical average over a physical operator $\hat{A}$. In an explicit way, we can rewrite Eq. (14.38) as

$$\frac{1}{m_x^*} = \frac{2}{N\hbar^2} \sum_k \frac{d^2 E_{SL}(k + k_0)}{dk^2} f_k \, , \tag{14.39}$$

where the factor of two comes from the spin degeneracy of electrons, $\hbar k_0 = P/N$, P is the center-of-mass momentum, $N = 2\sum_k f_k$, and f_k represents the non-equilibrium distribution of hot electrons in relative scattering motions. Using the effective Hamiltonian in Eq. (14.36) and the center-of-mass position operator defined in Eq. (14.35), we can calculate the center-of-mass velocity operator using the Heisenberg equation

$$\hat{V} = \frac{dX}{dt} = \frac{1}{i\hbar}[X, \mathcal{H}_{\text{eff}}] = \frac{1}{Ni\hbar}\sum_{j=1}^{N}[x_j, \mathcal{H}_{\text{eff}}]$$
$$= \frac{1}{N}\sum_{j=1}^{N}\frac{dE_{\text{SL}}(\hat{p}_j)}{dp_j}, \tag{14.40}$$

where $[\hat{A}, \hat{B}] \equiv \hat{A}\hat{B} - \hat{B}\hat{A}$. As a result, we get the draft velocity v_{d} of electrons in the center-of-mass frame through

$$v_{\text{d}} = \langle \hat{V} \rangle_{\text{av}} = \frac{2}{N}\sum_k v(k + k_0) f_k \tag{14.41}$$

with $v(k) = \hbar^{-1} dE_{\text{SL}}(k)/dk$ as the group velocity of Bloch electrons in the k state. Equation (14.41) gives rise to the nonlinear relationship between v_{d} and k_0. Furthermore, we obtain from Eq. (14.40) the force balance equation

$$\frac{dv_{\text{d}}}{dt} = \left\langle \frac{d\hat{V}}{dt} \right\rangle_{\text{av}} = \frac{1}{Ni\hbar}\left\langle \sum_{j=1}^{N}\left[\frac{dE_{\text{SL}}(\hat{p}_j)}{dp_j}, \mathcal{H}_{\text{eff}}\right]\right\rangle_{\text{av}}$$
$$= \frac{eF_0}{m_x^*} + A_{\text{i}}[k_0] + A_{\text{p}}[k_0], \tag{14.42}$$

where $A_{\text{i}}[k_0]$ and $A_{\text{p}}[k_0]$ are the ensemble-averaged resistive accelerations in the center-of-mass frame due to scattering by impurities (i) and phonons (ph), and are given by

$$A_{\text{i}}[k_0] = \left(\frac{N_{\text{i}}}{N}\right)\frac{2\pi}{\hbar}\sum_{k,k'}|U_{\text{i}}(|k' - k|)|^2\left[v(k' + k_0) - v(k + k_0)\right]$$
$$\times (f_k - f_{k'})\delta\left[E_{\text{SL}}(k' + k_0) - E_{\text{SL}}(k + k_0)\right], \tag{14.43}$$

$$A_{\text{p}}[k_0] = \frac{1}{N}\sum_{k,k'Q,\lambda}\left[v(k' + k_0) - v(k + k_0)\right]$$
$$\times\left[\Theta^{\text{abs}}_{k,k';Q\lambda}[k_0]N_{Q\lambda} - \Theta^{\text{em}}_{k,k';Q\lambda}[k_0](N_{Q\lambda} + 1)\right]. \tag{14.44}$$

Here, N_{i} is the number of ionized donors in the system, we only consider the acoustic-phonon scattering for low lattice temperatures T,

$$\begin{bmatrix}\Theta^{\text{abs}}_{k,k';Q\lambda}[k_0]\\ \Theta^{\text{em}}_{k,k';Q\lambda}[k_0]\end{bmatrix} = \frac{2\pi}{\hbar}|C_{Q\lambda}|^2$$
$$\times\left\{\begin{bmatrix}-f_{k'}(1 - f_k)\delta_{Q_x,k-k'}\delta\left[E_{\text{SL}}(k + k_0) - E_{\text{SL}}(k' + k_0) - \hbar\omega_{Q\lambda}\right]\\ f_{k'}(1 - f_k)\delta_{Q_x,k'-k}\delta\left[E_{\text{SL}}(k + k_0) - E_{\text{SL}}(k' + k_0) + \hbar\omega_{Q\lambda}\right]\end{bmatrix}\right.$$
$$\left.+\begin{bmatrix}(1 - f_{k'})f_k\delta_{Q_x,k'-k}\delta\left[E_{\text{SL}}(k' + k_0) - E_{\text{SL}}(k + k_0) - \hbar\omega_{Q\lambda}\right]\\ -(1 - f_{k'})f_k\delta_{Q_x,k-k'}\delta\left[E_{\text{SL}}(k' + k_0) - E_{\text{SL}}(k + k_0) + \hbar\omega_{Q\lambda}\right]\end{bmatrix}\right\}, \tag{14.45}$$

$\mathbf{Q} = (Q_x, \mathbf{Q}_\perp)$, $\mathbf{Q}_\perp = (Q_y, Q_z)$, and $\hbar\omega_{Q\lambda}$ are the wave vector and energy of acoustic phonons, and $\lambda = \ell, t$ are for longitudinal and transverse modes, respectively. Additionally, $N_{Q\lambda}$ represents the non-equilibrium distribution of hot phonons and is determined by

$$\frac{dN_{Q\lambda}}{dt} = \Gamma^{\text{em}}_{Q\lambda}[k_0](N_{Q\lambda}+1) - \Gamma^{\text{abs}}_{Q\lambda}[k_0]N_{Q\lambda} - \frac{N_{Q\lambda} - N_0(\omega_{Q\lambda})}{\tau_{Q\lambda}}, \tag{14.46}$$

where

$$\begin{aligned}
\begin{bmatrix} \Gamma^{\text{abs}}_{Q\lambda}[k_0] \\ \Gamma^{\text{em}}_{Q\lambda}[k_0] \end{bmatrix} &= \frac{2\pi}{\hbar}\sum_{k,k'} |C_{Q\lambda}|^2 \\
&\times \left\{ \begin{bmatrix} f_{k'}(1-f_k)\delta_{Q_x,k-k'}\delta\left[E_{\text{SL}}(k+k_0) - E_{\text{SL}}(k'+k_0) - \hbar\omega_{Q\lambda}\right] \\ f_{k'}(1-f_k)\delta_{Q_x,k'-k}\delta\left[E_{\text{SL}}(k+k_0) - E_{\text{SL}}(k'+k_0) + \hbar\omega_{Q\lambda}\right] \end{bmatrix} \right. \\
&= \left. \begin{bmatrix} (1-f_{k'})f_k\delta_{Q_x,k'-k}\delta\left[E_{\text{SL}}(k'+k_0) - E_{\text{SL}}(k+k_0) - \hbar\omega_{Q\lambda}\right] \\ (1-f_{k'})f_k\delta_{Q_x,k-k'}\delta\left[E_{\text{SL}}(k'+k_0) - E_{\text{SL}}(k+k_0) + \hbar\omega_{Q\lambda}\right] \end{bmatrix} \right\}.
\end{aligned} \tag{14.47}$$

Here, $\Gamma^{\text{abs}}_{Q\lambda}[k_0]$ is the absorption rate of phonons, $\Gamma^{\text{em}}_{Q\lambda}[k_0]$ is the emission rate of phonons, $N_0(\omega_{Q\lambda}) = [\exp(\hbar\omega_{Q\lambda}/(k_B T)) - 1]^{-1}$ is the Boson function for thermal-equilibrium acoustic phonons and $\tau_{Q\lambda}$ is the lifetime of acoustic phonons due to scattering by rough surfaces of a sample. The explicit expressions for $|U_{\text{i}}(q_x)|^2$ and $|C_{Q\lambda}|^2$ in Eqs. (14.43) and (14.45) will be given by Eqs. (14.54) and (14.57) below.

14.2.2 Boltzmann Scattering Equation

For the relative scattering motion of electrons, we still use k to represent the relative wave number of Bloch electrons under the restraint $|k| \le \pi/d$. Therefore, the microscopic equation for relative scattering motion of electrons can be written as [12, 13]

$$\begin{aligned}
\frac{dg_k}{dt} &= \mathcal{W}^{(\text{in})}_k[k_0](1-f_k) - \mathcal{W}^{(\text{out})}_k[k_0]f_k \\
&= \mathcal{W}^{(\text{in})}_k[k_0]\left[1 - f^{(0)}_k - g(k)\right] - \mathcal{W}^{(\text{out})}_k[k_0]\left[f^{(0)}_k + g(k)\right],
\end{aligned} \tag{14.48}$$

where $f_k = f^{(0)}_k + g(k)$, $f^{(0)}_k$ is the Fermi function for thermal-equilibrium electrons, $g(-\pi/d) = g(\pi/d)$ to ensure the periodicity, $\sum_k g(k) \equiv 0$ to ensure the conservation of electrons in the system,

$$\begin{bmatrix} \mathcal{W}^{(\text{in})}_k[k_0] \\ \mathcal{W}^{(\text{out})}_k[k_0] \end{bmatrix} = \begin{bmatrix} \mathcal{W}^{(\text{in})}_{k(1)} + \mathcal{W}^{(\text{in})}_{k(2)}[k_0] + \mathcal{W}^{(\text{in})}_{k(3)}[k_0] \\ \mathcal{W}^{(\text{out})}_{k(1)} + \mathcal{W}^{(\text{out})}_{k(2)}[k_0] + \mathcal{W}^{(\text{out})}_{k(3)}[k_0] \end{bmatrix} \tag{14.49}$$

are the total scattering-in (in) rate for electrons in the final k state and the total scattering-out (out) rate for electrons in the initial k state. In Eq. (14.49), $\mathcal{W}^{(\text{in})}_{k(j)}$ and

$\mathcal{W}_{k(j)}^{(\text{out})}$ for $j = 1, 2, 3$ represent the contributions from electron–electron, electron–impurity and electron–phonon scattering, respectively.

For the relative electron–electron scattering part in Eq. (14.49), we have

$$\begin{bmatrix} \mathcal{W}_{k(1)}^{(\text{in})} \\ \mathcal{W}_{k(1)}^{(\text{out})} \end{bmatrix} = \frac{2\pi}{\hbar} \sum_{k',k_1,k_1'}^{(k' \neq k_1')} |V_c(|k - k_1|)|^2 \begin{bmatrix} (1 - f_{k'}) f_{k_1} f_{k_1'} \\ f_{k'}(1 - f_{k_1})(1 - f_{k_1'}) \end{bmatrix}$$
$$\times\, \delta_{k_1'-k',k-k_1} \delta\left[E_{\text{SL}}(k) + E_{\text{SL}}(k') - E_{\text{SL}}(k_1) - E_{\text{SL}}(k_1')\right], \quad (14.50)$$

and the Coulomb interaction matrix between a pair of electrons is given by

$$|V_c(|q_x|)|^2 = \frac{1}{2} \left(\frac{e^2}{2\pi\epsilon_0\epsilon_r L_x}\right)^2 e^{-q_x^2 \ell^2}$$
$$\times \left[\int_0^{+\infty} d\eta \exp\left(-\frac{\eta^2}{2\alpha^2}\right) \frac{\mathcal{F}_0(|\eta|, |q_x|)}{\sqrt{\eta^2 + q_x^2}}\right]^2 . \quad (14.51)$$

Here, we assume that the quantum confinement for electrons in the z-direction is a parabolic potential, that is, $m^* \Omega_0^2 z^2/2$. Additionally, the quantum confinement in the y-direction is a square-well potential with a finite barrier height V_b, the ground-state wave function of a single dot in the x-direction is in a Gaussian form with a width 2ℓ, L_x is the length of QDS, ϵ_r is the average dielectric constant of host materials, $\alpha = \sqrt{m^* \Omega_0/\hbar}$, and the form factor in Eq. (14.51) is defined by

$$\mathcal{F}_0(|\eta|, |q_x|) = \int_{-\infty}^{+\infty} dy\, \psi_1^2(y) \int_{-\infty}^{+\infty} dy_1 \psi_1^2(y_1) e^{-\sqrt{\eta^2+q_x^2}|y-y_1|}, \quad (14.52)$$

where $\psi_1(y)$ is the ground-state wave function of electrons in a square well.

For the relative electron–impurity scattering part in Eq. (14.49), we have

$$\begin{bmatrix} \mathcal{W}_{k(2)}^{(\text{in})}[k_0] \\ \mathcal{W}_{k(2)}^{(\text{out})}[k_0] \end{bmatrix} = N_i \frac{4\pi}{\hbar} \sum_{k'}^{(k' \neq k)} |U_i(|k - k'|)|^2$$
$$\times \begin{bmatrix} f_{k'} \delta\left[E_{\text{SL}}(k + k_0) - E_{\text{SL}}(k' + k_0)\right] \\ (1 - f_{k'}) \delta\left[E_{\text{SL}}(k' + k_0) - E_{\text{SL}}(k + k_0)\right] \end{bmatrix}, \quad (14.53)$$

where the Coulomb interaction matrix between an electron and an ionized impurity atom is given by

$$|U_i(|q_x|)|^2 = \left(\frac{e^2}{2\pi\epsilon_0\epsilon_r L_x}\right)^2 \exp\left(-\frac{q_x^2 \ell^2}{2}\right)$$
$$\times \left[\int_0^{+\infty} d\eta \exp\left(-\frac{\eta^2}{4\alpha^2}\right) \frac{\mathcal{F}_1(\eta, q_x, y_0)}{\sqrt{\eta^2 + q_x^2}}\right]^2 . \quad (14.54)$$

The form factor in Eq. (14.54) is defined as

$$\mathcal{F}_1(\eta, q_x, y_0) = \int_{-\infty}^{+\infty} dy\, \psi_1^2(y) e^{-\sqrt{\eta^2+q_x^2}|y-y_0|} , \tag{14.55}$$

where y_0 is the position of a single δ-doping layer inside the square well perpendicular to the y-direction.

For the relative electron–phonon scattering part in Eq. (14.49), we have

$$\begin{aligned}
\begin{bmatrix} \mathcal{W}_{k(3)}^{(\text{in})}[k_0] \\ \mathcal{W}_{k(3)}^{(\text{out})}[k_0] \end{bmatrix} &= \frac{2\pi}{\hbar} \sum_{k'}^{(k'\neq k)} \sum_{Q,\lambda} |C_{Q\lambda}|^2 \begin{bmatrix} f_{k'} \\ (1-f_{k'}) \end{bmatrix} \\
&\times \left\{ N_{Q\lambda} \begin{bmatrix} \delta_{Q_x,k-k'} \delta\left[E_{\text{SL}}(k+k_0) - E_{\text{SL}}(k'+k_0) - \hbar\omega_{Q\lambda}\right] \\ \delta_{Q_x,k'-k} \delta\left[E_{\text{SL}}(k'+k_0) - E_{\text{SL}}(k+k_0) - \hbar\omega_{Q\lambda}\right] \end{bmatrix} \right. \\
&\left. + (N_{Q\lambda}+1) \begin{bmatrix} \delta_{Q_x,k'-k} \delta\left[E_{\text{SL}}(k+k_0) - E_{\text{SL}}(k'+k_0) + \hbar\omega_{Q\lambda}\right] \\ \delta_{Q_x,k-k'} \delta\left[E_{\text{SL}}(k'+k_0) - E_{\text{SL}}(k+k_0) + \hbar\omega_{Q\lambda}\right] \end{bmatrix} \right\} ,
\end{aligned} \tag{14.56}$$

where $\omega_{Q\lambda} = c_\lambda Q$ is the frequency of acoustic phonons in Debye's model, c_λ is the sound velocity, and the interaction matrix between an electrons and the lattice is given by

$$|C_{Q\lambda}|^2 = V_{Q\lambda}^2 \exp\left(-\frac{Q_z^2}{2\alpha^2}\right) \exp\left(-\frac{Q_x^2 \ell^2}{2}\right) \mathcal{F}_2(Q_y) . \tag{14.57}$$

In Eq. (14.57), the form factor is defined as

$$\mathcal{F}_2(Q_y) = \left| \int_{-\infty}^{+\infty} dy\, \psi_1^2(y) e^{iQ_y y} \right|^2 , \tag{14.58}$$

and the expressions for $V_{Q\lambda}^2$ with $\lambda = \ell$ (LA-phonon) and $\lambda = t$ (TA-phonon) are given by

$$V_{Q\ell}^2 = \frac{\hbar Q}{2\mathcal{V}\rho_i c_\ell} \left[D^2 + (eh_{14})^2 \frac{A_\ell(\mathbf{Q})}{Q^2} \right] , \tag{14.59}$$

$$V_{Qt}^2 = \frac{\hbar q}{2\mathcal{V}\rho_i c_\ell} (eh_{14})^2 \frac{A_t(\mathbf{Q})}{Q^2} , \tag{14.60}$$

where $\mathcal{V}$ is the sample volume, ρ_i is the ion-mass density, D is the deformation potential, and h_{14} is the piezoelectric field. Additionally, there are the structural factors for acoustic phonons

$$A_\ell(\mathbf{Q}) = \frac{36 Q_x^2 Q_y^2 Q_z^2}{Q^6} , \tag{14.61}$$

$$A_t(\mathbf{Q}) = \frac{2[Q^2(Q_x^2 Q_y^2 + Q_y^2 Q_z^2 + Q_z^2 Q_x^2) - 9 Q_x^2 Q_y^2 Q_z^2]}{Q^6} . \tag{14.62}$$

References

1 Huang, D.H., Lyo, S.K., and Gumbs, G. (2009) Bloch oscillations, dynamical localization, and optical probing of electron gases in quantum-dot superlattices in high electric fields. *Phys. Rev. B*, **79**, 155308.

2 Lyo, S.K. and Huang, D.H. (2003) Temperature-dependent magnetoconductance in quantum wires: Effect of phonon scattering. *Phys. Rev. B*, **68**, 115317.

3 Lyo, S.K. and Huang, D.H. (2001) Multisublevel magnetoquantum conductance in single and coupled double quantum wires. *Phys. Rev. B*, **64**, 115320.

4 Lyo, S.K. and Huang, D.H. (2002) Magnetoquantum oscillations of thermoelectric power in multisublevel quantum wires. *Phys. Rev. B*, **66**, 155307.

5 Lyo, S.K. and Huang, D.H. (2006) Effect of electron–electron scattering on the conductance of a quantum wire studied with the Boltzman transport equation. *Phys. Rev. B*, **73**, 205336.

6 Lyo, S.K. (1989) Magnetoquantum oscillations of the phonon-drag thermoelectric power in heterojunctions. *Phys. Rev. B*, **40**, 6458.

7 Huang, D.H. and Lyo, S.K. (2000) Suppression of impurity and interface-roughness back-scattering in double quantum wires: theory beyond the Born approximation. *J. Phys.: Condens. Matter*, **12**, 3383.

8 Lyo, S.K. (2008) Bloch oscillations and nonlinear transport in a one-dimensional semiconductor superlattice. *Phys. Rev. B*, **77**, 195306.

9 Huang, D.H., Lyo, S.K., Thomas, K.J., and Pepper, M. (2008) Field-induced modulation of the conductance, thermoelectric power, and magnetization in ballistic coupled double quantum wires under a tilted magnetic field. *Phys. Rev. B*, **77**, 085320.

10 Esaki, L. and Tsu, R. (1970) Superlattice and negative differential conductivity in semiconductors. *IBM J. Res. Dev.*, **14**, 61.

11 Winnerl, S., Schomburg, E., Grenzer, J., Regl, H.-J., Ignatov, A.A., Semenov, A.D., Renk, K.F., Pavel'ev, D.G., Koschurinov, Y., Melzer, B., Ustinov, V., Ivanov, S., Schaposchnikov, S., and Kop'ev, P.S. (1997) Quasistatic and dynamic interaction of high-frequency fields with miniband electrons in semiconductor superlattices. *Phys. Rev. B*, **56**, 10303.

12 Ziman, J.M. (1967) *Electrons and Phonons*, Oxford University Press, London, p. 298.

13 Ziman, J.M. (1972) *Principles of the Theory of Solids*, 2nd edn, Cambridge Press, p. 212.

15
Spontaneous and Stimulated Nonlinear Wave Mixing of Multi-excitons

Nonlinear optical signals are generated when a material system interacts with several laser beams. There are different types of signal classifications: spontaneous *vs.* stimulated, coherent *vs.* incoherent, and short *vs.* long range. Some signals scale like $\sim N$, whereas others scale like $\sim N^2$, where N is the number of active quantum dots (QDs). The various signals are usually treated using different formalisms, thus making it difficult to precisely establish their connection and often confusing with esoteric acronyms (CARS, HORSES etc.). The main reason for such a baffling plethora of nonlinear techniques is that it involves several interactions between the optical field and matter. Since both may be characterized by several parameters (transition dipole moments, energy levels, central frequencies, polarizations, arrival times, etc.), we have various experimental setups. In this chapter, we discuss the unified classification of those signals in accordance with the last (measured) interactions. Although this classification by itself is limited, it serves as a basis for further perturbative expansion, thus bringing all those spectroscopic techniques into perspective.

The semiclassical theory of nonlinear spectroscopy that assumes a classical optical field interacting with a quantum system has had great success in describing various coherent measurements [1–8]. The signals are calculated via the Maxwell equations with the polarization source created by the QDs and are written in terms of response functions and cannot describe spontaneous processes. The response function is a perturbative expansion of the polarization term driven by the coherent material response to the incoming fields. There are two tools commonly used to compute the nonlinear optical signals. The first uses nonlinear susceptibilities, conventionally represented by double sided Feynman diagrams for the density matrix. The second is given by Fermi's golden rule for optical transitions from its initial to final state and represented by single sided Feynman diagrams for the wave function. The semiclassical signal calculations require two steps. The QD's (primary source) driven field is a secondary source for the detected field which accounts for the propagation. Also, the signal is an interaction pattern between various quantum pathways (evolution of the QD confined electron density matrix by the interactions). This makes the signals hard to interpret. These difficulties and limitations can be overcome partially by using a fully quantum mechanical descrip-

Properties of Interacting Low-Dimensional Systems, First Edition. G. Gumbs and D. Huang.

tion (both optical field and matter). Corresponding formalism and its implications are the main subjects of this chapter.

As an example, let us consider a setup where two beams of frequency ω_1 and ω_2 generate a signal with frequency close to $\omega_1 + \omega_2$. Possible signals of this type are: sum frequency generation (SFG), hyper Rayleigh scattering (HRS), two-photon-induced fluorescence (TPIF) and hyper Raman (HRA). Such signals are used for various spectroscopic applications for probing QD energy levels and ultra-fast dynamical processes as well as in high resolution imaging and nonlinear microscopy. SFG, TPIF and HRS are commonly used for biomolecular and cell imaging. Some investigators have reported the simultaneous observation of two types of signals, for example, SFG plus TPF or SFG plus HRS in the same system.

The different types of signals are depicted in Figure 15.1. The primary classification is in distinguishing between stimulated (coherent) $S_{\mathrm{ST,coh}}$ and spontaneous S_{SP}. The latter are divided into incoherent $S_{\mathrm{SP,inc}}$, coherent short range $S_{\mathrm{SP,coh,sr}}$ and long range $S_{\mathrm{SP,coh,lr}}$. This gives, for the total signal,

$$S = S_{\mathrm{ST,coh}} + S_{\mathrm{SP,inc}} + S_{\mathrm{SP,coh,sr}} + S_{\mathrm{SP,coh,lr}} \,. \tag{15.1}$$

The optical signals are broadly classified as either stimulated where the signal is generated in the direction of an existing strong classical field or is spontaneous where the signal is generated in a new direction, that is, the detected mode is initially in the vacuum state. By treating the radiation field quantum mechanically, we can describe both spontaneous and stimulated processes [7–12].

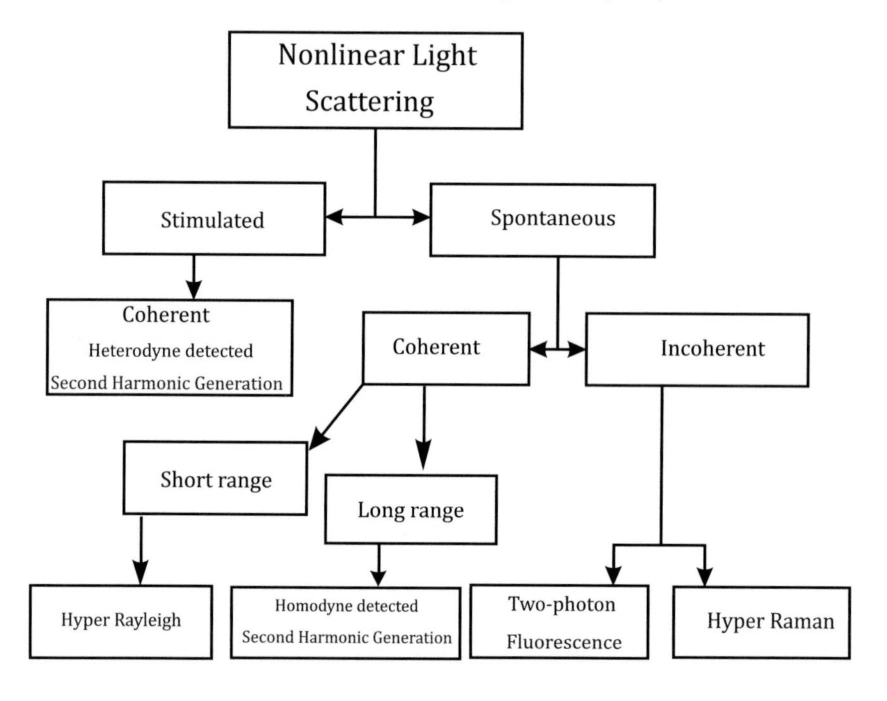

Figure 15.1 Classification of various wave mixing signals.

Classification of the signal as being coherent is well defined with respect to the driving field or incoherent where no such phase relation exists. Stimulated signals are coherent and scale as $\sim N$ and the field itself (both amplitude and phase) can be obtained via heterodyne detection. However, spontaneous signals can be either coherent or incoherent. The homodyne-detected coherent signal that is generated in a sample much larger than the optical wavelength is directional and scales as $\sim N^2$. However, short-range correlations can induce a Rayleigh or hyper Rayleigh scattering signal coming from pairs of close QDs. This signal is isotropic and scales as $\sim N$. Spontaneous incoherent signals are generated by QDs that emit independently. They scale as $\sim N$ and are denoted as spontaneous light emission (SLE). Such signals may be further classified as either Raman, hyper Raman or fluorescence.

The classification shown in Figure 15.1 is valid to all orders in the fields. We shall recast the possible signals using compact superoperator expressions that can be expanded in the optical field to generate specific signals. To the first order, we only have the coherent linear response that is self heterodyned or ordinary Rayleigh scattering. The simplest model that shows these signals is depicted in Figure 15.2, where the emitted signals are either at or in the vicinity of $\omega_1 + \omega_2$. For this model, the stimulated/coherent signal heterodyne detected is "sum frequency generation" (SFG). The spontaneous/coherent/long-range signal is homodyne detected SFG. The spontaneous/coherent/short-range signal is known in this case as hyper Rayleigh, and the spontaneous/incoherent signal is two-photon-induced light emission. This can further be classified as two-photon-induced florescence and hyper Raman. The ratio of the two depends on the pure dephasing rate. This separation is well documented and will not be repeated here.

Two detection modes are commonly used: heterodyne and homodyne. Semiclassically, in an $n + 1$ wave mixing measurement, n incoming waves interact with a QD to induce a polarization $\sim \langle V(\mathbf{r}, t)\rangle_{\{n\}}$ (the details and reasoning for such unusual notation will be provided in the next sections). This polarization serves as a source in the Maxwell equations for the signal field $\mathcal{E}_{n+1}(\mathbf{r}, t)$. If necessary, the polarization can be further orientationally averaged and summed over all the QDs. Note that this is a simplified description and more detailed analysis of the detection involves the propagation effects (see Appendix 15.A below for more details) [13–18].

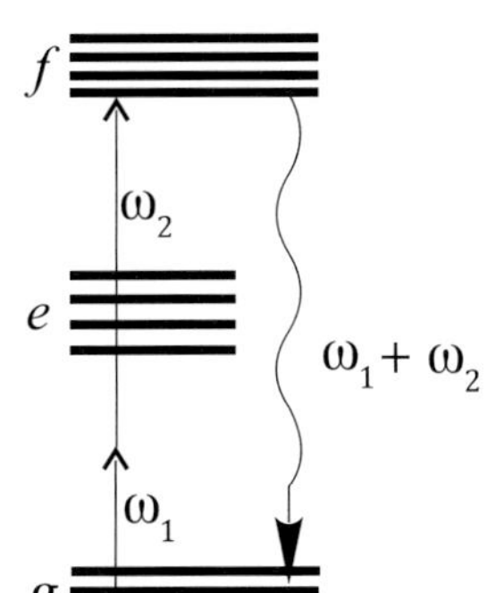

Figure 15.2 Level scheme for nonlinear two-photon-induced single photon emitted signals in the vicinity of $\omega_1 + \omega_2$.

Heterodyne signals, detected by interference with a heterodyne mode, give both the amplitude and phase of the nonlinear polarization. With the aid of the polarization, we obtain the electric field of the detected mode. For a collection of N QDs, a coherent signal is obtained by adding the amplitudes of all QDs and scales as $\sim \mathrm{Im}[N\langle V(\boldsymbol{r},t)\rangle_{\{n\}}\mathcal{E}^*_{n+1}(\boldsymbol{r},t)]$. Since the heterodyne-detected signal is phase sensitive, its maximum is generated along one of the possible 2^n phase matching directions, that is, $\Delta\boldsymbol{k} = \boldsymbol{k}_{\{n\}} - \boldsymbol{k}_{n+1} = 0$ with $\boldsymbol{k}_{\{n\}} = \sum_{j=1}^{n} \pm\boldsymbol{k}_j$, as illustrated in Figure 15.3a.

Homodyne detection is phase insensitive and measures the intensity of the scattered lighted $\sim |\langle V(\boldsymbol{r},t)\rangle_{\{n\}}|^2$ and can be either coherent or incoherent. The later is a sum of all individual QD's contributions $\sim N$, while the former is produced by QD pairs and scales as $\sim N(N-1)$. The coherence length is related to the optical phase variation between two QDs, $\Delta\boldsymbol{k}\cdot(\boldsymbol{r}_\alpha - \boldsymbol{r}_\beta)$. For a sufficiently large $\Delta\boldsymbol{k}$, the phase rapidly oscillates and the coherent part of the signal vanishes. Thus, coherent QD response thus shows up in the phase-matching direction $\Delta\boldsymbol{k} = 0$ and depends quadratically on the number of active QDs, that is, $\sim N^2$ for $N \gg 1$.

Inelastic processes, such as hyper-Raman scattering, are incoherent and do not produce a macroscopic electric field. Different QDs emit a signal independently with a random phase. One way to see this is by deviating from the semiclassical picture and looking at the joint state of the QD and detected mode field ($|\mathrm{mol},\mathrm{phot}\rangle$) at the end of the process: $|g,0\rangle + \alpha|g',1\rangle$ (see Figure 15.3b). Here, the scattered mode is initially in the vacuum state with the QD in state $|g\rangle$ and finally we have one emitted photon in the detected mode. The energy difference between the initial and final state of the QD is supplied by a linear combination of the incoming modes and the signal mode (it corresponds to $\omega_1 + \omega_2 - \omega_s$ in Figure 15.3b). The expectation value of the signal field mode [formally defined in Eq. (15.3)] with this state vanishes since $|g\rangle$ and $|g'\rangle$ are orthogonal to each other.

In contrast, parametric or elastic scattering processes are always phase matched with $\Delta\boldsymbol{k} = 0$. A state like $|g,0\rangle + \alpha|g,1\rangle$ gives a finite field amplitude. Hyper-Rayleigh and hyper-Raman scattering can be seen as elastic and inelastic counterparts of two-photon-induced fluorescence, that is, incoherent and not phase matched.

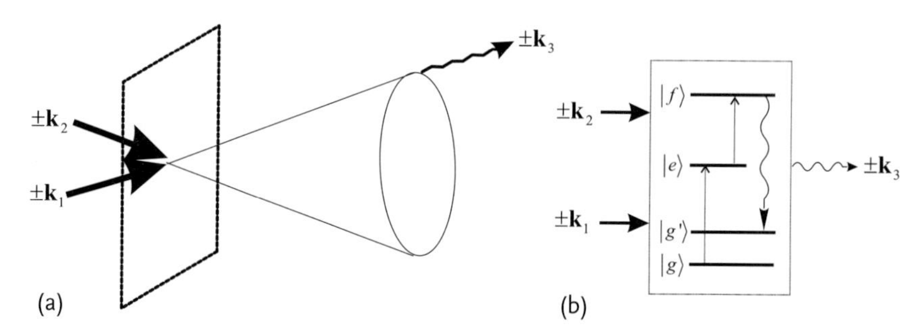

Figure 15.3 Three-wave mixing process with two classical and one quantum mode: (a) phase-matching condition; (b) QD level scheme.

The microscopic description of the signal treats both the QDs and the optical field quantum mechanically. This allows us to classify the signals by the initial state of the detected mode rather than by the detection method. If that mode initially has a finite number of photons, one has a stimulated (emission or absorption) process. If it is in the vacuum state, we have a spontaneous process. Stimulated processes, for example, a pump-probe, are heterodyne detected. We will mainly focus on spontaneous processes, but present the stimulated signals for completeness. Understanding the connection between the various signals is important for applications to nonlinear imaging. We show that the coherent part of the scattering should be classified according to the coherence range. Rayleigh and nonlinear light scattering are coherent processes involving pairs of QDs. However, they only probe short-range correlations and therefore scale as $\sim N$. When inter-QDs interactions become important, the density dependent part of the Rayleigh signal shows that this part becomes dominant in the vicinity of anomalous first order phase transitions and vanishes for ordinary first order transitions in dilute solutions of QDs.

In the following section, we calculate the signals using a quantum mechanical description of the field and we express it in a form suitable for expansion. It is based on non-equilibrium Green's functions that can be represented graphically in terms of close time path loop diagrams. Some signals scale with the single QD and others with QD pair distribution functions. To some extent, these are the analogue of the Maxwell equations in the fully quantum mechanical description. However, unlike the Maxwell equations that are deterministic, the QD distributions are statistical in origin. Then, we will survey statistical models for these distribution functions. For a solution of weakly interacting QDs or polymers, signatures of structural phase transitions are illustrated. The last section of this chapter summarizes our results and presents a comparison of various two-photon-induced signals based on the level scheme in Figure 15.3.

15.1 Spontaneous, Stimulated, Coherent and Incoherent Nonlinear Wave Mixing

We partition the optical field into its positive and negative frequency parts with $E(\boldsymbol{r}, t) + E^*(\boldsymbol{r}, t)$. The positive energy optical field operator at position $\boldsymbol{r}$ and time t is given by the sum of all optical modes including the detected $n + 1$ one, that is,

$$E(\boldsymbol{r}, t) = \sum_{j=1}^{n+1} \sqrt{\frac{\hbar\omega_j}{2\epsilon_0 \Omega}}\, a_j(t) e^{i\boldsymbol{k}_j\cdot\boldsymbol{r} - i\omega_j t} . \tag{15.2}$$

Here, a_j ($a_j^\dagger$) is the annihilation (creation) operator for the jth field mode of wave vector $\boldsymbol{k}_j$ and angular frequency ω_j, satisfying the bosonic commutation relation $[a_i, a_j^\dagger] = \delta_{i,j}$, ϵ_0 is the vacuum permittivity, and Ω is the field quantization volume.

The detector located at distance $|\boldsymbol{R}|$ from the sample (similar to the semiclassical case depicted in Figure 15.4), sees a spherical wave with

$$E_{n+1}(\boldsymbol{R}+\boldsymbol{r},t) \sim \sqrt{\frac{\hbar\omega_{n+1}}{2\epsilon_0 \Omega_{n+1}}}\, a_{n+1}(t) \left[\frac{e^{i\boldsymbol{k}_{n+1}\cdot(\boldsymbol{R}+\boldsymbol{r})-i\omega_{n+1}t}}{k_{n+1}|\boldsymbol{R}+\boldsymbol{r}|}\right]. \tag{15.3}$$

We avoid the details of the detector location by defining the plane wave signal mode in a local system of coordinates $E_{n+1}(\boldsymbol{r},t)$ by Eq. (15.2) apart from the summation. The large distance of the detector from the sample yields

$$E_{n+1}(\boldsymbol{R}+\boldsymbol{r},t) \sim E_{n+1}(\boldsymbol{r},t)\left[\frac{e^{i\boldsymbol{k}_{n+1}\cdot\boldsymbol{R}}}{k_{n+1}|\boldsymbol{R}|}\right].$$

From this point, we treat the signal mode as a plane wave. However, its fully defined spherical version becomes important when the signals are calculated semiclassically via the Maxwell equations (see Appendix 15.A).

We split the detected electric field into two terms, that is, $E_{n+1}(\boldsymbol{r},t) = \mathcal{E}_s(\boldsymbol{r},t) + E_s(\boldsymbol{r},t)$. The classical (coherent) part $\mathcal{E}_s(\boldsymbol{r},t)$ is given and is not affected by the interaction with matter and while $E_s(\boldsymbol{r},t)$ is initially (at $t=-\infty$) in its vacuum state and changes its state due to the field-matter coupling. The signal is defined as the change in the signal mode dimensionless intensity due to the coupling with the sample

$$S(t) = S_{\text{ST}}(t) + S_{\text{SP}}(t) = \left(\frac{2\epsilon_0}{\hbar\omega_s}\right) \times \left\{2\,\text{Re}\left[\int d^3\boldsymbol{r}\langle\mathcal{E}_s^*(\boldsymbol{r},t)E_s(\boldsymbol{r},t)\rangle\right] + \int d^3\boldsymbol{r}\langle E_s^\dagger(\boldsymbol{r},t)E_s(\boldsymbol{r},t)\rangle\right\}. \tag{15.4}$$

To calculate the expectation value of the optical field, the total Hamiltonian of the field-matter system must by specified. This is

$$H(t) = H_0 + H_{\text{int}}(t)\,. \tag{15.5}$$

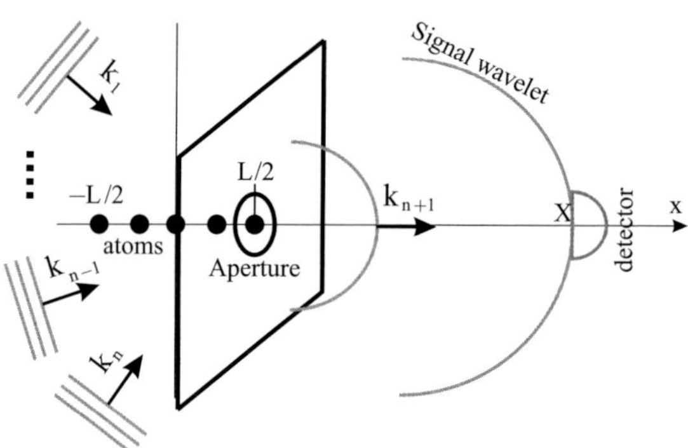

Figure 15.4 Semiclassical heterodyne detection of incoherent nonlinear signals.

Here, H_0 describes the sample and H_{int} represents its interaction with the optical modes. We assume that the sample consists of N identical QDs and these are defined by the positions $\boldsymbol{r}_\alpha$, energy eigenstates $|i\rangle$ and transition dipole moments $\mu_{i,j}$. We partition the dipole operator into excitation and de-excitation parts $V^\dagger(\boldsymbol{r}) + V(\boldsymbol{r})$ where

$$V(\boldsymbol{r}) = \sum_{\alpha=1}^{N} \delta(\boldsymbol{r} - \boldsymbol{r}_\alpha) \sum_{k>j} \mu_{j,k} |j\rangle\langle k| \,. \tag{15.6}$$

Making use of Eqs. (15.2) and (15.6), the radiation–matter interaction in the rotating wave approximation takes the form of

$$\begin{aligned} H_{\text{int}}(t) &= H_{\text{int}}^{(n+1)}(t) + H_{\text{int}}^{\{n\}}(t) \equiv \int d^3\boldsymbol{r}\, H_{\text{int}}(\boldsymbol{r}, t) \\ &= \int d^3\boldsymbol{r} \left[E_{n+1}(\boldsymbol{r}, t) V^\dagger(\boldsymbol{r}) + E_{n+1}^\dagger(\boldsymbol{r}, t) V(\boldsymbol{r}) \right] \\ &\quad + \int d^3\boldsymbol{r} \sum_{j=1}^{n} \left[E_j(\boldsymbol{r}, t) V^\dagger(\boldsymbol{r}) + E_j^\dagger(\boldsymbol{r}, t) V(\boldsymbol{r}) \right] . \end{aligned} \tag{15.7}$$

The stimulated and spontaneous parts of the signal are given by the two terms in Eq. (15.4) that can be calculated by the Heisenberg equations of motion involving only the detected mode. For the stimulated part, we have

$$\begin{aligned} \frac{d}{dt} \langle \mathcal{E}_s^*(\boldsymbol{r}, t) E_s(\boldsymbol{r}, t) \rangle &= \frac{i}{\hbar} \mathcal{E}_s^*(\boldsymbol{r}, t) \langle [H_{\text{int}}(t), E_s(\boldsymbol{r}, t)] \rangle \\ &= i \left(\frac{\omega_s}{2\epsilon_0} \right) \mathcal{E}_s^*(\boldsymbol{r}, t) \, \langle V(\boldsymbol{r}, t) \rangle \,. \end{aligned} \tag{15.8}$$

We used the fact that the coherent part of the detected mode is not affected by the interaction with the QDs. Also, $\langle \cdots \rangle$ denotes averaging over the radiation and matter degrees of freedom.

To perform this calculation, we introduce superoperators that facilitate bookkeeping of the various interactions. For an arbitrary operator A, these are defined as

$$\begin{aligned} A_{\text{L}} X &= AX \,, \quad A_{\text{R}} X = XA \,, \\ A_{\pm} &= \frac{1}{\sqrt{2}} (A_{\text{L}} \pm A_{\text{R}}) \,. \end{aligned} \tag{15.9}$$

The interaction Hamiltonian in Eq. (15.7) recast in the superoperator representation assumes the form of $\sqrt{2} H_{\text{int},-} = E_{\text{L}} V_{\text{L}}^\dagger + E_{\text{L}}^\dagger V_{\text{L}} - V_{\text{R}}^\dagger E_{\text{R}} - V_{\text{R}} E_{\text{R}}^\dagger$, where we suppressed the time and coordinate dependence of the superoperators.

The time evolution of the dipole superoperators can be calculated in the interaction picture using the bare QD Hamiltonian as a reference, leading to

$$\begin{aligned}\langle V_\nu(\boldsymbol{r},t)\rangle &= \mathrm{Tr}\left[V_\nu(\boldsymbol{r},t)\rho(t)\right] \\ &= \left\langle\left[e^{(i/\hbar)(H_{0,L}-H_{0,R})t}\,V_\nu(\boldsymbol{r})e^{-(i/\hbar)(H_{0,L}-H_{0,R})t}\right]\right\rangle \\ &= \left\langle \hat{T}\,V_\nu(\boldsymbol{r},t)\exp\left[-\frac{\sqrt{2}i}{\hbar}\int_{-\infty}^{t} d\tau \int d^3\boldsymbol{r}'\,H_{\mathrm{int},-}(\boldsymbol{r}',\tau)\right]\right\rangle . \end{aligned} \tag{15.10}$$

In this notation, the subscript ν denotes either L or R superoperator indices. Also, $\hat{T}$ is the time ordering operator in Liouville space that acts on a product of the following superoperators and rearranges them so that their time arguments increase from right to left.

Heterodyne detected $(n+1)$-wave mixing signals in a macroscopic $(N \gg 1)$ sample are generated along one of the 2^n directions given by combinations of the n incoming wave vectors $\boldsymbol{k}_{\{n\}} = \pm\boldsymbol{k}_1 \pm \boldsymbol{k}_2 \pm \ldots \pm \boldsymbol{k}_n$. This can be obtained by expanding the density operator $\rho(t)$ in Eq. (15.10) to the first order in the n incoming modes each interacting once with a single QD, and summing over all the QDs in the sample. Therefore, we get

$$\langle V(\boldsymbol{r},t)\rangle_{\{n\}} = \sum_{\alpha=1}^{N} \delta(\boldsymbol{r}-\boldsymbol{r}_\alpha)\langle V_L(t)\rangle_{\{n\}}e^{i\boldsymbol{k}_{\{n\}}\cdot\boldsymbol{r}} . \tag{15.11}$$

The subscript $\{n\}$ indicates that the averaging involves the density operator calculated by taking into account interactions of each of the incoming modes with a single QD. The signal $n+1$ mode is treated separately.

When Eq. (15.8) together with the initial condition $\langle E_s(\boldsymbol{r}, t=-\infty)\rangle = 0$ and the expansion in Eq. (15.11) are substituted into Eq. (15.4), we obtain the stimulated incoherent signal

$$S_{\mathrm{ST}}^{(n)}(t) = \frac{1}{\hbar}\,\mathrm{Im}\left[F_1(\Delta\boldsymbol{k})\int_{-\infty}^{t} d\tau \mathcal{E}_s^*(\tau)\langle V_L(\tau)\rangle_{\{n\}}\right] , \tag{15.12}$$

where the auxiliary function $F_1(\Delta\boldsymbol{k}) = \sum_\alpha e^{i\Delta\boldsymbol{k}\cdot\boldsymbol{r}_\alpha}$ carries all information about the macroscopic sample geometry as well as the distribution of QDs. This gives rise to phase matching and is a hallmark of heterodyne detected signals. Self-heterodyne signals, such as, pump-probe and stimulated Raman/hyper-Raman scattering, also fall into the stimulated signal category.

We now consider the spontaneous part of the signal in Eq. (15.4) that is obtained by solving the Heisenberg equation of motion

$$\begin{aligned}\frac{d}{dt}\langle E_s^\dagger(\boldsymbol{r},t)E_s(\boldsymbol{r},t)\rangle &= \frac{i}{\hbar}\left\langle\left[H_{\mathrm{int}}, E_s^\dagger(\boldsymbol{r},t)E_s(\boldsymbol{r},t)\right]\right\rangle \\ &= 2\,\mathrm{Im}\left[\left(\frac{\omega_s}{2\epsilon_0}\right)\langle E_s^\dagger(\boldsymbol{r},t)V(\boldsymbol{r},t)\rangle\right] \end{aligned} \tag{15.13}$$

with the initial condition $\langle E_s^\dagger E_s \rangle = 0$ at $t = -\infty$. The right-hand side of this equation can be factorized if the density operator is treated perturbatively with respect to the E_s part of the signal mode. To the first order, the spontaneous signal has the form

$$S_{\mathrm{SP}}^{(n)}(t) = \frac{2}{\hbar^2}\left(\frac{\hbar\omega_s}{2\epsilon_0\Omega}\right)\mathrm{Re}\left[\int d^3\boldsymbol{r}\int d^3\boldsymbol{r}' e^{-i\boldsymbol{k}_{n+1}\cdot(\boldsymbol{r}-\boldsymbol{r}')}\right.$$
$$\left.\times\int_{-\infty}^{t} d\tau \int_{-\infty}^{\tau} d\tau' e^{i\omega_{n+1}(\tau-\tau')}\langle\hat{\mathcal{T}}\, V_{\mathrm{L}}(\boldsymbol{r},\tau)\, V_{\mathrm{R}}^\dagger(\boldsymbol{r}',\tau')\rangle_{\{n\}}\right]. \quad (15.14)$$

When all interactions with the optical fields occur with the same QD, $\langle\hat{\mathcal{T}}\, V_{\mathrm{L}}(\boldsymbol{r},\tau)$ $V_{\mathrm{R}}^\dagger(\boldsymbol{r}',\tau')\rangle_{\{n\}}$ assumes the form $\langle\hat{\mathcal{T}}\, V_{\mathrm{L}}(\tau)\, V_{\mathrm{R}}^\dagger(\tau')\rangle_{\{n\}}\delta(\boldsymbol{r}-\boldsymbol{r}')/\Omega$ and we recover the incoherent signal in Eq. (15.14). Expanding it to the lowest order in the interactions with each of the incoming modes, we obtain

$$S_{\mathrm{SP,\,incoh}}^{(n)}(t) = \frac{2}{\hbar^2}\left(\frac{\hbar\omega_s}{2\epsilon_0\Omega}\right)\mathrm{Re}\left[F_1(0)\int_{-\infty}^{t} d\tau \int_{-\infty}^{\tau} d\tau'\right.$$
$$\left.\times\, e^{i\omega_{n+1}(\tau-\tau')}\left\langle\hat{\mathcal{T}}\, V_{\mathrm{L}}(\tau)\, V_{\mathrm{R}}^\dagger(\tau')\right\rangle_{\{n\}}\right]. \quad (15.15)$$

The incoherent $[F_1(0) = N]$ homodyne detected signal is phase insensitive. The n-photon induced fluorescence and hyper-Raman scattering are examples of such signals.

The coherent part of the spontaneous signal is generated when the optical modes are allowed to interact with all possible QD pairs in the sample. Interactions with different QDs are not time ordered and $\langle\hat{\mathcal{T}}\, V_{\mathrm{L}}(\boldsymbol{r},\tau)\, V_{\mathrm{R}}^\dagger(\boldsymbol{r}',\tau')\rangle$ can be factored into $\langle V_{\mathrm{L}}(\boldsymbol{r},\tau)\rangle\langle V_{\mathrm{R}}^\dagger(\boldsymbol{r}',\tau')\rangle$. By expanding the two factors to first order in each of the n incoming modes, we obtain the coherent part of the homodyne detected signal from

$$S_{\mathrm{SP,\,coh}}^{(n)}(t) = \frac{1}{\hbar^2}\left(\frac{\hbar\omega_s}{2\epsilon_0\Omega}\right)$$
$$\times\,\mathrm{Re}\left[F_2(\Delta\boldsymbol{k})\left|\int_{-\infty}^{t} d\tau e^{i\omega_{n+1}\tau}\langle V_{\mathrm{L}}(\tau)\rangle_{\{n\}}\right|^2\right], \quad (15.16)$$

where we have used the identity

$$\int_{-\infty}^{t} d\tau \int_{-\infty}^{\tau} d\tau' = \frac{1}{2}\int_{-\infty}^{t} d\tau \int_{-\infty}^{t} d\tau'\,. \quad (15.17)$$

The auxiliary function $F_2(\Delta\boldsymbol{k}) = \sum_{\alpha,\beta\neq\alpha} e^{i\Delta\boldsymbol{k}\cdot(\boldsymbol{r}_\alpha-\boldsymbol{r}_\beta)}$ is determined by the distribution function of QD pairs and the sample geometry. Equation (15.16) describes n-harmonic generation and hyper-Rayleigh scattering.

Equations (15.12), (15.15), and (15.16) constitute the starting formal expressions for the CTPL diagrammatic calculation of specific signals. In the following section, we will deal with the QD and QD pair distribution functions, that is, $F_1(\Delta \boldsymbol{k})$ and $F_2(\Delta \boldsymbol{k})$.

15.2
$n + 1$ Wave Mixing in QD Fluids and Polymer QDs Molecule Solutions

Now, we more closely examine the role of QD distributions in nonlinear wave scattering. We consider a system of N identical QDs given by hard spheres whose diameter a is smaller than the wavelength $\lambda = 2\pi c/\omega$ of the detected mode. The QDs are placed in a solvent and occupy the volume $L^3 \approx \Omega'$. Equations (15.12), (15.15), and (15.16) describe the scattering due to the solute. Three cases will be considered. First, we will look at the scattering from an ideal solute with no long-range order as depicted in Figure 15.5c. Second, we investigate the scattering from a solution

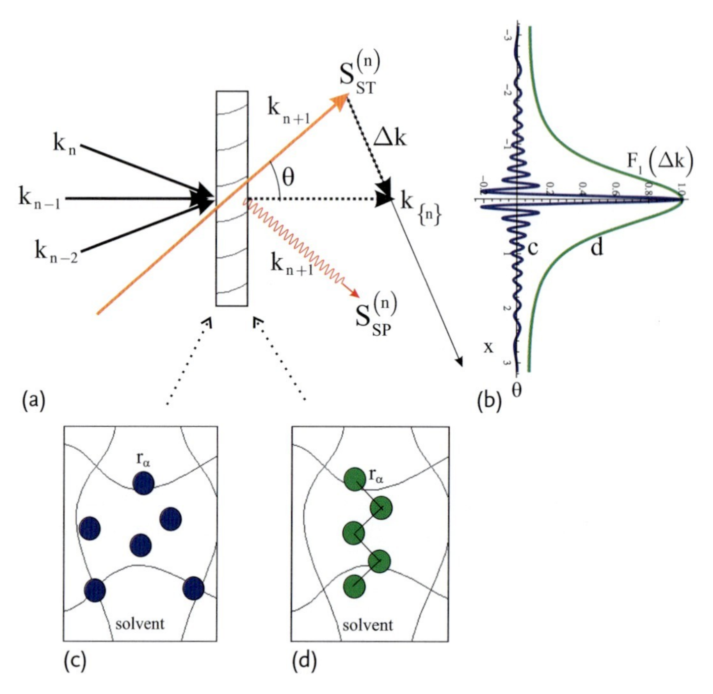

Figure 15.5 (a) Schematic of nonlinear wave mixing. θ is the phase matching angle between stimulated (thick red line)/spontaneous (wavy line) and a linear combination of the incoming modes (doted line). Panel (b) shows the angular distribution of the stimulated signal from a solute of non-interacting (blue rapidly oscillating curve) and polymer QDs molecule (smooth green curve). Panels (c) and (d) represent a solute made of non-interacting QDs and polymer QDs molecules correspondingly. The parameters used are: $L/\lambda = 10$, $N = 100$ and $b/\lambda = 0.01$. $F_1(\Delta k)$ is shown normalized to the number of QDs.

of threadlike-polymer bonded QDs (see Figure 15.5d). Finally, we discuss a real (includes long-range interactions) QD solution close to a phase transition point. Furthermore, these cases will be classified as shown in Figure 15.1.

We shall assume that the phase $\Delta \boldsymbol{k} \cdot \boldsymbol{r}$ does not change appreciably within the sample, that is, $L|\Delta \boldsymbol{k}| \ll 1$. Recall that in the opposite limit, the problem can be treated in the continuum limit, where the Maxwell equations self-consistently connect the polarization $\langle V(\boldsymbol{r}, t)\rangle_{\{n\}}$ and the induced electric field $E_{n+1}(\boldsymbol{r}, t)$. In the latter case, the signal is calculated in two steps. First, the atoms acting as the primary sources induce the field at the aperture. This field serves as the secondary source for the signal that is calculated using the propagator formalism. Both semiclassical and quantum approaches yield the same result as shown in the Appendix 15.A.

15.2.1 Stimulated and Spontaneous Incoherent Signals

The foundation for this type of signal lies within Eqs. (15.12)–(15.15). Both are determined by the QD distribution and are the subject of investigation in this subsection.

The probability of finding a solute QD in the volume $d^3\boldsymbol{r}$ centered at the position $\boldsymbol{r}$ is given by $F_1(\boldsymbol{r}) d^3\boldsymbol{r}/\Omega'$. The QD distribution function is normalized so that its average value in the sample volume Ω' is the total number of QDs: $(N/\Omega') \int_{\Omega'} d^3\boldsymbol{r} F_1(\boldsymbol{r}) = N$. Converting the summation over a large number of independent QD coordinates $\boldsymbol{r}_\alpha$ to an integration over $\boldsymbol{r}$, we obtain

$$F_1(\Delta \boldsymbol{k}) = \frac{N}{\Omega'} \int_{\Omega'} d^3\boldsymbol{r} F_1(\boldsymbol{r}) e^{i\Delta \boldsymbol{k}\cdot\boldsymbol{r}} . \tag{15.18}$$

More generally, the QD distribution function must include internal QD degrees of freedom and rotational averaging, though it is beyond the scope of this discussion.

Ideal QD solute fluids

Let us specify $\Delta \boldsymbol{k}$ running along the $\hat{\boldsymbol{x}}$-direction as shown in Figure 15.5a. In the absence of long-range order, that is, $F_1(\boldsymbol{r}) = 1$ (see Figure 15.5c), a straightforward calculation of the integral in Eq. (15.18) yields

$$F_{1,\text{fluid}}(\Delta \boldsymbol{k}) = N \mathcal{P}_{\text{fluid}}(\theta) \tag{15.19}$$

with the polarization angular distribution:

$$\mathcal{P}_{\text{fluid}}(\theta) = \operatorname{sinc}\left(\frac{2\pi L}{\lambda_{n+1}} \sin\frac{\theta}{2}\right) . \tag{15.20}$$

Here, $\operatorname{sinc}(x) \equiv \sin x/x$, $\lambda_{n+1} = 2\pi c/\omega_{n+1}$, θ is the angle between the detected $\boldsymbol{k}_{n+1}$ mode and a linear combination of the incoming modes $\boldsymbol{k}_{\{n\}}$.

Solute of QDs bonded into a thread-like polymer

We now consider polymers in Figure 15.5d. The QD probability distribution can be calculated using the theory of random walks described as

$$\frac{N}{\Omega'}F_1(\mathbf{r})d^3\mathbf{r} = \frac{2}{N}\sum_j\sum_{i>j} W_{i,j}(r)d^3\mathbf{r}\,, \tag{15.21}$$

$$W_{i,j}(r) = \left(\frac{3}{2\pi b^2|i-j|}\right)^{3/2} \exp\left(\frac{-3r^2}{2b^2|i-j|}\right)\,. \tag{15.22}$$

Here, the characteristic length b depends on the polymer geometry. The quantity $W_{i,j}(r)d^3\mathbf{r}$ stands for the probability of finding the jth QD at the distance r from the ith QD within the volume of $d^3\mathbf{r}$. Switching the summation in Eq. (15.21) to an integration and substituting Eq. (15.21) into Eq. (15.18), we obtain

$$F_{1,\mathrm{poly}}(\Delta\mathbf{k}) = N\mathcal{P}_{\mathrm{poly}}(\theta, N)\,, \tag{15.23}$$

$$\mathcal{P}_{\mathrm{poly}}(\theta, N) = \frac{2}{U(N,\theta)}\left[e^{U(N,\theta)} - 1 + U(N,\theta)\right]\,, \tag{15.24}$$

$$U(N,\theta) = \frac{8\pi^2}{3}\left(\frac{b^2 N}{\lambda_{n+1}^2}\right)\sin^2\frac{\theta}{2}\,. \tag{15.25}$$

Equation (15.12) together with Eq. (15.19) or Eq. (15.23) give the stimulated signal which is peaked in the direction $\Delta\mathbf{k} = 0$. Long-range order breaks the linear-N dependence of the signal from ideal solutions.

15.2.2
Spontaneous Coherent Signal

Spontaneous coherent signals given by Eq. (15.16) are determined by the Fourier transform of the QD *pair* distribution function

$$F_2(\Delta\mathbf{k}) = \frac{N(N-1)}{2\Omega'^2}\int_{\Omega'} d^3\mathbf{r}_\alpha \int_{\Omega'} d^3\mathbf{r}_\beta\, F_2(\mathbf{r}_\alpha, \mathbf{r}_\beta)e^{i\Delta\mathbf{k}\cdot(\mathbf{r}_\alpha - \mathbf{r}_\beta)}\,. \tag{15.26}$$

Here, $[N(N-1)/2\Omega'^2]F_2(\mathbf{r}_\alpha, \mathbf{r}_\beta)d^3\mathbf{r}_\alpha d^3\mathbf{r}_\beta$ is the joint probability of the QDs in the pair between $(\mathbf{r}_\alpha, \mathbf{r}_\beta)$ and $(\mathbf{r}_\alpha + d\mathbf{r}_\alpha, \mathbf{r}_\beta + d\mathbf{r}_\beta)$. The pair distribution function is normalized so that when integrated over the sample, it gives the total number of QD pairs, that is,

$$\frac{N(N-1)}{2\Omega'^2}\int_{\Omega'} d^3\mathbf{r}_\alpha \int_{\Omega'} d^3\mathbf{r}_\beta\, F_2(\mathbf{r}_\alpha, \mathbf{r}_\beta) = \frac{N(N-1)}{2}\,. \tag{15.27}$$

It can be further partitioned as

$$F_2\left(\mathbf{r}_\alpha, \mathbf{r}_\beta\right) = F_1\left(\mathbf{r}_\alpha\right)F_1\left(\mathbf{r}_\beta\right) + g_2\left(\mathbf{r}_\alpha, \mathbf{r}_\beta\right)\,. \tag{15.28}$$

The interaction function g_2, a measure of the effect of inter-QD interactions, measures the deviation of the pair distribution function $F_2(\mathbf{r}_\alpha, \mathbf{r}_\beta)$ from a product of single QD distributions $F_1(\mathbf{r}_\alpha)F_1(\mathbf{r}_\beta)$.

Long-range coherence

When the first term in Eq. (15.28) is substituted into Eq. (15.26), we obtain the long-range coherent spontaneous signal

$$F_2(\Delta \boldsymbol{k}) = N(N-1)\mathcal{P}^2_{\text{fluid}}(\theta) \,. \tag{15.29}$$

For linear scattering ($n = 1$), the signal given by Eq. (15.29) is indistinguishable from the incident beam. However, the signal is distinct for nonlinear scattering with non-collinear beam geometry. A similar result holds for a collection of N' polymers each made of N QDs. In the absence of long-range order between polymers, we can use Eq. (15.29) with $N \to N'$.

Short-range coherence

Short-range coherent spontaneous signals are described by Eq. (15.16). The distance between the QDs involved in the light–matter interaction is restricted by $g_2(\boldsymbol{r}_\alpha, \boldsymbol{r}_\beta)$ so that $|\boldsymbol{r}_\alpha - \boldsymbol{r}_\beta|/\lambda_{n+1} \ll 1$. The exponent in the QD pair distribution function in Eq. (15.26) can be approximated by unity. We begin by considering a solution of hard sphere QDs of diameter a and the volume per solute QD, that is, $\pi a^3/6 = \Omega'/N = v$. In this case, we have

$$g_2(r_{\alpha,\beta}) = \begin{cases} 0 & \text{for } r_{\alpha,\beta} > a \\ -1 & \text{for } r_{\alpha,\beta} \le a \end{cases} \,, \tag{15.30}$$

where $\boldsymbol{r}_{\alpha,\beta} = \boldsymbol{r}_\alpha - \boldsymbol{r}_\beta$. Using the relation $(1/\Omega')\int_{\Omega'} d^3\boldsymbol{r}_\alpha \int_{\Omega'} d^3\boldsymbol{r}_\beta g_2(\boldsymbol{r}_\alpha, \boldsymbol{r}_\beta) = \int_{\Omega'} d^3\boldsymbol{r}_{\alpha,\beta} g_2(r_{\alpha,\beta})$, we obtain

$$F_{2,\text{fluid}}(\Delta \boldsymbol{k}) = -\frac{N-1}{2} \,. \tag{15.31}$$

The short-range interaction for a collection of N' polymer bonded QDs each comprised of N segments is calculated in a molecular context, that is,

$$F_{2,\text{poly}}(\Delta \boldsymbol{k}) = \frac{N^4}{v'^2} X \mathcal{P}^2_{\text{poly}}(\theta, N) \,, \tag{15.32}$$

where $v' = \Omega'/N'$ is the volume per single polymer QD molecule, and X describes the average short-range interaction between the segments of two thread-like polymer QD molecules.

To discuss the validity of the hard sphere model Eq. (15.31), we impose three limitations on the solute QDs and their interactions. First, the solute is treated as a non-ideal gas of classical molecules capable of undergoing a thermodynamic phase transition. Second, the pair interaction potential falls off with the fourth or higher power of the distance. Third, the total potential energy of the system is representable as the sum of pair potentials that only depend on the separation. Unlike the polymer QD molecules discussed previously, the solute molecules are not chemically bound.

The deviation of the solute from the ideal gas is described by the fugacity Z normalized in density v^{-1} units:

$$Z[v] = \frac{1}{v} \exp\left(-\sum_{\ell \geq 1} \beta_\ell v^{-\ell}\right). \tag{15.33}$$

The irreducible integrals β_ℓ are defined so that for the ideal gas $\beta_\ell \to 0$. We rewrite Eq. (15.33) in its differential form and get the pressure as

$$p = \frac{\partial}{\partial v} \ln Z[v] = \frac{1}{v}\left[\sum_{\ell \geq 1} \ell \beta_\ell v^{-\ell} - 1\right]. \tag{15.34}$$

The pressure p of the gas above the solvent also shows the deviation from the ideal gas that can be formally written with irreducible integrals as

$$\left(\frac{\partial p}{\partial Z}\right)_T = \frac{N k_B T}{Z v} \tag{15.35}$$

$$\left(\frac{\partial p}{\partial \Omega'}\right)_T = -\frac{N k_B T}{\Omega'^2}\left(1 - \sum_{\ell \geq 1} \ell \beta_\ell v^{-\ell}\right), \tag{15.36}$$

where k_B is the Boltzmann constant and T is the temperature. Using the generalized form of the grand partition function, Eq. (15.33), as well as connection between the cluster and irreducible integrals, it has been shown that

$$\frac{1}{2\Omega'} \int_{\Omega'} d^3 \boldsymbol{r}_{\alpha,\beta} g_2(\boldsymbol{r}_{\alpha,\beta}) = \frac{Z^2 v^2}{2 k_B T \Omega'} \left(\frac{\partial^2 p}{\partial Z^2}\right)_T. \tag{15.37}$$

Substituting Eq. (15.35) into Eq. (15.37) and utilizing Eq. (15.34), we obtain

$$\frac{1}{2\Omega'} \int_{\Omega'} d^3 \boldsymbol{r}_{\alpha,\beta} g_2(\boldsymbol{r}_{\alpha,\beta}) = -\frac{v}{2\Omega'}\left(1 - \frac{1}{1 - \sum_{\ell \geq 1} \ell \beta_\ell v^{-\ell}}\right). \tag{15.38}$$

Combining with Eqs. (15.28) and (15.26), we find that

$$\begin{aligned} F_{2,\text{fluid}}(\Delta \boldsymbol{k}) &= -\frac{N-1}{2}\left(1 - \frac{1}{1 - \sum_{\ell \geq 1} \ell \beta_\ell v^{-\ell}}\right) \\ &= -\frac{N-1}{2}\left[1 + \frac{N k_B T}{\Omega'^2}\left(\frac{\partial p}{\partial \Omega'}\right)_T^{-1}\right]. \end{aligned} \tag{15.39}$$

This equation confirms that the short-range coherent spontaneous signal vanishes for an ideal solution. It also indicates that short-range coherent signals from the solute in the absence of the strong van der Waals forces is not phase sensitive and depends on the solute density ν^{-1}. Therefore, it represents Rayleigh ($n = 1$) and hyper Rayleigh ($n > 1$) scattering.

The calculation of the irreducible integrals β_ℓ from first principles is extremely complicated. Therefore, we will only discuss the role of $\sum_\ell \beta_\ell \nu^{-\ell}$. The phase transitions are characterized by divergence of the fugacity density series in Eq. (15.33) on the real axis at $T = T_c$. Hence, $\sum_\ell \beta_\ell \nu^{-\ell}$ either diverges (first order transitions) or becomes unity (anomalous first order transitions). In the first case, at the singularity, $\sum_\ell \beta_\ell \nu^{-\ell}$ increases and reaches unity at some temperature T_0 lower than the temperature at which the singularity moves into the complex plane at T_c. Close to T_0, a slight change in the partial volume of the solute dos not change with pressure and the second term in Eq. (15.39) dominates the short-range coherent spontaneous signal.

An anomalous first-order transition occurs in the temperature range $T_0 < T < T_c$. We can then neglect the second term in Eq. (15.39) and the signal reduces to that obtained from the rigid spheres model in Eq. (15.31). The $(N-1)/2$ factor signifies that only QD pairs composed of nearby QDs contribute to the short-range coherence in this case.

15.3 Application to Two-Photon-Induced Signals

In the above sections, we presented a unified microscopic description of $n+1$ wave mixing processes. The nonlinear signal is defined as the change in the intensity of the detected mode due to the other n optical modes. The signal is formally expressed in terms of polarization superoperators that are calculated by the Heisenberg equations of motion for the field (stimulated signal) or the field intensity (spontaneous signal). We identified four types of signals that can be calculated by utilizing standard statistical quantities, namely, the QD and QD pair distribution functions. These results can be summarized as follows:

$$S^{(n)}(t) = S^{(n)}_{\mathrm{ST}}(t) + S^{(n)}_{\mathrm{SP,\,icoh}}(t) + S^{(n)}_{\mathrm{SP,\,coh,\,lr}}(t) + S^{(n)}_{\mathrm{SP,\,coh,\,sr}}(t)\,, \tag{15.40}$$

$$S^{(n)}_{\mathrm{ST}}(t) = \frac{N}{\hbar}\,\mathrm{Im}\left\{\left[\begin{array}{c}\mathcal{P}_{\mathrm{fluid}}(\theta)\\ \mathcal{P}_{\mathrm{poly}}(\theta, N)\end{array}\right]\int_{-\infty}^{t} d\tau\,\mathcal{E}_s^*(\tau)\langle V_{\mathrm{L}}(\tau)\rangle_{\{n\}}\right\}, \tag{15.41}$$

$$S^{(n)}_{\mathrm{SP,\,incoh}}(t) = \left(\frac{N\omega_s}{\hbar\epsilon_0\Omega}\right)\mathrm{Re}\left\{\int_{-\infty}^{t} d\tau\int_{-\infty}^{\tau} d\tau' e^{i\omega_{n+1}(\tau-\tau')}\right.$$
$$\left.\times\left\langle\hat{\mathcal{T}}\,V_{\mathrm{L}}(\tau)V_{\mathrm{R}}^{\dagger}(\tau')\right\rangle_{\{n\}}\right\}, \tag{15.42}$$

$$S^{(n)}_{\mathrm{SP,coh,lr}}(t) = \left(\frac{\omega_s}{2\hbar\epsilon_0\Omega}\right) N(N-1)\mathcal{P}^2_{\mathrm{fluid}}(\theta) \times \left| \int_{-\infty}^{t} d\tau e^{i\omega_{n+1}\tau} \langle V_{\mathrm{L}}(\tau)\rangle_{\{n\}} \right|^2 , \tag{15.43}$$

$$S^{(n)}_{\mathrm{SP,coh,sr}}(t) = \left(\frac{\omega_s}{2\hbar\epsilon_0\Omega}\right) \left| \int_{-\infty}^{t} d\tau e^{i\omega_{n+1}\tau} \langle V_{\mathrm{L}}(\tau)\rangle_{\{n\}} \right|^2 \times \left[\frac{-\frac{N-1}{2\,\mathrm{Re}\left\{1-\left(1-\sum_{\ell\geq 1}\ell\beta_\ell \nu^{-l}\right)^{-1}\right\}}}{\mathrm{Re}\left\{\frac{N^2}{\nu'}\mathcal{P}_{\mathrm{poly}}(\theta,N) + \frac{N^4}{\nu'^2} X\mathcal{P}^2_{\mathrm{poly}}(\theta,N)\right\}} \right] . \tag{15.44}$$

Here, S_{ST} represents the stimulated heterodyne detected signal. This includes self-heterodyne detected techniques such as pump-probe and stimulated hyper-Raman scattering. The remaining terms describe spontaneously generated signals that can be homodyne detected. Also, for multi-photon induced fluorescence, $S_{\mathrm{SP,incoh}}$ is incoherent, phase insensitive and scales like $\sim N$. The quantity $S_{\mathrm{SP,coh,lr}}$ describes the coherent response of all possible QD pairs $\sim N(N-1)/2$. These types of linear signals coincide with the incoming beam. Nonlinear signals include hyper-Raman scattering sum/difference frequency generation. $S_{\mathrm{SP,coh,sr}}$ denotes a short-range coherent spontaneous signal. Identical oriented polymer QD molecules give a directed phase matched signal and the extent of phase-matching depends on the polymer size, internal structure and interaction between the polymers. The random-walk model suggests that the signal consists of two terms in the QD density $1/\nu'$ expansion.

We have further investigated the nonlinear scattering from the solute represented by the non-ideal gas described by the osmotic pressure, density and fugacity. The signal is phase-insensitive and can be recast into an infinite series of the QD density $\sum_\ell \beta_\ell \nu^{-\ell}$. The exact calculation of the necessary irreducible integrals is a difficult challenge. We discussed two limiting cases of ordinary and anomalous first order transitions and compared them to the hard sphere model. Such signals are both phase-insensitive and depend on the QD density, and we associate them with Rayleigh and hyper-Rayleigh scattering.

Equations (15.40) provide the starting point for the superoperator CTPL expansion of the nonlinear polarization. CTPL's are rule based diagrammatic techniques (see the Appendix 15.B) [13–18]. We shall illustrate this for stationary nonlinear spontaneous signals. We consider signals generated by two incoming classical fields: $\mathcal{E}_1 e^{-i\omega_1 t}$ and $\mathcal{E}_2 e^{-i\omega_2 t}$ around two-photon resonances $\omega_3 \approx \omega_1 + \omega_2$. The QDs are described by the three level ladder system: $|g\rangle, |g'\rangle$, $|e\rangle$ and $|f\rangle$, shown in Figure 15.3b. The ground state is a manifold that contains at least two levels $\{|g\rangle, |g'\rangle\}$. This model allows the Brillouin scattering. Then, the Stock shifted pattern generates an acoustic wave. This, in turn, lifts the degeneracy of the QD ground state and modifies the density dependent prefactor for the short-range coherent signals. In some cases, the acoustic wave can also reflect the incoming modes via Brag diffraction, thereby increasing the power of the generated signal. The Brillouin scattering is considered to be a type of Raman scattering.

The incoherent signal in Eq. (15.42) gives rise to hyper-Raman and two-photon-induced fluorescence which can be distinguished by including dephasing processes. This goes beyond the scope here. Since all incoming modes are classical fields, the stationary signals ($t \to \infty$) can be expressed in terms of nonlinear susceptibil-

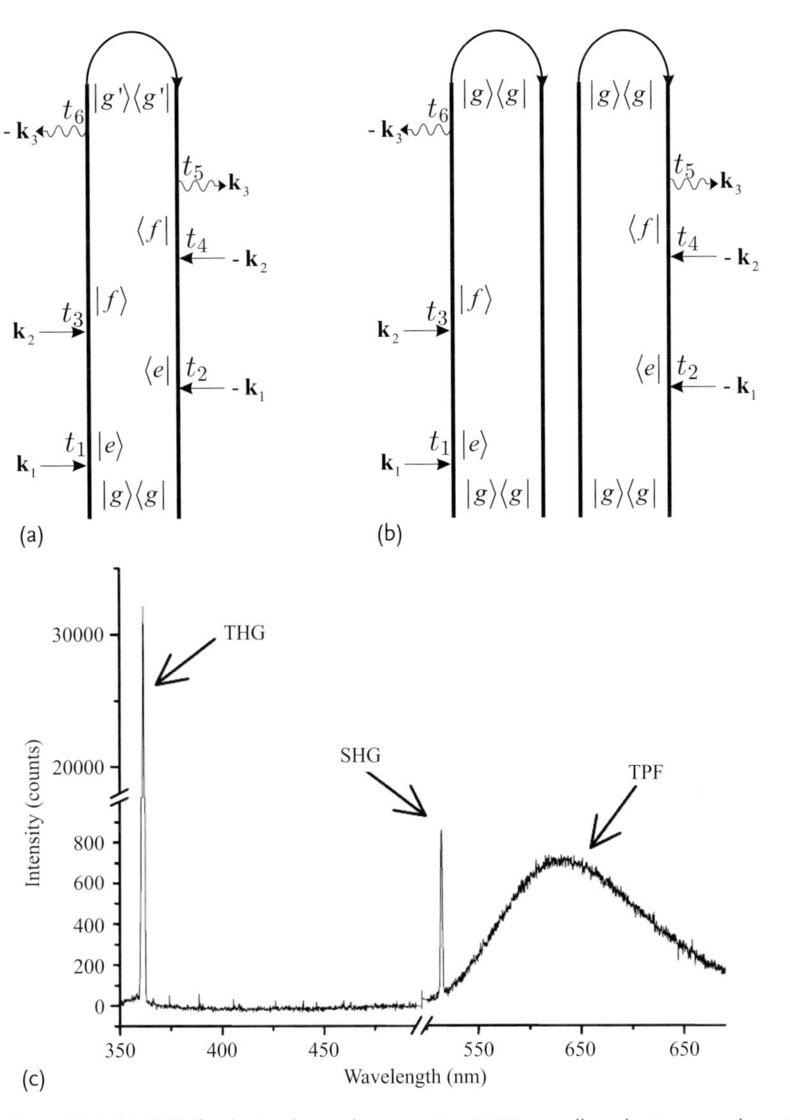

Figure 15.6 (a) CTPL for the incoherent hyper-Raman and two-photon-induced fluorescence (TPIF); (b) CTPL for the long-range coherent homodyne detected sum frequency generation (SFG) as well as short-range coherent hyper-Rayleigh; and (c) Spectra measured from PMMA polymers of oriented DCM.

ities using the CTPL shown in Figure 15.6a. We have

$$
\begin{aligned}
&S_{\text{HRAM, TPIF}}(-\omega_3;\omega_2,\omega_1) \\
&= 2N\,\text{Re}\left[\chi^{(5)}_{LR,---}(-\omega_3;\omega_3,-\omega_2,\omega_2,-\omega_1,\omega_1)\right]|\mathcal{E}_1|^2|\mathcal{E}_2|^2\,,
\end{aligned}
\tag{15.45}
$$

where the susceptibility is given in the mixed representation (L/R for the generated mode, and $+,-$ for the classical incoming modes) and can be written in terms of the Green's function $G(\omega) = \hbar/(\hbar\omega - H_0 + i\hbar\gamma)$, where γ is a dephasing rate

$$
\begin{aligned}
&\chi^{(5)}_{LR,---}(-\omega_3;\omega_3,-\omega_2,\omega_2,-\omega_1,\omega_1) = \frac{1}{\epsilon_0\Omega'}\sum_p \frac{i^5}{5!\hbar^5} \\
&\quad\times \langle g|VG^\dagger(\omega_g+\omega_1)VG^\dagger(\omega_g+\omega_1+\omega_2)V^\dagger G^\dagger(\omega_g+\omega_1+\omega_2-\omega_3) \\
&\quad\times VG(\omega_g+\omega_1+\omega_2)V^\dagger G(\omega_g+\omega_1)V^\dagger|g\rangle\,.
\end{aligned}
\tag{15.46}
$$

In this notation, p stands for permutations of the incoming field within each branch of the loop diagram. Expanding Eq. (15.46) in QD energy levels ω_{eg}, ω_{ef}, $\omega_{fg'}$ and the corresponding transition dipole moments μ_{eg}, μ_{ef}, $\mu_{fg'}$, we obtain

$$
\begin{aligned}
&\chi^{(5)}_{LR,---}(-\omega_3;\omega_3,-\omega_2,\omega_2,-\omega_1,\omega_1) \\
&= \frac{1}{\epsilon_0\Omega'}\sum_p \frac{i^5}{5!\hbar^5}\sum_{g,g'} \frac{|\mu_{eg}\mu_{ef}\mu_{fg'}|^2}{[(\omega_1-\omega_{eg})^2+\gamma^2](\omega_1+\omega_2-\omega_{fg}+i\gamma)} \\
&\quad\times \frac{1}{(\omega_1+\omega_2-\omega_{fg'}-i\gamma)(\omega_1+\omega_2-\omega_3-\omega_{gg'}-i\gamma)}\,.
\end{aligned}
\tag{15.47}
$$

The long-range coherent signal in Eq. (15.43) for our model is a homodyne-detected sum frequency generation (SFG) with

$$
\begin{aligned}
&S_{\text{SFG}}(-\omega_3;\omega_2,\omega_1) = N(N-1)|\mathcal{E}_1|^2|\mathcal{E}_2|^2 \\
&\quad\times |\mathcal{P}_{\text{fluid}}(\theta)\delta(\omega_3-\omega_2-\omega_1)\chi^{(2)}_{L,--}(-\omega_3;\omega_2,\omega_1)|^2\,.
\end{aligned}
\tag{15.48}
$$

This susceptibility can be calculated from the CTPL in Figure 15.6b,

$$
\begin{aligned}
&\chi^{(2)}_{L,--}(-\omega_3;\omega_2,\omega_1) \\
&= \frac{1}{\epsilon_0\Omega'}\sum_p \frac{i^2}{2!\hbar^2}\langle g|VG(\omega_g+\omega_1+\omega_2)V^\dagger G(\omega_g+\omega_1)V^\dagger|g\rangle \\
&= \frac{1}{\epsilon_0\Omega'}\sum_p \frac{i^2}{2!\hbar^2}\sum_g \frac{\mu_{ge}\mu_{ef}\mu_{fg}}{(\omega_1-\omega_{eg}+i\gamma)(\omega_1+\omega_2-\omega_{gf}-i\gamma)}\,.
\end{aligned}
\tag{15.49}
$$

The short-range coherent signal in Eq. (15.44) for our model gives the density dependent hyper-Rayleigh (HRAY) scattering

$$
\begin{aligned}
&S_{\text{HRAY}}(-\omega_3; \omega_2, \omega_1) \\
&= \left[\begin{array}{c} -\dfrac{N-1}{2\,\mathrm{Re}\{1-\left(1-\sum_{\ell\geq 1} \ell\beta_\ell \nu^{-\ell}\right)^{-1}\}} \\ \mathrm{Re}\{\frac{N^2}{\nu'}\mathcal{P}_{\text{poly}}(\theta, N) + \frac{N^4}{\nu'^2} X \mathcal{P}^2_{\text{poly}}(\theta, N)\} \end{array} \right] |\mathcal{E}_1|^2 |\mathcal{E}_2|^2 \\
&\quad \times \left| \delta(\omega_3 - \omega_2 - \omega_1) \chi^{(2)}_{L,--}(-\omega_3; \omega_2, \omega_1) \right|^2 .
\end{aligned}
\tag{15.50}
$$

In Figure 15.6c, we present the experimental spontaneously generated signal from a polymer solute. The SFG signal has a sharp resonance, as expected from the delta function in Eq. (15.48), while the TPIF signal is broadened and covers the range of $\omega_{g',g}$ in accordance with Eq. (15.45). The hyper-Raylcigh signal in Eq. (15.50) has the same resonance as SFG since both are determined by the square of the second-order susceptibility.

Note that all the signals discussed above are generated by classical incoming fields, and can be calculated using the standard susceptibility without the CTPL formalism. The power of the present microscopic treatment is its ability to predict signals generated by non-classical incoming modes. Furthermore, even though in this chapter we neglected the QD orientational degrees of freedom, they play an important role in distinguishing between SFG and HRAY processes. Taking that into account, we add a superscript to the transition dipole moment μ^{α}_{jk}, which indicates its orientation with respect to the αth component of the optical field. The intensity of the signals in Eqs. (15.48) and (15.50) is then proportional to

$$
\left\langle \left(\mu^{\alpha'}_{g'e'}\right)^* \left(\mu^{\beta'}_{e'f'}\right)^* \left(\mu^{\gamma'}_{f'g'}\right)^* \mu^{\alpha}_{ge} \mu^{\beta}_{ef} \mu^{\gamma}_{fg} \right\rangle_r
\tag{15.51}
$$

where primed and unprimed indices denote two different QDs in the QD pair and $\langle\cdots\rangle_r$ denotes rotational averaging. For such long-range coherent signals such as SFG, correlation between the two QDs in a pair is negligible and Eq. (15.51) can be factorized as $|\langle \mu^{\alpha}_{ge}\mu^{\beta}_{ef}\mu^{\gamma}_{fg}\rangle_r|^2$. In isotropic media, this is zero due to orientation fluctuations, thus only leaving short-range coherent signals HRAY.

Appendix 15.A
Semiclassical vs. Quantum Field Derivation of Heterodyne Detected Signals

In this appendix, we calculate the heterodyne detected incoherent nonlinear signal from a linear chain of QDs that interact with $n+1$ classical optical fields. The chain lies between $-L/2$ and $L/2$ along the x-axis. The heterodyne detected signal is given by the electric field of the signal mode at $x = X$ far from the sample, as shown in Figure 15.4.

We shall demonstrate equivalence of the semiclassical and quantum approaches. We begin with the semiclassical nature of the signal field. The calculation is composed of two steps. First, we calculate the electric field on the auxiliary object (aperture) via the Maxwell equations with the optical field driven by the nonlinear polarization of the atomic *primary* sources. Second, the aperture serves as the point *secondary* source of a spherical signal wave that is calculated with the propagator formalism.

For $k_{\{n\}}L \gg 1$, the sample is treated as a continuous medium. The incoming waves create a nonlinear polarization wave along the sample with

$$P_{\{n\}}(x,t) = P_n(t)e^{ik_{\{n\}}x - i\omega_{\{n\}}t}\,, \tag{15.A.1}$$

where $P_n(t)$ is slowly varying, that is, $|(\partial/\partial t)P_{\{n\}}(t)| \ll |\omega_{\{n\}}P_{\{n\}}(t)|$. This polarization is the principal source of the generated mode whose electric field is

$$E_{n+1}(x,t) = \mathcal{E}_{n+1}(x,t)e^{ik_{n+1}x - i\omega_{n+1}t}\,, \tag{15.A.2}$$

and $\mathcal{E}_{n+1}(t)$ is the slowly varying field amplitude (in space and time).

The electric field of the generated mode and the polarization induced by the incoming modes are related through the Maxwell's equations

$$\left[-\frac{\partial^2}{\partial x^2} + \frac{k_{n+1}^2}{\omega_{n+1}^2}\frac{\partial^2}{\partial t^2}\right]E_{n+1}(x,t) = -\frac{1}{\epsilon_0 c^2}\frac{\partial^2}{\partial t^2}P_{\{n\}}(x,t)\,. \tag{15.A.3}$$

Substituting Eqs. (15.A.1) and (15.A.2) into Eq. (15.A.3) and using the slowly varying amplitude approximation for the generated polarization, we have

$$ik_{n+1}\frac{\partial}{\partial x}\mathcal{E}_{n+1}(x,t) = -\frac{\omega_{\{n\}}^2}{2\epsilon_0 c^2}P_{\{n\}}(t)e^{i\Delta kx - i(\omega_{\{n\}}-\omega_{n+1})t}\,. \tag{15.A.4}$$

At the beginning of the illuminated region, the amplitude of the generated mode vanishes so that $\mathcal{E}_{n+1}(-L/2,t) = 0$. Using this condition and integrating Eq. (15.A.4) over the sample, we obtain the generated mode at the aperture

$$\begin{aligned} E_{n+1}\left(\frac{L}{2},t\right) &= -i\frac{\omega_{\{n\}}^2 L}{2\epsilon_0 k_{n+1}c^2}P_{\{n\}}(t) \\ &\quad\times \operatorname{sinc}\left(\frac{\Delta kL}{2}\right)e^{ik_{n+1}L/2 - i\omega_{\{n\}}t}\,. \end{aligned} \tag{15.A.5}$$

The signal field is given by Fresnel diffraction from a point-like secondary source that corresponds to a single Huygens' wavelet

$$\begin{aligned} E_{n+1}(X,t) &= -\frac{i}{k_{n+1}X}E_{n+1}\left(\frac{L}{2},t\right)e^{ik_{n+1}(X-L/2)} \\ &= -\frac{L}{2\epsilon_0 X n^2(\omega_{n+1})}P_{\{n\}}(t)\operatorname{sinc}\left(\frac{\Delta kL}{2}\right)e^{ik_{n+1}X - i\omega_{\{n\}}t}\,. \end{aligned} \tag{15.A.6}$$

Here, $n(\omega)$ is the refractive index of the sample and the factor of $1/X$ accounts for the spherical nature of the Huygens' wavelet.

Let us consider the signal obtained from a quantum description of the field. For this, each atom is the primary and sole source of the signal wave. The signal wave is given by the interference from the Huygens' wavelets in Eq. (15.3). Using Eq. (15.7), we obtain the equation of motion for the photon annihilation operator, that is,

$$\begin{aligned} \frac{d}{dt} a_{n+1}(t) &= \frac{i}{\hbar}\left[H_{int}^{(n+1)}, a_{n+1}(t)\right] \\ &= \frac{i}{\hbar}\int dx \sqrt{\frac{\hbar\omega_{n+1}}{2\epsilon_0 \Omega_{n+1}}} \langle V(x,t)\rangle_{\{n\}} e^{-ik_{n+1}x + i\omega_{n+1}t} . \end{aligned} \tag{15.A.7}$$

We shall integrate the equation of motion (15.A.7) under the following conditions

1. The expectation value of the polarization operator is given by Eq. (15.11);
2. Initially, the polarization $\langle V(x,-\infty)\rangle_{\{n\}}$ is zero;
3. The polarization has a slowly varying temporal amplitude so that

$$\int_{-\infty}^{t} d\tau \langle V(\tau)\rangle_{\{n\}} = P_{\{n\}}(t)\left[\frac{\exp(-i\omega_{\{n\}}t)}{-i\omega_{\{n\}}}\right]. \tag{15.A.8}$$

From Eq. (15.3), we obtain the signal optical field

$$E_{n+1}(X,t) = \frac{-N\omega_{n+1}}{2\epsilon_0 \Omega_{n+1} L \omega_{\{n\}}} \frac{e^{ik_{n+1}X - i\omega_{\{n\}}t}}{k_{n+1}X} P_{\{n\}}(t) \int_{-L/2}^{L/2} dx e^{i\Delta k x} . \tag{15.A.9}$$

Using the resonant condition $\omega_{n+1} - \omega_{\{n\}} \approx 0$, we finally obtain

$$\begin{aligned} E_{n+1}(X,t) = &-\frac{Nc}{2\epsilon_0 \Omega_{n+1}\omega_{n+1} n(\omega_{n+1}) X} P_{\{n\}}(t) \\ &\times \operatorname{sinc}\left(\frac{\Delta k L}{2}\right) e^{ik_{n+1}X - i\omega_{n+1}t} . \end{aligned} \tag{15.A.10}$$

By comparing Eq. (15.A.10) with Eq. (15.A.6), we see that the two approaches give the same results, except for the factors $L/n(\omega_{n+1})$ versus $Nc/\Omega_{n+1}\omega_{n+1}$ that depend on the model and arise from the single signal mode approximation. The heterodyne signal is obtained by treating the heterodyne wave as a spherical wave emitted by the aperture [that brings the Goui phase factor $(i/k_{n+1}X)$]: $\mathcal{E}_s(X,t) = (i/k_{n+1}X)\mathcal{E}_s(L,t)$ that brings up the Goui phase factor $(i/k_{n+1}X)$. Substituting the above equation along with Eq. (15.A.10) (or Eq. (15.A.6)) into the signal Eq. (15.4), we have

$$S_{\text{HET}} \sim \operatorname{Im}\left\{\langle V(\mathbf{r},t)\rangle_{\{n\}} \mathcal{E}_{n+1}^{*}(\mathbf{r},t)/X^2\right\} . \tag{15.A.11}$$

Formally, we apply the Goui phase twice in Eq. (15.4) for propagating the signal and the heterodyne part. This leads to an overall unimportant prefactor $(1/k_{n+1}X)^2$. We can skip the propagation steps and use Eq. (15.A.5) directly $E_{n+1} \sim iP_{\{n\}}$. This is the standard semiclassical procedure.

Appendix 15.B
Generalized Susceptibility and Its CTPL Representation

In this appendix, we provide the formalism for the generalized susceptibilities used in the Section 15.3. The basis of this is the superoperator non-equilibrium Green's function (SNGF). The SNGFs of nth order are defined as traces of time ordered products of such superoperators:

$$\langle \hat{T} A_{+}(t) \underbrace{A_{+}(t_n) \cdots A_{+}(t_{n-m+1})}_{m} \underbrace{A_{-}(t_{m-n}) \cdots A_{-}(t_1)}_{n-m} \rangle \,, \quad \text{where} \quad m = 0\,, \ldots, n \,.$$

The SNGFs may contain an arbitrary number of "+" and "−" superoperators. The chronologically last superoperator must be a + one, otherwise, the SNGF vanishes (which is a consequence of the permutational invariance of the trace and superoperators definition). More specifically, the material $\mathbb{V}$ and optical $\mathbb{E}$ SNGF's are defined by

$$\mathbb{V}^{(n)}_{L,\nu_n \cdots \nu_1}(\tau, t_n, \ldots, t_1) = \langle \hat{T} V'_L(\tau) V'_{\nu_n}(t_n) \cdots V'_{\nu_1}(t_1) \rangle \,, \tag{15.B.1}$$

$$\mathbb{E}^{(n)}_{\nu_n \ldots \nu_1}(t_n, \ldots, t_1) = \langle \hat{T} E'_{\bar{\nu}_n}(t_n) \cdots E'_{\bar{\nu}_1}(t_1) \rangle \,, \tag{15.B.2}$$

where the subscript ν is the superoperator index that depends on the representation, $V'_\nu = V_\nu + V^\dagger_\nu$ and the net field operators. SNGFs of the form

$$\mathbb{V}^{(m)}_{+\underbrace{-\cdots-}_{m}}$$

give causal ordinary response functions of mth order. The material SNGF of the form

$$\mathbb{V}^{(m)}_{+\underbrace{+\cdots+}_{m}}$$

represent mth moment of QD's fluctuations. The material SNGF of the form

$$\mathbb{V}^{(m)}_{+\underbrace{+\cdots+}_{m'}\underbrace{-\cdots-}_{m-m'}}$$

indicates changes in m'th moment of QD's fluctuations induced by $m - m'$ light–matter interactions. In the other representation, the material SNGF

$$\mathbb{V}^{(m)}_{L\underbrace{L\cdots L}_{n}\underbrace{R\cdots R}_{m-n}}$$

represents a *Liouville space pathway* with n interactions from the left (i.e., with the ket) and $m-n$ interactions from the right (i.e., with the bra).

The average material field in Eqs. (15.40)–(15.44) can be written in terms of the SNGF's, defined above, as

$$\langle V_L(\tau)\rangle_{\{n\}} = \frac{i^n}{n!\hbar^n} \sum_{\nu_n} \cdots \sum_{\nu_1} \int_{-\infty}^{\infty} dt_n \cdots \int_{-\infty}^{\infty} dt_1 \\ \times \Theta(\tau) \mathbb{V}^{(n+1)}_{L,\nu_n\cdots\nu_1}(\tau, t_n, \ldots, t_1) \mathbb{E}^{(n)}_{\bar{\nu}_n\cdots\bar{\nu}_1}(t_n, \ldots, t_1) , \qquad (15.B.3)$$

where $t_n, \cdots, t_1$ are the light–matter interaction times for the incoming modes. The factor $\Theta(\tau) = \prod_{i=1}^{n} \theta(\tau - t_i)$ guarantees that the τ is the last light–matter interaction with the detected mode which has been accounted for separately. The indices $\bar{\nu}_j$ are the conjugates to ν_j and are defined as follows: The conjugate of $+(R)$ is $-(L)$ and vice versa; Equation (15.B.3) implies that the excitations in the material are produced by fluctuations in the optical field and vice versa; and we use the mixed representation in order to separate classical incoming ($\pm$ representation) from the quantum detected (L, R representation) optical modes.

Equations (15.46) and (15.49) have been obtained by recasting material SNGFs in Eq. (15.B.3) into the form of generalized susceptibilities. Those are formally defined in the frequency domain by performing a multiple Fourier transform, that is,

$$\chi^{(n)}_{L,\nu_n\cdots\nu_1}(-\omega_{n+1}; \omega_n, \cdots, \omega_1) = \int_{-\infty}^{\infty} d\tau \cdots \int_{-\infty}^{\infty} dt_1 \Theta(\tau) e^{i(\omega_n t_n + \cdots + \omega_1 t_1)} \\ \times \delta(-\omega_{n+1} + \omega_n + \cdots + \omega_1) \mathbb{V}^{(n)}_{L,\nu_n\cdots\nu_1}(\tau, t_n, \cdots, t_1) . \qquad (15.B.4)$$

The SNGF

$$\chi^{(n)}_{\underbrace{+-\cdots-}_{n}}(-\omega_{n+1}; \omega_n, \cdots, \omega_1)$$

(with one "+" and all the others as "−" indices) are the nth order nonlinear susceptibilities or causal response functions. Others can be interpreted similarly to their time domain counterparts in Eq. (15.B.1).

The generalized susceptibilities, written in terms of L, R superoperators, can be represented by close-time path loop (CTPL) diagrams introduced by the Schwinger–Keldysh many-body theory. The following rules are used to construct the diagrams:

1. Time runs along the loop clockwise from bottom left to bottom right;
2. The left branch of the loop represents the "ket", while the right represents the "bra";
3. Each interaction with a field mode is represented by an arrow line either on the right (R-operators) or on the left (L-operators);
4. The field is marked by dressing the lines with arrows, where an arrow pointing to the right (left) represents the field annihilation (creation) operator $E_\alpha(t)$ $[E^\dagger_\alpha(t)]$;

5. Within the RWA, each interaction with the field $E_\alpha(t)$ annihilates the photon and is accompanied by the operator $V_\alpha^\dagger(t)$ that leads to exciting a state represented by "ket" and de-exciting a state represented by "bra", respectively. Arrows pointing "inwards" (i.e., pointing to the right on a ket and to the left on a bra) consequently cause absorption of a photon by exciting the system, whereas arrows pointing "outwards" (i.e., pointing to the left on a bra and to the right on a ket) represent exciting the system by photon emission;
6. The observation time t is fixed and is always the last. By convention, it occurs at the left most. This can be achieved by a reflection of all interactions through the central line between kets and bras that corresponds to taking the complex conjugate of the original correlation function;
7. The loop translates into an alternating product of interactions (arrows) and periods of free evolutions (vertical solid lines) along the loop;
8. Since the loop time goes clockwise along the loop, periods of free evolution on the left branch amount to propagating forward in real time with the propagator give by the retarded Green's function G. Whereas evolution on the right branch corresponds to the backward propagation (advanced Green's function $G^\dagger$);
9. The frequency arguments of the various propagators are cumulative, that is, they are given by the sum of all "earlier" interactions along the loop. Additionally, the ground state frequency is added to all arguments of the propagators;
10. The Fourier transform of the time-domain propagators adds an additional factor of i or $(-i)$ for each retarded or (advanced) propagator;
11. The overall sign of the SNGF is given by $(-1)^{N_R}$, where N_R represents the number of R superoperators.

References

1. Mukamel, S. (1995) *Principle of Nonlinear Optical Spectroscopy*, Oxford University Press, New York.
2. Mukamel, S. and Hanamura, E. (1986) Four-wave mixing using partially coherent fields in systems with spatial correlations. *Phys. Rev. A*, **33**, 1099.
3. Andrews, D.L. and Allcock, P. (2002) *Optical Harmonics in Molecular Systems*, Wiley-VCH Verlag GmbH, Weinheim.
4. Glauber, R.J. (2007) *Quantum Theory of Optical Coherence: Selected Papers and Lectures*, Wiley-VCH Verlag GmbH, Weinheim.
5. Scully, M.O. and Zubairy, M.S. (1997) *Quantum Optics*, Cambridge University Press.
6. Bloembergen, N. (1996) *Nonlinear Optics*, World Scientific.
7. Denk, W., Strickler, J.H., and Webb, W.W. (1990) Two-photon laser scanning fluorescence microscopy. *Science*, **248**, 73.
8. Mertz, J. (2004) Nonlinear spectroscopy: new techniques and applications. *Curr. Opin. Neurobiol.*, **14**, 610.
9. Maker, P.D. (1970) Spectral broadening of elastic second-harmonic light scattering in liquids. *Phys. Rev. A*, **1**, 923.
10. Terhune, R.W., Maker, P.D., and Savage, C.M. (1965) Measurements of nonlinear light scattering. *Phys. Rev. Lett.*, **14**, 681.
11. Marx, C.A., Harbola, U., and Mukamel, S. (2008) Nonlinear optical spectroscopy of single, few, and many molecules: Nonequilibrium Green's function QED approach. *Phys. Rev. A*, **77**, 022110.

12 Clays, K. and Persoons, A. (1991) Hyper-Rayleigh scattering in solution. *Phys. Rev. Lett.*, **66**, 2980.

13 Roslyak, O. and Mukamel, S. (2009) A unified description of sum frequency generation, parametric down conversion and two-photon fluorescence. *Mol. Phys.*, **107**, 265.

14 Roslyak, O., Marx, C.A., and Mukamel, S. (2009) Nonlinear spectroscopy with entangled photons: Manipulating quantum pathways of matter. *Phys. Rev. A*, **79**, 033832.

15 Roslyak, O. and Mukamel, S. (2009) Photon entanglement signatures in difference-frequency-generation. *Opt. Express*, **17**, 1093.

16 Roslyak, O., Marx, C.A., and Mukamel, S. (2009) Generalized Kramers-Heisenberg expressions for stimulated Raman scattering and two-photon absorption. *Phys. Rev. A*, **79**, 063827.

17 Roslyak, O. and Mukamel, S. (2009) Multidimensional pump-probe spectroscopy with entangled twin-photon states. *Phys. Rev. A*, **79**, 063409.

18 Rahav, S., Roslyak, O., and Mukamel, S. (2009) Manipulating stimulated coherent anti-Stokes Raman spectroscopy signals by broad-band and narrow-band pulses. *J. Chem. Phys.*, **131**, 194510.

16
Probing Excitons and Biexcitons in Coupled QDs by Coherent Optical Spectroscopy

Up to this point, we have considered the frequency resolved spectra [1–3]. Let us now turn to the time resolved two-dimensional (2D) spectroscopy that provides a new tool for studying coupled excitons in molecular aggregates, photosynthetic complexes, semiconductor quantum wells and quantum dots (QDs). In these experiments, the system is subjected to three temporally well separated femtosecond pulses propagating in the directions $\boldsymbol{k}_1$, $\boldsymbol{k}_2$ and $\boldsymbol{k}_3$, and centered at times τ_1, τ_2 and τ_3, as shown in Figure 16.1. 2D signals of two coupled quantum wells were recently simulated using a free-carrier model, neglecting Coulomb interactions and many-body effects such as biexcitons. In this chapter, we calculate these signals for a system of two coupled QDs (dimer).

The QD molecule is described by a tight-binding two-band Hamiltonian for confined electrons and holes taking into account inter-dot electron and hole hopping along with monopole-monopole and dipole–dipole Coulomb interactions. We employ realistic parameters for the calculations. The Hamiltonian is then expressed in terms of e-h pair variables and truncated at the two-exciton manifold [4–7]. In this chapter, we confine our attention to two types of heterodyne detected signals S_I and S_III generated in the phase-matching directions $k_\text{I} = -k_1 + k_2 + k_3$ (photon echo), and $k_\text{III} = k_1 + k_2 - k_3$ (double quantum coherence), respectively. Both are recorded as a function of time delays $t_1 = \tau_2 - \tau_1$, $t_2 = \tau_3 - \tau_2$, and $t_3 = t - \tau_3$, where t is the detection time. The 2D spectra, obtained by a Fourier transform of S_I with respect to t_1 and t_3, $S_\text{I}(\Omega_3, t_2, \Omega_1)$ and of S_III with respect to t_2 and t_3, $S_\text{III}(\Omega_3, \Omega_2, t_1)$, reveal exciton correlations and two-exciton resonances. The signals are obtained by numerically solving the nonlinear exciton equations (NEE) that account for exciton-exciton interactions and their quasi-bosonic nature. To classify biexciton states in terms of their single exciton constituents, we analyze both the S_I and S_III signals using the corresponding sum-over-states (SOS) expressions. The roles of different coupling mechanisms at various inter-dot distances d will be investigated. The analysis shows that, at short distances, electron delocalization contributes to the creation of new biexcitonic peaks in the spectra, while exciton hopping shifts the two-exciton peaks.

The chapter is organized as follows: The first section presents our model Hamiltonian for two coupled quantum dots; the second section discusses the variation of the absorption spectra and single exciton eigenstates with inter-dot distance; in the

Properties of Interacting Low-Dimensional Systems, First Edition. G. Gumbs and D. Huang.

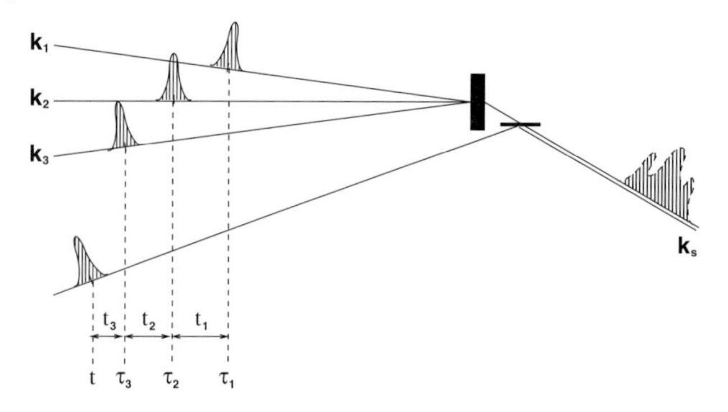

Figure 16.1 The pulse sequence in a four-wave-mixing experiment: the system is excited by three pulses propagating in directions $\boldsymbol{k}_1$, $\boldsymbol{k}_2$, and $\boldsymbol{k}_3$, and centered at times τ_1, τ_2, and τ_3, respectively. The signal is heterodyne detected in the phase-matching direction $\boldsymbol{k}_s$ at time t.

third section, we present the photon echo and double quantum coherence spectra and identify biexcitonic contributions; the fourth section is a brief chapter summary and the last section consists of appendices 16.A–16.C.

16.1 Model Hamiltonian for Two Coupled Quantum Dots

We consider a system of two vertically stacked, lens-shaped InGaAs/GaAs QD's of height 2 nm and radius 10 nm aligned along the z-axis. It is described by the tight-binding two-band Hamiltonian

$$H_{\mathrm{QD}} = H_0 + H_C \,. \tag{16.1}$$

Here, H_0 is the Hamiltonian for non-interacting electrons and holes and H_{C} is the contribution from the Coulomb interaction. Due to the quantum confinement, the conduction and valence bands split into discrete atomic-like levels. Each dot in InGaAs has two spin-degenerate electron states with angular momentum quantum numbers $(J, M) = (1/2, \pm 1/2)$ and two heavy hole states $(3/2, \pm 3/2)$. The free carrier Hamiltonian has the form

$$H_0 = \sum_{m_1,n_1} t^{(1)}_{m_1,n_1} \hat{c}^\dagger_{m_1} \hat{c}_{n_1} + \sum_{m_2,n_2} t^{(2)}_{m_2,n_2} \hat{d}^\dagger_{m_2} \hat{d}_{n_2} \,, \tag{16.2}$$

where the indices 1 (2) correspond to electrons (holes) in the conduction (valence) band. The subscripts, for example, $m_1 = (R_{m_1}, \sigma_{m_1})$, describe the QD position $R = \{T, B\}$ ("T" for top and "B" for bottom) and the spin projection σ of the carrier ($\uparrow, \downarrow$ for up and down spin projection, respectively). The creation and annihilation operators of an electron (hole) at site R_{m_1} (R_{m_2}) with spin z-component σ_{m_1} (σ_{m_2}) are $\hat{c}^\dagger_{m_1}$ and $\hat{c}_{m_1}$ ($\hat{d}^\dagger_{m_2}$ and $\hat{d}_{m_2}$). These satisfy the Fermionic algebra

$$\{\hat{c}_{m_1}, \hat{c}^\dagger_{n_1}\} \equiv \hat{c}_{m_1}\hat{c}^\dagger_{n_1} + \hat{c}^\dagger_{n_1}\hat{c}_{m_1} = \delta_{m_1,n_1} \,, \tag{16.3}$$

$$\{\hat{d}_{m_2}, \hat{d}^\dagger_{n_2}\} = \delta_{m_2 n_2} \,. \tag{16.4}$$

All other anti-commutators are zero. The diagonal elements of $t^{(1)}$ and $t^{(2)}$ give the electron and hole energies, while off-diagonal elements represent carrier hopping in the conduction and valence bands, respectively. The Coulomb interaction is given by

$$\begin{aligned} H_C = &\frac{1}{2}\sum_{m_1,n_1} V^{\text{ee}}_{m_1,n_1} \hat{c}^\dagger_{m_1} \hat{c}^\dagger_{n_1} \hat{c}_{n_1} \hat{c}_{m_1} + \frac{1}{2}\sum_{m_2,n_2} V^{\text{hh}}_{m_2 n_2} \hat{d}^\dagger_{m_2} \hat{d}^\dagger_{n_2} \hat{d}_{n_2} \hat{d}_{m_2} \\ &- \sum_{m_1,n_2} V^{\text{eh}}_{m_1 n_2} \hat{c}^\dagger_{m_1} \hat{d}^\dagger_{n_2} \hat{d}_{n_2} \hat{c}_{m_1} \\ &+ \sum_{\substack{m_1,m_2,n_1,n_2 \\ R_{m_1}=R_{m_2}\neq R_{n_1}=R_{n_2}}} V^F_{m_1 m_2, n_1 n_2} \hat{c}^\dagger_{m_1} \hat{d}^\dagger_{m_2} \hat{d}_{n_2} \hat{c}_{n_1} \,. \end{aligned} \tag{16.5}$$

The first three terms are monopole–monopole contributions of electron–electron, hole–hole and electron–hole interactions, respectively. Pseudo-potential calculations, including strain and realistic band structure, have been performed for this system. The on-site energies, hopping parameters and Coulomb interaction energies have been reported as a function of the inter-dot distance. Electrons that tunnel at short ($\lesssim$ 8 nm) distances create delocalized bonding and anti-bonding one-particle states. However, the heavy holes remain localized, even at shorter distance ($\lesssim$ 8 nm), and their energies are lowered with decreasing distance. This is due to the high inter-dot barrier for the heavy holes that suppresses hole tunneling (the heavy hole/electron effective mass ratio is $m_{\text{hh}}/m_e \approx 0.4/0.06 \approx 6$) as well as to the effect of strain and band structure. We will use the parameters from, as listed in Table 16.1, and assume $V^{\text{ee}} = V^{\text{hh}} = -V^{\text{eh}}$. The last term in Eq. (16.5) describes the electrostatic dipole–dipole interactions between the charge distributions in the QDs which induce for exciton hopping

$$\begin{aligned} V^F_{m_1 m_2, n_1 n_2} = &\frac{1}{4\pi\epsilon_0\epsilon_b r^3_{mn}} \left[\boldsymbol{\mu}_{m_1,m_2} \cdot \boldsymbol{\mu}_{n_1,n_2} \right. \\ &\left. - \frac{3(\boldsymbol{\mu}_{m_1,m_2} \cdot \boldsymbol{r}_{mn})(\boldsymbol{\mu}_{n_1,n_2} \cdot \boldsymbol{r}_{mn})}{r^2_{mn}} \right], \end{aligned} \tag{16.6}$$

where ϵ_b is the dielectric constant of the host material, $\boldsymbol{\mu}_{m_1,m_2}$ is the inter-band dipole moment at site $R_{m_1} = R_{m_2} \equiv R_m$ and $r_{mn} = |R_m - R_n|$ is the distance between sites m and n. Equation (16.6) has been shown to be satisfactory for direct gap semiconductor quantum dots of radius 0.5–2.0 nm even when they are almost in contact.

The interaction with the optical field is described by the Hamiltonian in the rotating wave approximation

$$H_{\text{int}}(t) = -\left[\boldsymbol{E}(t) \cdot \hat{\boldsymbol{V}}^\dagger + \text{h.c.}\right], \tag{16.7}$$

Table 16.1 Parameters for a system of two vertically stacked quantum dots labeled as "T" (top) and "B" (bottom). Electron and hole on-site energies (E^e and E^h), tunneling couplings (t^e and t^h) as well as e–h Coulomb interaction elements (V^{eh}) vs. inter-dot distance d.

Parameter (meV)	Distance dependence d (nm)
E_T^e	$1450 - 436d^{-1} + 3586d^{-2} - 7382d^{-3}$
E_B^e	$1449 - 452d^{-1} + 3580d^{-2} - 6473d^{-3}$
E_T^h	$167 + 129d^{-1} - 2281d^{-2} + 6582d^{-3}$
E_B^h	$163 + 274d^{-1} - 3780d^{-2} + 9985d^{-3}$
t_e	$-255 \exp(-d/2.15)$
t_h	$-4.25 \exp(-d/3.64)$
V_{BB}^{eh}	$-29.0 + 7.98/d$
V_{TT}^{eh}	$-29.6 + 19.6/d$
V_{BT}^{eh}	$-99.1/\sqrt{d^2 + 3.72^2}$
V_{TB}^{eh}	$-98.5/\sqrt{d^2 + 4.21^2}$

where $\hat{V} = \sum_{m_1,m_2} \boldsymbol{\mu}_{m_1,m_2} \hat{d}_{m_2} \hat{c}_{m_1}$ is the dipole moment annihilation operator and $\boldsymbol{E}(t)$ is the negative frequency part of the optical field. The interband dipole moment $\boldsymbol{\mu}_{m_1,m_2}$ at the site $R_{m_1} = R_{m_2}$ is given by

$$\boldsymbol{\mu}_{\uparrow,\uparrow} = \frac{\mu}{\sqrt{2}}(\hat{x} - i\hat{y}) , \qquad \boldsymbol{\mu}_{\downarrow,\downarrow} = \frac{\mu}{\sqrt{2}}(\hat{x} + i\hat{y}) , \tag{16.8}$$

where $\hat{x}$ and $\hat{y}$ are the polarization directions of the optical pulses.

The total Hamiltonian for the QD molecule-light system is

$$H = H_{QD} + H_{int}(t) . \tag{16.9}$$

The Hamiltonian in Eq. (16.9) can be transformed into the excitonic representation by introducing the electron–hole pair operators

$$\hat{B}^{\dagger}_{m_1,m_2} = \hat{c}^{\dagger}_{m_1} \hat{d}^{\dagger}_{m_2} , \quad \hat{B}_{m_1,m_2} = \hat{d}_{m_2}, \hat{c}_{m_1} . \tag{16.10}$$

The main steps of this transformation are given in Appendix 16.A. The transformed Hamiltonian will be used in the following sections to study the one and two exciton properties.

16.2 Single-exciton Manifold and the Absorption Spectrum

Fourier calculated the absorption spectrum by transforming the linear response obtained by Eq. (16.B.7) and it is given by the energy loss term

$$\propto \sum_e \mathrm{Im}\left[\frac{|\mu_{eg}|^2}{\hbar\omega - E_e + i\hbar\gamma_{eg}}\right], \tag{16.11}$$

where the subscript "e" denotes the single exciton excited state with energy E_e and γ_{eg} is the dephasing rate. We will use a dephasing rate of $\gamma = 0.05$ meV obtained from the measured line-widths in a similar system. The single exciton block of the Hamiltonian has four fourfold spin-degenerate eigenstates. Spin does not affect the absorption, thus we neglect it in this section, but include it in the next section where we discuss the nonlinear response (since two-exciton states may be formed by single excitons with different spin).

The single exciton eigenstates $|\alpha\rangle$, $|\beta\rangle$, $|\gamma\rangle$ and $|\delta\rangle$ are expanded as per the following basis (see Figure 16.2)

$$\begin{aligned} |a\rangle &= \hat{c}_B^\dagger \hat{d}_B^\dagger |g\rangle\,, \quad |c\rangle = \hat{c}_B^\dagger \hat{d}_T^\dagger |g\rangle\,, \\ |b\rangle &= \hat{c}_T^\dagger \hat{d}_T^\dagger |g\rangle\,, \quad |d\rangle = \hat{c}_T^\dagger \hat{d}_B^\dagger |g\rangle\,, \end{aligned} \tag{16.12}$$

where "B" ("T") denotes the bottom (top) QD, and $|g\rangle$ is the ground state. Also, $|a\rangle$ and $|b\rangle$ describe direct (intra-dot) excitons, while $|c\rangle$ and $|d\rangle$ are indirect (inter-dot) excitons.

The dependence of the calculated absorption spectra on inter-dot distance d is presented in Figure 16.3. For a large separation ($d = 17$ nm), the two QDs are uncoupled, therefore, indirect excitons are optically forbidden. Because of slight asymmetry for the QDs (see Table 16.1), there are two peaks corresponding to direct excitons $|\alpha\rangle = |a\rangle$ and $|\beta\rangle = |b\rangle$. When the distance is decreased, the absorption peaks are blue-shifted and their splitting is reduced. This is due to the decreasing hole energies as well as the electron tunneling that leads to carrier delocalization. As shown in Figure 16.3, exciton splitting is minimized at $d = 8.6$ nm, where the single exciton eigenstates are approximately given by

$$\begin{aligned} |\alpha\rangle &\approx |a\rangle + |b\rangle\,, \quad |\gamma\rangle \approx |c\rangle\,, \\ |\beta\rangle &\approx |a\rangle - |b\rangle\,, \quad |\delta\rangle \approx |d\rangle \end{aligned} \tag{16.13}$$

with energies $E_\alpha \approx E_\beta < E_\gamma < E_\delta$. The QD localized excitons for large separation d now becomes strongly entangled bonding and anti-bonding excitons. Indirect excitons also appear at the same time. These are blue-shifted since their binding energy is smaller than that of the direct excitons due to the reduced Coulomb attraction.

At shorter distances, two lower excitons anti-cross, while the contribution of the two upper ones in the absorption spectrum become stronger. The exciton eigenstates now form bonding and anti-bonding states

$$\begin{aligned} |\alpha\rangle &\approx |b\rangle + |c\rangle\,, \quad |\gamma\rangle \approx |b\rangle - |c\rangle\,, \\ |\beta\rangle &\approx |a\rangle + |d\rangle\,, \quad |\delta\rangle \approx |a\rangle - |d\rangle \end{aligned} \tag{16.14}$$

with the energies satisfying $E_\alpha < E_\beta < E_\gamma < E_\delta$. Therefore, the electron is delocalized, but the hole is not. This is attributed to the smaller effective mass of the electron as well as to the more complicated valence band structure and the effect of strain that favors hole localization.

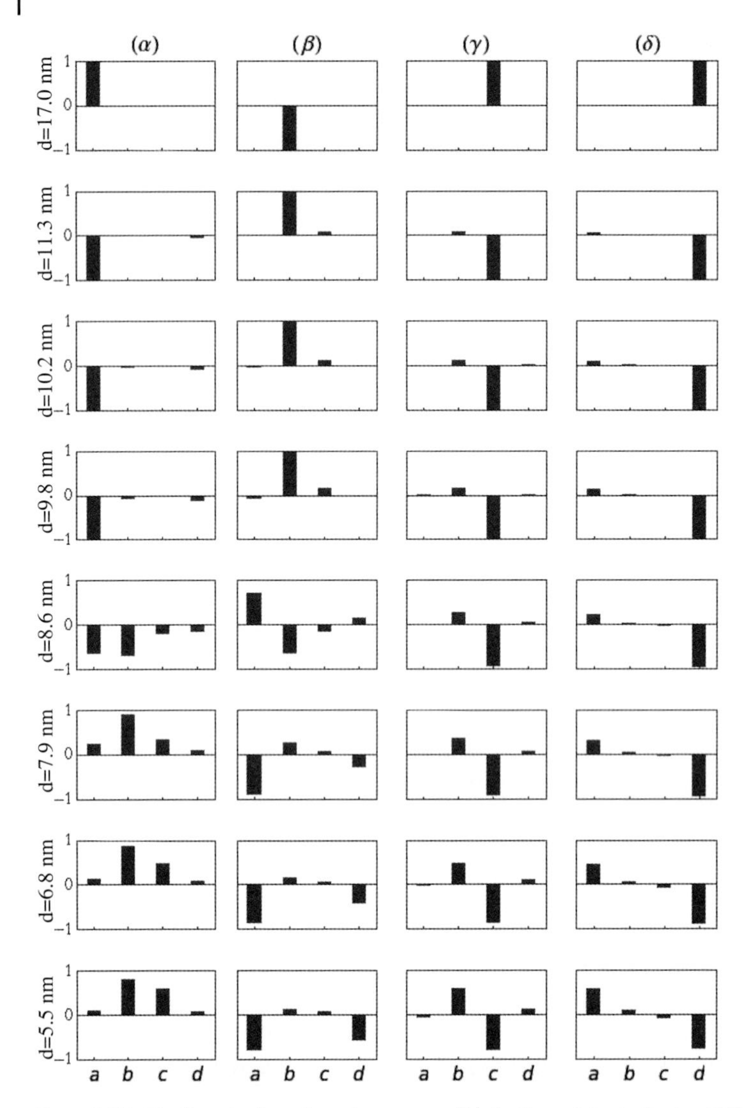

Figure 16.2 The four single exciton eigenstates $|\alpha\rangle$, $|\beta\rangle$, $|\gamma\rangle$ and $|\delta\rangle$ as a function of inter-dot distance d, expanded in the e–h pair basis $|a\rangle$ (both particles in the bottom QD), $|b\rangle$ (both particles in the top QD), $|c\rangle$ (electron in the bottom and hole in the top QD) and $|d\rangle$ (electron in the top and hole in the bottom QD).

The effect of dipole–dipole interactions for the shortest distance, $d = 5.5$ nm, is shown in Figure 16.4. The peak positions are the same but their intensity changes. The effect of dipole–dipole interactions is more pronounced in the nonlinear response and is discussed in the next section.

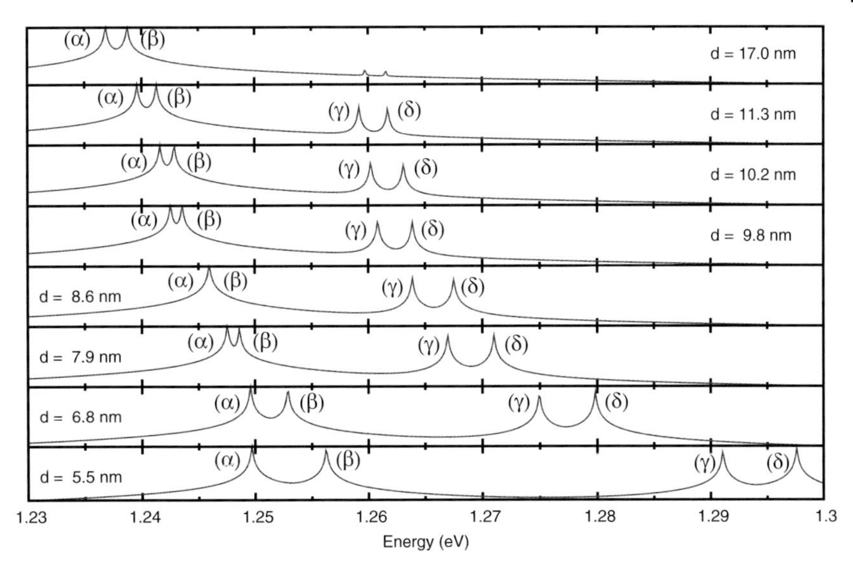

Figure 16.3 Linear absorption spectra (log scale) for various inter-dot distances d.

16.3
Two-exciton Manifold and the 2D Spectra

We now consider two-exciton states and their energies E_f. The optical response is calculated with the use of the nonlinear exciton equations (NEE) for the single and two-particle variables $\langle \hat{B}_m \rangle$ and $\langle \hat{B}_m \hat{B}_n \rangle$ (see Appendix 16.B). The 2D spectra in Eqs. (16.C.2) and (16.C.9) are calculated in terms of the single-exciton Green's functions and the exciton scattering matrix (see Appendix 16.C).

Simulations were carried out using the spectron package. However, it is more convenient to analyze them using the alternative sum-over-states expressions

$$S_{\mathrm{I}}^{\nu_4\nu_3\nu_2\nu_1}(\Omega_3, t_2 = 0, \Omega_1) = i \sum_{e,e'} \frac{\mu_{ge'}^{\nu_1}}{\hbar\Omega_1 + E_{e'} + i\hbar\gamma_{e'e}} \times \left[\sum_f \frac{\mu_{eg}^{\nu_2}\mu_{fe}^{\nu_3}\mu_{e'f}^{\nu_4}}{\hbar\Omega_3 - E_f + E_{e'} + i\hbar\gamma_{fe'}} - \frac{\left(\mu_{eg}^{\nu_2}\mu_{e'g}^{\nu_3} + \mu_{e'g}^{\nu_2}\mu_{eg}^{\nu_3}\right)\mu_{ge}^{\nu_4}}{\hbar\Omega_3 - E_e + i\hbar\gamma_{eg}} \right], \tag{16.15}$$

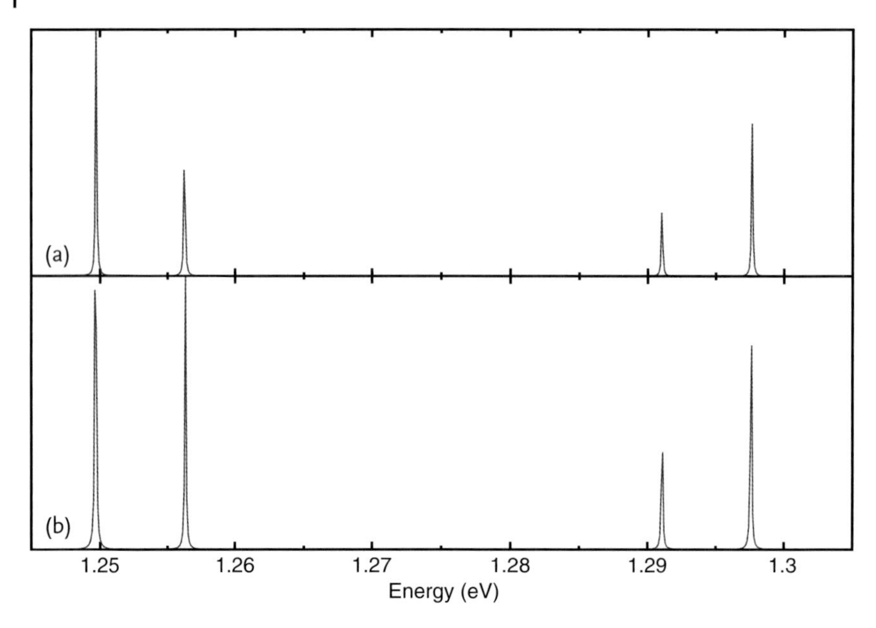

Figure 16.4 Absorption spectrum with (a) dipole–dipole interactions and (b) without dipole–dipole interactions for the shortest inter-dot distance $d = 5.5$ nm.

$$S_{\mathrm{III}}^{\nu_4\nu_3\nu_2\nu_1}(\Omega_3, \Omega_2, t_1 = 0) = i \sum_{e,e',f} \frac{\mu_{eg}^{\nu_1}\mu_{fe}^{\nu_2}}{\hbar\Omega_2 - E_f + i\hbar\gamma_{fg}} \times \left[\frac{\mu_{ge'}^{\nu_4}\mu_{e'f}^{\nu_3}}{\hbar\Omega_3 - E_f + E_{e'} + i\hbar\gamma_{fe'}} - \frac{\mu_{ge'}^{\nu_3}\mu_{e'f}^{\nu_4}}{\hbar\Omega_3 - E_{e'} + i\hbar\gamma_{e'g}} \right], \qquad (16.16)$$

where g denotes the ground state, e and e' are single excitons and f is a doubly excited state. Also, γ_{eg}, γ_{fe} and γ_{fg} are the corresponding dephasing rates.

The k_{I} signal has resonances at single exciton energies on the Ω_1-axis, at $\Omega_1 = -E_{e'}$ and two types of resonances along Ω_3: at $\Omega_3 = E_e$ and $\Omega_3 = E_f - E_{e'}$. Therefore, the diagonal $\Omega_3 = -\Omega_1 = E_e$ shows single exciton peaks that are similar to linear absorption, while the off-diagonal cross peaks reveal exciton coherence and biexciton contributions. To see what excitons contribute to the formation of each biexciton, we examine the double-sided Feynman diagrams that represent the sequences of interactions with the optical fields and the state of the excitonic density matrix during the intervals between interactions. For the k_{I} technique, there are three diagrams shown in Figure 16.5: ground state bleaching (a), stimulated emission (b) and excited state absorption (c). The two-exciton states f formed by excitons e and e' only show up in Figure 16.5c. Thus, a cross peak at $\Omega_1 = -E_{e'}$ and $\Omega_3 = E_f - E_{e'}$ indicates that exciton e' contributes to that biexciton.

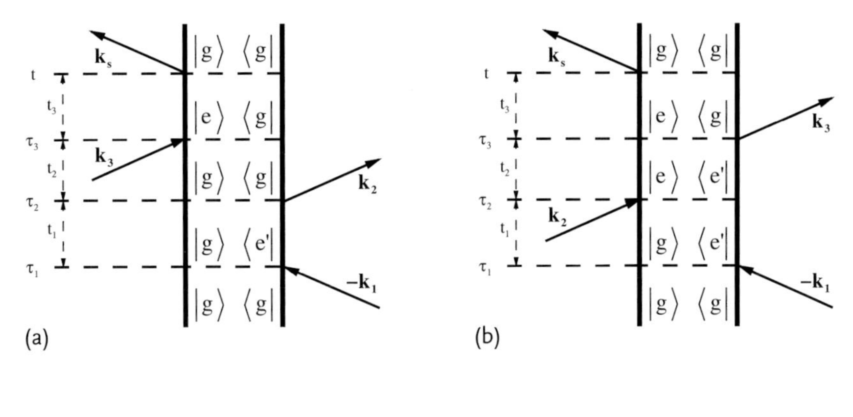

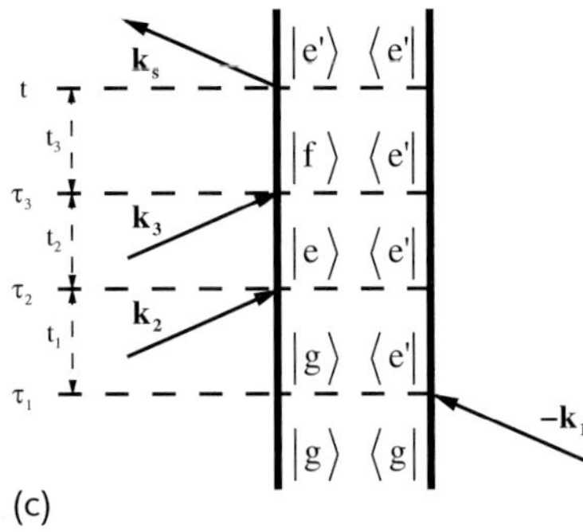

Figure 16.5 Double-sided Feynman diagrams for the k_{I}-technique: ground state bleaching (a), stimulated emission (b), and excited state absorption (c).

The k_{III} spectrum provides complementary information. As in k_{I}, the resonances along Ω_3 are at single exciton energies $\Omega_3 = E_{e'}$ and at $\Omega_3 = E_f - E_{e'}$. Along Ω_2, though, the signal directly reveals two-exciton energies $\Omega_2 = E_f$. There are two corresponding Feynman diagrams shown in Figure 16.6 that both describe excited state absorption. The two-exciton state is formed by excitons e and e', and results in cross peaks at $\Omega_2 = E_f$ and $\Omega_3 = E_{e'}$ Figure 16.6a or $\Omega_2 = E_f$ and $\Omega_3 = E_f - E_{e'}$ Figure 16.6b. We thus obtain information about the contributing single exciton states.

The variation of the k_{I} signals with d is displayed in Figure 16.7 and 16.8, and shows the k_{III} signals. At $d = 17\,\mathrm{nm}$, direct excitons $|\alpha\rangle = |a\rangle$ and $|\beta\rangle = |b\rangle$ result in two diagonal peaks at E_α and E_β in the k_{I} spectrum. Biexcitons can be formed either from excitons within the same QD or from excitons in different QDs. Because of the slight asymmetry between the dots, there are two same-dot biexcitons (at the top or bottom QD) that create two closely spaced peaks, labeled A, in the k_{III} spectrum (see Figure 16.8a) at $E_{f_1} = 2360\,\mathrm{meV}$ and $E_{f_2} = 2363\,\mathrm{meV}$. In k_{I} (see Figure 16.7a), they create two cross peaks at $\Omega_3 = E_{f_1} - E_\alpha$ and $E_{f_2} - E_\beta$. The absence of cross peaks at $\Omega_3 = E_{f_1} - E_\beta$ and $E_{f_2} - E_\alpha$ shows that biexciton f_1 (f_2) is formed by two $|a\rangle$-type ($|b\rangle$-type) single excitons localized in the bottom (top)

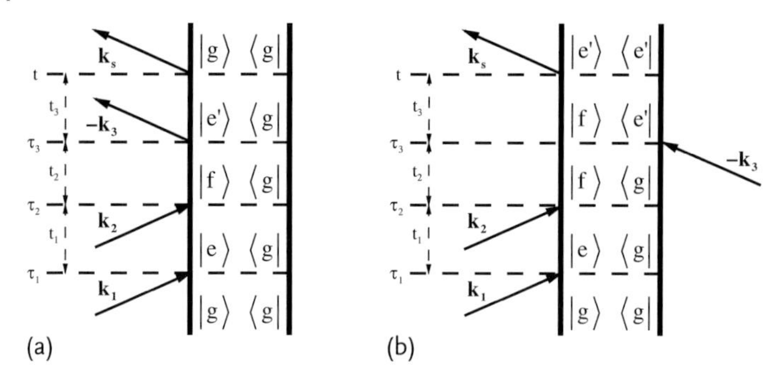

Figure 16.6 Double-sided Feynman diagrams for the k_{III}-technique with (a) cross peaks at $\Omega_2 = E_f$ and $\Omega_3 = E_{e'}$, or (b) $\Omega_2 = E_f$ and $\Omega_3 = E_f - E_{e'}$.

QD. The third biexciton, labeled B, has energy $E_{f_3} = 2453$ meV and is higher than the other two. As shown in the k_{I} spectrum, it creates peaks at both $\Omega_3 = E_{f_3} - E_\alpha$ and $E_{f_3} - E_\beta$, indicating that it is formed by one exciton on each QD ($|\alpha\rangle$ and $|\beta\rangle$).

When $d = 10.2$ nm, the indirect excitons $|\gamma\rangle = |c\rangle$ and $|\delta\rangle = |d\rangle$ show themselves up on the diagonal (Figures 16.7b and 16.8b). In the k_{III} spectrum, there are three bound biexcitons, labeled A and B, while the remaining cross peaks are due to unbound two-exciton states at energies $2E_\alpha$, $E_\alpha + E_\beta$ and $2E_\beta$. The first two biexcitons (A) are formed from excitons within the same QD, as for $d = 17$ nm. The third biexciton (B) is formed primarily from the exciton $|\beta\rangle = |b\rangle$ and a small contribution from $|\gamma\rangle = |c\rangle$, which is an indication of electron delocalization.

The 2D spectra at the critical distance of $d = 8.6$ nm, where the two lower excitons become degenerate, are shown in Figures 16.7c and 16.8c. The k_{III} spectrum has four biexciton peaks at energies $E_{f_1} = 2381$ meV and $E_{f_2} = 2383$ meV (labeled A), $E_{f_3} = 2457$ meV (B), and $E_{f_4} = 2523$ meV (C). The remaining peaks, at energies $E_e + E_{e'}$ $(e, e' = \alpha, \beta, \gamma, \delta)$, are due to unbound two-exciton states. The k_{I} spectrum shows that biexcitons A and B are mostly formed by the first two excitons which are now delocalized and may not be attributed to a single QD. The excitons are represented by bonding and anti-bonding orbitals $|a\rangle \pm |b\rangle$. The last cross peak C consists mostly of the two higher single exciton states, the indirect excitons $|c\rangle$ and $|d\rangle$.

At shorter distances ($d = 6.8$ nm and $d = 5.5$ nm), the electrons become delocalized and the spectra get even richer, as shown in Figures 16.7d,e and 16.8d,e. Four sets of biexcitonic peaks now appear in k_{III}. When $d = 5.5$ nm, we observe eight biexcitonic peaks at $E_{f_{1,2}} = 2407, 2411$ meV (A), $E_{f_{3,4,5}} = 2424, 2440, 2449$ meV (D), $E_{f_6} = 2490$ meV (B) and $E_{f_{7,8}} = 2560, 2574$ meV (C). Looking at region (I) of the k_{I} spectrum (see Figure 16.7b), we see that biexcitons f_1, f_2 of group A and f_3 of group D are formed mainly from the bonding orbitals $|\alpha\rangle = |b\rangle + |c\rangle$ and $|\beta\rangle = |a\rangle + |d\rangle$. Similarly, regions (I) and (II) show that f_4 and f_5 of group D forms mostly from the anti-bonding ones $|\gamma\rangle = |b\rangle - |c\rangle$ and $|\delta\rangle = |a\rangle - |d\rangle$. Region (II) also suggests that all single exciton states significantly contribute to the formation of biexcitons B and C.

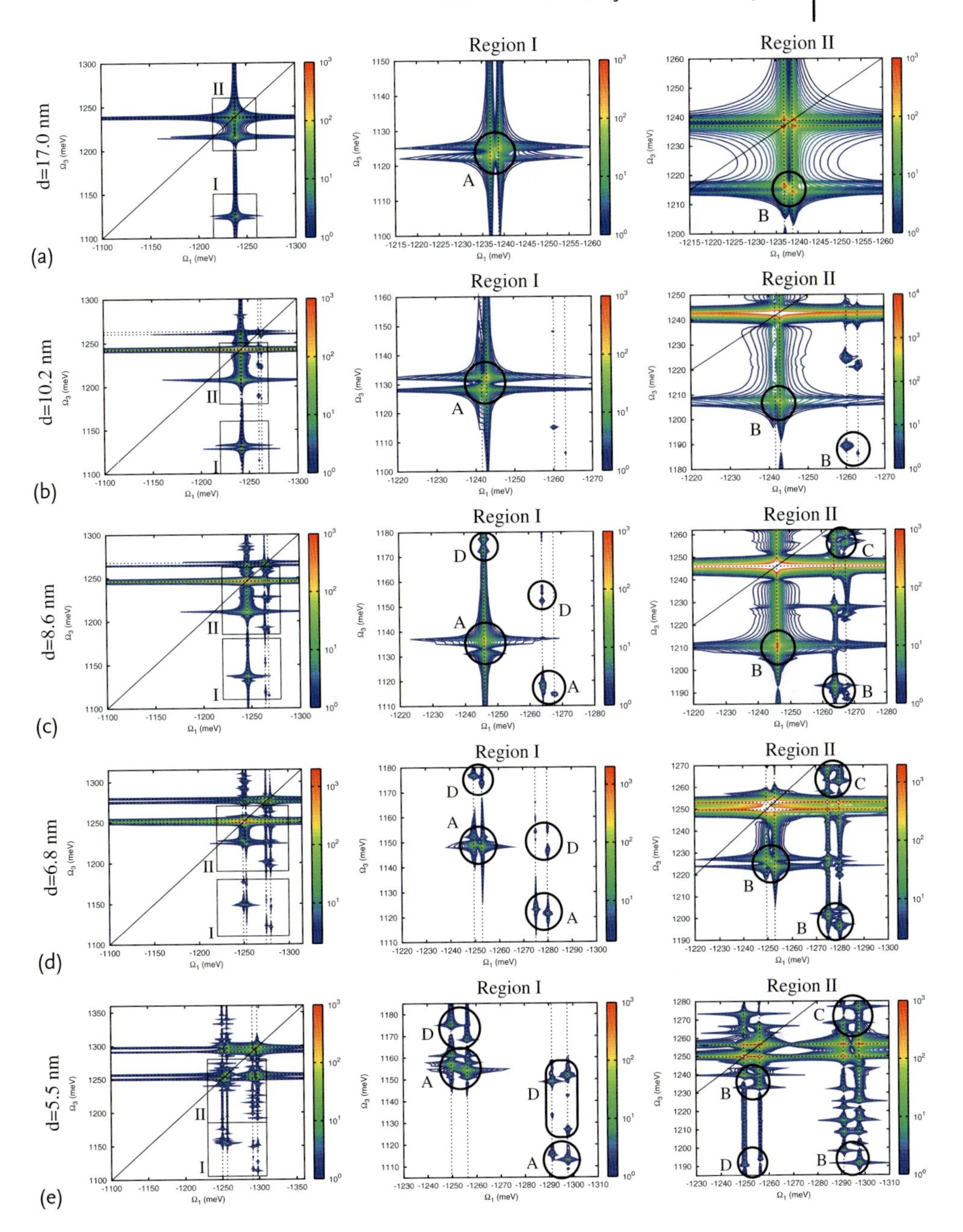

Figure 16.7 Absolute value of the k_{I} signal for $xxxx$ polarization at several inter-dot distances d. Regions designated I and II are magnified in the center and right columns, respectively. Single exciton energies are marked with dashed lines and biexciton peaks are circled. Rows (a), (b), (c), (d), and (e) correspond to $d = 17.0$, 10.2, 8.6, 6.8 and 5.5 nm, respectively.

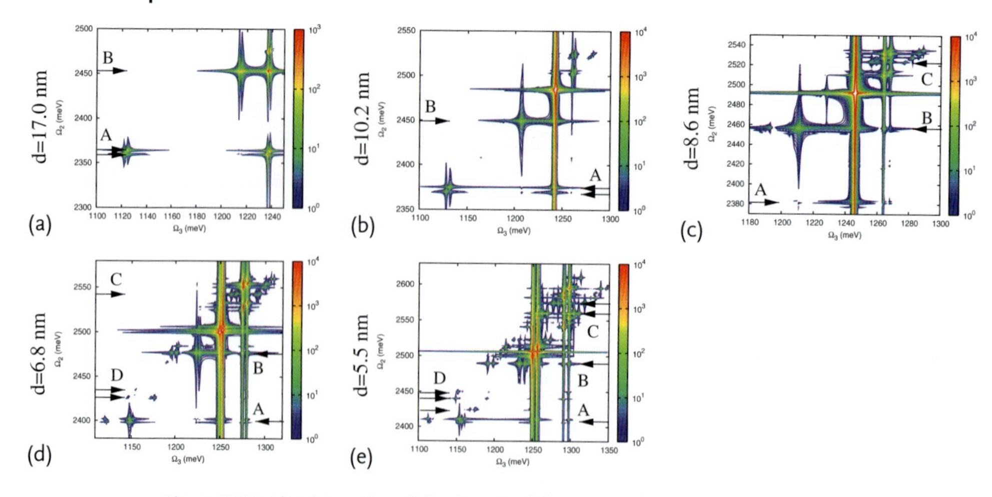

Figure 16.8 Absolute value of the k_{III} signal for *xxxx* polarization at various inter-dot distances *d*. Arrows mark biexciton contributions. Labels (a), (b), (c), (d) and (e) correspond to $d =$ 17.0, 10.2, 8.6, 6.8 and 5.5 nm, respectively.

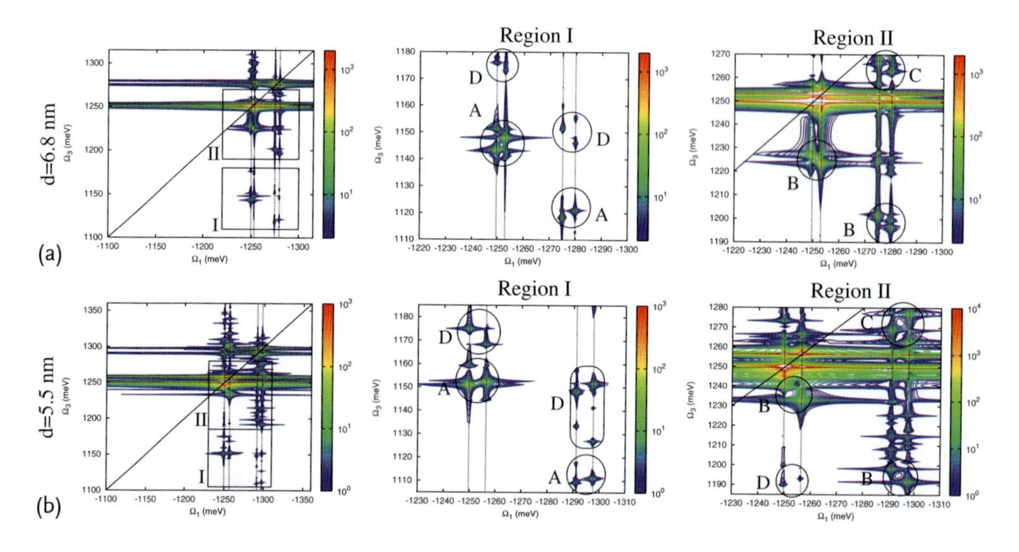

Figure 16.9 Absolute value of the k_{I} signal for *xxxx* polarization without dipole–dipole interactions (compare Figure 16.7d,e). Rows (a) and (b) correspond to $d = 6.8$ and 5.5 nm, respectively.

In Figure 16.9, we display the k_{I} spectra calculated by neglecting dipole–dipole interactions at short distances, $d = 6.8$ and 5.5 nm (compare this with Figure 16.7d,e). The lowest excitonic peak on the diagonal becomes stronger. Biexcitons A, formed by bonding orbitals, red shifts while the remaining biexcitonic peaks remain at the same locations.

16.4 Summary

The single and double excitons in two coupled quantum dots and their dependence on inter-dot distance were considered in this chapter. Both carrier tunneling and exciton migration via dipole–dipole interactions were included. The absorption spectra were used to classify the single exciton states in terms of localized e–h pairs and the 2D spectra in directions $k_{\text{I}} = -k_1 + k_2 + k_3$ and $k_{\text{III}} = k_1 + k_2 - k_3$ were calculated by means of nonlinear exciton equations. Analysis of these spectra using the corresponding sum-over-states expressions allows the classification of biexcitons according to their single exciton components. At large distances, only direct excitons are active, and two types of biexcitons are formed, either in the same or different QDs. At shorter distances, we also obtained biexcitons created from indirect excitons. At distances where electron inter-dot tunneling takes place, additional biexcitonic peaks appear. These peaks are thus a signature of electron delocalization. The exciton hopping becomes significant only at short distances where it affects the intensity of the excitonic peaks and shifts the biexciton energies.

Appendix 16.A Transformation of the Electron–Hole Hamiltonian Using Excitonic Variables

Introducing the electron–hole pair operators in Eq. (16.10), Chernyak and Mukamel expressed the Hamiltonian in Eq. (16.9), as well as the commutation relations of these operators, in terms of an infinite series of normally ordered operators $\hat{B}^\dagger$ and $\hat{B}$. Since the Hamiltonian conserves the number of particles, each term contains an equal number of creation ($\hat{B}^\dagger$) and annihilation ($\hat{B}$) operators. For calculating the third-order response, the Hamiltonian can be truncated at fourth order and yields

$$\begin{aligned} H = & \sum_{m,n} h_{m,n} \hat{B}^\dagger_m \hat{B}_n + \sum_{mn,kj} U_{mn,kl} \hat{B}^\dagger_m \hat{B}^\dagger_n \hat{B}_k \hat{B}_j \\ & - \sum_m \left[\boldsymbol{\mu}^*_m \cdot \boldsymbol{E}(t) \hat{B}_m + \text{h.c.} \right], \end{aligned} \tag{16.A.1}$$

where electron–hole pairs are denoted with composite indices $m = (m_1, m_2)$ and $n = (n_1, n_2)$. The quantities $h_{m,n}$ and $U_{mn,kl}$ of the effective Hamiltonian can be determined by successively comparing the matrix elements of Hamiltonians in Eqs. (16.9) and (16.A.1) order by order in the space of one, two, three, and so on electron–hole pair excitations. This is possible because normally ordered N creation and N annihilation operators do not contribute in the subspaces of $N - 1$ and smaller number of excitations. Thus, in the one electron–hole pair excitation subspace, we obtain

$$\begin{aligned} h_{m,n} = & t^{(1)}_{m_1,n_1} \delta_{m_2,n_2} + t^{(2)}_{m_2 n_2} \delta_{m_1,n_1} - V^{\text{eh}}_{m_1,m_2} \delta_{m_1,n_1} \delta_{m_2,n_2} \\ & + V^F_{m_1 m_2, n_1 n_2} \delta_{R_{m_1},R_{m_2}} \delta_{R_{n_1},R_{n_2}} \left(1 - \delta_{R_{m_1},R_{n_1}}\right). \end{aligned} \tag{16.A.2}$$

Here, diagonal elements ($m = n$) describe the electron–hole pair energy given as the sum of electron and hole kinetic energies reduced by the electron–hole Coulomb attraction. Off-diagonal elements ($m \neq n$) describe electron, hole or exciton hopping between adjacent sites.

Similarly, to describe the one and two-electron–hole pair subspace, we need the quadratic term

$$U_{mn,kj} = \bar{U}_{mn,kj} + F_{mn,kj} , \tag{16.A.3}$$

where

$$\begin{aligned}\bar{U}_{mn,kj} &= \frac{1}{4}\Big[V^{\text{ee}}_{m_1,n_1}\delta_{m_1 j_1}\delta_{n_1,k_1}\delta_{m_2,k_2}\delta_{n_2,j_2} \\ &+ V^{\text{hh}}_{m_2,n_2}\delta_{m_1,k_1}\delta_{n_1,j_1}\delta_{m_2,j_2}\delta_{n_2,k_2}\Big] \\ &- \frac{1}{4}\Big[t^{(1)}_{m_1,k_1}\delta_{m_2 k_2}\delta_{n_1,j_1}\delta_{n_2,j_2} + t^{(2)}_{m_2,k_2}\delta_{m_1,k_1}\delta_{n_1,j_1}\delta_{n_2,j_2} \\ &+ t^{(1)}_{n_1,j_1}\delta_{m_1,k_1}\delta_{m_2,k_2}\delta_{n_2,j_2} + t^{(2)}_{n_2,j_2}\delta_{m_1,k_1}\delta_{m_2,k_2}\delta_{n_1,j_1}\Big] .\end{aligned} \tag{16.A.4}$$

We further define the matrix F by the equation

$$F_{mn,kj} + F_{mn,jk} - 2\sum_{p,q} P_{mn,pq}F_{pq,kj} = 0 , \tag{16.A.5}$$

where P is tetradic matrix defined as

$$P_{mn,pq} = \frac{1}{2}\delta_{m_1,q_1}\delta_{m_2,p_2}\delta_{n_1,p_1}\delta_{n_2,q_2} + \frac{1}{2}\delta_{m_1,p_1}\delta_{m_2,q_2}\delta_{n_1,q_1}\delta_{n_2,p_2} . \tag{16.A.6}$$

The U matrix is invariant to the addition of any matrix F that satisfies Eq. (16.A.5) and thus, is not uniquely defined. This freedom arises since Hamiltonians in Eqs. (16.9) and (16.A.1) are only required to coincide in our physically relevant subspace of one and two e–h pair excitations, but can differ in higher manifolds. Similarly, the commutation relation of the electron–hole particle operators is expanded in terms of a series of normally ordered operators $\hat{B}^\dagger$ and $\hat{B}$. For the third-order response, this can be truncated at quadratic order, that is,

$$\left[\hat{B}_m, \hat{B}^\dagger_n\right] = \delta_{m,n} - 2\sum_{p,q} P_{mp,nq}\hat{B}^\dagger_p\hat{B}_q , \tag{16.A.7}$$

where $\delta_{m,n} = \delta_{m_1,n_1}\delta_{m_2,n_2}$. The P matrix is responsible for the deviation from boson statistics.

Appendix 16.B
The Nonlinear Exciton Equations

We calculate the nonlinear optical using the Heisenberg equation of motion for the electron–hole operator $\hat{B}_m$ that can be obtained from Eqs. (16.A.1) and (16.A.7):

$$i\hbar\frac{d}{dt}\left\langle\hat{B}_m\right\rangle = \sum_n h_{m,n}\langle\hat{B}_n\rangle + \sum_{n,k,j} V_{mn,kj}\left\langle\hat{B}_n^\dagger\hat{B}_k\hat{B}_j\right\rangle - \boldsymbol{\mu}_m^*\cdot\boldsymbol{E}(t) + 2\sum_{n,k,j}\boldsymbol{\mu}_n^*\cdot\boldsymbol{E}(t)P_{mk,nj}\left\langle\hat{B}_k^\dagger\hat{B}_j\right\rangle . \tag{16.B.1}$$

The last term on the right-hand side is known in the context of the simpler semiconductor Bloch equations as phase-space filling. The second term describes the exciton-exciton interactions where V is given by

$$\begin{aligned} V_{mn,,kj} &= 2U_{mn,kj} - 2\sum_p P_{mn,pk}h_{p,j} - 2\sum_{p,q} P_{mn,pq}U_{pq,kj} \\ &= 2\bar{U}_{mn,kj} - 2\sum_p P_{mn,pk}h_{p,j} - 2\sum_{p,q} P_{mn,pq}\bar{U}_{pq,kj} \,. \end{aligned} \tag{16.B.2}$$

Note that it is independent of the matrix F defined in Eq. (16.A.5). In a similar fashion, for the two-particle variable $\langle\hat{B}_m\hat{B}_n\rangle$, we obtain to the second order in the optical field

$$i\hbar\frac{d}{dt}\left\langle\hat{B}_m\hat{B}_n\right\rangle = \sum_{k,j} h^{(2)}_{mn,kj}\left\langle\hat{B}_k\hat{B}_j\right\rangle - \left[\boldsymbol{\mu}_m^*\cdot\boldsymbol{E}(t)\left\langle\hat{B}_n\right\rangle + \left\langle\hat{B}_m\right\rangle\boldsymbol{\mu}_n^*\cdot\boldsymbol{E}(t)\right] + 2\sum_{k,j}\boldsymbol{\mu}_j^*\cdot\boldsymbol{E}(t)P_{mn,kj}\left\langle\hat{B}_k\right\rangle , \tag{16.B.3}$$

where

$$h^{(2)}_{mn,kj} = h_{n,j}\delta_{m,k} + h_{m,k}\delta_{n,j} + V_{mn,kj} \,. \tag{16.B.4}$$

The diagonal elements of $h^{(2)}$ represent two electron–hole pair energies, while $V_{mn,mn}$ describes the biexciton binding energy. To see this, we employ the definition of the electron–hole pair operators in Eq. (16.10) and the anti-commutation relations in Eqs. (16.3) and (16.4), in conjunction with Eqs. (16.A.2)–(16.A.6) to rewrite the V matrix in the form

$$\begin{aligned} V_{mn,kj} &= V^{\text{ee}}_{m_1,n_1}\delta_{m_1,j_1}\delta_{n_1k_1}\delta_{m_2,k_2}\delta_{n_2,j_2} \\ &+ V^{\text{hh}}_{m_2n_2}\delta_{m_1,k_1}\delta_{n_1,j_1}\delta_{m_2,j_2}\delta_{n_2,k_2} \\ &+ \frac{1}{2}\left(V^{\text{F}}_{m_1n_2,k_1j_2}\delta_{n_1,j_1}\delta_{m_2,k_2} + V^{\text{F}}_{m_1n_2,j_1k_2}\delta_{n_1,k_1},\delta_{m_2,j_2}\right) \\ &+ \frac{1}{2}\left(V^{\text{F}}_{n_1m_2l_1k_2}\delta_{m_1k_1}\delta_{n_2l_2} + V^{\text{F}}_{n_1m_2,k_1j_2}\delta_{m_1,j_1}\delta_{n_2,k_2}\right) \\ &- \frac{1}{2}\left(V^{\text{eh}}_{m_1,n_2} + V^{\text{eh}}_{n_1,m_2}\right)\delta_{m_1,k_1}\delta_{n_1,j_1}\delta_{m_2,k_2}\delta_{n_2,j_2} \\ &- \frac{1}{2}\left(V^{\text{eh}}_{m_1,n_2} + V^{\text{eh}}_{n_1,m_2}\right)\delta_{m_1,j_1}\delta_{n_1,k_1}\delta_{m_2,j_2}\delta_{n_2,k_2} \,. \end{aligned} \tag{16.B.5}$$

This expression clearly shows that V describes Coulomb interactions between the particles that constitute the two pairs.

Neglecting incoherent exciton transport, we can make the factorization

$$\left\langle \hat{B}_k^\dagger \hat{B}_j \right\rangle = \left\langle \hat{B}_k^\dagger \right\rangle \left\langle \hat{B}_j \right\rangle , \quad \text{and} \quad \left\langle \hat{B}_n^\dagger \hat{B}_k \hat{B}_j \right\rangle = \left\langle \hat{B}_n^\dagger \right\rangle \left\langle \hat{B}_k \hat{B}_j \right\rangle . \tag{16.B.6}$$

Expanding the equations of motion in powers of the optical field $E(t)$ and defining $B_m = \langle \hat{B}_m \rangle$, $Y_{m,n} = \langle \hat{B}_m \hat{B}_n \rangle$, we finally obtain the nonlinear exciton equations (NEE)

$$i\hbar \frac{d}{dt} B_m^{(1)} = \sum_n h_{m,n} B_n^{(1)} - \boldsymbol{\mu}_m^* \cdot \boldsymbol{E}(t) , \tag{16.B.7}$$

$$\begin{aligned} i\hbar \frac{d}{dt} Y_{m,n}^{(2)} &= \sum_n h_{mn,kj}^{(2)} Y_{k,j}^{(2)} - \left[\boldsymbol{\mu}_m^* \cdot \boldsymbol{E}(t) B_n^{(1)} + \boldsymbol{\mu}_n^* \cdot \boldsymbol{E}(t) B_m^{(1)} \right] \\ &\quad + 2 \sum_{k,j} \boldsymbol{\mu}_j^* \cdot \boldsymbol{E}(t) P_{mn,kj} B_k^{(1)} , \end{aligned} \tag{16.B.8}$$

$$\begin{aligned} i\hbar \frac{d}{dt} B_m^{(3)} &= \sum_n h_{m,n} B_n^{(3)} + \sum_{n,k,j} V_{mnkj} (B_n^{(1)})^* Y_{k,j}^{(2)} \\ &\quad + 2 \sum_{n,k,j} \boldsymbol{\mu}_n^* \cdot \boldsymbol{E}(t) P_{mk,nj} (B_k^{(1)})^* B_j^{(1)} . \end{aligned} \tag{16.B.9}$$

Appendix 16.C
The 2D Signals

The NEE may be solved using the single-exciton Green's functions and the exciton scattering matrix. The polarization is then expressed in terms of the response function

$$\begin{aligned} V^{\nu_4}(t) &= \int dt_1 \int dt_2 \int dt_3 \, S^{\nu_4\nu_3\nu_2\nu_1}(t_3, t_2, t_1) E^{\nu_3}(t - t_3) \\ &\quad \times E^{\nu_2}(t - t_3 - t_2) E^{\nu_1}(t - t_3 - t_2 - t_1) . \end{aligned} \tag{16.C.1}$$

The response function in the $k_{\mathrm{I}} = -k_1 + k_2 + k_3$ direction is given by

$$\begin{aligned} & S_{\mathrm{I}}^{\nu_4\nu_3\nu_2\nu_1}(\Omega_3, t_2, \Omega_1) \\ &= 2i \sum_{e_1,e_2,e_3,e_4} \mu_{e_1}^{\nu_1} (\mu_{e_2}^{\nu_2})^* (\mu_{e_3}^{\nu_3})^* \mu_{e_4}^{\nu_4} I_{e_1}^*(t_2) I_{e_2}(t_2) I_{e_1}^*(-\Omega_1) I_{e_4}(\Omega_3) \\ &\quad \times \Gamma_{e_4,e_1,e_2,e_3}(\Omega_3 + E_{e_1} + i\gamma_{e_1}) \mathcal{G}_{e_3 e_2}^0 (\Omega_3 + E_{e_1} + i\gamma_{e_1}) , \end{aligned} \tag{16.C.2}$$

where ν_1, ν_2, ν_3 and ν_4 are the polarizations of the optical pulses, e_1, e_2, e_3 and e_4 label eigenstates of the single exciton block of the Hamiltonian with energies E_e and dephasing rates γ_e,

$$I_e(t) \equiv \langle e | \hat{G}(t) | e \rangle = -i\theta(t) e^{-iE_e t - \gamma_e t} , \tag{16.C.3}$$

$$I_e(\omega) \equiv \langle e|\hat{G}(\omega)|e\rangle = (\hbar\omega - E_e + i\hbar\gamma_e)^{-1}\,, \tag{16.C.4}$$

and $G(t)$ is the single exciton Green's function. The Fourier transform of $G(t)$ and its inverse one are defined as

$$G(\omega) = \int dt e^{i\omega t} G(t)\,, \tag{16.C.5}$$

$$G(t) = \int \frac{d\omega}{2\pi} e^{-i\omega t} G(\omega)\,. \tag{16.C.6}$$

Finally, we have

$$\begin{aligned}\mathcal{G}^0_{e_2,e_1}(\omega) &\equiv \langle e_1 e_2|\hat{\mathcal{G}}^0(\omega)|e_1 e_2\rangle \\ &= \frac{1}{\hbar\omega - E_{e_2} - E_{e_1} + i\hbar(\gamma_{e_2} + \gamma_{e_1})}\,,\end{aligned} \tag{16.C.7}$$

which is the Green's function that represents non-interacting two-excitons and the scattering matrix is given by

$$\Gamma(\omega) = \left[\mathbb{I} - V\mathcal{G}^0(\omega)\right]^{-1} V\mathcal{G}^0(\omega)\,(\mathbb{I} - P)\,[\mathcal{G}^0(\omega)]^{-1} - P\left[\mathcal{G}^0(\omega)\right]^{-1}\,, \tag{16.C.8}$$

where V is given in Eq. (16.B.5), and $\mathbb{I}$ is the tetradic identity matrix in the two-exciton space.

We also find the response in the $k_{\text{III}} = k_1 + k_2 - k_3$ direction is given by

$$\begin{aligned}&S^{\nu_4\nu_3\nu_2\nu_1}_{\text{III}}(\Omega_3, \Omega_2, t_1) \\ &= 2\sum_{e_1,e_2,e_3,e_4} (\mu^{\nu_1}_{e_1})^*(\mu^{\nu_2}_{e_2})^*\mu^{\nu_3}_{e_3}\mu^{\nu_4}_{e_4} I^*_{e_1}(t_1)\, I_{e_4}(\Omega_3)\, I^*_{e_3}(\Omega_2 - \Omega_3) \\ &\quad\times \left[\Gamma_{e_4,e_3,e_2,e_1}(\Omega_3 + E_{e_3} + i\gamma_{e_3})\mathcal{G}^0_{e_2,e_1}(\Omega_3 + E_{e_3} + i\gamma_{e_3})\right. \\ &\quad\left. - \Gamma_{e_4,e_3,e_2,e_1}(\Omega_2)\mathcal{G}^0_{e_2,e_1}(\Omega_2)\right].\end{aligned} \tag{16.C.9}$$

References

1 Mukamel, S. (1995) *Principle of Nonlinear Optical Spectroscopy*, Oxford University Press, New York.

2 Bloembergen, N. (1996) *Nonlinear Optics*, World Scientific, Singapore.

3 Marx, C.A., Harbola, U., and Mukamel, S. (2008) Nonlinear optical spectroscopy of single, few, and many molecules: Nonequilibrium Green's function QED approach. *Phys. Rev. A*, **77**, 022110.

4 Roslyak, O. and Mukamel, S. (2009) Photon entanglement signatures in difference-frequency-generation. *Opt. Express*, **17**, 1093.

5 Roslyak, O., Marx, C.A., and Mukamel, S. (2009) Generalized Kramers-Heisenberg expressions for stimulated Raman scattering and two-photon absorption. *Phys. Rev. A*, **79**, 063827.

6 Roslyak, O., Marx, C.A., and Mukamel, S. (2009) Nonlinear spectroscopy with entangled photons: Manipulating quantum pathways of matter. *Phys. Rev. A*, **79**, 033832.

7 Roslyak, O. and Mukamel, S. (2009) Multidimensional pump-probe spectroscopy with entangled twin-photon states. *Phys. Rev. A*, **79**, 063409.

17
Non-thermal Distribution of Hot Electrons

17.1
Introduction

The use of a Boltzmann transport equation with a drift term [1] under a time-dependent electric field can be justified only within the limit of $\nu_f \tau_p \ll 1$, where ν_f is the frequency of the external field and τ_p is the momentum-relaxation time of carriers. This approach can no longer be physically justified [2–5] for an incident electromagnetic field with $\nu_f \geq 1\,\text{THz}$, where $\tau_p = 1\,\text{ps}$ is assumed. This is because electrons are expected to be spatially localized when the time period $(1/\nu_f)$ of the electromagnetic field becomes equal to or shorter than the momentum-relaxation time (τ_p) of electrons. As a result, no drift of electrons can occur under such an electromagnetic field.

When the motion of electrons is separated into center-of-mass and relative motions, the incident electromagnetic field is found to be coupled only to the center-of-mass motion in the dipole approximation but not to the relative motion of electrons [6–9]. This generates an oscillating drift velocity in the center-of mass motion, but the time-average value of this drift velocity also remains zero as described above. This oscillating drift velocity will, however, affect the electron–phonon and electron–impurity interactions. whereas the thermodynamics of electrons are determined by the relative motion of electrons [9–11]. This includes the scattering of electrons with impurities, phonons, and other electrons.

When the incident electromagnetic field is spatially uniform, electrons in bulk GaAs can not directly absorb incident photons through an intraband transition [9]. Because both impurity atoms and lattice ions do not move with the electron center-of-mass, electrons inside a drifting system feel that impurities and ions are oscillating against them due to the Galilean principle of relative motion. This leads to impurity- and phonon-assisted photon absorption in the system.

The relative scattering motion of electrons can not be fully described by a simple energy-balance equation [7, 8, 12]. The oscillating drift velocity is accounted for, but the thermal effect of pair scattering on the distribution of electrons is not included in this equation. The peak structures separated by multiples of the photon energy in the distribution of electrons is also absent in this equation. This leads to an inaccurate estimate of the effective electron temperature.

Properties of Interacting Low-Dimensional Systems, First Edition. G. Gumbs and D. Huang.

The relative scattering motion of electrons can be described very well by a Boltzmann scattering equation (Boltzmann transport equation without a drift term) [9, 10]. The effect of incident optical field is reflected in the impurity- and phonon-assisted photon absorption through modifying the scattering of electrons with impurities and phonons. This drives the distribution of electrons away from the thermal-equilibrium distribution to a non-thermal one. At the same time, the electron temperature increases with the strength of the incident electromagnetic field, creating hot electrons.

In this chapter, we will establish a Boltzmann scattering equation for an accurate description of the relative scattering motion of many electrons interacting with an intense optical field by including both the impurity- and phonon-assisted photon absorption processes as well as the Coulomb scattering between two electrons. We will study the thermodynamics of hot electrons by calculating the effective electron temperature as a function of both the amplitude of the optical field and the incident photon energy.

17.2 Boltzmann Scattering Equation

The non-thermal distribution n_k for electrons in a conduction band satisfies the Boltzmann scattering equation [9, 10]

$$\frac{d}{dt} n_k = \mathcal{W}_k^{(\mathrm{in})} (1 - n_k) - \mathcal{W}_k^{(\mathrm{out})} n_k \,, \tag{17.1}$$

where k is a wave number of electrons, n_k is the non-thermal distribution of electrons, and a Markovian process is assumed for the dynamics of electron scattering. In the above equation,

$$\mathcal{W}_k^{(\alpha)} = \mathcal{W}_k^{(\alpha)(\mathrm{im})} + \mathcal{W}_k^{(\alpha)(\mathrm{ph})} + \mathcal{W}_k^{(\alpha)(\mathrm{c})} \,, \tag{17.2}$$

where $\mathcal{W}_k^{(\alpha)}$ with $\alpha = in$ or out represents scattering-in/scattering-out rate for electrons staying in/leaving the k state. The superscripts im, and ph, c represent the impurity, phonon, and Coulomb scattering of electrons.

The electron scattering-in/scattering-out rates due to impurities are given by [9]

$$\begin{aligned} \mathcal{W}_k^{(\mathrm{in})(\mathrm{im})} &= N_\mathrm{I} \frac{2\pi}{\hbar} \sum_q \left| U^{\mathrm{im}}(q) \right|^2 \sum_{M=-\infty}^{\infty} J_{|M|}^2(\mathcal{M}_q) \\ &\quad \times \left[n_{k-q} \delta(\varepsilon_k - \varepsilon_{k-q} - M\hbar\Omega_{\mathrm{op}}) + n_{k+q} \delta(\varepsilon_k - \varepsilon_{k+q} + M\hbar\Omega_{\mathrm{op}}) \right] , \end{aligned} \tag{17.3}$$

$$\begin{aligned} \mathcal{W}_k^{(\mathrm{out})(\mathrm{im})} &= N_\mathrm{I} \frac{2\pi}{\hbar} \sum_q \left| U^{\mathrm{im}}(q) \right|^2 \sum_{M=-\infty}^{\infty} J_{|M|}^2(\mathcal{M}_q) \\ &\quad \times \left[(1 - n_{k+q}) \delta(\varepsilon_{k+q} - \varepsilon_k - M\hbar\Omega_{\mathrm{op}}) \right. \\ &\quad \left. + (1 - n_{k-q}) \delta(\varepsilon_{k-q} - \varepsilon_k + M\hbar\Omega_{\mathrm{op}}) \right] , \end{aligned} \tag{17.4}$$

where $J_M(x)$ is the mth order first-kind Bessel function, $\varepsilon_k = \hbar^2 k^2/2m^*$ is the kinetic energy of an electron in a conduction band, m^* is the effective mass of electrons, $n_\mathrm{I} = N_\mathrm{I}/\mathcal{V}$ is the impurity concentration with sample volume $\mathcal{V}$ and $\hbar\Omega_\mathrm{op}$ ($\Omega_\mathrm{op} = 2\pi\nu_\mathrm{f}$) is the incident photon energy,

$$\mathcal{M}_q = \frac{eqE_\mathrm{op}}{\sqrt{2}m^*\Omega_\mathrm{op}^2} \tag{17.5}$$

is for non-polarized incident optical field, E_op is the amplitude of the incident optical field and the impurity scattering potential is

$$U^\mathrm{im}(q) = \frac{Ze^2}{\epsilon_0\epsilon_\mathrm{r}\left(q^2 + Q_\mathrm{s}^2\right)\mathcal{V}} \,. \tag{17.6}$$

Here, ϵ_r is the average dielectric constant of GaAs, Z is the charge number of an ionized impurity atom, $Q_\mathrm{s}^2 = (e^2/\epsilon_0\epsilon_\mathrm{r})(m^*/\pi^2\hbar^2)(3\pi^2 n_\mathrm{3D})^{1/3}$ represents the static Thomas–Fermi screening effect [13] and n_3D is the concentration of conduction electrons in a bulk GaAs.

The electron scattering-in/scattering-out rates due to phonons, including phonon-assisted photon absorption, are given by [9]

$$\begin{aligned} \mathcal{W}_k^{(\mathrm{in})(ph)} &= \frac{2\pi}{\hbar}\sum_{q,\lambda}\left|C_{q\lambda}\right|^2 \sum_{M=-\infty}^{\infty} J_{|M|}^2(\mathcal{M}_q) \\ &\times \left[n_{k-q}N_{q\lambda}\delta(\varepsilon_k - \varepsilon_{k-q} - \hbar\omega_{q\lambda} - M\hbar\Omega_\mathrm{op})\right. \\ &\left. + n_{k+q}(N_{q\lambda}+1)\delta(\varepsilon_k - \varepsilon_{k+q} + \hbar\omega_{q\lambda} + M\hbar\Omega_\mathrm{op})\right], \end{aligned} \tag{17.7}$$

$$\begin{aligned} \mathcal{W}_k^{(\mathrm{out})(\mathrm{ph})} &= \frac{2\pi}{\hbar}\sum_{q,\lambda}\left|C_{q\lambda}\right|^2 \sum_{M=-\infty}^{\infty} J_{|M|}^2(\mathcal{M}_q) \\ &\times \left[(1 - n_{k+q})N_{q\lambda}\delta(\varepsilon_{k+q} - \varepsilon_k - \hbar\omega_{q\lambda} - M\hbar\Omega_\mathrm{op})\right. \\ &\left. + (1 - n_{k-q})(N_{q\lambda}+1)\delta(\varepsilon_{k-q} - \varepsilon_k + \hbar\omega_{q\lambda} + M\hbar\Omega_\mathrm{op})\right]. \end{aligned} \tag{17.8}$$

Here, $\hbar\omega_q$ represents the phonon energy. The phonons are assumed to be in thermal equilibrium with an external heat bath at a fixed temperature T. The distribution of phonons is a Bose–Einstein function given by

$$N_{q\lambda} = \frac{1}{\exp\left(\frac{\hbar\omega_{q\lambda}}{k_\mathrm{B}T}\right) - 1} \,. \tag{17.9}$$

For optical phonon scattering, we find from the Fröhlich electron–phonon coupling [14] (with $\lambda = \mathrm{LO}$)

$$\left|C_{q\mathrm{LO}}\right|^2 = \left(\frac{\hbar\omega_\mathrm{LO}}{2\mathcal{V}}\right)\left(\frac{1}{\epsilon_\infty} - \frac{1}{\epsilon_\mathrm{s}}\right)\frac{e^2}{\epsilon_0\left(q^2 + Q_\mathrm{s}^2\right)} , \tag{17.10}$$

where ω_{LO} is the frequency of dominant longitudinal-optical (LO) phonon modes at high temperatures, ϵ_∞ and ϵ_{s} are the high-frequency and static dielectric constants of GaAs. For acoustic phonon scattering, we find from the deformation-potential approximation [9] (with $\lambda = \ell, t$)

$$\left|C_{q\ell}\right|^2 = \frac{\hbar\omega_{q\ell}}{2\rho_{\mathrm{i}} c_\ell^2 \mathcal{V}} \left[D^2 + \frac{9}{32q^2}(eh_{14})^2\right] \left(\frac{q^2}{q^2 + Q_s^2}\right)^2 , \tag{17.11}$$

$$\left|C_{qt}\right|^2 = \frac{\hbar\omega_{qt}}{2\rho_{\mathrm{i}} c_t^2 \mathcal{V}} \frac{13}{64q^2}(eh_{14})^2 \left(\frac{q^2}{q^2 + Q_s^2}\right)^2 , \tag{17.12}$$

where $\lambda = \ell, t$ corresponds to one longitudinal and two transverse acoustic-phonon modes, c_ℓ and c_t are the sound velocities for these modes, ρ_i is the ion mass density, D is the deformation-potential coefficient and h_{14} is the piezoelectric constant. Applying the Debye model to low-energy acoustic phonons, we get $\omega_{q\lambda} = c_\lambda q$ with $\lambda = \ell, t$.

The electron scattering-in/scattering-out rates due to the Coulomb interaction between electrons are given by [15, 16]

$$\begin{aligned} \mathcal{W}_k^{(\mathrm{in})(\mathrm{c})} &= \frac{2\pi}{\hbar} \sum_{k',q} \left|V^{\mathrm{c}}(q)\right|^2 (1 - n_{k'}) n_{k-q} n_{k'+q} \\ &\quad \times \delta\left(\varepsilon_k + \varepsilon_{k'} - \varepsilon_{k-q} - \varepsilon_{k'+q}\right) , \end{aligned} \tag{17.13}$$

$$\begin{aligned} \mathcal{W}_k^{(\mathrm{out})(\mathrm{c})} &= \frac{2\pi}{\hbar} \sum_{k',q} \left|V^{\mathrm{c}}(q)\right|^2 n_{k'} (1 - n_{k-q})(1 - n_{k'+q}) \\ &\quad \times \delta\left(\varepsilon_{k-q} + \varepsilon_{k'+q} - \varepsilon_k - \varepsilon_{k'}\right) , \end{aligned} \tag{17.14}$$

where the Coulomb scattering potential is

$$V^{\mathrm{c}}(q) = \frac{e^2}{\epsilon_0 \epsilon_{\mathrm{r}} \left(q^2 + Q_{\mathrm{s}}^2\right) \mathcal{V}} . \tag{17.15}$$

Coulomb scattering of electrons is a relative motion between electrons. As a result, the external optical field does not directly couple to it.

We know that the electron temperature T_{e} is generally a reflection of the magnitude of the average kinetic energy of all electrons, even in a non-equilibrium state. Therefore, we can formally define an effective electron temperature at each moment by employing a quasi-equilibrium Fermi–Dirac function ($T_{\mathrm{e}} \neq T$) in the equation of the average kinetic energy of electrons [9]

$$\int_0^{+\infty} \frac{k^4 dk}{1 + \exp\left[\frac{\varepsilon_k - \mu(T_{\mathrm{e}})}{k_{\mathrm{B}} T_{\mathrm{e}}}\right]} = \int_0^{+\infty} n_k k^4 dk , \tag{17.16}$$

where the chemical potential $\mu(T_{\mathrm{e}})$ can be determined for given T_{e} and n_{3D} by

$$\frac{1}{\pi^2} \int_0^{+\infty} \frac{k^2 dk}{1 + \exp\left[\frac{\varepsilon_k - \mu(T_{\mathrm{e}})}{k_{\mathrm{B}} T_{\mathrm{e}}}\right]} = n_{\mathrm{3D}} . \tag{17.17}$$

For a thermal equilibrium distribution of electrons, the electron temperature equals the lattice temperature. On the other hand, the electron temperature for a quasi-equilibrium distribution of electrons is not the same as the lattice temperature. However, the functional form of the quasi-equilibrium distribution is assumed to be the Fermi–Dirac function with the electron temperature determined by the additional energy-balance equation [9, 17, 18]. For a general transient or steady-state non-thermal distribution of electrons, there is no simple quantum-statistical definition for the electron temperature in all ranges. However, at high electron temperatures, we can still define an *effective* electron temperature [19] through the Fermi–Dirac function according to Eq. (17.17) with the conservation of the total number of electrons.

The Fermi–Dirac function describes the quantum statistics of degenerate electrons at low electron temperatures and high electron densities. When either the electron temperature is high or the electron density is low, the Fermi–Dirac function reduces to the Maxwell–Boltzmann distribution for nondegenerate electrons with conservation of the total number of electrons. In the nondegenerate case, the average kinetic energy of electrons is proportional to the electron temperature. The numerically calculated distribution of electrons in this chapter is not the Fermi–Dirac function. We only use the Fermi–Dirac function to define an effective electron temperature in the high-temperature range in Eq. (17.16) by equating the numerically-calculated average kinetic energy of electrons with that of the Fermi–Dirac function for the same number of electrons.

In earlier work [17, 18], the temperature of plasma electrons was defined through a simplified energy-balance equation, including inelastic scattering and joule heating of electrons. The inelastic scattering of electrons was treated within a perturbation theory and the electron current was calculated using a Drude-type model including displacement current. The electron temperature was found to be independent of time when an optical field was applied to the plasma. However, in this chapter, the distribution of electrons is calculated exactly and the effective electron temperature is obtained based upon this calculated non-thermal distribution. The heating from the incident optical field is included through the phonon-assisted photon absorption process.

17.3
Numerical Results for Effective Electron Temperature

In the numerical calculations below, we chose GaAs as the host material. For GaAs, we have taken parameters [13] as follows: $m^* = 0.067 m_0$ with free-electron mass m_0; $n_{3D} = 1 \times 10^{18}\,\text{cm}^{-3}$; $\hbar\omega_{LO} = 36\,\text{meV}$; $\epsilon_r = 12$, $\epsilon_s = 13$; $\epsilon_\infty = 11$, $\rho_i = 5.3\,\text{g/cm}^3$; $c_\ell = 5.14 \times 10^5\,\text{cm/s}$; $c_t = 3.04 \times 10^5\,\text{cm/s}$; $D = -9.3\,\text{eV}$, $h_{14} = 1.2 \times 10^7\,\text{V/cm}$; and $T = 300\,\text{K}$. Other parameters, such as E_{op} and $\hbar\Omega_{op}$, are given directly in the figure captions. In the numerical calculations, we only show results for steady-state cases.

Shown in Figure 17.1 is the calculated effective electron temperature T_e as a function of the amplitude E_{op} of an incident electromagnetic field at $T = 300\,\text{K}$ for $\hbar\Omega_{op} = 25\,\text{meV}$ (solid curve) and $\hbar\Omega_{op} = 45\,\text{meV}$ (dashed curve). From this figure, we find that T_e increases with increasing E_{op} because the rate of leading assisted photon absorption with $M \neq 0$ is proportional to E_{op}^2/Ω_{op}^4. The increase of E_{op} for fixed Ω_{op} implies that the enhancement of photon absorption increases the electron temperature by pushing up average electron kinetic energy. Moreover, we find that the increase of T_e with E_{op} is reduced when the photon energy $\hbar\Omega_{op}$ changes from 25 meV (solid curve) to 45 meV (dashed curve). This is because the rate of leading assisted photon absorption with $M \neq 0$ is inversely proportional to photon energy.

The decrease of T_e with $\hbar\Omega_{op}$ can be seen even more clearly in Figure 17.2 where the calculated effective electron temperature T_e is presented as a function of the incident photon energy $\hbar\Omega_{op}$ at $T = 77\,\text{K}$ for $E_{op} = 25\,\text{kV/cm}$ (dashed curve) and $E_{op} = 50\,\text{kV/cm}$ (solid curve). Also from Figure 17.2, we find that T_e decreases with increasing $\hbar\Omega_{op}$, initially, and levels out at the given lattice temperature $T = 77\,\text{K}$ finally (solid curve). As described above, the rate of leading assisted photon absorption is proportional to E_{op}^2/Ω_{op}^4. The increase of Ω_{op} for fixed E_{op} means the reduction of photon absorption. Again, from Figure 17.2, we can see the initial drop of T_e is a result of the decreased assisted photon absorption. The final leveling-out of T_e is attributed to electrons and the lattice gradually approaching a thermal-equilibrium state due to complete suppression of assisted photon absorption. This correct asymptotic behavior was not seen from the numerical result of the energy-balance equation [12]. When the incident electromagnetic field changes from 50 to 25 kV/cm (dashed curve), the initial drop of T_e is partially suppressed and the asymptotic approach of T_e to T occurs at an even lower photon energy as expected.

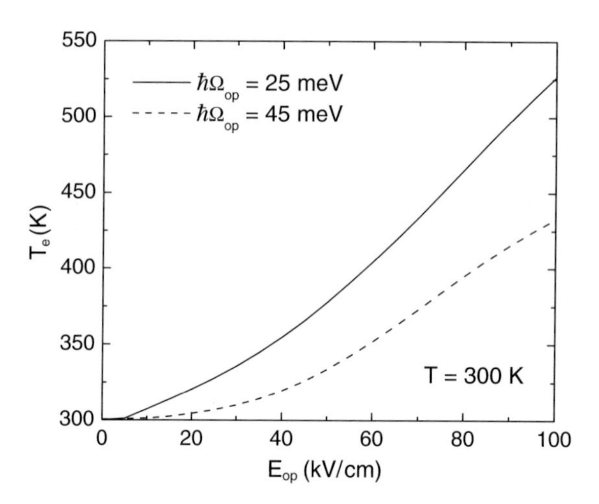

Figure 17.1 Calculated effective electron temperature T_e for a bulk GaAs as a function of the amplitude E_{op} of incident electromagnetic field at $T = 300\,\text{K}$ with $\hbar\Omega_{op} = 25\,\text{meV}$ (solid curve) and $\hbar\Omega_{op} = 45\,\text{meV}$ (dashed curve). The other parameters are given in the text.

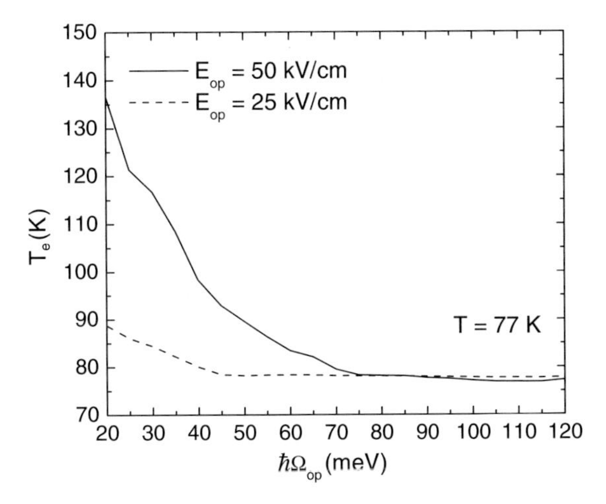

Figure 17.2 Calculated effective electron temperature T_e for a bulk GaAs as a function of the energy $\hbar\Omega_{op}$ of incident photons at $T = 77$ K with $E_{op} = 25$ kV/cm (dashed curve) and $E_{op} = 50$ kV/cm (solid curve). The other parameters are given in the text.

17.4 Summary

In conclusion, a Boltzmann scattering equation has been established for the accurate description of relative scattering motion of electrons interacting with an incident optical field by including the impurity- and phonon-assisted photon absorption as well as the Coulomb scattering between two electrons. The effective electron temperature has been calculated as functions of both the incident-field amplitude and the photon energy. The use of a transport equation with a drift term for such optical-field frequencies has been found not physically justified.

The effect of pair scattering is sometimes included as a homogeneous broadening in a phenomenological theory. However, the effect of pair scattering in this chapter is found to be inhomogeneous that depends on the wave number of electrons. Solving a Boltzmann scattering equation beyond the relaxation-time approximation is a difficult numerical procedure. The exact solution in this chapter provides a testing tool for the justification of simplified theories, for example, relaxation-time approximation and linearized Fokker–Planck equation in certain parameter ranges.

The impurity- and phonon-assisted photon absorption processes create multiple small peaks on the high-energy tail of the Fermi–Dirac distribution. The multiple peaks predicted by our theory were completely ignored in the simple energy-balance equation [12]. The occurrence of these high-energy peaks are attributed to electrons that have predominantly been scattered-out of low-energy states below the Fermi energy that can be seen from terms with $M \neq 0$ in Eqs. (17.3) and (17.7). At the same time, the combination of the electrons that have been scattered-out from below the Fermi energy and the electrons that have been scattered-out from the band edge leave an oscillation of the electron distribution there. From the conservation of electron number in the conduction band, we expect that the electron distribution below the Fermi energy will drop as distribution spreads towards high-energy states. This can be equivalently viewed as an increase of an effective electron temperature through assisted photon absorption. The effect of pair scattering was zero in the energy-balance equation due to conservation of total energy. However, when the effect of pair scattering is included here, the multiple peaks both above and below the Fermi energy are found to be reduced in size and broadened. As a price, the overall electron distribution above/below the Fermi energy is enhanced/suppressed, leading to an even higher electron temperature through pair scattering. This analysis shows that the simple energy-balance equation [12] will generally lead to an inaccurate estimation of the electron temperature.

Although the appearance of the multiple peaks in the distribution of electrons can be foreseen through the phonon-assisted photon absorption process, the relative strength of these peaks are very hard to predict. The relative strength of peaks at high electron kinetic energies turns out to be extremely important for understanding the laser damage of the semiconductor material through impact ionization of valence electrons [11] and for hot-electron transport [9].

References

1 Ziman, J.M. (1964) *Principles of the Theory of Solids*, 1st edn, Cambridge University Press, London, pp. 179–186.

2 Zhao, X.-G., Georgakis, G.A., and Niu, Q. (1997) Photon assisted transport in superlattices beyond the nearest-neighbor approximation. *Phys. Rev. B*, **56**, 3976.

3 Xu, W. and Zhang, C. (1997) Nonlinear transport in steady-state terahertz-driven two-dimensional electron gases. *Phys. Rev. B*, **55**, 5259.

4 Yan, W.-X., Bao, S.-Q., Zhao, X.-G., and Liang, J.-Q. (2000) Theory of dynamical conductivity of interacting electrons. *Phys. Rev. B*, **61**, 7269.

5 Romanov, Y.A., Mourokh, L.G., and Horing, N.J.M. (2003) Negative high-frequency differential conductivity in semiconductor superlattices. *J. Appl. Phys.*, **93**, 4696.

6 Ting, C.S., Ying, S.C., and Quinn, J.J. (1976) Theory of dynamical conductivity of interacting electrons. *Phys. Rev. B*, **14**, 4439.

7 Lei, X.L. and Ting, C.S. (1985) A new approach to non-linear transport for an electron–impurity system in a static electric field. *J. Phys. C: Solid State Phys.*, **18**, 77.

8 Lei, X.L. and Ting, C.S. (1985) Green's-function approach to nonlinear electronic transport for an electron–impurity-phonon system in a strong electric field. *Phys. Rev. B*, **32**, 1112.

9 Huang, D.H., Apostolova, T., Alsing, P.M., and Cardimona, D.A. (2004) High-field transport of electrons and radiative effects using coupled force-balance and Fokker-Planck equations beyond the relaxation-time approximation. *Phys. Rev. B*, **69** 075214.

10 Kaiser, A., Rethfeld, B., Vicanek, H., and Simon, G. (2000) Microscopic processes in dielectrics under irradiation by subpicosecond laser pulses. *Phys. Rev. B*, **61**, 11437.

11 Apostolova, T., Huang, D.H., Alsing, P.M., McIver, J., and Cardimona, D.A. (2002) Effect of laser-induced antidiffusion on excited conduction electron dynamics in bulk semiconductors. *Phys. Rev. B*, **66**, 075208.

12 Lei, X.L. (1998) Balance-equation approach to hot-electron transport in semiconductors irradiated by an intense terahertz field. *J. Appl. Phys.*, **84**, 1396.

13 Lyo, S.K. and Huang, D.H. (2002) Magnetoquantum oscillations of thermoelectric power in multisublevel quantum wires. *Phys. Rev. B*, **66**, 155307.

14 Fröhlich, H. and Paranjape, B.V. (1956) Dielectric breakdown in solids. *Proc. Phys. Soc., Sect. B*, **69**, 21.

15 Lindberg, M. and Koch, S.W. (1988) Effective Bloch equations for semiconductors. *Phys. Rev. B*, **38**, 3342.

16 Moloney, J.V., Indik, R.A., Hader, J., and Koch, S.W. (1999) Modeling semiconductor amplifiers and lasers: from microscopic physics to device simulation. *J. Opt. Soc. Am. B*, **11** 2023.

17 Ginzburg, V.L. and Gurevich, A.V. (1960) Nonlinear phenomena in a plasma located in an alternating electromagnetic field. *Sov. Phys. Usp.*, **3**, 115.

18 Bass, F.G. and Gurevich, Y.G. (1971) Nonlinear theory of the propagation of electromagnetic waves in a solid-state plasma and in a gaseous discharge. *Sov. Phys. Usp.*, **14** 113.

19 Huang, D.H., Alsing, P.A., Apostolova, T., and Cardimona, D.A. (2005) Effect of photon-assisted absorption on the thermodynamics of hot electrons interacting with an intense optical field in bulk GaAs. *Phys. Rev. B*, **71**, 045204.

Index

Properties of Interacting Low-Dimensional Systems, First Edition. G. Gumbs and D. Huang.